Computational Vision

This is a volume in
COMPUTER SCIENCE AND SCIENTIFIC COMPUTING

Werner Rheinboldt and Daniel Siewiorek, editors

Computational Vision

Harry Wechsler

Computer Science Department
George Mason University
Fairfax, Virginia

9-25-90
To Jayshree
with best wishes for success
Harry Wechsler

ACADEMIC PRESS, INC.
Harcourt Brace Jovanovich, Publishers
Boston San Diego New York
London Sydney Tokyo Toronto

Copyright © 1990 by Academic Press, Inc.
All rights reserved.
No part of this publication may be reproduced or
transmitted in any form or by any means, electronic
or mechanical, including photocopy, recording, or
any information storage and retrieval system, without
permission in writing from the publisher.

ACADEMIC PRESS, INC.
1250 Sixth Avenue, San Diego, CA 92101

United Kingdom Edition published by
ACADEMIC PRESS LIMITED
24–28 Oval Road, London NW1 7DX

Library of Congress Cataloging-in-Publication Data

Wechsler, Harry.
    Computational vision/Harry Wechsler.
        p.   cm.—(Computer science and scientific computing)
    Includes bibliographical references.
    ISBN 0-12-741245-X (alk. paper)
    1. Computer vision. I. Title. II. Series.
TA1632.W43 1990
006.3'7—dc20   89–17867
                                                    CIP

Printed in the United States of America
90 91 92 93   9 8 7 6 5 4 3 2 1

To my parents and my wife, Michele

In the beginning God created the heaven and the earth. Now the earth was unformed and void, and darkness was upon the face of the deep; and the spirit of God hovered over the face of the waters. And God said: 'Let there be light.' And there was light. And God saw the light, that it was good; and God divided the light from the darkness. And God called the light Day, and the darkness He called Night. And there was evening and there was morning, one day.

Contents

The young alchemist. (Engraving by Major, T., after D. Teniers.) Reprinted courtesy of the National Library of Medicine.

Introduction

The first rule was never to accept anything as true unless I recognized it to be certainly and evidently such: that is, carefully to avoid all precipitation and prejudgement, and to include nothing in my conclusions unless it presented itself so clearly and distinctly to my mind that there was no reason or occasion to doubt it.

The second was to divide each of the difficulties which I encountered into as many parts as possible, and as might be required for an easier solution.

The third was to think in an orderly fashion when concerned with the search for truth, beginning with the things which were simplest and easiest to understand, and gradually and by degrees reaching toward more complex knowledge, even treating, as though ordered, materials which were not necessarily so.

The last was, both in the process of searching and in reviewing when in difficulty, always to make enumerations so complete, and reviews so general, that I would be certain that nothing was omitted.

From *Discourse on Method and Meditations* by Rene Descartes

It was during the Renaissance when scientists and artists like Leonardo da Vinci began to explore the wonders of the human body. Knowledge exploded and new worlds were dreamed of and even discovered. Indeed, people would become accustomed to think of and search for the unknown in revolutionary ways.

As the twentieth century closes, and we prepare for the next millenium, we still seek to understand our origins and to decide what our future quests should be. We will replay the Renaissance, turn our attention inward, and try to understand ourselves. Our scientists look to decipher the intertwined mysteries of the genome and those of the mind. It is our deep desire for knowledge and new challenges that has driven us so far, and will continue to do so, well into the future. The thirst to understand the mind goes much beyond engineering and future technologies. It is an intellectual and spiritual challenge as well.

In a spiritual way, vision is the art of seeing the impossible and charging forward. Aristotle finds it worth mentioning that, "all men by nature desire to know. An indication of this is the delight we take in our senses, for even apart from their usefulness they are loved for themselves and above all others the sense of sight. For not only with a view to action, but even when we are not going to do anything we prefer seeing to everything else. The reason is that this, most of all senses, makes us know and brings to light many differences between things." Vision is essential if we want to engage our environment in any meaningful way.

I have always been intrigued by the way we think and perceive. Note that I have chosen to hedge on *vision* as *computational* rather than *computer*, because I believe that we compute but not necessarily in the same way that most of the present computers do. This is why the book is titled *Computational Vision*. The thrust of the book provides an integrated theory on how computational vision might work and be implemented.

This book is the result of work by many people before me and of my own undertakings and thoughts. The result of a synergetic approach, it seeks to draw from many scientific endeavors and to link human and machine vision. Another goal was to make the book self-contained. To that end I include fundamental concepts, definitions, and some examples.

The underlying challenge for computational vision is that the sheer complexity of the visual task has been mostly ignored by current approaches. Most visual tasks are complex because they are undercon-strained—or in mathematical terms are ill-posed; however, the visual system displays robustness despite stringent constraints on what can be accomplished and with relatively limited computational resources. Consequently, I present an integrated goal-driven theory of distributed computation over space, time, and function. Distributed computation goes beyond mere processing to include image representations and strategies.

Distributed strategies represent active perception and its symbiotic link to the environment. They are functional if flexible processing is allowed, or exploratory when we engage the environment to interpret it better. The visual system casts most visual tasks as minimization problems and solves them using distributed computation and enforcing nonaccidental, natural constraints. The constraints are either pre-wired or evolve through system adaptation and learning. The computational vision theory could then be represented as

$$
\begin{aligned}
CV &= PDC \\
&= PDR + PDP + PDS \\
&= PDR + PDP + AP \\
&= PDR + PDP + (FAP + EAP),
\end{aligned}
$$

where the abbreviations stand for computational vision (CV), parallel distributed computation (PDC), parallel distributed representation (PDR), parallel distributed processing (PDP), parallel distributed strategies (PDS), active perception (AP), functional active perception (FAP), and exploratory active perception (EAP). A multilevel architecture, which implements such a computational vision theory, is suggested throughout the book and is graphically displayed in Fig. 6.1.

Our computational vision theory is embedded within a general model, the Perception-Control-Action (PCA) cycle, Fig. I.1. The theory, discussed mostly in Sections 6.1., 8.2., and 10.2., recognizes the fact that perception is *directed* and primed under the guidance of memory structures called schemata. Such structures are the natural outgrowth of both nature and nurture. Vision, just one of several sensory capabilities, is embedded within a cycle that consists of control, planning and adaptation. Anticipation is an internal means to implement active perception strategies and leads to robustness and reduced computational loads. Acting, essential to robust behavior, connects mental processing units and links an autonomous system displaying intelligence to its surrounding environment. Be it for need or curiosity, active perception harnesses a rich and complex world and stimulates thinking. People and the environment are one ecosystem. They interact and influence each other, and in the ensuing process they transform and remodel themselves.

The book starts with an introductory chapter on computational theory. The second and fourth chapters consider in great detail the

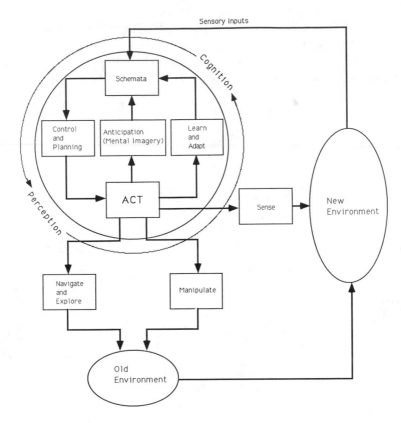

Figure I.1
The Perception-Control-Action (PCA) Cycle

corresponding cojoint representations of space and spatial-frequency, and the parallel and distributed processing (PDP). I have tried to show the evolutionary development of specific concepts, such as scale space, and the motivation behind them. I have also tried to suggest analogies, whenever possible, such as the ones connecting PDP models to physical systems, and to emphasize their relevance in generating new ideas.

Having alluded so far only to spatial distribution and how it can alleviate the computational burden, I move on to the temporal distribution aspect—related to active perception and the complex interaction between the viewer and his environment—in Chapter 6. We, the observers, are continuously searching and looking around. It turns out that active perception not only helps to alleviate the computational burden, but is also beneficial because we discover new things and enrich ourselves. The objects we recognize almost effortlessly come in many forms and shapes. Thus, the third chapter on invariance discusses how to reduce their countless appearances to a manageable task. The seventh chapter then considers how the object recognition process might actually proceed.

We do not know where perception ends and cognition begins. We also have a difficult time defining intelligence and how it is actually implemented. Still, a book on vision would be incomplete without Chapter 8, which covers adaptive behavior and spatial cognition and reasoning. Adaptation is one of the most important ways nature copes with complexity. The ninth chapter considers appropriate parallel and distributed hardware architectures that can implement distributed (visual) computation.

The Epilogue is a deserved respite from so many mathematical considerations. It takes us into the world of plastic arts and pure thought. Artists have tried over the centuries to represent in a faithful way the world surrounding us. What can we learn from them, and what conclusions could we draw? We also recall the concept of perception, and what directed perception means and can offer for facilitating the workings of the visual system. And finally, we consider where we go from here and how.

There are many people who made this book possible and to whom I extend much gratitude. I feel deeply indebted to Norman Caplan from NSF who supported my early research on multiscale representations and helped launch what has been a very fruitful academic career. Many thanks go to my former doctoral students Lowell Jacobson, Todd Reed, and Lee Zimmerman. Eitan Gurari (Chapter 1), Behrooz and Behzad Kamgar-Parsi (Chapter 4), Lee Zimmerman (Chapter 7) and Brad Kjell (Chapter 9) read parts of the manuscript and provided useful feedback. My wife, Michele, had enough patience and perseverance to edit the manuscript and helped convey my thoughts in clear and concise language. Kathy Gilette did a wonderful job typing the many revisions of the manuscript, and Jerzy Bala helped create many of the graphics. Jan Lazarus enthusiastically directed me to the art reproductions, here courtesy of the National Library of Medicine.

Harry Wechsler
Washington, January 1990

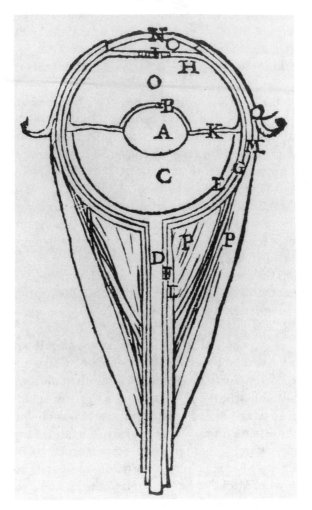

Eye. (From Vesalius, A., De humani corporis fabrica, 1543.) Reprinted courtesy of the National Library of Medicine.

1

Computational Complexity

The limitless Aleph, the microcosm of the alchemists and Kabalists, is one of the points in space that contains all other points. The only place on earth where all places are seen from every angle, each standing clear, without any confusion or blending. I saw the ℵ from every point and angle, and in the ℵ I saw the earth and in the earth the ℵ and in the ℵ the earth; and I felt dizzy and wept, for my eyes had seen that secret and conjectured object whose name is common to all men but which no man has looked upon—the unimaginable universe.

Reprinted with permission from *The Aleph,* by Jorge Luis Borges, ©1968, E.P. Dutton, Inc.

This book provides an integrated theory on the workings and implementation of computational vision. The sheer complexity and robustness of the visual task provides the unifying theme—achieving robust performance is constrained by relatively limited computational resources. Thus, we focus on those characteristics that make vision computationally feasible. I have chosen to focus on *vision* as *computational* rather than *computer*, because I believe that we indeed compute but not necessarily in the same way as most computers today. Thus, this book is entitled *Computational Vision* and appropriately begins with an introductory chapter on computational theory.

I consider intelligent behavior as an information processing task and then address specific descriptive levels where tasks of varying complexity are understood. Marr (1982) argued that the specific levels are function or strategy, process (representation and algorithm), and mechanism (implementation). Basic computational theory, the first level, specifies the task, its appropriateness, and the implementation strategy. Next, the representation and algorithm specify the computational approach in terms of input and output representations, *and* the corresponding transformations. Last, the hardware specifies the actual implementation. Consequently, the task determines the mixture of representations and algorithms, and a good match between the three levels is the ultimate goal.

According to Simon (1982), "All mathematical derivation can be viewed simply as change of representation, making evident what was previously true but obscure. This view can be extended to all of problem solving—solving a problem then means representing (transforming) it so as to make the solution transparent. In other words, knowledge has to be made explicit if and when needed."

Vision is a sequence of transformations, some of them carried out in parallel, that captures some invariant aspect of the world surrounding us. The invariance helps us to recognize and move around to perform safely, despite image variability. Ullman (1980) similarly states that "the role of processing is not to create information but to extract it, integrate it, make it explicit and usable."

The temptation to use the transformational/computational paradigm of the digital computer as an analogy to brain function is easily understood and explained by Edelman (1982) as "science's attempt to compare the brain, in various epochs, to the most prevalent machines of those times. To Sigmund Freud, the metaphor was a hydraulic pump to handle what has been called psychic energy." Consequently, by using the term *computational* rather than *computer* vision (CV) in this volume, we avoid introducing a theoretical bias and allow for alternative paradigms.

Computational complexity issues are broad and pervasive in the development of a perceptual theory that models perception computationally. McClelland and Rumelhart (1986) claim that the time and space requirements of any cognitive theory are important determinants of the theory's (biological) plausibility. According to Tsotsos (1987), "complexity satisfaction provides a major source of constraint on the solution of the problem. Much past work in computer vision, motivated by Marr's philosophy, has tacitly assumed that the language of continuous mathematics is equivalent to the language of computation. Mathematical modeling is not equivalent to computational modeling. There are still issues of representation, discretization, sampling, numerical stability, and computational complexity (at least) to contend with."

We suggest throughout the book a new computational paradigm for visual perception based on critical constraints with which any system, including the human visual system (HVS), has to cope. The constraints fall within three classes—representation, algorithm and strategy, and fault-tolerant behavior. Following the complexity issue discussion, we explore additional constraints, and possible solutions to cope with the constraints are dealt with in later chapters.

1.1. Computation

We will review major results related to computation and its complexity. The interested reader can choose among several excellent books (Garey and Johnson, 1979; Gurari, 1989). We consider first a Turing machine as a way to formally characterize the concept of computability. The study of computations ("Turing") looks at the bounds placed on resources (time and space) available for computation. Classes of computation and their interrelationship are defined and treated, respectively, according to the resources needed for computation. We will then consider some specific perceptual problems and their associated complexity. We continue by introducing the concepts of probabilistic and parallel computation. The concluding remarks point out the need for distributed and parallel computation in representations and processing.

1.1.1. Computability

The Turing machine M is an abstract computing machine that consists of a finite state control, an input tape, a read only input head, m auxiliary work tapes for some $m \geq 0$, and a read-write work tape head for each auxiliary work tape. The state transition table determines each move of M according to the present state of M, the symbol under the input head, and the symbols under the heads of the auxiliary work tapes. Each move of M consists of changing the state of M, changing the symbol under the head of each auxiliary tape, and relocating each head by at most one position in any direction. Initially,

M is assumed to have its input $a_1 \ldots a_n$ stored in the input tape between a left endmarker $\not{c}$ and a right endmarker \$. In addition, the input head is assumed to be located at the start of the input, the auxiliary work tapes are assumed to contain just blank symbols B, and M is assumed to accept a given input, if on such an input it can reach a distinguished state of M, called an accepting state. M can be modified to compute a function by designating one of the auxiliary work tapes to be a one-way write only tape. The resulting machine is then called a Turing transducer.

Many characterizations have been offered for capturing the notion of computability, and they include Turing transducers and programs. These characterizations, however, are equivalent. The equivalency is the basis for the Church thesis, which states that a function is computable (respectively, partially computable) if and only if it is computable (respectively, partially computable) by a Turing transducer. By the Church thesis, a decision problem is partially decidable if and only if there is a Turing machine that accepts those instances of the problem that have the answer "yes." Similarly, the problem is decidable if and only if there is a Turing machine that accepts exactly those instances that have the answer "yes" and halts also on all instances that have the answer "no."

1.1.2. Formal Languages

The previous mathematical system offers a useful abstraction for simulating and analyzing the behavior of programs. The Chomsky hierarchy of formal languages (Hopcroft and Ullman, 1979) considers the corresponding underlying grammars, and the machines that can accept the languages they generate. A grammar G is formally defined as a quadruple $\langle V, T, P, S \rangle$ where V and T are (finite) sets of variable symbols and terminal symbols, respectively, such that $V \cap T = \phi$, P is a set of production rules, S is the start symbol, and $S \in V$. There are four classes of grammars according to the constraints imposed on the set P of productions. Regular grammars (RG) (type three) are defined by $P_3 = \{A \to \omega_1 B, A \to \omega_2 | A, B \in V$ and ω_1 and ω_2 are (possibly empty) string of terminals}. Context-free grammars (CFG) (type two) are defined as $P_2 = \{A \to \alpha | A \in V$ and α is a (possibly empty) string of terminal and nonterminal symbols}. The context-sensitive grammars (CSG) (type one) are defined as $P_1 = \{\alpha \to \beta |$ the length of the string β is greater or equal to that of string α, i.e., $|\beta| \geq |\alpha|\}$. CSGs can be alternatively defined in their normal form as $\alpha = \alpha_1 A \alpha_2$, $\beta = \alpha_1 \gamma \alpha_2$, and γ is not the empty symbol ε, i.e., replacement is legal only in the context (α_1, α_2). Finally, phrase structure grammars (PSG) (type zero) are defined as $P_0 = \{\alpha \rightarrow \beta |,\ \alpha$ and β are arbitrary strings, and $\alpha \neq \varepsilon\}$.

Each formal language is implicitly defined by a corresponding grammar and a machine playing the role of a (acceptor) parser. The type zero language has as its parser the Turing machine. The classes of languages that are accepted by linear bounded automata (LBA), pushdown automata (PA), and finite state automata (FSA) are exactly the languages of types one, two, and three, respectively, i.e., the CSL, CFL, and RL. The LBA is a Turing machine (TM) that allows only linear space, the PA is a TM with a one-way input tape and one auxiliary tape that behaves like a pushdown store. The hierarchy defined by Chomsky specifies the languages of type $i = 0, 1, 2, 3$, where level zero is the highest. The higher the language in the hierarchy, the more powerful (i.e., the wider the scope of the language) but the more difficult it is to implement a corresponding parser.

Formal languages have been used in syntactic pattern recognition (Fu, 1982) by attempting to impose natural (even semantic) constraints among visual primitives. The parser built for the resulting grammar is then a recognizer for a particular class of objects. It is interesting to note some further relationships between the four types of languages discussed so far, and those languages with which one is more familiar. CFG are equivalent to BNF (Backus normal form), while most programming languages (PL) can be described by deterministic context-free languages (DCFL) such that CFL $\supset$ DCFL $\supset$ RL. DCFLs are equivalent to LR(k) left-to-right at most k look-ahead grammars. (ALGOL is a LR(1).) Regular languages are the most efficiently parsed, but they do not allow embedding, and as such could not be useful for contextual image recognition. As it is for linguistics, however, the transformational grammars, i.e., the syntax of the language, represent only an abstract, idealized theory of competence. The actual performance is further predicated on additional factors, such as the semantics of the situation and/or specific pragmatics, including goals, and beliefs. Including such factors in the actual treatment of image understanding lies beyond our current grasp and goal. The increased synaptic capacity resulting from the interactions of several factors, however, might in fact alleviate the complexity of computation.

Integration of factors labeled as data fusion or multisensory data integration permeates computations characteristic of the visual cortex.

The finiteness of memory and the restricted access to it are constraints that set the boundaries of the capabilities of finite state machines and pushdown automata. In the case of Turing machines, however, none of the memory constraints are significant, because they can all be removed and still buy no more definition power from the machines. Yet there are languages that Turing machines cannot accept. The intuitive explanation for this phenomenon is that each Turing machine is a description of a language, which itself has a description. While each language is a collection of strings, and there are an uncountably infinite number of languages, a Turing machine has only infinitely many possible descriptions. Consequently, there are more languages than Turing machines.

1.1.3. Resource-Bounded Computation

The consideration given to algorithms so far assumed the ideal case, where no limit is placed on the amount of resources, such as time and space. Such idealization allowed the development of some useful, but still incomplete, results about algorithms and generated an approach for identifying unsolvable problems. It provides no hint, however, of the feasibility of solving those problems that are solvable. A natural outgrowth of the study of unrestricted computation is of computation bounded by the amount of resources allocated.

The time and space requirements of a given algorithm depend on the algorithm itself and the executing agent. Each agent, or processing element (PE) has its own set of primitive data items and primitive operations. Each primitive data item of a given agent requires some fixed amount of (memory) space. Similarly, each primitive operation of a given agent requires some fixed amount of execution time. As a result, each computation of a given algorithm requires some "αs" space and some "βt" time, where s and t depend only on the program, and α and β depend only on the agent. α represents the packing power of the agent in use, while β represents the speed of the agent and the simulation power of its operations. Since different agents differ in their constants α and β, and since the study of computation aims at the development of a general theory, then one can, with no loss of generality, restrict the analyses of time and space to behavioral analyses. These analyses provide the required accuracy only up to some linear factor from the time and the memory requirements of the actual agents. Such analyses can be carried out by employing models of computing machines like the Turing machine and/or random access machines (RAM) (abstractions of von Neumann computers). In general, the time and space required by RAMs for finding a solution to a problem at a given instance increase with the length of the representation of the instance. Consequently, the time and space requirements of computing machines are specified by functions of the input lengths.

In the case of Turing machines, the transition rules are assumed to be the primitive operations, and the symbols are assumed to be the primitive data items. Each move is assumed to take one unit of time. The time that a computation takes is assumed to equal the number of moves made during the computation. The space that a computation requires is assumed to equal the number of locations visited in the (auxiliary work) tape, which has the maximal such number. A Turing machine M is said to be a $T(n)$ time-bounded Turing machine, or of *time complexity* $T(n)$, if the computation of M on all possible inputs x of length n takes no more than $T(n)$ time. M is said to be of *polynomial time complexity*, if $T(n)$ is a polynomial in n. Similar definitions can be obtained for space complexity as well. Based on such definitions, the Church thesis is refined next, even though its refinement cannot be proven correct. (However, here too, one can intuitively be convinced of the statement's correctness, by showing the existence of mappings between the different classes of computation models where the result is invariant.)

The sequential computation thesis. A function is computable (respectively, partially computable) by an algorithm A only if it is computable (respectively, partially computable) by a deterministic Turing transducer that satisfies the following conditions: A on a given input has a computation that takes $T(n)$ time and $S(n)$ space only if on such an input the Turing transducer has a computation that takes $p(T(n))$ time and $p(S(n))$ space, where $p(\ )$ is some fixed polynomial that does not depend on the input.

Based on the previous definitions complexity classes can be defined as follows. DTIME($T(n)$) denotes the class of languages that have time

complexity $O(T(n))$. NTIME($T(n)$) will denote the class of languages
that have nondeterministic (i.e., one has a choice of transition rules
at each step) time complexity $O(T(n))$. Similarly one can define
DSPACE($S(n)$) and NSPACE($S(n)$), respectively.

P will denote the class of membership problems for the languages in
$\bigcup_{p(n)}$ DTIME($p(n)$), where $p(n)$ stands for a polynomial in n. NP
will denote the class of membership problems for the languages
in $\bigcup_{p(n)}$ NTIME($p(n)$). EXPTIME denotes the class of membership
problems for the languages in $\bigcup_{p(n)}$ DTIME($2^{p(n)}$). PSPACE denotes
the class of membership problems for the languages in
$\bigcup_{p(n)}$ DSPACE($p(n)$). NLOG and DLOG denote the class of membership
problems for the languages in NSPACE(log n) and DSPACE(log n),
respectively.

Not all problems are equal with respect to time complexity, in that
some require more time than others to be solved. Besides showing the
existence of a time hierarchy, one could also show lower bounds on the
time complexity of some problems. Deriving these lower bounds is of
special interest in the identification of intractable problems, i.e., of
problems that require impractical amounts of resources for solving
them. Such an identification can save a considerable amount of futile
effort invested in trying to efficiently solve intractable problems. In
general, an algorithm is considered tractable if it is of polynomial time
complexity, because then its time requirements grow slowly with the
input length. On the other hand, a problem of exponential time
complexity is considered intractable because its time requirements
grow quickly with the length and therefore can be practically solved
only for small inputs.

Identification of (in)tractable problems employs polynomial time-
bounded reductions. Thus, one is interested in identifying "easy"
(regarding reductions) intractable problems for complexity classes C.
A problem K is said to be an NP *hard* problem with respect to
polynomial time reductions, or just an NP *hard* problem, if every
problem in NP is polynomially time reducible to K. The problem K is
said to be NP *complete* (NPC) if it is an NP *hard* problem in NP. If the
class C under consideration is that of NP, $K \in$ NP and for every
$K' \in$ NP, $K' \rightarrow K$ (i.e., K is reducible to K') then K is NPC. It follows that
if K is NPC, $K' \in$ NP, and $K \rightarrow K'$ then K' is also NPC. What was needed
was to get started by showing that a specific problem is NPC. Cook's
(1971) theorem was first in establishing the NPC for a specific problem,

that of the satisfiability of Boolean functions in conjunctive normal forms.

The relationships between complexity classes are shown in Fig. 1.1. By definition $NP \supseteq P$, and one can show that $EXPTIME \supseteq NP$, and $EXPTIME \supseteq P$. Whether P and NP are properly contained in NP and EXPTIME, respectively, is not known. Consequently, the importance of NP arises from the refinement that it may offer to the boundary between tractability and intractability of problems. In particular, if $NP \neq P$, as is widely conjectured now, then NP is likely to provide some of the easiest problems, namely the NPC, for proving the intractability of new problems by means of reducibility. On the other hand, if $NP = P$, then many important problems, which received considerable attention for many decades without any success in finding efficient algorithms for them, will become solvable in polynomial time.

It is interesting to study the limitations of some subclasses of P. The motivation might be from a theoretical point of view, as in the case of the subclass NLOG of P, or from a practical point of view as in the case of the subclass U_NC (Uniform_NC) of the problems in P that can be solved by efficient parallel programs. (See Sections 1.3.) A problem K is said to be a P hard problem if every problem in P is log space reducible to K. The problem is said to be P *complete* if it is a P *hard* problem in P. By the previous definitions, the P *complete* problems are the hardest problems in P. An example of P *complete* problems is the emptiness problems for context-free grammars (CFG).

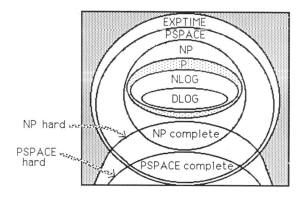

Figure 1.1.
Relationships between some complexity classes.

Polynomial space complexity studies can be undertaken as well. One can show that $\text{NTIME}(T(n)) \subseteq \text{DSPACE}(T(n))$, and so PSPACE contains NP. Furthermore, PSPACE is contained in EXPTIME. Such containments propose PSPACE for a study similar to that done for NP. Such a study will be relevant for the case that $NP = P$, for the same reasons that the study has been important for NP in the first place. Should NP turn out to be different from P, then the study of PSPACE might provide some insights into the factors that increase problem complexity.

Next we consider several combinatorial problems and discuss their complexity and relevance to computational vision. Given a finite system of linear equations and inequalities in real variables, finding a feasible solution, if one exists at all, is known as linear programming (LP). LP is known to be in P and is relevant to problems encountered in geometric spatial reasoning, where a consistent interpretation of imperfect or incomplete data is sought. Graph theoretic approaches are used for clustering and attempt to capture the relationship among objects' components. Examples are the minimum spanning tree (MST) and the clique. The MST for a graph $G = (V, E)$ of vertices V and links E is a subgraph of G, which is a tree that contains all vertices of G. MST for a weighted graph is a spanning tree of minimal weight (think of connecting a number of cities by a minimum length network of roads). MST is used for taxonomy and clustering, and its characteristic graph stands for the internal structure characterizing a particular class of objects. The complexity of MST algorithms is of the order $O(|V|^2)$ and as such $\text{MST} \in P$. MST has been used, among other things, to classify the traces left by nuclear particles in bubble-chamber analysis through matching (graph isomorphism) MSTs. Graph isomorphism and linear programming are open problems regarding their exact complexity and are sometimes considered of intermediate complexity labeled as NPI, where $\text{NPI} = \text{NP} - (P \cup \text{NPC})$. The Steiner tree (ST) problem, to be defined next, which is a generalization of MST, is however an NP *complete* problem. An instance of ST is a connected graph $G(V, E)$, a subset of vertices $v \subseteq V$, a length function $l(e) > 0$, and a positive integer k. The question to be answered by ST is if there is a tree $T(V^*, E^*)$ in G such that $v \subseteq V^* \subseteq V$, $E^* \subseteq E$ and $\sum_{e \in E^*} l(e) \leq k$? (Clearly if $v = V$ then $\text{ST} = \text{MST}$.) Cliques are completely connected sets of nodes, which are not subsets of larger completely connected sets. The problem of detecting the existence of cliques of a given size is an NP *complete* problem. Clique characteristics, once detected, can be useful in locating subparts, as would be needed in the assembly of mechanical parts. Another graph problem that we return to later is the well-known travel salesman problem (TSP), which is an NP *complete* problem. Polyhedral scene labeling problems have been shown to be inherently NPC (Kirousis and Papadimitriou, 1985).

Early work on complexity problems related to computational vision was done by Gurari and Wechsler (1982). They addressed the difficulties involved in the segmentation of pictures, where segmentation is defined in the following manner (Pavlidis, 1977):

Definition 1. Let X denote a grid of sample points for a given picture. Let Y be a subset of X containing at least two points. Then a *uniformity predicate* $P(Y)$ is one that assigns the value *true* or *false* to Y depending only on properties of the picture value function f for the points of Y. Furthermore, P has the properties that if Z is a nonempty subset of Y, then $P(Y) = $ true $\Rightarrow P(Z) = $ true.

Definition 2. A segmentation of the grid X for a uniformity predicate P is a partition of X into disjoint nonempty subsets $\{X_i\}_{i=1}^n$ such that

(a) $\bigcup_{i=1}^n X_i = X$;

(b) $\forall 1 \leq i \leq n$, X_i is connected and $P(X_i) = $ *true*; and

(c) P is *false* on the union of any number of adjacent members of the partition.

The results shown by Gurari and Wechsler (Theorem 1) state that there exists no algorithm that decides for any arbitrarily given predicate P whether or not P is a uniformity predicate, and that (Theorem 2) there exists no algorithm that decides for any arbitrarily given uniformity predicate P whether or not the predicate induces more than one segment in any image.

A thorough discussion of planning, geometry, and complexity, as in robot motion, is presented by Schwartz *et al.* (1987) and more recently by Sharir (1989). It is also worth noting that the problem of planning robot motion revolves around PSPACE complete (PSC), i.e., the solution requires a number of steps, exponential in the size of the input needed to define the specific problem. (PSC problems can be defined in a similar way to NPC.)

1.2. Probabilistic Computation

So far complexity analysis was considered according to the programs' worst case behavior. Programs whose worst behavior is good are obviously the most desirable programs for solving a given problem. In many instances one might also be satisfied with programs that on the average behave nicely as long as no more efficient solution is available to solve the problem in its general form. In fact, one might even be satisfied with programs even when they have a small probability to provide wrong answers. Programs obtained by introducing instructions that make random choices are called probabilistic (Gurari, 1989). The need to accept average good behavior fits with the principle of bounded rationality as introduced by Simon (1982) and explained by the limited amount of resources available toward solving a specific problem. Furthermore, many algorithms in computational vision such as those based on Markov/Gibbs random fields (Geman and Geman, 1984) include the concept of choice (see Section 4.2.3.). Last but not least, random synaptic connections seem to be quite characteristic of the neural network substrate and evolutionarily motivated.

Many probabilistic algorithms fall within two classes—the Las Vegas and Monte Carlo types. Las Vegas probabilistic algorithms have three possible outputs: "yes," "no," and "maybe." The answers "yes" and "no" are always correct. The probability of maybe is at most a constant less than one half. The Monte Carlo algorithms have only two possible answers, yes and no. The answer yes is always correct, and the answer no is wrong with probability of at most some constant that is smaller than one half.

Randomization can be an important programming tool, and its power stems from the ability to choose through the use of appropriate random assignment instructions of the form $x = random(S)$, where S can be any finite set. The execution of such instructions takes $|v| + \log|S|$ time, where $|v|$ is assumed to be the length of the representation of the value v chosen from S, and $|S|$ is assumed to denote the cardinality of S. A probabilistic program is then said to have an *expected time complexity* $\bar{t}(x)$ on input x if $\bar{t}(x) = \sum_{i=0}^{\infty} p_i(x) \cdot i$. The function $p_i(x)$ is the probability that the program requires i units of time to compute on x. The program is said to have an expected time complexity $\bar{T}(n)$ if $\bar{T}(|x|) \geq \bar{t}(x)$ for each x. At the price of erring, one can still attempt to reduce the complexity of solving a specific problem. As an example, brute force algorithms for solving the nonprimality (COMPOSITE) problems take exponential time. If one is ready to accept errors, however, polynomial time algorithms are available.

The ability to make "random choices" can be viewed as a degeneration of the ability to make "nondeterministic choices." In the nondeterministic case, upon the execution of each nondeterministic instruction, a choice must be made between options. Some options might be "good," and some might be "bad." The choice must be for a good option, whenever such an option exists. The problem with nondeterministic choices is that, in general, they do not seem to be made in an efficient way. In the random case, the options that arise during the execution of a probabilistic instruction are similar to those that arise in the nondeterministic case. No restriction is now made, however, on the nature of the chosen option. Instead, each of the good and bad options is assumed to have equal probability of being chosen. Consequently, the unbiasness among the different options enables efficient execution of choices.

The burden of increasing the probability of obtaining good choices is placed on the programmer. If the programmer is nature and the goal is self-organization and adaptation, then one considers the so-called genetic algorithms. (See Sections 8.4.8. and 9.4.2. for a brief discussion on genetic algorithms and their possible parallel implementation on the BUTTERFLY machine.) Much computational effort is spent on search and information retrieval. Gordon (1988) considers probabilistic and genetic algorithms for document retrieval and shows how competing descriptions are associated with a document and changed over time by a genetic algorithm, according to the queries used and the relevant judgments made during retrieval. Intelligent and adaptive search helps resolve the computational burden encountered when attempting to retrieve information from large data depositories. One has to seek adaptive probabilistic programs, which have both polynomial time complexity and bounded error probability. The adaptive aspect overcomes some shortcomings of the pure probabilistic methods, including independence assumptions, high computational cost, and lack of reliable feedback.

1.3.   Parallel Computation

The constraints introduced so far concerning the complexity of algorithms seem to suggest that distributed input representations and computations might be required. Distributed representations are considered in Chapter 2, parallel and distributed processing (PDP) is addressed in Chapter 4, and parallel hardware architectures are presented in Chapter 9. One should keep in mind, however, that PDP for its own sake is not a panacea for reducing computational complexity. Specifically, assume that the sequential portion of a given algorithm requires a portion f of the total computation. Then, the informal (Amdahl) speed factor $s = 1/[f + (1 - f)p] \to 1/f$, where p is the number of processors allocated to a computation, which contains an inherently sequential subcomputation. Clever sequential algorithms (maybe of the genetic type) perform nearly as well as a naive algorithm that explores all paths in parallel, because the latter one will execute the same operations as the sequential one only with some of the operations carried out in parallel.

Theoretical results show that the space requirements of sequential computations and the time requirements of parallel computations are polynomially related. Many abstract models of parallel machines have been offered in the literature. Unlike the case for the abstract models of sequential machines, however, the way to relate the different abstract models of parallel machines is not obvious. Therefore, the lowest common denominator of such models, i.e., their hardware representations, seem to be natural choices when analyzing the models for the resources that they require.

The representations are considered in terms of undirected acyclic graphs, called *combinatorial Boolean circuits* (circuits). Each node in a circuit is assumed to have an indegree no greater than 2, and an outdegree of unbounded value. Each node of indegree 0 is labeled either with a variable name, or with the constant 0, or with the constant 1. Each node of indegree 1 is labeled with the Boolean function ¬. Each node of indegree 2 is labeled either with the Boolean function ∧, or with the Boolean function ∨.

Each node of indegree greater than 0 is called a *gate*. A gate is said to be a *not gate* if it is labeled with ¬, an *and gate* if it is labeled with ∧, and an *or gate* if it is labeled with ∨. Nodes that are labeled with variable names are called input nodes. Nodes of outdegrees 0 are called output nodes. A node that is labeled with 0 is called a constant node 0. A node that is labeled with 1 is called a constant node 1.

A circuit with n input nodes and m output nodes, computes a function $f: \{0, 1\}^n \to \{0, 1\}^m$ in the obvious way.

The *size* of a circuit is the number of gates in the circuit. The *depth* of a circuit is the number of gates in the longest path from an input node to an output node.

$C = (c_0, c_1, c_2, \ldots)$ is said to be a *family of circuits* if c_n is a circuit with n input nodes for each $n \geq 0$. The family is said to have *depth complexity* $D(n)$ if $D(n) \geq$ (depth of c_n) for all $n \geq 0$.

A family $C = (c_0, c_1, c_2, \ldots)$ is said to compute a function $f: \{0, 1\}^* \to \{0, 1\}^*$ if for each $n \geq 0$ the circuit c_n computes the function $f_n: \{0, 1\}^n \to \{0, 1\}^k$ for some $k \geq 0$ that depends on n. f_n is assumed to be a function that satisfies $f_n(x) = f(x)$ for each x in $\{0, 1\}^n$.

A function f is said to be of *size complexity* $Z(n)$ if it is computable by a family of circuits of size $Z(n)$. The function f is said to have *depth complexity* $D(n)$ if it is computable by a family of circuits that has depth complexity $D(n)$.

A family $C = (c_0, c_1, c_2, \ldots)$ of circuits of size complexity $Z(n)$ is said to be a uniform family of circuits if an $O(\log Z(n))$ space-bounded deterministic Turing transducer can compute $\{(1^n, c_n) | n \geq 0\}$. The size of circuits is a major resource for parallel computations, as time is for sequential computation. The size complexity of the uniform family of circuits is polynomially related to the time of sequential complexity.

Sequential computations are considered feasible only if they are polynomially time-bounded. Similarly, families of circuits are considered feasible only if they are polynomially size-bounded. As a result, parallelism has no apparent major impact on problems that are unsolvable in polynomial time. On the other hand, for those problems that are solvable in polynomial time, parallelism is key when it can provide a significant speedup. One such class of problems is the class that consists of the problems that can be solved by uniform families of circuits that have, simultaneously, polynomial size complexity and *polylog*($= O(\log^i n)$ for some $i \geq 0$) depth complexity. This class of problem is denoted U_NC, Uniform-NC. U_NC is a subclass of problems that can be solved polynomially in deterministic time. If the inclusion U_NC $\subset P$ is proper as it is conjectured now, then P *complete* problems are examples of problems that are excluded from U_NC, i.e., there are *P complete* problems, as discussed in Section 1.1.3., that cannot be solved efficiently in parallel.

These results show that parallelism does not increase the class of problems that can be feasibly solved. The wisdom of the story is that only by decreasing the sequential part "*f*," i.e., by designing truly parallel algorithms, can one speedup the computation. The parallel cytoarchitecture characteristic to the cortex (Hubel and Wiesel, 1962) is one indication that the human visual system reached such a conclusion a long time ago. Studies (Abu-Mostafa, 1986) indicate that PDP models may provide speed at the expense of an exponential growth of the parallel network. Sequential computation involves a number of steps (time complexity), memory (space complexity), and algorithm length (Kolmogorov complexity). PDP models, when applied to the same problem, must iterate (time complexity) and include state variable-attractors (space complexity) and synaptic connections (Kolmogorov complexity). The analysis performed by Abu-Mostafa shows that for N (neural) elements, the number of iterations is $O(N^2)$ and the synaptic capacity is $O(N^3)$ bits. Thus, the capacity of the synaptic connections grows much faster than the number of neurons, due to the strong connectivity among the processing elements. As a consequence, PDP models would be quite appropriate for long algorithms. If the problem is very demanding in space, however, then the number of neurons has to be large, and the capacity must follow suit (even if the original Kolmogorov complexity is modest), resulting in inefficiency and waste of wetware.

This discussion indicates that different complexities are related, and that their interactions should be fully analyzed. Another aspect suggested by Abu-Mostafa (1986) for further study is the role of analog computation and continuous-time processing on the complexity of the algorithm, and the relationship between random algorithms and large databases of information, which could make PDP models useful due to their large synaptic capacity.

Much recent research is indeed dedicated to both the analysis and the potential of PDP models for alleviating complexity issues. Bruck and Goodman (1987) discuss such issues in the context provided by problems defined as co-NP, where co-NP is the complement of the languages that are in NP. The relationship between NP and co-NP given the conjecture $\widetilde{NP} \neq co\text{-}\widetilde{NP}$ is stronger than the one relating P and NP, i.e., ($P \neq$ NP). It is obvious that P is closed under complementation, i.e., $P = \text{co-P}$, so $\text{NP} \neq \text{co-NP} \rightarrow P \neq \text{NP}$; although it might be the case that $P \neq \text{NP}$ even though $\text{NP} = \text{co-NP}$. The results given by Bruck and Goodman (1987) state that (Theorem 1) a network with polynomial (in the size of the input) number of neurons cannot solve an NP-*hard* problem even if it operates for an exponential length of time (unless $\text{NP} = \text{co-NP}$ and (Theorem 2) a network with polynomial (in the size of the input) number of neurons, which always gets to an ε-approximate solution for the TSP does not exist unless $P = \text{NP}$.

1.4. Conclusions

We considered in this chapter the abstract concepts of computation and corresponding complexities and commented on them as relevant constraints with which the visual system must cope. It seems appropriate to conclude by suggesting that one way of coping with such constraints is to employ truly distributed and hierarchical representations and processing algorithms and a good match between representation and algorithm. It has also been shown that distributed networks of computation trade space for time, and that a major advantage in using them could be their inherently large synaptic capacity. Additional ways of dealing with complexity are suggested in later chapters and they include invariant image representation, active perception, and adaptation. The concept of computation can go beyond the traditional, sequential von Neuman architecture, and one could even contemplate optical implementations. As suggested earlier, however, such abstractions are still qualified and predicted on our present knowledge and on the ubiquitous computer. Novel analog and/or biological computational approaches, yet unknown to us, could still surprise us and provide an alternative to overcome the computational burden. Finding such approaches will indeed lead to a revolution in building perceptual systems.

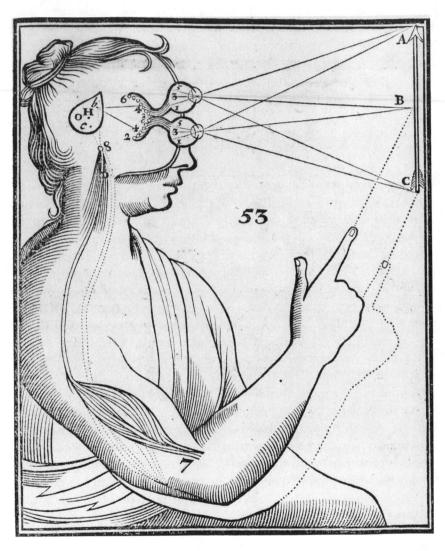

Diagram of figure's vision of arrow through eye to brain. (From Descartes, R., Tractatus de homine . . . , 1677.) Reprinted courtesy of the National Library of Medicine.

2

Distributed and Multiscale Representations

"What a curious feeling!" said Alice. "I must be shutting up like a telescope." And so it was indeed: she was now only ten inches high, and her face brightened up at the thought that she was now the right size for going through the little door into that lovely garden. First, however, she waited for a few minutes to see if she was going to shrink any further: she felt a little nervous about this: "for it might end, you know," said Alice, "in my going out altogether, like a candle. I wonder what I should be like then?"

From *Alice in Wonderland* by Lewis Carroll

Mathematical derivations are problem-solving processes that formally prove previously obscured truths. One such problem-solving process, computational vision (CV), seeks to produce useful descriptions of visual input to allow an artificial or natural system to safely negotiate its environment. Vision itself is a sequence of transformations—some of them carried out in parallel—which captures invariant aspects of the world surrounding us. Invariance is what allows us to recognize and to move and maneuver safely, despite image variability.

CV is a multidisciplinary and synergetic effort. It brings together signal processing, psychophysics (a branch of psychology that measures the relationship between perceived and physical attributes of a stimulus), neurophysiology, and cognitive sciences, among others, to elucidate the mysteries of the human visual system (HVS) and build artificial visual systems (AVS). This chapter opens our attempt to build

19

a foundation and theory for how the HVS and AVS could function. We
start by considering the flow of information as it makes its journey
from the retina toward the striate cortex (see Fig. 2.1.) or equivalently
from a CCD camera toward some "artificial" brain. Initially, the flow
will be unidirectional (bottom up), notwithstanding the evidence and
the need for top-down priming processes (see Chapter 6) and multisensory data integration (see Chapter 8).

We begin with that aspect of perception called low level or "early
vision." Much of this chapter is concerned with multiresolution (or
scale space) and distributed schemes of image representations. Our
ultimate goal is to demonstrate that distributed cojoint representations of space and spatial frequency are ideally suited for low-level
image representations. Such computations are (probably) preceded by
derivations to achieve at least geometrical invariance to scale and
rotation changes and possibly to perspective distortions. (This is
discussed in Chapter 3.) Such representations address the complexity
issue by being distributed and seeming to provide optimal resolution.

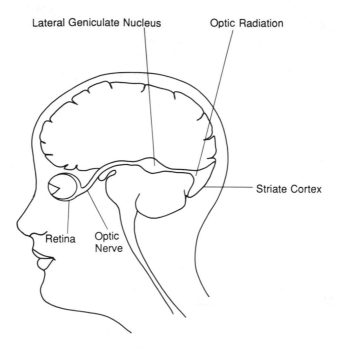

Figure 2.1.
Flow of information from the retina to the primary visual cortex.

Resolution, according to Haralick (1986), is a generic term that describes how well a system, process, component, or image can reproduce an isolated object consisting of separate but closely spaced objects or lines. The *limiting resolution* is defined in terms of the smallest dimension of the target or object that can be discriminated or observed. Resolution may then be a function of an object's spatial position, as well as the structure of its "primitive" elements and clustering characteristics.

2.1. Frequency Analysis

The input to the retina or to a CCD camera is a mere signal, and this is why we start by briefly reviewing the fundamental signal processing (SP) concepts of linear systems and frequency analysis.

2.1.1. Linear Systems

Consider the functions f and g of one or more temporal and/or spatial variables. (For simplicity we usually consider 1D functions even though the optical array input is 2D.) A system produces an output h for a given input f, if its operation is defined in terms of some transformation T, known as the transfer function, such that $h = Tf$. Assuming that two inputs f_1 and f_2 produce two outputs, h_1 and h_2, the system is said to be linear if

$$T(af_1 + bf_2) = aT(f_1) + bT(f_2) = ah_1 + bh_2. \qquad (2\text{-}1)$$

Furthermore, the same system is shift invariant, if for a shifted input the output shifts by the same amount. Assuming that the shift is given by α, the shift-invariance property is

$$T[f(t - \alpha)] = h(t - \alpha). \qquad (2\text{-}2)$$

We are interested in defining a relationship linking the input signal to the system output. The relationship for a shift-invariant linear system is expressed in terms of the convolution $\otimes$ of two functions g and f as

$$h = g \otimes f, \qquad (2\text{-}3a)$$

and it is given as

$$h(\alpha) = \int g(\alpha - t)f(t)dt, \qquad (2\text{-}3b)$$

where the function g, characteristic of the system, is called the impulse response function. Convolution is the measure of overlap between $g(t)$, reflected about the origin, and shifted by α to yield $g(\alpha - t)$ and $f(t)$. One can define the corresponding 2D digital convolution as

$$h(i, j) = \sum_m \sum_n g(i - m, j - n)f(m, n), \qquad (2\text{-}3c)$$

where g, the impulse response, assumes the form of a mask template, as those used for edge detection or image smoothing. Summation is finite and is carried out only for those shifts (i, j) that ensure overlap between g and f. For two arrays of sizes $(m_1 \times n_1)$ and $(m_2 \times n_2)$, the resulting convolution array is of size $(m_1 + m_2 - 1) \times (n_1 + n_2 - 1)$.

Examples of convolutions abound, especially where the system output is a blurred image of the input. If an object moves faster than the exposure time of the camera, then the image is blurred (as a result of subsampling and low-pass filtering). Assume that the moving object undergoes linear translation only. The image of the object is given by $f(x, y)$, the true picture value function of spatial variables x and y, and E is the exposure or effective integration time of the camera. Then, what the camera records is the output $h(x, y)$ given as

$$h(x, y) = \int_{-E/2}^{E/2} f[x - \alpha(t), y - \beta(t)]dt, \qquad (2\text{-}4)$$

where $\alpha(t)$ and $\beta(t)$ stand for the motion in the x and y directions, respectively. Equivalently, there is an impulse response function $g(x, y)$, such that $h = g \otimes f$, and the effects of linear motion can thus be modeled as a shift-invariant linear system.

The impulse distribution $\delta(t)$, known also as the Dirac function, plays a major role in linear systems, and it is defined next. $\delta(t)$ is infinite at $t = 0$, zero elsewhere, and its integral is given as

$$\int_{-\infty}^{\infty} \delta(t)dt = 1. \qquad (2\text{-}5a)$$

The impulse distribution enjoys the *sifting property*, given as

$$f(t_0) = \int \delta(t - t_0)f(t)dt. \tag{2-5b}$$

The function, $g(t)$, defined in Eq. (2-3b) as the system's impulse response, is the output a system yields in response to an impulse as input, i.e., $g = g \otimes \delta$.

2.1.2.   Fourier Transform

The Fourier transform (FT) can be expressed and derived in several ways. The radial FT and the corresponding inverse transform are given by the pair $\langle F(\omega), f(t) \rangle$ as defined:

$$F(\omega) = \int_{-\infty}^{\infty} f(t)\exp(-j\omega t)dt \tag{2-6a}$$

$$f(t) = \frac{1}{2\pi} \int_{-\infty}^{\infty} F(\omega)\exp(j\omega t)d\omega, \tag{2-6b}$$

where $\exp(j\omega) = \cos(\omega) + j\sin(\omega)$, and $j = \sqrt{-1}$.

The FT as defined is a special case of the Laplace transform (LT)

$$\mathscr{L}(f) = F(s) = \int_{0}^{\infty} f(t)\exp(-st)dt, \tag{2-7}$$

where $f(t)$ is a causal signal (i.e., $f(t) = 0$ for $t < 0$), $s = a + j\omega$, and a a convergence factor. The FT is nothing more than the LT of the function $f(t) = \exp(-at)f(t)$ evaluated on the imaginary axis. If $f(t)$ is absolutely integrable, then $F(s) - F(\omega)$ for $s = j\omega$ as $a \to 0$.

One can use the radial FT and the sifting property of the impulse distribution to show that $FT(1) = 2\pi\delta(\omega)$. Assume a function $f(t)$ such that its FT is given by $F(\omega) = 2\pi\delta(\omega)$. Then, using the inverse FT

$$f(t) = \frac{1}{2\pi} \int_{-\infty}^{\infty} 2\pi\delta(\omega)\exp(j\omega t)d\omega$$

$$= \frac{1}{2\pi} \int_{-\infty}^{\infty} 2\pi\delta(\omega - 0)\exp(j\omega t)d\omega$$

$$= \exp(0) = 1.$$

In a similar fashion one can show that $FT[\exp(j\omega_0 t)] = 2\pi\delta(\omega - \omega_0)$.

The FT is a linear operator $\mathscr{F}$, such that $F(\omega) = \mathscr{F}[f]$ and $\mathscr{F}$ $[af_1 + bf_2] = a\mathscr{F}(f_1) + b\mathscr{F}(f_2)$, where f and f_2 are functions, and a and b are constants. One can evaluate the *FT* for functions that are not absolutely integrable, such that $f(t) = \cos(\omega_0 t)$, by expressing $\cos(\omega_0 t)$ as

$$\cos(\omega_0 t) = \tfrac{1}{2}[\exp(j\omega_0 t) + \exp(-j\omega_0 t)]$$

and following with

$$F(\omega) = \tfrac{1}{2}\{\mathscr{F}[\exp(j\omega_0 t)] + \mathscr{F}[\exp(-j\omega_0 t)]\}$$
$$= \pi[\delta(\omega - \omega_0) + \delta(\omega + \omega_0)].$$

Another alternative for defining the *FT* is to consider nonperiodic and continuous signals. The cyclic 2D *FT* and its inverse are defined as the pair $\langle F(u,v), f(x,y)\rangle$ given by

$$F(u, v) = \int\!\!\!\int_{-\infty}^{\infty} f(x, y)\exp[-j2\pi(ux + vy)]dxdy \qquad (2\text{-}8a)$$

$$f(x, y) = \int\!\!\!\int_{-\infty}^{\infty} F(u, v)\exp[j2\pi(ux + vy)]dudv. \qquad (2\text{-}8b)$$

The resulting FT is nonperiodic and continuous. (Note that a nonperiodic signal yields a continuous representation, and that a periodic signal yields a discrete representation, and vice versa, i.e., a continuous signal yields a nonperiodic representation and a discrete signal yields a periodic representation.) In the sequel, we use both the radial and cyclic FT.

The Hartley transform (HT) is another method of decomposing and analyzing a given function in terms of sinusoids. Whereas the FT involves real and imaginary numbers and a complex sum of sinusoidal functions, the HT involves only real numbers and a real sum of sinusoidal functions (Bracewell, 1989). The HT is defined as

$$HT(u, v) = \int\!\!\!\int_{-\infty}^{\infty} f(x, y)[\cos 2\pi(ux + vy) + \sin 2\pi(ux + vy)]dxdy. \qquad (2\text{-}8c)$$

The difference between the HT and the FT is a factor of $(-j)$ in front of the sine function and leads to significant savings in both storage and computation. The fast Hartley transform (FHT) is similar to the fast Fourier transform (FFT) (see Section 9.1.) but performs much faster because it requires only real arithmetic. The recursive nature of the FHT enables one to penetrate the next higher order FHT from two identical lower order FHT and also to derive the discrete Fourier transform (DFT) (see Section 2.1.3.; Hou, 1987).

We record several FT properties. First, by the *shift theorem* (see Section 3.1.1.), shifting a function does not change the magnitude of its FT. By far the most important property is the *convolution theorem*, which states that convolution in one domain (such as the spatial one) corresponds to multiplication in the other domain (such as the frequency one). Formally,

$$\mathscr{F}[f(x, y) \otimes g(x, y)] = F(u, v)G(u, v), \qquad (2\text{-}9a)$$

where $F(u, v) = \mathscr{F}[f(x, y)]$ and $G(u, v) = \mathscr{F}[g(x, y)]$. The *similarity theorem* (see also Eq. 3-4), which is related to the uncertainty principle, states that improving resolution (i.e., shrinking) in one domain leads to poorer resolution (i.e., expanding) in the other domain. The theorem is formally stated as

$$\mathscr{F}[f(\alpha x, \beta y)] = \frac{1}{|\alpha\beta|} F\left(\frac{u}{\alpha}, \frac{v}{\beta}\right). \qquad (2\text{-}9b)$$

The last result can be useful in finding the cyclic FT of functions such as the Gaussian. Assume the 1D case for simplicity and observe that the FT of $f(t) = \exp(-\pi t^2)$ is given by $F(u) = \exp(-\pi u^2)$. The Gaussian function is given as $f(t) = \exp(-t^2/2\sigma^2)$, where σ is the standard deviation. Assume that $\alpha = (2\pi\sigma^2)^{-1}$ and then the FT of the Gaussian is given as $\mathscr{F}[f(t)] = (2\pi\sigma^2)^{1/2} \exp[-u^2/2(\hat{\sigma})^2]$, i.e., the FT of a Gaussian of standard deviation σ and unit amplitude is another Gaussian of standard deviation $\hat{\sigma} = (2\pi\sigma)^{-1}$ and amplitude $(2\pi\sigma^2)^{1/2}$.

The power spectrum is given as 〈amplitude/magnitude, phase〉. Assume that the FT is written as $F(\omega) = \text{Re}[F(\omega)] + j\,\text{Im}[F(\omega)]$ and define the 〈magnitude, phase〉 pair as 〈P, ψ〉 where

$$P(\omega) = |F(\omega)|$$

$$\psi(\omega) = \tan^{-1}\left\{\frac{\text{Im}[F(\omega)]}{\text{Re}[F(\omega)]}\right\}. \qquad (2\text{-}10a)$$

Recovery of the original image function can be accomplished using the inverse FT. However, either the magnitude or phase (Oppenheim and Lim, 1981) alone could be used to approximately reconstruct the original image. Furthermore, localized spectral information as it pertains to cojoint representations, yet to be introduced, could further enhance the reconstruction process.

The energy E of a function $f(t)$ is defined as

$$E = \int_{-\infty}^{\infty} |f(t)|^2 dt \tag{2-10b}$$

assuming that $|f(t)|^2$ is absolutely integrable, i.e., $E < \infty$. It is easy to show that the FT preserves the energy and to write the Rayleigh's theorem:

$$E = \int_{-\infty}^{\infty} |F(\omega)|^2 d\omega = \int_{-\infty}^{\infty} |f(t)|^2 dt. \tag{2-10c}$$

One can also define the 1D power spectrum corresponding to the HT (Eq. 2-8c) as the pair $\langle P, \psi \rangle$ where

$$\langle P, \psi \rangle = \langle \{[HT(-u)]^2 + [HT(u)]^2\}^{1/2},$$
$$\tan^{-1}[HT(-u)/HT(u)] + 45° \rangle \tag{2-10d}$$

Note that the HT does not obey Rayleigh's Theorem given by Eq. (2-10c) (Bracewell, 1978).

Two functions of great interest in visual processing are autocorrelation and correlation. The correlation R_{gf} of two functions $g(t)$ and $f(t)$ given as $g \circledast f$, is

$$R_{gf}(\alpha) = g \circledast f = \int_{-\infty}^{\infty} g(\alpha + t)f(t)dt. \tag{2-11a}$$

The correlation $\circledast$ is similar to the convolution $\otimes$, defined in Eq. (2-3b), except that now the function $g(t)$ is not reflected about the origin. If $g \equiv f$ then the correlation simplifies to auto-correlation $R_f(\alpha)$, which is defined as

$$R_f(\alpha) = f \circledast f = \int_{-\infty}^{\infty} f(\alpha + t)f(t)dt. \tag{2-11b}$$

The correlation and auto-correlation functions measure the relative matching of two functions for different shifts given by α. The auto-correlation function $R_f(\alpha)$ has a maximum at $\alpha = 0$, and thus the periodicity of a function could be tested based on the peaks, if any, of $R_f(\alpha)$. (A function $f(t)$ is periodic with period T if and only if $f(t) = f(t + nT)$.) It is easy to show that the FT of the auto-correlation function of $f(t)$ is the power spectrum of $f(t)$. Specifically,

$$P(\omega) = \mathscr{F}[R_f(\alpha)] = \mathscr{F}[f(t) \circledast f(t)] = \mathscr{F}[f(t) \otimes f(-t)] = F(\omega)F(-\omega)$$
$$= F(\omega)F^*(\omega) = |F(\omega)|^2, \tag{2-11c}$$

where we used the convolution theorem, and the fact that for real functions the FT is Hermitian, i.e., $F(\omega) = F^*(-\omega)$, and F^* is the complex conjugate of F.

Linear systems theory aids system identification. A system is fully characterized by either the transfer function or its impulse response because of the convolution theorem. Assume that pairs $\langle f(t), F(\omega) \rangle$, $\langle g(t), G(\omega) \rangle$, and $\langle h(t), H(\omega) \rangle$ correspond to the input signal, $\langle$impulse response, transfer function$\rangle$, and output signal, respectively. For linear systems $h = g \otimes f$, or equivalently, $H(\omega) = G(\omega)F(\omega)$. To identify a system means to uncover either $G(\omega)$ or $g(t)$. The power spectra for the input and output signals are usually known, and if one can assume that $F(\omega) \neq 0$, it follows that $G(\omega) = H(\omega)/F(\omega)$, and $g(t) = \mathscr{F}^{-1}[G(\omega)]$. We considered in Eq. (2-4) a system characterized by motion blur and given in terms of unknown velocities $\alpha(t)$, $\beta(t)$. The blur was caused over integration time E, and the recorded output had a cyclic spectra given as $H(u, v)$. It is easy to see that

$$H(u, v) = \int\!\!\!\int_{-\infty}^{\infty} h(x, y)\exp[-j2\pi(ux + vy)]dxdy$$

$$= \int_{-E/2}^{E/2} \int\!\!\!\int_{-\infty}^{\infty} f[x - \alpha(t), y - \beta(t)]\exp[-j2\pi(ux + vy)]dxdydt$$

$$= \int_{-E/2}^{E/2} \int\!\!\!\int_{-\infty}^{\infty} f(\xi, \eta)\exp[-j2\pi(u\xi + v\eta)]d\xi d\eta$$

$$\times \exp\{-j2\pi[u\alpha(t) + v\beta(t)]\}dt$$

$$= F(u, v)G(u, v),$$

where $G(u, v) = \int_{-E/2}^{E/2} \exp\{-j2\pi[u\alpha(t) + v\beta(t)]\}dt$. We seek to deblur the image, i.e., to recover the original spatial function $f(x, y)$ by removing the effects due to motion. If motion can be appropriately characterized in terms of a moving object's velocity (see Section 5.1.2. for optical flow derivation) then $G(u, v)$ can be estimated, and the original function $f(x, y)$ is recovered solving the Fredholm integral using inverse filtering methods (Andrews and Hunt, 1977).

We conclude this section by considering two widely used image processing operations—differentiation and Laplacian. Both are associated with edge detection, i.e., finding transitions in the picture value function, and/or emphasizing the contrast of a picture. We explain this last effect in terms of the transfer function, while the case of optimal edge detection is discussed in Section 4.2.1. The partial differential operators are $(\partial f/\partial x)$ and $(\partial f/\partial y)$ for a given function $f(x, y)$. Assuming bounded functions (over space and/or time) and using integration by parts, one can derive that

$$\mathscr{F}\left[\frac{\partial f}{\partial x}\right] = j2\pi u F(u, v) \text{ and } \mathscr{F}\left[\frac{\partial f}{\partial y}\right] = j2\pi v F(u, v), \qquad (2\text{-}12)$$

where $F(u, v) = \mathscr{F}(f)$. Using this last result and defining the Laplacian operator $(\nabla^2 f)$ as

$$\nabla^2 f = \frac{\partial^2 f}{\partial x^2} + \frac{\partial^2 f}{\partial y^2}, \qquad (2\text{-}13\text{a})$$

one derives that

$$\mathscr{F}[\nabla^2 f] = -(2\pi)^2(u^2 + v^2)F(u, v). \qquad (2\text{-}13\text{b})$$

An alternative proof of Eqs. (2-12) and (2-13) can be obtained using Eq. (2-7), which states that FT is a special case of LT, and that $\mathscr{L}(f') = sF(s) - f(0)$ and $\mathscr{L}(f'') = s^2F(s) - sf(0) - f'(0)$.

We see that applying differential operators and/or the Laplacian is equivalent to a system characterized by transfer functions $H(u, v)$ given as $\langle j2\pi u, j2\pi v \rangle$ and $-(2\pi)^2(u^2 + v^2)$, respectively. Those functions are *high-pass* filters, i.e., they emphasize the high frequencies at the expense of the low frequencies, which explains why the contrast of an image "sharpens" when applying such differential operators.

2.1.3. The Discrete Fourier Transform

The Z transform (ZT) of a sequence $f(n)$ is defined as

$$F(z) = \sum_{m=0}^{\infty} f(m)z^{-m}. \qquad (2\text{-}14)$$

If one makes the substitution $z = r[\exp(j\omega)]$ and $r = 1$, i.e., z is a complex variable of m samples on the unit circle, then the FT is obtained as

$$F(\omega) = \sum_{m=0}^{\infty} f(m)\exp(-j\omega m). \tag{2-15}$$

Finally, if one assumes (discrete and) finite duration sequences $f(m)$, and if W_N is the Nth root of unity, i.e., if $W_N = \sqrt[N]{1} = \exp(-j2\pi/N)$, the discrete Fourier transform (DFT) is defined as $F(n)$, where

$$F(n) = \sum_{m=0}^{N-1} f(m)\exp[-j2\pi nm/N] = \sum_{m=0}^{N-1} f(m)W_N^{nm}. \tag{2-16a}$$

The inverse DFT is defined as

$$f(m) = \frac{1}{N}\sum_{n=0}^{N-1} F(n)W_N^{-nm}. \tag{2-16b}$$

As with the FT, one can define the power spectrum (magnitude and phase) for the DFT as well. First one can show that for real sequences $f(m)$, $0 \le m \le 2M - 1$, $N = 2M$, $F(M + k) = F^*(M - k)$. Then, using the DFT definition, one can write the Parseval identity as

$$\sum_{m=0}^{2M-1} f^2(m) = \frac{1}{2M}\sum_{n=0}^{2M-1} |F(n)|^2, \tag{2-17}$$

which is the discrete equivalent of the Rayleigh's theorem given in Eq. (2-10c). Define then $F^2(0)/N^2$ as the dc power, and $\langle F(n), \tan^{-1} \text{Im}[F(n)]/\text{Re}[F(n)]\rangle$ as the $\langle$amplitude, phase spectrum$\rangle$ pair $\langle P(n), \psi(n)\rangle$. Note that $F(n) = |F(n)|\exp[j\psi(n)]$.

There are two major issues to face when dealing with sampled sequences, which result from sampling m data items at equal intervals of time and space. The Shannon *sampling theorem* sets the minimum rate for sampling a bandwidth-limited continuous signal so it can be uniquely recovered from its samples. The theorem states that the sampling frequency f_N must be at least twice the bandwidth of the signal f_s (i.e., the largest frequency in the signal). The sampling frequency is called the Nyquist frequency and $f_N = 2f_s$. If the sampling theorem is not obeyed, aliasing results. This occurs when a continuous signal is sampled at such a rate that the translated versions of its spectrum (created by the sampling process) overlap in the frequency domain. This prevents recovery of the (continuous) original signal from the (discrete) sampled signal. To avoid aliasing, one must sample at twice the rate of the highest frequency in the continuous signal, i.e., the sampling must be done at the Nyquist rate.

Let us now consider the effects of digital masks using transfer functions. Assume two discrete 1D masks $g_1(i) = [1\ 2\ 1]$ and $g_2(i) = [-1\ 2\ -1]$. Using Equation (2-15), the corresponding transfer functions defined as $G(\omega) = \Sigma_{i=-N}^{N} g(i)\exp(-j\omega i)$ are found as $G_1(\omega) = 2 + 2\cos(\omega)$ and $G_2(\omega) = 2(1 - \cos(\omega))$. If $\omega \in (-\pi/2, \pi/2)$, then the masks $g_1(i)$ and $g_2(i)$ are low- and high-pass (impulse response) filters. Using discrete convolution, as defined in Eq. (2-3c), one finds that the 2D mask $g(i, j)$ corresponding to the original 1D mask $g_1(i)$ is given as

$$g(i, j) = g_1^T(i) \otimes g_1(i) = \begin{bmatrix} -1 & 2 & -1 \\ -2 & 4 & -2 \\ -1 & 2 & -1 \end{bmatrix},$$

and that the transfer function $G(\omega_1, \omega_2)$ is given as $G(\omega_1, \omega_2) = G(\omega_1)G(\omega_2) = 4(1 + \cos\omega_1)(1 - \cos\omega_2)$, and it corresponds to a high-pass filter, i.e., high frequencies are emphasized at the expense of lower frequencies.

Another example is that of averaging and then analytically evaluating the filtering effect. Assume a finite-impulse response (FIR) model of order $k = 1$, i.e.,

$$y(n) = \tfrac{1}{2}[x(n) + x(n - 1)] \qquad n \geq 0.$$

Then using the ZT, one readily obtains $Y(z) = \tfrac{1}{2}[X(z) + z^{-1}X(z)]$, where $Y(z) = ZT(y)$ and $X(z) = ZT(x)$, and that the transfer function is given by $G(z) = Y(z)/X(z) = 1/2(1 + z^{-1})$. Remembering that $z = \exp(j\omega)$ and that $\omega T = 2\pi f/f_s = \pi v$, where T is the period of the sequence, $f_s = (1/T) = 2f_N$, and $0 \leq v \leq 1$, one can then obtain the transfer function in terms of the gain ($|G|$), and phase response (ψ):

$$|G(v)|^2 = \tfrac{1}{2}[1 + \cos(\pi v)] \text{ and } \psi(v) = -\pi v/2.$$

It is clear that the gain is characteristic to a low-pass filter as one would expect for averaging (i.e., contrast is deemphasized).

Finally, we consider again the Laplacian operator defined in Eq. (2-13a). Using finite differences, one can write that $\partial/\partial x[f(x, y)] = f(x + 1, y) - f(x, y)$, and that $\partial^2/\partial x^2[f(x, y)] = f(x + 1, y) - 2f(x, y)$

$+ f(x - 1, y)$. The $(\nabla^2 f)$ Laplacian operator can then be written as

$$\nabla^2 f \cong \frac{\partial^2}{\partial x^2} [f(x, y)] + \frac{\partial^2}{\partial y^2} [f(x, y)]$$

$$= \frac{f(x + 1, y) - 2f(x, y) + f(x - 1, y)}{4}$$

$$+ \frac{f(x, y + 1) - 2f(x, y) + f(x, y - 1)}{4},$$

if $\partial^2/\partial y^2 [f(x, y)]$ is derived similarly to $\partial^2/\partial x^2 [f(x, y)]$. If $(\nabla^2 f)$ is equal to zero, then a discrete (digital) mask for the Laplacian results, and it is given by

$$\nabla^2 f = \begin{bmatrix} 0 & -1 & 0 \\ -1 & 4 & -1 \\ 0 & -1 & 0 \end{bmatrix}. \tag{2-18a}$$

If eight-neighbor rather than four-neighbor connectivity is used, one could easily derive the Laplacian as

$$\nabla^2 f = \begin{bmatrix} -1 & -1 & -1 \\ -1 & 8 & -1 \\ -1 & -1 & -1 \end{bmatrix}. \tag{2-18b}$$

Setting the Laplacian equal to zero amounts to smoothing an image, and this can be corroborated from the newly found value of $f(x, y)$ as given by

$$f(x, y) = \tfrac{1}{4}[f(x + 1, y) + f(x - 1, y) + f(x, y + 1) + f(x, y - 1)].$$

The corresponding ZT is given by $G(z_1, z_0) = \tfrac{1}{2}[(z_1 + z_1^{-1})/2 + (z_2 + z_2^{-1})/2]$. By setting z equal to $\exp(j\omega)$ one obtains a low-pass filter given by $G(\omega_1, \omega_2) = 1/2[\cos(\omega_1) + \cos(\omega_2)]$.

2.2. Scale Space

Scale space representations and analysis extend from the simple idea, which can be visually tested, that different characteristics of an image reveal themselves at different levels of resolution. This section presents basic multiscale concepts and leads to cojoint representations, which are discussed in Section 2.3.

2.2.1. *Heat Equation*

Chapter 4 is largely dedicated to the analogy between physical systems and PDP models of visual processing. The heat equation provides a similar rationale for the multiscale concept (Hummel, 1987).

Assume $f(x, y)$ to be a function of spatial variables x and y. The heat equation seeks a function $u(x, y, t)$ such that after t units of time, the initial function $f(x, y)$ will diffuse to $u(x, y, t)$. Formally, one seeks to solve the following equation

$$\frac{\partial u}{\partial t} = \nabla^2 u \tag{2-19}$$

$$u(x, y, 0) = f(x, y)$$

where the last equality plays the role of a boundary condition, and the ∇^2 is the Laplacian operator. The solution to the heat equation is obtained by blurring $f(x, y)$ with Gaussians of increasing width. Blurring is equivalent to convolution, and one thus writes

$$u(x, y, t) = \iint\limits_{R^2} G(x', y', t)f(x - x', y - y')dx'dy' \tag{2-20a}$$

or

$$u(x, y, t) = G(x, y, t) \otimes f(x, y), \tag{2-20b}$$

where $G(x, y, t) = 1/4\pi t \exp[-(x^2 + y^2)/4t]$. For any $t > 0$, $G(x, y, t)$ is a Gaussian centered at $(0, 0)$ with standard deviation $\sigma = \sqrt{2t}$, and if $t = 0$, then $G(x, y, t)$ is an impulse distribution that sifts out $u(x, y, 0)$ (see Eq. (2-5b)). The time t is equivalent to the level l within a pyramid (see Section 2.2.2.), and the scale space can be defined in terms of spatial location (x, y) and width (σ) of the blurring Gaussian. Note that the Gaussian itself is a solution to the heat equation, i.e., it obeys the equation $\partial G/\partial t = \nabla^2(G)$, and that one can also write

$$\frac{\partial G}{\partial t} = \lim_{\alpha \to 0} \frac{G(x, y, t + \alpha) - G(x, y, t)}{\alpha}. \tag{2-21}$$

One observes that the difference of Gaussians (DOG) approximates the

$\nabla^2(G)$ operator, i.e., as the difference in the widths of the two Gaussians approaches zero, the DOG approximates the Laplacian of a Gaussian. The DOG is a filter that can derive a multiscale representation. Assume that we again consider the function $f(x, y)$, and that based on the properties of differential operators and convolution, one can write $\hat{f}(x, y, t)$, the filtered image, as any of the following expressions:

$$\hat{f}(x, y, t) = \nabla^2(G) \otimes f$$
$$= G \otimes \nabla^2(f)$$
$$= \nabla^2(G \otimes f). \qquad (2\text{-}22)$$

Thus, the multiscale representation can be obtained by either (i) filtering the original image with the Laplacian of a Gaussian (or alternatively with the DOG), (ii) convolving the Laplacian of the original image with Gaussians of increasing width, or (iii) taking the Laplacian of the images resulting from convolving the original image with Gaussians of increasing width.

If the domain of integration R^2 in Eq. (2-20a) is infinite, then $\hat{f}(x, y, t) \to 0$ as $t \to \infty$, i.e., eventually the heat dissipates. One can consider alternative formulations of the heat equation, like the Dirichlet boundary problem, where heat diffuses within a cylinder (Hummel, 1987). Assume that the base of the cylinder is given by B and its (boundary) walls by ∂B. Then, the boundary conditions can be rewritten as $u(x, y, 0) = f(x, y)$ if $(x, y) \in B$, and $u(x, y, t) = f(x, y)$ if $(x, y) \in \partial B$ and $t \geq 0$. The multiscale representation can be potentially used for reconstructing the original image, as

$$f(x, y) = -\int_0^L \hat{f}(x, y, t)dt + u(x, y, L), \qquad (2\text{-}23)$$

where $L = \infty$, or L is the height of the cylinder given by B and ∂B. This last result follows from the fundamental theorem of integral calculus $\int_a^b f(x)dx = F(b) - f(a)$, where $F'(x) = f(x)$ with the substitutions $\partial u/\partial t = \hat{f}$, and $u(x, y, 0) = f(x, y)$.

We demonstrated earlier that the FT of a Gaussian is also a Gaussian, but of different amplitude and width. If an image is blurred with a Gaussian (like in DOG) then the Gaussian acts like a transfer function, and it represents a band-limiting (of some width) operator in the frequency domain. Therefore, the multiscale representation is a multichannel representation in the frequency domain where a channel corresponds to some specific bandwidth.

Discrete versions corresponding to the heat equations can be simulated and solved (see the WARP parallel systolic array in Section 9.2.3.). As an example, Hummel (1987) approximates the 1D heat Eq. (2-19) by

$$u(i, t + 1) - u(i, t) = \tfrac{1}{4}u(i + 1, t) - \tfrac{1}{2}u(i, t) + \tfrac{1}{4}u(i - 1, t), \quad (2\text{-}24a)$$

where $u(i, 0) = f(i)$. One can define a corresponding discrete Gaussian kernel $G(i, t)$ as

$$G(i, t) = \frac{1}{4^t}\begin{bmatrix} 2t \\ i + t \end{bmatrix}, \qquad (2\text{-}24b)$$

where $[^n_t] = \frac{n!}{t!(n-t)!}$, $[^n_t] = 0$ for $t < 0$ or $t > n$, and $[^n_t] = [^{n-1}_{t-1}] + [^{n-1}_{t}]$. It is easy to verify that the kernel $G(i, t)$ as given by Eq. (2-24b) satisfies the discrete heat equation as given by Eq. (2-24a), and that the filtered data is obtained as $u(i, t) = G \otimes f$. The recursive solution easily follows, and it is given by

$$u(i, t + 1) = \tfrac{1}{4}u(i - 1, t) + \tfrac{1}{2}u(i, t) + \tfrac{1}{4}u(i + 1, t). \qquad (2\text{-}24c)$$

2.2.2. *Pyramids*

The heat equation discussed previously suggests multiscale representations. Indeed, a vast body of research can be considered within this context. Rosenfeld and Thurston (1971) realized early the advantages of simultaneous use of different size operator masks to increase sensitivity to edges of variable resolution and also to withstand noise. Soon thereafter, the fact that masks of different sizes corresponded to the spatial-frequency channels became apparent.

An efficient structure for implementing multiscale representation and computation was labeled as *pyramid*. Assuming a picture-value function $f(x, y)$ defined over a digital grid, one can define the pyramidal structure as a collection of subsampled images connected by some mapping transformation. The first level of the pyramid, labeled as $f_0(x, y) = f(x, y, 0) = f(x, y)$, is of the same size as the original input ($N \times N$) and sits at the bottom. The level l of a pyramid corresponds to the time t used in the heat equation. Successive layers are defined using the mapping T such that $T: f_l \to f_{l+1}$ takes place, where $T: f(x, y, l) \to f(x, y, l + 1)$. Eventually, the top of the pyramid is reached, one pixel only. One example of a mapping is that of averaging 2×2 arrays, and from our earlier discussion on the relationship between masks and filtering, it corresponds to low-pass filtering. The resulting pyramidal data structure (Samet, 1989a; 1989b), known as a quad tree, stands for a multichannel system where finer details, corresponding to high spatial frequency components, tend to disappear as one climbs the pyramid. (The phenomena of visual defocusing parallels such behavior.) Burt and Adelson (1983) considered the implementation of the pyramidal structure as the Laplacian of successively blurred images. A pyramid of such blurred images is known as the Gaussian pyramid, and, as we already saw in Eqs. (2-21) and (2-22), the Laplacian used in the Gaussian pyramid is equivalent to the DOG of the same pyramid.

The three basic steps in implementing such an approach are blur (or to reduce), expand (to make two levels of the pyramid compatible in size), and difference (to subtract). The blur is achieved using convolution masks, which obey prespecified constraints: (i) normalization, (ii) symmetry, (iii) unimodality, and (iv) equal contribution to the next level. Specifically, for a convolution mask $k(m, n)$, known also as a generating kernel, those constraints can be written as

(i) $\quad \displaystyle\sum_{m=0}^{K-1} \sum_{n=0}^{K-1} k(m, n) = 1;$

(ii) $\quad k(m, n) = k(K - 1 - m, n) = k(m, K - 1 - n)$
$$= k(K - 1 - m, K - 1 - n); \hspace{3cm} (2\text{-}25)$$

(iii) $\quad 0 \le k(m, n) \le k(p, q) \text{ for } m \le p < \dfrac{K}{2}, n \le q < \dfrac{K}{2}; \text{ and}$

(iv) $\quad \displaystyle\sum_{i=0}^{K-1} \sum_{j=0}^{K-1} k(m + 2i, n + 2j) = \tfrac{1}{4} \quad \text{for } m, n = 0, 1,$

where K is the size of the kernel. From the viewpoint of signal processing, one can still question if there is some optimal way for deriving the corresponding levels of the pyramid in terms of the transfer function and corresponding filtering effects. Meer *et al.* (1987) responded by describing the construction of image pyramids as a two-dimensional decimation process. The optimality of the kernel masks is related to minimal information loss after the reduction in resolution, and they correspond to an ideal low-pass filter. An ideal low-pass filter is specified by the transfer function $K(\omega_1, \omega_2) = \{1$ for $0 \le |\omega_1| < a$, $0 \le |\omega_2| < b$, and 0 elsewhere}. The rectangular passband is the region where $K(\omega_1, \omega_2) = 1$. Assuming that the generating kernels and their corresponding Fourier transforms are given by $k(m, n)$ and $K(\omega_1, \omega_2)$, respectively, the similarity to an ideal low-pass filter can be expressed in terms of the following two spectral indexes:

(a)   The antialiasing (see Section 2.1.3.) index η indicates the relative amount of spectrum folded back (aliased into the spectral domain of interest $(0, \pi/2)$) after the sampling rate reduction

$$\pi = \frac{\iint_0^{\pi/2} |K(\omega_1, \omega_2)| d\omega_1 d\omega_2}{\iint_0^{\pi} |K(\omega_1, \omega_2)| d\omega_1 d\omega_2}. \tag{2-26a}$$

(b)   The filtering index μ indicates the amount of deviation from the ideal low-pass filter in the spectral domain of interest $(0, \pi/2)$

$$\mu = \iint\limits_0^{\pi/2} |1 - |K(\omega_1, \omega_2)|| d\omega_1 d\omega_2. \tag{2-26b}$$

An optimal kernel with good antialiasing characteristics has η close to one. The filtering index μ must be as small as possible in order to yield a filter that is maximally flat in the passband. The kernels corresponding to unweighted averaging of pixels yield for 2×2 masks, the pair $(\eta, \mu) = (0.5, 0.467)$, and for 4×4 masks, the pair $(\eta, \mu) = (0.598, 1.579)$. As such, both kernels have poor antialiasing characteristics and deviate from the ideal response. The ideal low-pass filter is not physically realizable because the corresponding kernel given as

$$k(m, n) = \frac{\sin m \dfrac{\pi}{2}}{m\pi} \cdot \frac{\sin n \dfrac{\pi}{2}}{n\pi}$$

(for $m, n = 0, \pm 1, \pm 2, \ldots$) is one of infinite extent. One must be satisfied with a physically realizable low-pass filter, where in the passband, the magnitude remains close to one, in the transition band, the magnitude drops fast to zero, and in the stopband, the magnitude remains close to zero. Such a filter is called the equiripple one and is characterized by the passband (δ_p) and stopband (δ_s) ripples; for a given tolerance, the length of the filter increases as the transition band narrows. The optimal sequence for a desired frequency response described by the passband (0 to θ_p), transitionband (θ_p to θ_s), and stopband (θ_s to π) (i.e., the pair $\langle \theta_p, \theta_s \rangle$), the tolerated ripples (δ_p, δ_s) and the length of the sequence are obtained by using optimization methods (McClellan *et al.*, 1973). Levels obtained using such an approach and corresponding experimental results from Meer *et al.* (1987) are shown in Fig. 2.2.

The multiscale and pyramid concepts are widely used. Characteristics of such efforts are the peaks and ridges in the difference of low-pass transforms (DOLP) (Crowley and Parker, 1984) (a DOG-like representation) as the underlying representation for shape. The wavelet representation introduced by Mallat (1989) suggests that, for efficiency reasons, successive layers of the pyramid should include only the additional details, which are not already available at preceding levels. To avoid redundancy in the stored information to be analyzed, one should seek only the difference of information and process it accordingly. Such an approach is akin to differential coding. Furthermore, the wavelet functions belong to $L^2(R)$ (which includes all square integrable functions such as the Gabor functions) and they form an orthonormal basis, while the Gabor functions do not. Another approach, which parallels that of a multichannel (pyramid) system, is that of the fractals, discussed next.

2.2.3. *Fractals*

Much credit on the work on fractals goes to Mandelbrot (1982). Mandelbrot coined *fractal* from the Latin adjective *fractus*, which corresponds to the Latin verb *frangere*, meaning to break or create irregular fragments. A theoretical fractal is self-similar under all magnifications and property changes are limited. Earlier work on natural imagery led Pentland (1983) to conclude that textures (see Section 2.4.1.) are fractal-like. Even though texture synthesis based on

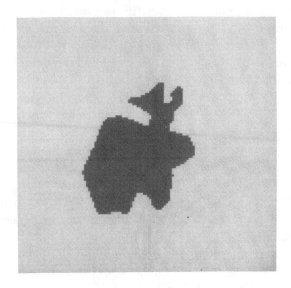

(a)

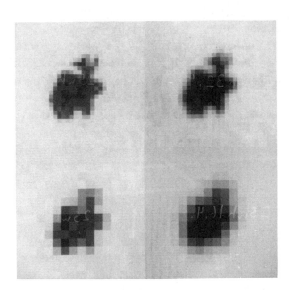

(b)

Figure 2.2.
Pyramid levels.

fractals for Lucas's Star Wars-type movies is quite successful, the conjecture made by Pentland has been refuted by Peleg *et al.* (1984). Originally, Mandelbrot reported that for many coastlines, the length $\mathscr{L}$, which is a function of the yardstick ε being used, is given as

$$\mathscr{L}(\varepsilon) \approx L\varepsilon^{1-\mathscr{D}}, \tag{2-27a}$$

where L is the true length, and $\mathscr{D}$ is the fractal dimension. (For a straight line, $\mathscr{D} = 1$.) Assuming a fractal wave, $\mathscr{D}$ is independent of ε, and the plot of $\mathscr{L}(\varepsilon)$ vs. ε on a log-log scale yields a straight line of slope $(1 - \mathscr{D})$. The area of fractal surfaces has been reported by Mandelbrot to behave like

$$\mathscr{A}(\varepsilon) \approx A\varepsilon^{2-\mathscr{D}}. \tag{2-27b}$$

One can easily be tempted to use changes in the measured areas vs. changing scale as a fractal signature for texture analysis. The 2D fractal dimension $\mathscr{D}$ can be estimated as

$$\mathscr{D} = T + (1 - H) \tag{2-27c}$$

assuming that the two-dimensional power spectrum $P(\omega)$ can be approximated by

$$P(\omega) \cong \omega^{-2(H+1)}, \tag{2-27d}$$

and that T is the topological dimension (i.e., for 2D the fractal dimension is $\mathscr{D} = 3 - H$). The suggestion made by Peleg *et al.* (1984) is that although textures are usually not fractals for the entire scale range, the change in those measurements is still helpful in characterizing texture. "Frequency" information about texture can then be inferred in the spatial domain without using the frequency domain. Specifically, assume that the (surface) area is defined using a blanket of thickness 2ε, which is defined by its upper and lower surfaces, U_ε and B_ε, respectively. Initially, $U_0(x, y) = B_0(x, y) = f(x, y)$, where $f(x, y)$ is the picture value function, and as ε increases, more and more textural details are suppressed. For $\varepsilon = 1, 2, 3, \ldots$

$$U_\varepsilon(x, y) = \max\left\{U_{\varepsilon-1}(x, y) + 1, \max_R U_{\varepsilon-1}(k, l)\right\} \tag{2-28a}$$

$$B_\varepsilon(x, y) = \min\left\{B_{\varepsilon-1}(x, y) - 1, \min_R B_{\varepsilon-1}(k, l)\right\}, \tag{2-28b}$$

where $R = \{(k, l); \text{distance } [(k, l), (x, y)] \leq 1\}$. Define then

$$V_\varepsilon = \sum_{x,y} [U_\varepsilon(x, y) - B_\varepsilon(x, y)] \qquad (2\text{-}28\text{c})$$

$$A(\varepsilon) = \frac{V_\varepsilon - V_{\varepsilon-1}}{2}. \qquad (2\text{-}28\text{d})$$

Note that the denominator used in Eq. (2.28d) is 2 rather than 2ε. The characteristic signature $S(\varepsilon)$ for a given image texture is then obtained by fitting a straight line through the points $[\log(\varepsilon - 1), \log(A(\varepsilon - 1))]$, $[\log(\varepsilon), \log(A(\varepsilon))]$, and $[\log(\varepsilon + 1), \log(A(\varepsilon + 1))]$. For fractals $S(\varepsilon)$ should be equal to $(2 - \mathscr{D})$, but Peleg *et al.* found this wasn't the case for most textures. Classification and discrimination could still be implemented using a distance d defined as

$$d(\alpha, \beta) = \sum_\varepsilon [S_\alpha(\varepsilon) - S_\beta(\varepsilon)]^2 \log \frac{\varepsilon + \frac{1}{2}}{\varepsilon - \frac{1}{2}}, \qquad (2\text{-}28\text{e})$$

where S_α and S_β are the signatures corresponding to textures α and β, respectively.

Since the fractal dimension alone is not sufficient to characterize natural textures, Keller *et al.* (1989) introduce the concept of *lacunarity*, which captures the second-order statistics of fractal surfaces. The lacunarity Λ measures the discrepancy between the actual and expected areas of the fractal set and is defined (Mandelbrot, 1982) as

$$\Lambda = E[(M/E(M) - 1)^2] \qquad (2\text{-}28\text{f})$$

Lacunarity, which is small when the texture is fine and large when the texture is coarse, leads to improved segmentation of natural textures if used together with the fractal dimension (Keller *et al.*, 1989).

2.2.4. Human Visual System

Frequency analysis and linear systems are basic tools for signal processing. Much experimental work in both psychophysics and neurophysiology suggests that analogous tools work within the human visual system (HVS).

The contrast sensitivity function (CSF) is the graph of the physical contrast needed to perceive a grating vs. the grating's spatial frequency. The minimum level of perceived physical contrast is called the threshold, and the contrast sensitivity is its inverse. An experiment, suggested by Campbell and Robson (1968) (informally described in Fig. 2.3.), could involve two gratings, a sine and a square wave, each of them of 10 cycles/degree frequency. If the HVS performs some type of frequency analysis, then the perceived sine wave should remain unchanged, while the square wave is decomposed into a fundamental at 10 cycles/degree and an infinite series of harmonics with the second and third at 20 and 30 cycles/degrees, respectively. The CSF graph (Fig. 2.3b.) shows, however, that the HVS is not very sensitive to such harmonics and, consequently, both the sine and square wave look the same. The CSF, also known as the *modulation transfer function* (MTF), has been mostly analyzed for human subjects at rest. Steinman *et al.* (1985) consider the case when compensatory eye movements are needed to maintain the retinal image position of attended visual targets. They report that head movement, with its concomitant retinal image motion, produced a need for a modest increase in contrast at high spatial frequencies and reduced the need for contrast at low spatial frequencies. This is one of the many cases where motion is beneficial to perception.

Insight concerning the inner workings of the HVS is provided by "aftereffect illusions," which humans experience after being exposed to prolonged, constant stimulation. Such illusions allow the researcher to correlate neurophysiological to psychophysical findings and thus to explain the mechanisms of visual perception. Experiments involving aftereffect illusions include a period of prolonged adaptation to a stimulus (possibly a grating of specific attributes such as width and orientation). The adaptation period is preceded by an original test stimulus and followed by the same test stimulus once again. The difference between the final perceived appearance of the test stimulus and its original appearance can be attributed to fatigue caused by the adaptation period and can thus indicate to what the HVS is sensitive. The observed sensitivity to width and/or orientation in the case of grating experiments seems again to suggest that the HVS includes mechanisms responsible for frequency analysis. Decreased sensitivity is evident in the post adaptation CSF, and the observed "drops" correspond to those frequencies the subject has been exposed to during the adaptation period.

Assuming that the HVS performs some type of frequency analysis, the question remains as to the size of the support area where such

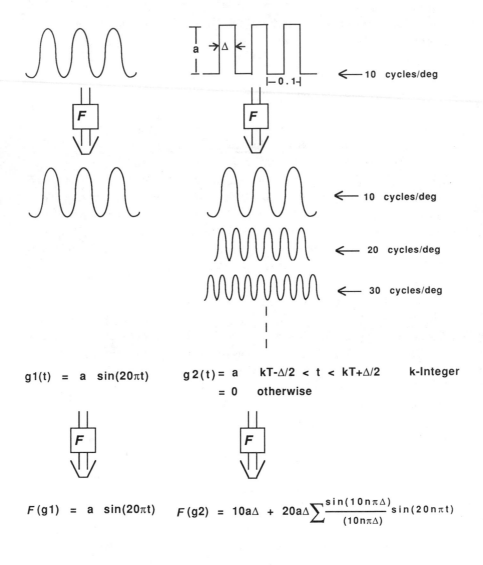

$$g1(t) = a \ \sin(20\pi t) \qquad g2(t) = a \quad kT-\Delta/2 < t < kT+\Delta/2 \qquad k\text{-Integer}$$
$$= 0 \quad \text{otherwise}$$

$$F(g1) = a \ \sin(20\pi t) \qquad F(g2) = 10a\Delta + 20a\Delta \sum \frac{\sin(10n\pi\Delta)}{(10n\pi\Delta)} \sin(20n\pi t)$$

(a)   Spatial Decomposition of Gratings using Fourier Series

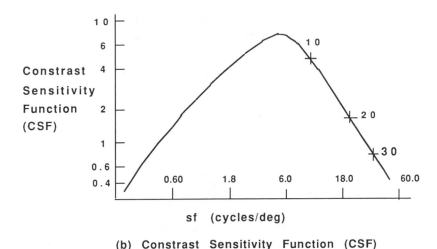

(b) Constrast Sensitivity Function (CSF)

Figure 2.3.
Evidence for the HVS performing frequency analysis.

analysis is performed. The size relates to resolution, which Section 2.3. covers. We have already noted, however, some results related to this issue. Gagalowicz (1981), Julesz and Bergen (1983), and Caelli (1985) offer strong evidence to the effect that visual perception is a local process. The relatively poor performance exhibited by early frequency analysis methods (which were global in nature) (Connors and Harlow, 1980; Weszka et al., 1976) can thus be easily explained.

Hubel and Wiesel (1962) defined simple cells in the visual (striate) cortex as early relay stations, orientation sensitive, where the receptive field (RF) is made up of *on* (excitatory)/*off* (inhibitory) regions. They then showed that a simple cell's response to bars and edges could be predicted from its corresponding receptive fields. Such experiments led to the unfounded belief that simple cells act as edge detectors. The use of sine-wave gratings as visual test stimuli showed that the HVS performs some kind of frequency analysis. Additional research, such as that of Wilson and Bergen (1979), made the concept of a multichannel frequency analysis system well established. The results of Hubel and Wiesel (1962) could be easily accounted for by a multichannel system. The degree to which a simple cell's response to bars could be predicted from its response to sine-wave gratings was quantitatively studied by several groups including DeValois et al. (1979).

There is growing interest in the RF profiles and their definition. If we remember the heat equation, the pyramid approach is only a rough solution. From Eq. (2-22) we remember that one of the solutions to the heat equation is filtering the original image with the Laplacian of a Gaussian (or alternately with the DOG). Koenderink and van Doorn (1987) suggest that the DOG-like solution is the equivalent of an appropriate RF and that convolving such RF with the original image yields the exact partial derivatives of the retinal illuminance blurred to a specified degree. Remember that convolution is equivalent to filtering and that a multichannel approach could result if such RF are thought of as multichannel filters. The 1D RF look like Gabor functions (to be briefly discussed next and then again in Section 2.3.7.) and in the frequency domain the corresponding bandpass filter is given as $\phi_n(\omega, t) = (j\omega)^n \exp(-\omega^2 t)$, where n is the order of the derivative. The bandwidth is approximately constant and asymptotically equal to $\Delta\omega = 1/(2\sqrt{t})$. The amplitude ratios are independent of position, while the phase angles are proportional to position. The spatial form of the RF is (asymptotically) a sine or cosine (for odd and even orders, respectively, and thus, a phase difference of exactly $\pi/2$) of spatial (circular) frequency $\omega_n = \sqrt{[(n + 1)/2]/t}$ and modulated by a Gaussian envelope of halfwidth $4\sqrt{t}$. The width of the RF depends on the differentiation order and is given as $\Delta x \approx 2\sqrt{t}$, i.e., it is almost independent of the order. Hence, the uncertainty relation (see Section 2.3.) for order n, given as $\Delta x \Delta\omega \geq 1$, holds. This discussion easily extends to the 2D case.

Porat and Zeevi (1988) and Clark and Bovik (1989) worked on the Gabor representation for early vision. A distributed architecture, made up of multiple spatially and spectrally localized RF and defined as Gabor filters, yields an early low-level representation of the visual input. The 2D Gabor functions can be represented as

$$h(x, y) = g(x', y')\exp[2\pi j(u_0 x + v_0 y)] = g(x', y')\exp(2\pi j\omega_0 x'), \quad (2\text{-}29a)$$

where

$$(x', y') = (x \cos \theta_0 + y \sin \theta_0, -x \sin \theta_0 + y \cos \theta_0), \theta_0 = \tan^{-1}(v_0/u_0),$$

are rotated coordinates, and

$$g(x, y) = \frac{1}{2\pi\sigma_x\sigma_y} \exp\left[-\frac{(x/\sigma_x)^2 + (y/\sigma_y)^2}{2}\right], \quad (2\text{-}29b)$$

where $g(x, y)$ is a 2D Gaussian scaled by widths σ_x and σ_y, with width to length ratio $\gamma = \sigma_x/\sigma_y \geq 1$. Hence, $h(x, y)$ is a complex sinusoidal grating modulated by a 2D Gaussian with major axis oriented at an angle θ_o from the horizontal and of 2D center frequency (u_o, v_o) or radial frequency $\omega_o = (u_o^2 + v_o^2)^{1/2}$ (Clark and Bovik, 1989). The frequency response, i.e., the transfer function of $h(x, y)$ from Eq. (2-29a) is given by

$$H(u, v) = \exp\{-2\pi^2[(u' - \omega_o)^2\sigma_x^2 + (v'\sigma_y)^2]\}, \qquad (2\text{-}29c)$$

where $(u', v') = (u \cos \theta_o + v \sin \theta_o, -u \sin \theta_o + v \cos \theta_o)$. The transfer function corresponds to a Gaussian of width to length aspect ratio $1/\lambda$ oriented at an angle θ_o from the horizontal u axis. Note that $h(x, y)$ is complex and has real (cosine) and imaginary (sine) components given by

$$h_c(x, y) = g(x', y')\cos(2\pi\omega_o x')$$
$$h_s(x, y) = g(x', y')\sin(2\pi\omega_o x'). \qquad (2\text{-}29d)$$

The (transfer) response is identical for both $h_c(x, y)$ and $h_s(x, y)$ except for a phase difference of $\pi/2$. (The $\pi/2$ difference, known as the phase quadrature, was shown by Pollen and Ronner (1981) to characterize neighboring simple cells found in the cortex.) The distributed architecture we alluded to before can now be implemented in terms of RF filters given as $h(x, y)$ or alternately as $H(u, v)$. The distributed architecture thus yields a filtered image, where large values are expected over those (local) areas characterized by spatial frequency characteristics to which the filter is also tuned. Watson (1987) suggests a similar architecture, called the cortex transform, which is conceptually modeled after the Gabor-type RF discussed previously, yields a distributed representation in terms of both spatial and spectral localization and can be efficiently implemented.

2.2.5. Zero Crossings

The lowest-order isotropic (orientation-independent) operator is the Laplacian. From Section 2.2.1. and the Eq. (2-22) we remember that one of the solutions to the heat equation can be obtained by taking the Laplacian of the images resulting from convolving the original image with Gaussians of increasing width. Thus, the Laplacian can be used to detect intensity changes at different resolutions corresponding to appropriate levels of blurring. The corresponding $\nabla^2 G$ filter can be approximated by DOG (as we saw earlier), and neurophysiological studies seemed to suggest that the ratio of inhibitory to excitatory areas within the RF is about 1:1.6. Those findings prompted Marr (1982) to advance the concept of a primal sketch made up of zero crossings. Zero crossings make explicit the relevant information about a 2D image, primarily the intensity changes and their geometrical distribution and organization. Formally, zero crossings are thought of in the context of the heat equation and are those (x, y, t) for whom $\hat{f}(x, y, t)$, the solution to the heat equation (Eq. (2-22)) is zero. One argument Marr brought in support of his primal sketch theory was that the zero crossings could yield a discrete, symbolic representation that probably incurs no loss of information. The argument, disputed by Daugman (1983; 1988b), rests on a theorem stating that a one-octave band-pass signal can be completely reconstructed (up to an overall multiplicative constant) from its zero crossings.

Daugman (1983) reported impressive results, which cast great doubts about the validity of assuming zero crossing as an initial, early low-level image representation. He researched the relationship between the organizational principle of a 2D anisotropic spatial filter (elongation, concatenation of subunits, or differential operators mediated by lateral inhibition) and the resulting general consequences for spatial frequency and orientation selectivity. Examples brought by Daugman to disprove the feasibility of zero crossings included (i) the observation that the subharmonic functions have no zero crossings, but that they are still visible to the human eye; and (ii) that the zero crossings are linear in their operation, while the visual system is not. One familiar member of the subharmonic functions class is the sum of a sine-wave grating plus a parabola: $f(x, y) = \sin(\omega x) + Ax^2$, where $A > \omega^2/2$. Its Laplacian is $-\omega^2 \sin(\omega x) + 2A$, so this image, which humans easily detect visually, would be invisible to a visual system that operates according to Marr's proposed bank of Laplacian filters followed by zero-crossing detection. Another example would consider two different functions such as $f_1(x, y) = \sin(x)$, and $f_2(x, y) = \sin[x(2 + \cos x)]$. The corresponding filtered image $\hat{f}(x, y, t)$ has identical zero crossings at all levels of resolution, even that the original functions can be visually discriminated.

Daugman (1988b) showed additional examples of simple information-processing operations that are apparent in pattern and motion analysis and cannot be treated using zero crossings. A simple example will consider a class of textures whose 2D FT contains energy at only seven distinct (u, v) locations and is given by

$$F(u, v) = L_o \delta(u, v) + \frac{1}{2} \delta(u - u_o, v) + \frac{1}{2} \delta(u + u_o, v)$$

$$+ \frac{m}{4} \delta(u - u_o, v - v_o) + \frac{m}{4} \delta(u + u_o, v + v_o)$$

$$+ \frac{m}{4} \delta(u - u_o, v + v_o) + \frac{m}{4} \delta(u + u_o, v - v_o).$$

The mean luminance L_o ensures positive values for image intensities, and m satisfies the inequality

$$|m| < \frac{u_o^2}{u_o^2 + v_o^2}.$$

The corresponding family of images $f(x, y)$, using the FT of $\cos(\omega)$ derived in Section 2.2.2., is given by

$$f(x, y) = L_o + \cos(u_o x) + \frac{m}{2} [\cos(u_o x + v_o y) + \cos(u_o x - v_o y)].$$

The Laplacian of the Gaussian operator $\nabla^2(G)$ can be implemented by first computing the Laplacian of the family of images $f(x, y)$ and convolving the result with Gaussian of arbitrary widths σ. One can derive that

$$-\nabla^2 G_\sigma \otimes f(x, y) = -G_\sigma(x, y) \otimes \nabla^2 f(x, y)$$

$$= \exp(-u_o^2 \sigma^2) u_o^2 \cos(u_o x)$$

$$\times \left[1 + m \frac{u_o^2 + v_o^2}{u_o^2} \exp(-v_o^2 \sigma^2) \cos(v_o y) \right].$$

The term in brackets is purely positive because of the constraint on m given before. Then, the only zero crossings are those of the cosine term, which depend on u_o, but not on v_o, and the zeros are also independent of the scale σ. Thus, one can generate a family of image textures (corresponding to different values of v_o) and yet all images (and even their superposition) have identical zero crossings. The experiments performed by Daugman (1988b) on perceptual capabilities offer strong evidence against zero crossings as a model of early image representation in human vision.

Based on the before mentioned, a multichannel frequency system seems well founded and what remains open is the exact form such a system might assume. We address this very question in the next section, where we treat in great detail the issue of cojoint spatial/ spatial-frequency (s/sf) representations. The results are summarized in the concluding remarks (Section 2.3.11.) and again in the section dedicated to texture (Section 2.4.). Readers who may be inclined to skim the section on a first reading will miss the discussion on redefining resolution within the solid mathematical framework provided by signal processing theory. Strong arguments are brought forth in support of distributed cojoint s/sf representations for low-level visual representations.

2.3.   Cojoint Spatial/Spatial-Frequency Representations

We first explore current interest in cojoint signal representations for visual processing and shape representation problems. Cojoint resolution, as it pertains to joint energy and pseudo-energy signal representations, is then discussed. Notable cojoint representations are reviewed and cast into a common analytical framework, previously introduced by Cohen (1966), which permits a quantitative description of the relationship between cojoint bilinear signal representations. We close with a comparative analysis of the high cojoint resolution, which these representations provide. The following discussion follows on Jacobson and Wechsler (1988).

2.3.1.   Benefits of Cojoint Representations

Cojoint spatial/spatial-frequency representations have emerged as an important vision research issue largely from empirical studies suggesting that they are fundamental to visual information encoding in the cortex of all mammals, including our own human species (Campbell and Robson, 1968; Blakemore and Campbell, 1969; Campbell *et al.*, 1969; Sachs *et al.* 1971; Maffei and Fiorentini, 1973, 1976; and Movshon *et al.* 1978a, 1978b); for a more recent review, see Pollen and

Ronner (1983). Moreover, the major known classes of neurons, leading from the retina to the early stages of visual cortical processing (retinal ganglion, LGN, simple and complex cells—see Fig. 2.1.), have filtering properties characterized by a monotonic progression toward increased cojoint spatial/spatial-frequency resolution.

Cojoint spatial/spatial-frequency representations are primarily useful because they improve pattern *separability* relative to pure space or spatial-frequency representations. Representational separability is a central problem in pattern recognition and signal detection theory, and, not surprising, it should be an important component of texture, shape, and object recognition representations. The separability offered by cojoint representations can also be exploited in optical flow computations and, by extrapolation, may benefit stereo and other visual computations.

To illustrate the notion of separability as it pertains to image representation, consider the image formed by scanning a page of newsprint, as in an OCR (optical character readers) application. The gray-scale pattern associated with each letter is separated (i.e., has spatially disjoint suport) from the gray-scale patterns of all other letters on the page. One could, therefore, successfully employ, for example, normalized correlation between the image and a set of letter templates for recognition—assuming that, for each positioning of the template, a new energy normalization is performed over the image subregions covered by the support region of the template.

Given the same hypothetical image of a page of newsprint, suppose that one computes its 2D Fourier power spectrum. Then, the spectral energy arising from each letter on the page is greatly dispersed across the frequency plane, and normalized correlation between a pattern template (say the Fourier power spectrum of one letter) and the transform of the page is no longer practical because the individual pattern spectra, whose superposition comprises the spectrum of the page, share a common region of support.

In the example, separability was evident in the space domain but not in the spatial-frequency domain. Now consider Fig. 2.4., which schematically depicts an image composed of two orthogonal gratings. Whereas the two gratings overlap in their spatial support everywhere in the 2D image plane, their signatures in the spatial-frequency domain are disjoint. Therefore, given a spatial-frequency domain template together with a spatial-frequency domain image representation, a normalized correlation method would offer a viable approach to single-grating signature recognition.

The two previous examples indicate that, for some visual patterns of interest, separability is most readily obtained in the spatial domain, whereas, for others, separability is most evident in the spatial-frequency domain. A multitude of other interesting patterns exist that are poorly separable in either the space or spatial-frequency domain. Nearly complete energy segregation, however, is achieved in the cojoint space/spatial-frequency domain.

Representations of different object classes, which have signatures with disjoint support, generally enhance pattern recognition. As illustrated, cojoint space/spatial-frequency representations greatly enhance the separability of object signatures whenever energy in these signatures is segregated in the space domain only, the spatial-frequency domain only, or in the joint space and spatial-frequency domains. Space/spatial-frequency representations are used in a theory for form

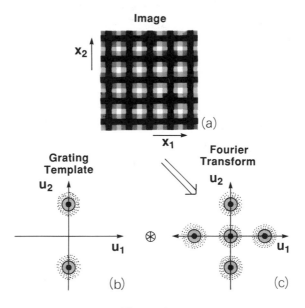

Figure 2.4.
(a) Image of two superimposed sinusoidal gratings. (b) Spectral template of a single grating. (c) Spectrum of the image. Note common-support superposition of gratings in space but separability of their associated spectra.

analysis, which provides separability of the objects' signature and allows multiplicative image components such as reflectance and illumination to separate (Section 3.1.2.).

2.3.2. Cojoint Uncertainty

The cojoint spatial/spatial-frequency uncertainty of a signal is typically defined in terms of the signal's effective spatial extent and effective bandwidth. Such a definition underlies a proof—popularly attributed to Gabor (1946), but proved earlier by Weyl (1932)—that a family of functions obtained by multiplying a complex exponential by a Gaussian having real variance uniquely yields the smallest cojoint space-bandwidth product permitted by the uncertainty principle of Fourier analysis. The difference of Gaussians (DOG) and Gabor representations use linear filters, which are Gaussians and Gaussian-weighted complex exponentials, respectively. Thus, on the basis of the minimum uncertainty property of modulated Gaussians, DOG and Gabor representations provide optimal resolution when compared with alternative cojoint representations (Marr and Hildreth, 1979; Porat and Zeevi, 1988). Such claims, however, are based upon a serious misconception about cojoint resolution as it applies to cojoint representations. Specifically, it is incorrect to infer that high cojoint localization of a *linear filter* necessarily implies high cojoint resolution of a *cojoint representation* computed using that filter. The cojoint resolution of a cojoint representation is properly measured directly from the cojoint representation itself. It is not immediately clear, however, how one might extend familiar notions of *signal uncertainty* to define the cojoint resolution of each of the possible *representations* of a signal. A quantitative definition of something one might refer to as *representational uncertainty* has been lacking. Beyond discussing various proposed cojoint signal representations, we propose a possible definition for representational uncertainty, the inverse of which would measure the cojoint resolution of a cojoint signal representation.

As discussed, signal uncertainty is commonly defined as the effective width of the signal times the effective bandwidth of its Fourier transform (see Daugman (1985) who generalized this concept to two-dimensional signals). Given an N-D signal $f(x)$ whose Fourier transform is denoted by $F(u)$, one can define the effective width along each dimension of the signal as

$$\Delta x_i = \sqrt{\frac{\int (x_i - \bar{x}_i)^2 f(x)f^*(x)dx}{\int f(x)f^*(x)dx}}, \qquad (2\text{-}30a)$$

and the effective bandwidth along each dimension of the signal's Fourier transform as

$$\Delta u_i = \sqrt{\frac{\int (u_i - u_i)^2 F(u)F^*(u)du}{\int F(u)F^*(u)du}}, \qquad (2\text{-}30b)$$

where "$*$" denotes complex conjugation. The uncertainty product is then given by $\Pi_{i=1}^{N} \Delta x_i \Delta u_i$, which must satisfy the uncertainty inequality

$$\prod_{i=1}^{N} \Delta x_i \Delta u_i \geq \left(\frac{1}{2}\right)^N. \qquad (2\text{-}31)$$

As seen from the definitions, the space-bandwidth product is formed from the variances of the *marginal energy distributions* associated with the signal and its Fourier transform. Note that u_i denotes here angular frequency. If we had chosen to use cyclic frequencies $v = u/(2\pi)$, then the inequality (2-31) would read $\Pi_{i=1}^{N} \Delta x_i \Delta u_i \geq (2(2\pi))^{-N}$ as it appears in Daugman (1985).

An alternative definition of uncertainty employs the concept of *entropy* along with a specially chosen joint energy distribution associated with the signal. (The *entropy H* of a source is defined as the average information provided by a source image f, i.e., $H(f) = -\Sigma p_k \log p_k$, where p_k is the probability of a specific image f_k.) Suppose that, given an N-D signal, $f(x)$, one can define an "intrinsic" cojoint energy distribution, $P_f(x, u)$, with the properties expected of a probability density function of continuous type, namely,

$$P_f(x, u) \geq 0 \qquad \forall x, u \text{ and } \int P_f(x, u)dxdu = 1, \qquad (2\text{-}32)$$

and, which has as its marginal densities, the marginal energy distributions of the signal $f(x)$ and its Fourier transform $F(u)$. One is then led to consider the entropy, $H(P_f)$, of the distribution function, $P_f(x, u)$, as given by

$$H(P_f) = -\iint P_f(x, u)\log[P_f(x, u)]dxdu. \qquad (2\text{-}33)$$

Note that, whereas the *uncertainty product* of inequality (Eq. (2-31)) is a measure of the "spread about the mean" of marginal energy distributions, the *cojoint signal entropy*, as defined, is in some sense a measure of the "informational" uncertainty of a signal's joint energy distribution.

Leipnik (1959) presents a related definition for entropic uncertainty, which can be formulated in terms of *marginal* distributions; this companion definition leads to an *entropic uncertainty inequality*, which is similar to the conventional variance-based uncertainty principle given by formulae (2-30) and (2-31). Generalizing Leipnik's formulation to N-dimensional signals, one defines *entropic uncertainties* $\tilde{\Delta}x_1$ and $\tilde{\Delta}u_i$ respectively as

$$\tilde{\Delta}x_i = \exp\left\{-\iint P_f(x, u)\log[P_{x_i}(x_i)]dxdu\right\} \qquad (2\text{-}34)$$

and

$$\tilde{\Delta}u_i = \exp\left\{-\iint P_f(x, u)\log[P_{u_i}(u_i)]dxdu\right\}, \qquad (2\text{-}35)$$

which correspond to the exponentiated entropies of the marginal densities $P_{x_i}(x_i)$ and $P_{u_i}(u_i)$, of the joint density function $P_f(x, u)$. (Note that to keep the notation simple, the subscript f has been dropped in going from joint to marginal densities; recall that marginal densities depend on intrinsic joint energy distribution, $P_f(x, u)$ of the signal $f(x)$.) The entropic uncertainty product $\Pi_{i=1}^{N} \tilde{\Delta}x_i\tilde{\Delta}u_i$ must then satisfy (Leipnick, 1959) the uncertainty inequality

$$\prod_{i=1}^{N} \tilde{\Delta}x_i\tilde{\Delta}u_i \geq (\pi e)^N. \qquad (2\text{-}36)$$

Leipnik (1959) proved, in the one-dimensional case, that the minimum is in fact attained uniquely by any complex exponential multiplied by a Gaussian with real-valued variance (a Gabor function). Weyl (1932) obtained the same result using the more familiar form of uncertainty inequality in Eq. (2-31). Note the similarity of the *entropic uncertainties* in Eqs. (2-34) and (2-35) to the conventional (variance) uncertainties in Eqs. (2-30a) and (2-30b) and the similarity between the corresponding uncertainty principles in inequalities in Eqs. (2-36) and (2-31).

Suppose that one defines the *joint entropic uncertainty* in terms of the previously defined entropy, $H(P_f)$, of the joint density function $P_f(x, u)$ as

$$\Omega_f = \exp\{H(P_f(x, u))\}. \tag{2-37}$$

It is now shown that this definition of uncertainty is equivalent to that of formulae (2-34), (2-35), and (2-36) whenever the signal energy distribution function $P_f(x, u)$ is separable, that is, whenever

$$P_f(x, u) = \prod_{i=1}^{N} P_{x_i}(x_i) P_{u_i}(u_i). \tag{2-38}$$

Using Eq. (2-38) in Eq. (2-33),

$$H(P_f) = -\iint P_f(x, u) \log\left(\prod_{i=1}^{N} P_{x_i}(x_i) P_{u_i}(u_i) \right)$$

$$= -\iint P_f(x, u) \sum_{i=1}^{N} (\log P_{x_i}(x_i) + \log P_{u_i}(u_i))$$

$$= \sum_{i=1}^{N} \left(-\iint P_f(x, u) \log P_{x_i}(x_i) dx du \right.$$

$$\left. -\iint P_f(x, u) \log P_{u_i}(u_i) dx du \right). \tag{2-39}$$

Therefore, using this result in Eq. (2-37),

$$\Omega_f = \exp\left\{ \sum_{i=1}^{N} \left(-\iint P_f(x, u) \log P_{x_i}(x_i) dx du \right. \right.$$

$$\left. \left. -\iint P_f(x, u) \log P_{u_i}(u_i) dx du \right) \right\}$$

$$= \prod_{i=1}^{N} \exp\left\{ -\iint P_f(x, u) \log P_{x_i}(x_i) dx du \right\}$$

$$\times \exp\left\{ -\iint P_f(x, u) \log P_{u_i}(u_i) dx \, du \right\}$$

$$= \prod_{i=1}^{N} \tilde{\Delta} x_i \tilde{\Delta} u_i. \tag{2-40}$$

Therefore, for separable joint density functions $P_f(x, u)$, the *joint entropic uncertainty* (Eq. (2-37)) is equivalent to the *entropic uncertainty* (formulae (2-34), (2-35), and (2-36)) which was defined earlier in terms of the entropies of the marginal density functions of $P_f(x, u)$.

Note that definitions of uncertainty in terms of *marginal* energy distributions, whether using variance (inequality (2-31)) or entropic (Eq. (2-30)) measures are precise characterizations of signal uncertainty only when a signal is separable. Leipnik (1959) recognized a more general definition of cojoint signal uncertainty to meaningfully characterize the uncertainty of a broader class of signals having nonseparable cojoint energy distributions.

Such motivations led Leipnik to use an entropic measure based on a particular joint energy distribution function of a signal having marginal distributions equal to those of the signal's power function $|f(x)|^2$ and frequency power function $|F(u)|^2$. Wigner (1932) had previously proposed an energy distribution function, now popularly referred to as the *Wigner distribution* (WD), which has most of the desired properties, and Leipnik (1960) employed this distribution function with the joint entropy measure (Eq. (2-33)) to prove that the minimum entropy signals are modulated Gaussians with complex-valued variances.

To summarize, uncertainty products based on *marginal distributions* are most useful for the cojoint localization of simple separable signals, such as separable Gaussians or rectangular impulses. Given nonstationary signals that have nonseparable joint energy distributions, especially signals with long duration and broad bandwidth, the overall "uncertainty" is best measured, however, using a definition such as Eq. (2-37), which deals with the *cojoint energy distribution* of the signal.

Whereas this section has focused on proposed methods for measuring the intrinsic uncertainty of a *signal*, our primary goal is to better understand cojoint resolution as it pertains to arbitrary joint energy *representations* of a signal. The definition of the *cojoint entropic* uncertainty of a signal suggests that an entropic uncertainty measurement be applied to arbitrary cojoint representations of a signal in order to quantify cojoint representational resolution. By comparing the relative uncertainties of various representations of the same signal, one can quantify the relative cojoint resolutions of these signal representations. Such a resolution measure will ultimately depend as much on the signal as on the particular manner used to transform the signal into the cojoint domain of representation. Signal and signal transformation interaction will be the topic of the next several sections discussing the Wigner distribution, the Gabor representation, the difference of Gaussians (DOG) representation, and computable approximations to the Wigner distribution.

2.3.3. General Spatial/Spatial-Frequency Signal Representations

In the following sections, we discuss popular linear and bilinear image representations. An analytic framework introduced by Cohen (1966) will be used to formulate the relationships between the bilinear (power) spectra associated with these different signal representations. Given a square-integrable function of N dimensions, $f(x) = f(x_1, x_2, \ldots, x_N)$, we seek a general expression for a class of $2N$-dimensional cojoint energy representations that includes the energy spectra associated with popular image representations, such as the difference of Gaussians (DOG) and Gabor representations. Cohen (1966) introduced an appropriate class of joint energy representations for one-dimensional signal analysis. A multidimensional variant of this defining relation is given by

$$E_f(x, u; \Phi) = \frac{1}{(2\pi)^N} \iiint \Phi(\xi, \alpha) f\left(\rho + \frac{1}{2}\alpha\right) f^*\left(\rho - \frac{1}{2}\alpha\right)$$

$$\times e^{j(\xi x - u\alpha - \xi\rho)} d\rho \, d\xi \, d\alpha, \tag{2-41}$$

where $x^T = (x_1, x_2, \ldots, x_N)$ is a point in the N-dimensional signal space, $u^T(u_1, u_2, \ldots, u_N)$ is a point in the N-dimensional angular-frequency space, and $\Phi(\xi, \alpha)$ is a kernel that, when specified, defines a particular member of the previous class of cojoint energy representations and is independent of x and u.

Equation (2-41) can be re-expressed in two alternative forms. Following the convention of Claasen and Mecklenbrauker (1980, Part III), the first of these is

$$E_f(x, u; \Phi) = \frac{1}{(2\pi)^N} \iint \Phi(\xi, \alpha) A_f(\xi, \alpha) e^{j(\xi x - u\alpha)} d\xi \, d\alpha, \tag{2-42}$$

where $A_f(\xi, \alpha)$ is closely related to the ambiguity function, from the radar field (Woodward, 1953), and is defined as

$$A_f(\xi, \alpha) = \int f(x + \tfrac{1}{2}\alpha) f(x - \tfrac{1}{2}\alpha) e^{-j(\xi x)} dx. \tag{2-43}$$

A second expression for $E_f(x, u; \phi)$ is obtained from Eq. (2-42), by the convolution theorem, as

$$E_f(x, u; \Phi) = \frac{1}{(2\pi)^N} \iint \phi(x - x', u - u') W_f(x', u') dx' du', \quad (2\text{-}44)$$

where

$$\phi(x, u) = \frac{1}{(2\pi)^N} \iint \Phi(\xi, \alpha) e^{j(\xi x - u\alpha)} d\xi d\alpha, \quad (2\text{-}45)$$

and $W_f(x, u)$ is the Wigner distribution of $f(x)$ defined by

$$W_f(x, u) = \frac{1}{(2\pi)^N} \iint A_f(\xi, \alpha) e^{j(\xi x - u\alpha)} d\xi d\alpha. \quad (2\text{-}46)$$

To summarize, either Eq. (2-42) or (2-44) defines a family of cojoint representations $E_f(x, u; \Phi)$ with each member corresponding to a particular choice for the kernel $\Phi(\xi, \alpha)$. Relation (2-42) states that each member of the family $E_f(x, u; \Phi)$ can be viewed as the *Fourier transform of a weighted ambiguity-related function*, where the representation-specific weighting function is $\Phi(\xi, \alpha)$. Alternatively, Eq. (2-44) shows that any member of the family $E_f(x, u; \Phi)$ can be regarded as a *linearly filtered version of the Wigner distribution*, where the representation-specific impulse response of the filter is $\phi(x, u)$, related to $\Phi(\xi, \alpha)$ by Eq. (2-45).

In Section 2.3.2. we showed that the resolution of a cojoint representation is determined by a mixture of filter and signal. For the class of representations encompassed by $E_f(x, u; \Phi)$, Eqs. (2-42) and (2-46) make this interdependence explicit, namely, these equations include a filter-specific term ($\Phi(\xi, \alpha)$ or $\phi(x, u)$, respectively) and a signal-specific term ($A_f(\xi, \alpha)$ or $W_f(x, u)$, respectively).

Several cojoint image energy representations of widespread interest to vision researchers belong to the family $E_f(x, u; \Phi)$. Definitions (2-42) and (2-46), therefore, allow us to compare the cojoint resolution capabilities of these image representations within a unified framework, making explicit the signal-filter interdependence. With this goal in mind, we derive the particular kernels $\Phi(\xi, \alpha)$ (or, equivalently, $\phi(x, u)$) that produce representations such as the Gabor energy spectrum.

First we will discuss the Wigner distribution $W_f(x, u)$ and the ambiguity-related function $A_f(\xi, \alpha)$, which are central to relations (2-42) and (2-46) but not too familiar to the vision community.

2.3.4. *Wigner Distribution*

2.3.4.1. Background

The Wigner distribution (WD) is a bilinear (quadratic) signal representation introduced in 1932 within the context of quantum mechanics. The WD later received the attention of Ville (1948), who advocated its application to communication theory. During the following three decades, the WD received only the infrequent attention of physicists and mathematicians. Its successful application to problems in optics in the late 1970s triggered a wave of interest by researchers from diverse disciplines who had a common interest in cojoint representations. For example, the WD has made possible the reconciliation of many aspects of Fourier and ray optics (Bastiaans, 1978, 1980), has been employed in time-varying filtering design (Boudreaux-Bartels, 1983), and has been used for the analysis of various one-dimensional waveforms, including general detection and classification tasks (Kumar and Carroll, 1983). Its computation (Bartelt, 1980; Athale *et al.*, 1982; and Bamler and Glunder 1983a,b) has also been investigated and is further discussed in Chapter 9. A more comprehensive discussion of the WD can be found in the classic series of articles by Claasen and Mecklenbrauker (1980).

Jacobson and Wechsler (1982a,b, 1983) first suggested using the WD for 2D image processing and an approximation of the WD for the representation of shape and texture information in images. They formulated a unique theory for invariant visual pattern recognition in which the WD plays a central role. Jacobson and Wechsler (1987) later employed the WD in a theory for computing the optical flow of time-varying images, which is discussed in Section 5.1.2.

2.3.4.2. Definition

In Section 2.3.3, we defined the WD as the Fourier transform of the ambiguity-related function $A_f(\xi, \alpha)$. Like the function $A_f(\xi, \alpha)$, the WD can also be defined in terms of the original signal $f(x)$ or its Fourier transform $F(u)$.

The radial Fourier transform pair (see Eq. (2-6)) is defined again, using the substitution u for w, as

$$F(u) = \mathscr{F}\{f(x)\} = \int f(x)e^{-j(u \cdot x)}dx \tag{2-47}$$

$$f(x) = \mathscr{F}^{-1}\{F(u)\} = \frac{1}{(2\pi)^N} \int F(u)e^{j(u \cdot x)}dx \tag{2-48}$$

Given an arbitrary N-dimensional signal $f(x)$, defined for $x = (x_1, x_2, \ldots, x_N)^T$, $-\infty < x_i < \infty$, $i = 1, 2, \ldots, N$, the WD of $f(x)$ is given by

$$W_f(x, u) = \int f(x + \tfrac{1}{2}\alpha)f^*(x - \tfrac{1}{2}\alpha)e^{-j(u \cdot \alpha)}d\alpha, \qquad (2\text{-}49)$$

$$= \int r_f(x, \alpha)e^{-j(u \cdot \alpha)}d\alpha$$

where

$$r_f(x, \alpha) = f(x + \tfrac{1}{2}\alpha)f^*(x - \tfrac{1}{2}\alpha). \qquad (2\text{-}50)$$

The WD is also defined in terms of the signal's Fourier transform $F(u)$ as

$$W_f(x, u) = \frac{1}{(2\pi)^N}\int F(u + \tfrac{1}{2}\xi)F^*(u - \tfrac{1}{2}\xi)e^{j(\xi \cdot x)}d\xi$$

$$= \frac{1}{(2\pi)^N}\int R_f(\xi, u)e^{j(\xi \cdot x)}d\xi, \qquad (2\text{-}51)$$

where

$$R_f(\xi, u) = F(u + \tfrac{1}{2}\xi)F^*(u - \tfrac{1}{2}\xi). \qquad (2\text{-}52)$$

By Eqs. (2-42) and (2-46), or alternatively, by Eqs. (2-44) and (2-45), the WD is of the class $E_f(x, u; \Phi)$ corresponding to the choice $\Phi(\xi, \alpha) = 1$ as summarized in Table 2.1.

Table 2.1. Wigner distribution.

$W_f(x, u) = E_f(x, u; \Phi_W)$ where $\begin{cases} \Phi_W(\xi, \alpha) = 1, \text{ or} \\ \psi_W(\lambda, u) - \mathscr{F}_\xi^{-1}\mathscr{F}_\alpha[\Psi_W(\xi, \alpha)] = \dfrac{1}{(2\pi)^N}\delta(x)\delta(u). \end{cases}$

2.3.4.3. Properties

The WD has the following important properties:

1. P1: $W_f(x, u)$ is a strictly real-valued function; finite duration (or bandwidth) signals have a Wigner representation with the same property.

2.   P2: For real $f(x)$, $W_f(x, u) = W_f(x, -u)$.

3.   P3: $\dfrac{1}{(2\pi)^N} \displaystyle\int W_f(x, u)du = |f(x)|^2$.

4.   P4: $\displaystyle\int W_f(x, u)dx = |F(u)|^2$.

5.   P5: If $g(x) = f(x - x_0)$, then $W_g(x, u) = W_f(x - x_0, u)$ (shift property).

6.   P6: If $g(x) = f(x)\exp[j(u_0, x)]$, then $W_g(x, u) = W_f(x, u - u_0)$ (modulation property).

7.   P7:   If $f(x) = g(x) \otimes h(x)$,   then $W_f(x, u) = W_g(x, u) \underset{x}{\otimes} W_h(x, u)$

        (convolution property).

8.   P8: If $f(x) = g(x)h(x)$, then $W_f(x, u) = \dfrac{1}{(2\pi)^N} W_g(x, u) \underset{u}{\otimes} W_h(x, u)$

        (windowing property).

2.3.4.4. Discussion

Property (1) implies that the WD, unlike the Fourier transform, has no associated phase. Yet, the WD, like the Fourier transform, is a reversible transformation. (Actually, given its WD, a function can only be recovered to within a minus sign.) That phase information is implicit in the WD is important since, as Oppenheim and Lim (1981) have shown, much information is contained in the Fourier phase function of typical signals. They found that Fourier phase information is especially relevant to the encoding of spatial information in nonstationary images. In fact, according to Eq. (2-51), Fourier phase information is specifically employed to obtain the spatial dependence of the spectral information encoded by the Wigner distribution.

Though the WD has an obvious interpretation as defining a local power spectrum at each point x of the signal, the analogy with the familiar power spectrum is not complete since the WD can in fact attain negative values. Nonetheless, as with other cojoint representations of the $E_f(x, u; \Phi)$ class, defined in Eq. (2-41), averages of the WD always yield positive values (see DeBruijn (1967), for further discussion of this inherent uncertainty principle).

Properties (3) and (4) state that the projection of the Wigner distribution onto the subspaces x or u alone yields the instantaneous signal energy or the power spectrum, respectively. (Recall the Rayleigh's theorem from Section 2.1.2.) This is no accident as the term (Wigner) *distribution* implies. Wigner originally sought a joint probability distribution over the position and momentum of a particle. Any possible representation must have the correct marginal probability distributions. Although there are many possible representations satisfying this requirement (all belonging, incidentally, to the class $E_f(x, u; \Phi)$), the simplest of these is the Wigner distribution. Wigner (1979) proved that there is no bilinear cojoint representation having the correct marginal distributions that is also nonnegative definite.

Properties (5) and (6) imply that the WD translates in x or u, as does the signal and its Fourier transform, respectively. These two properties, therefore, highlight a major difference between the WD and the ambiguity function. Whereas the WD translates in response to shifts of a signal and its spectrum, the ambiguity related function $A_f(\xi, \alpha)$ (to be discussed shortly) is left unmodified except for a linear phase change. Recall that the WD and the ambiguity-related function $A_f(\xi, \alpha)$ form a Fourier transform pair—Eq. (2-46).

Finally, by properties (7) and (8), the WD of the convolution or product of two signals is given by the signal domain or frequency domain convolution, respectively, of the separate WDs of the two signals.

2.3.4.5. Examples

One-dimensional signals. In the next examples, the one-dimensional time and frequency domain coordinates are respectively denoted as x and u.

Example A. Impulse distribution
Signal:

$$f(x) = \delta(x - \kappa). \tag{2-53}$$

Wigner distribution:

$$W_f(x, u) = \int \delta(x - \kappa + \tfrac{1}{2}\alpha)\delta(x - \kappa - \tfrac{1}{2}\alpha)e^{-ju\alpha}d\alpha = \delta(x - \kappa). \tag{2-54}$$

The WD of an impulse distribution has nonzero energy only at instant κ, where the associated spectrum is white.

Example B. One-dimensional chirp
Signal:

$$f(x) = e^{j\kappa x^2/2}, \ \kappa \text{ constant.} \tag{2-55}$$

Wigner distribution:

$$W_f(x, u) = \int e^{j\kappa(x + \alpha/2)^2/2} e^{-j\kappa(x - \alpha/2)^2/2} e^{-ju\alpha} d\alpha = 2\pi\delta(u - \kappa x). \tag{2-56}$$

At any instant x, the spectral energy in the WD is concentrated at the instantaneous angular frequency κx.

Example C. One-dimensional Gaussian
Signal:

$$f(x) = e^{-\kappa x^2}. \tag{2-57}$$

Wigner distribution:

$$W_f(x, u) = \int e^{-\kappa(x + \alpha/2)^2} e^{-\kappa(x - \alpha/2)^2} d\alpha = (2\pi/\kappa)^{1/2} e^{-2\kappa x^2} e^{-(1/2\kappa)u^2}. \tag{2-58}$$

The WD of a 1D Gaussian is a 2D Gaussian in the time-frequency plane. Note that the areas of the $1/e$ ellipses in the $x - u$ plane are always equal to $(2/\sqrt{2\kappa})(2\sqrt{2\kappa})\pi = 4\pi$, irrespective of the choice for k.

   Two-dimensional signals.   In the next examples, the coordinates of the 2D spatial domain and the 2D spatial-frequency domain are given respectively as $x^T = (x_1, x_2)$ and $u^T = (u_1, u_2)$.

Example D. Impulse distribution in the two-dimensional plane
Signal:

$$f(x) = \delta(x_1 - \kappa_1, x_2 - \kappa_2). \tag{2-59}$$

Wigner distribution:

$$W_f(x, u) = \delta(x_1 - \kappa_1, x_2 - \kappa_2). \tag{2-60}$$

Example E. Pair of two-dimensional superposed sinusoids
Signal:

$$f(x) = \cos(\kappa_1 x_1) + \cos(\kappa_2 x_2). \tag{2-61}$$

Wigner distribution:

$$W_f(x, u) = 32\pi^2[\cos(2\kappa_1 x_1) + \cos(2\kappa_2 x_2)]\delta(u_1, u_2)$$
$$+ 16\pi^2[\delta(u_1 - \kappa_1, u_2) + \delta(u_1 + \kappa_1, u_2) + \delta(u_1, u_2 - \kappa_2)$$
$$+ \delta(u_1, u_2 + \kappa_2)] + 32\pi^2 \cos(\kappa_1 x_1 + \kappa_2 x_2)[\delta(u_1 - \tfrac{1}{2}\kappa, u_2 + \tfrac{1}{2}\kappa)$$
$$+ \delta(u_1 + \tfrac{1}{2}\kappa, u_2 - \tfrac{1}{2}\kappa)] + 32\pi^2 \cos(\kappa_1 x_1 - \kappa_2 x_2)$$
$$\times [\delta(u_1 - \tfrac{1}{2}\kappa, u_2 - \tfrac{1}{2}\kappa) + \delta(u_1 + \tfrac{1}{2}\kappa, u_2 + \tfrac{1}{2}\kappa)]. \qquad (2\text{-}62)$$

The spatially modulating components of the spectrum are due to the modulation of the signal envelope (and hence signal energy) with changing position, whereas the nonmodulating components reflect the spatially constant energy in the pure sinusoidal frequencies.

Example F. Two-dimensional Gaussian
Signal:

$$f(x) = e^{-(\kappa_1 x_1^2 + \kappa_2 x_2^2)}. \qquad (2\text{-}63)$$

Wigner distribution:

$$W_f(x, u) = \frac{2\pi}{\sqrt{\kappa_1 \kappa_2}} \exp[-2(\kappa_1 x_1^2 + \kappa_2 x_2^2)]\exp[-(u_1^2/\kappa_1 + u_2^2/\kappa_2)/2]. \qquad (2\text{-}64)$$

The expression is obtained by using the result for a 1D Gaussian, and by understanding that the integral defining $W_f(x, u)$ is separable.

Numerically computed examples.

Example G. Wigner distribution of an annulus.
In this example, 2D spectra from the Wigner distribution of an image are shown at three different image points. The image was represented on a 128×128 grid and consisted of a smooth annular bright region that was centered at (70, 70) and had a radius of nine pixels. Figures 2.5, 2.6, and 2.7, depict (a) 2D slices of the function $r_f(x, \alpha)$ (Eq. (2-50)) at the indicated spatial points, and (b) 2D slices of the Wigner distribution $W_f(x, u)$ at the same spatial points—obtained by Fourier transforming the corresponding correlation functions. Note how the spectra vary with spatial position.

2.3.5. Ambiguity Function

2.3.5.1. Background

The function $A_f(\xi, \alpha)$ is closely related to the ambiguity function, first used to characterize the tradeoff between velocity and range

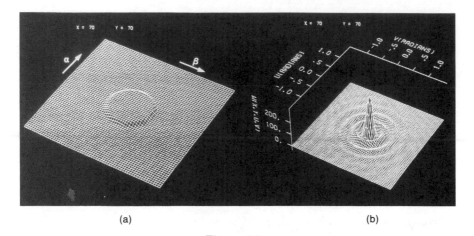

(a)   (b)

Figure 2.5.
(a) The local correlation function, and (b) the corresponding WD at the center (70, 70) of
an image of an annular brightness profile.

resolution, and which is encountered when designing radar signal
processors (Woodward, 1953). The ambiguity function has since become an invaluable tool in the radar field and has also found applications in other fields including sonar, radio astronomy, geophysics,
optics, and communications.

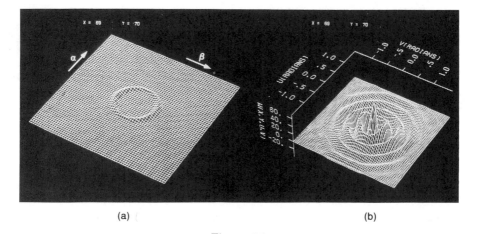

(a)   (b)

Figure 2.6.
(a) The local correlation function, and (b) the corresponding WD at point (69, 70) one
pixel from the center of the annulus.

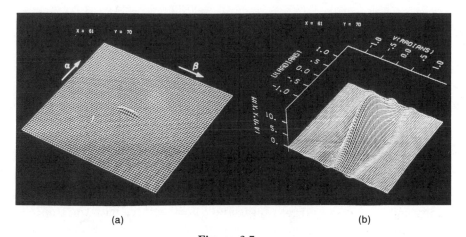

(a)   (b)

Figure 2.7.
(a) The local correlation function, and (b) the corresponding WD at point (61, 70) on the
edge of the annulus.

2.3.5.2. Definition

Like the WD, the ambiguity-related function $A_f(\xi, \alpha)$ can be defined
either in terms of a signal $f(x)$ (see Eq. 2-43) or its Fourier transform
$F(u)$

$$A_f(\xi, \alpha) = \int F(u + \tfrac{1}{2}\xi)F^\star(u - \tfrac{1}{2}\xi)e^{ju\cdot\alpha}du = \frac{1}{(2\pi)^N}\int R_f(\xi, u)e^{ju\cdot\alpha}du, \quad (2\text{-}65)$$

where $R_f(\xi, u)$ is as defined in Eq. (2-52).

2.3.5.3. Discussion

Compare definitions (2-43) and (2 65) of the function $A_f(\xi, \alpha)$ with
those for the Wigner distribution, Eqs. (2-49) and (2-51), respectively.
Note that the function $A_f(\xi, \alpha)$ and the Wigner distribution are both
obtained by Fourier transforming either of two possible functions,
which we have defined respectively as $r_f(x, \alpha)$ and $R_f(\xi, u)$. The definitions of the function $A_f(\xi, \alpha)$ and the Wigner distribution are actually
N–D Fourier transform pairs, which we had used in Section 2.3.3. to
define the WD (see Eq. (2-46)). We refrain, therefore, from explicitly
listing any properties of the function $A_f(\xi, \alpha)$, since they can easily be
deduced from the combined properties of the WD (Section 2.3.4.) and
the Fourier transform. In response to translations of a signal $f(x)$ or its
Fourier transform $F(u)$, the WD translates identically with $f(x)$ or $F(u)$

(properties (5) and (6) of the WD), whereas the function $A_f(\xi, \alpha)$ merely changes by a linear phase factor. The function $A_f(d, c)$ can not be considered a cojoint pseudo-energy spectrum over the independent variables of the signal and its Fourier transform, x and u, respectively —an interpretation that fits the WD. How then does one interpret the significance of the function $A_f(\xi, \alpha)$? As is well known to radar experts, the ambiguity function is the squared modulus of the correlation between the complex envelope of the signal $f(x)$ and all-time and frequency-shifted versions of itself. As such, the (ξ, α)-plane corresponds to all combinations of time lag α and frequency shift ξ where, in the radar field, α is associated with the two-way signal propagation delay and hence with target range, and ξ is associated with the Doppler frequency shift (see Section 5.2.) or, equivalently, with the down-range component of the target velocity. Therefore, the ambiguity-related function $A_f(\xi, \alpha)$ is by itself not a member of the cojoint representations $E_f(x, u; \Phi)$.

2.3.6. Spectrogram (Periodogram)

2.3.6.1. Background

The *spectrogram*, or equivalently, *periodogram*, is a real-valued joint representation that has long been used to represent speech and other acoustic signals in the time-frequency plane. The real-valued spectrogram has a computational precursor known as the finite support Fourier transform.

The generalization of the spectrogram for 4D spatial/spatial-frequency representations has been applied to vision problems including texture representation (Bajcsy and Lieberman, 1976), fractal dimension calculation (Pentland, 1983), and translation-invariant object recognition (Cavanagh, 1984, 1985).

2.3.6.2. Definition

The procedure for computing the spectrogram of an N–D signal $f(x)$ is briefly defined as follows:

Step 1. Multiply the signal $f(x')$ by a window $h(x')$ that is centered on some arbitrary point x to obtain a windowed signal $f_x(x')$:

$$f_x(x') = f(x')h(x - x'). \tag{2-66}$$

Step 2. Take the Fourier transform of the windowed signals $f_x(x')$:

$$F_x(u) = \int f_x(x') e^{-u \cdot x'} dx'. \tag{2-67}$$

Step 3. Compute the squared modulus of the Fourier transform $F_x(u)$ to yield the spectrogram $S_f(x, u)$ at a single point x:

$$S_f(x, u) = |F_x(u)|^2. \tag{2-68}$$

Step 4. To obtain the complete spectrogram, repeat the procedure for all points x in the signal domain.

2.3.6.3. Spectrogram as a Smoothed WD

The spectrogram is equivalently obtained by convolving the Wigner distribution of the signal $f(x)$ with the WD of the window function $h(x)$ and by rewriting Eq. (2-68) as follows:

$$S_f(x, u) = |F_x(u)|^2 = \int W_{f_x}(x', u)dx' \qquad \text{(by property (4) of the WD)}$$

$$= \frac{1}{(2\pi)^N} \iint W_f(x', u')W_h(x - x', u - u')dx'du', \tag{2-69}$$

where in the last step we have used the windowing property (8) of the WD along with the defining relation (2-66) for $f_x(x')$.

Table 2.2. Spectrogram.

$$S_f(x, u) = E_f(x, u; \Phi_s) \text{ where} \begin{cases} \Phi_s(\delta, \alpha) = A_h(\xi, \alpha), \text{ or} \\ \phi_s(x, u) = W_h(\xi, \alpha) \end{cases}$$

Comparing this result with Eq. (2-44), the spectogram falls into Cohen's (1966) general class of cojoint representations. The corresponding kernel in Cohen's definition is given by $\phi(x, u) = W_h(x, u)$, namely, the Wigner distribution of the window function $h(x)$.

Equation (2-69) leads to an interesting observation, namely, that the Wigner distributions W_f and W_h of the signal and window, respectively, play essentially symmetric roles. Whether a window $h(x)$ is being used to compute the spectrogram of $f(x)$ or conversely, a "window" $f(x)$ is being used to compute the spectrogram of a "signal" $h(x)$, is ambiguous. Accordingly, it may seem surprising that the spectrogram has proved useful. The convention of choosing a tapered analysis window with an approximately low-pass Fourier spectrum helps to assure, however, that the cojoint energy variation in the spectrogram, which carries the desired information, primarily reflects the cojoint energy changes that are intrinsic to the signal.

2.3.6.4. Finite-Support Fourier Transform as a Cross-Wigner Distribution

The spectrogram is equivalent to the convolution between the respective WDs of the signal and the window. There similarly exists a close signal-window relationship between the finite-support Fourier transform and the *cross Wigner distribution*. This relationship is derived next.

The *cross-WD*, $W_{fg}(x, u)$, of two functions $f(x)$ and $g(x)$, is defined by

$$W_{fg}(x, u) = \int f(x + \tfrac{1}{2}\alpha)g^*(x - \tfrac{1}{2}\alpha)e^{-j(u \cdot \alpha)}d\alpha. \tag{2-70}$$

The WD is therefore more precisely referred to as the *auto-WD* because of the *cross-WD* that results when $g \equiv f$.

The definition for the finite support Fourier transform (Eq. (2-67)) can be rewritten as

$$F_x(u) = \int f_x(x')e^{-ju \cdot x'}dx' \tag{2-71}$$

$$= \int f(x')h(x - x')e^{-ju \cdot x'}dx', \tag{2-72}$$

where we have substituted for $f_x(x')$ using Eq. (2-66). After further substituting $x' = \tfrac{1}{2}x + \tfrac{1}{2}\alpha$ and recalling that x is an arbitrary point, this becomes

$$F_x(u) = \tfrac{1}{2}e^{-ju \cdot x/2} \int f(\tfrac{1}{2}x + \tfrac{1}{2}\alpha)h(\tfrac{1}{2}x - \tfrac{1}{2}\alpha)e^{-ju \cdot \alpha/2}d\alpha \tag{2-73}$$

$$= \tfrac{1}{2}e^{-ju \cdot x/2} W_{fh^*}(\tfrac{1}{2}x, \tfrac{1}{2}u). \tag{2-74}$$

Therefore, within a linear phase factor, the finite-support Fourier transform is equivalent to a cross-WD between the signal and the conjugate window.

Note the symmetry between the signal and window in Eq. (2-73). As with the spectrogram (see Eq. (2-69)), functions $f(x)$ and $h(x)$ play ambiguous roles in the definition of the finite-support Fourier transform. Equation (2-73) remedies this problem by choosing a window $h(x) = f^*(x)$. That is, one can use the signal as its own window in Eq. (2-73). The finite-support Fourier transform then becomes essentially equivalent to the definition of the auto-WD. Generally, the signal may not be of finite support, but in practice, one would window the signal with a finite portion of itself, resulting in the so-called *pseudo Wigner distribution* (PWD) computation, which we explore in depth later.

2.3.6.5. The Spectrogram as the Square Modulus of a Cross-WD

Equations (2-74) and (2-68) taken together readily yield a new expression that relates the spectrogram to the cross-WD of $f(x)$ and $h(x)$:

$$S_f(x, u) = |F_x(u)|^2 = |\tfrac{1}{2}e^{-ju \cdot x/2} W_{fh*}(\tfrac{1}{2}x, \tfrac{1}{2}u)|^2 = \tfrac{1}{4}|W_{fh*}(\tfrac{1}{2}x, \tfrac{1}{2}u)|^2. \quad (2.75)$$

The previous formula complements Eq. (2-69), which related the spectrogram to the auto-WDs of $f(x)$ and $h(x)$. By using the definition of the crosss-WD (Eq. 2-68) and performing a couple of variable substitutions, one can indeed show that Eq. (2-75) reduces to Eq. (2-69).

2.3.6.6. Discussion

The findings reviewed in this section underscore the strong interrelationship between the finite support Fourier transform, the spectrogram, the auto-WD, and the cross-WD. Table 2.2 summarizes the most important result, namely, that the spectrogram is a special case of Cohen's (1966) general class of cojoint representations, and that it is equivalent to the Wigner distribution of the signal $f(x)$ convolved (smoothed) by the Wigner distribution of the window function $h(x)$.

2.3.7. Gabor Representation

The *Gabor energy spectrum* is a special spectrogram case that uses Gaussian support windows in its computation. It has a linear, complex-valued precursor that is commonly referred to as the *Gabor representation* (Marcelja, 1980; Daugman, 1985). As mentioned earlier, the Gabor spectrum and representation have received considerable attention in the vision community because research strongly suggests that such a representation is computed in the visual cortex of mammals. To account for the functional benefit of such a representation, researchers have emphasized that 2D filters used to compute the Gabor representation have optimal cojoint localization in space and spatial frequency (Marcelja, 1980; Daugman, 1985; Porat and Zeevi, 1988). A casual reading of such reports can easily leave one with the impression that the Gabor representation of a signal (like the filters used to compute it) must therefore have the highest possible cojoint resolution. As discussed in Section 2.3.2., however, such a claim is untenable. While the Gabor representation enjoys the benefits of a linear transform, this very property prevents it from accounting for the known nonlinear properties exhibited by the human visual system. This section highlights previous work involving the Gabor spectrum.

2.3.7.1. Background

Gabor (1946) developed a discrete joint time-frequency signal representation, referred to by him as the *information diagram*, which is obtained by computing the coefficients or *logons* of a signal expansion in terms of *time*-shifted and frequency-modulated Gaussian functions or *elementary signals*. For Gabor, each logon occupied a rectangular *cell* in the time-frequency plan of his information diagram. Each logon provided a discrete quantum of information about the signal that is independent of the logon in any other cell, and it was, therefore, important to select the elementary signals to be highly concentrated jointly in time and frequency. Gabor chose to provide his elementary signals with *Gaussian envelopes* since, as earlier shown by Weyl (1932), the Gaussian has the smallest time-bandwidth product of any function. The Gaussian, in fact, achieves the smallest time-bandwidth product permitted by the *uncertainty principle* of Fourier analysis as previously described in Section 2.3.2.

A continuous-domain variation of Gabor's discrete time-frequency representation are used in the derivations next, appropriately generalized for use with N-D signals. This section examines the power spectrum that results from taking the square modulus of the complex-valued Gabor representation. This power function is important for our discussion on cojoint resolution.

2.3.7.2. Definition

The Gabor power spectrum can be computed just as the spectrogram in Steps 1–4 of Section 2.3.6., where we now take the analysis window function $h(x)$ as a Gaussian function $h_G(x)$. For simplicity, we assume that the Gaussian is separable. Thus, the window $h_G(x)$ is given by

$$h_G(x) = \exp\left\{ -\sum_{i=1}^{N} \kappa_i x_i^2 \right\}. \tag{2-76}$$

2.3.7.3. Gabor Spectrum as a Smoothed Wigner Distribution

By Eq. (2-69), the Gabor power spectral representation $G_f(x, u)$ is given by

$$G_f(x, u) = \frac{1}{(2\pi)^N} \iint W_f(x', u') W_{h_G}(x - x', u - u')dx'du', \tag{2-77}$$

where the smoothing kernel (the Wigner distribution) of $h_G(x)$ is given by

$$W_{h_G}(x, u) = \left(\prod_{i=1}^{N} \sqrt{2\pi/\kappa_i} \right)\exp\left\{ -2\sum_{i=1}^{N} \kappa_i x_i^2 \right\}\exp\left\{ -\frac{1}{2}\sum_{i=1}^{N} u_i^2/\kappa_i \right\} \tag{2-78}$$

as can easily be derived by using the result for the WD of a 1D Gaussian (Example C of Section 2.3.4.5. along with the separability of $h_G(x)$ as defined in Eq. (2-78).)

Table 2.3. Gabor spectrum.

$G_f(x, u) = E_f(x, u; \Phi_G)$, where

$$\Phi_G(\xi, \alpha) = \left(\prod_{i=1}^{N} \sqrt{\pi/(2\kappa_i)} \right)\exp\left\{ -\frac{1}{8}\sum_{i=1}^{N} \frac{\xi_i^2}{\kappa_i} \right\}\exp\left\{ -\frac{1}{2}\sum_{i=1}^{N} \kappa_i \alpha_i^2 \right\}, \text{ or}$$

$$\phi_G(x, u) = \left(\prod_{i=1}^{N} \sqrt{2\pi/\kappa_i} \right)\exp\left\{ -2\sum_{i=1}^{N} \kappa_i x_i^2 \right\}\exp\left\{ -\frac{1}{2}\sum_{i=1}^{N} u_i^2/\kappa_i \right\}$$

It is clear from Eq. (2-77) that the Gabor spectrum is equivalent to the Wigner distribution convolved with a $2N$-D Gaussian function. Note that the width of the Gaussian along any axis x_i of the spatial domain is inversely proportional to the width along the corresponding spectral dimension u_i. Therefore, as is well known, there is an inherent tradeoff in resolution along corresponding axes x_i and u_i when using the Gabor representation; this holds true, of course, with the more general spectrogram as well.

Using the definition of $h_G(x)$ (in Eq. (2-76)) along with Table 2.3., we conclude that the Gabor spectrum is a member of Cohen's general class $E_f(x, u; \Phi)$ of cojoint representations with kernels as specified in Table 2.3.

2.3.8. *Difference of Gaussians*

2.3.8.1. Background

Current interest in the difference-of-Gaussians (DOG) representation is rooted in physiological discoveries of past decades. Kuffler (1952) discovered the antagonistic *center-surround* organization of the receptive fields of cat retinal ganglion cells. Subsequent studies led to the discovery of similar tuning properties for cells in the lateral geniculate nucleus of the cat. Rodieck (1965) later proposed that a class of retinal ganglion cells, denoted *x-cells*, have receptive fields corresponding to the difference of two Gaussians. This model for the antagonistic center-surround organization of both retinal ganglion cells and LGN cells has been popularized by Marr (1982), who incorporated the DOG representation into his theory for early visual processing.

Marr and Hildreth (1979) advocated presmoothing an image with Gaussians of different sizes. One then computes the Laplacian of the Gaussian smoothed images, from which zero-crossing coordinates are determined to obtain a *primal sketch* of the original image. The Laplacian of a Gaussian is well approximated by a difference of Gaussians. Marr and Hildreth have justified the special status of the Gaussian filters in their theory for edge detection on the grounds that the Gaussian function has the minimum cojoint variance attainable by any filter.

Note that whereas the Gabor representation employs orientation-selective filters to provide a 4D cojoint representation of a 2D image, the DOG representation uses isotropic filters and hence provides a 2D cojoint image (pyramid) representation. The power function that results from computing the modulus of any single level of a DOG image pyramid is equivalently obtained by smoothing the Wigner distribution of the same image with a kernel whose form is derived next. Understanding the DOG power function is useful for defining later the cojoint resolution of a *representation*.

2.3.8.2. Definition

For generality, we assume N-dimensional DOG filters $h_D(x)$ that are applied to N-dimensional signals $f(x)$. The filter associated with one resolution level of the DOG representation, therefore, has the following form:

$$h_D(x) = h_1(x) - h_2(x) = A \, \exp\left\{ -\sum_{i=1}^{N} a_i x_i^2 \right\}$$

$$- B \, \exp\left\{ -\sum_{i=1}^{N} b_i x_i^2 \right\}. \tag{2-79}$$

Spatial convolution of this filter with the signal $f(x)$ leads to an N-D real-valued representation whose associated power, denoted $D_f(x)$, is given as

$$D_f(x) = |f(x) \otimes h_D(x)|^2 = |f(x) \otimes [h_1(x) - h_2(x)]|^2. \tag{2-80}$$

2.3.8.3. DOG Power Spectrum as Smoothed Wigner Distribution

Note that the DOG filter $h_D(x)$ may be regarded as the difference between two DC (unmodulated) Gabor filters with differently sized and weighted Gaussian windows $h_1(x)$ and $h_2(x)$, respectively. The DOG power function $D_f(x)$ is, therefore, equivalent to the "zero-frequency slice" out of a "Difference of Gabor power spectrum"—a spectrogram (see Eq. (2-68)) that employs a DOG window function $h(x) = h_D(x)$ in Eq. (2-66). The resulting Eq. (2-69) establishes a relation between the Wigner distribution and a single resolution level of the DOG power representation defined by Eq. (2-77). Recall that Eq. (2-69) expresses a sliding-window power spectral analysis as a smoothed version of the Wigner distribution, where the smoothing kernel is the Wigner distribution of the window function used in computing the Fourier power spectra. Using this result with the DOG window and extracting only the zero-frequency part, gives

$$D_f(x) = \frac{1}{(2\pi)^N} \iint W_f(x', u') W_{h_D}(x - x', u - u') dx' du'|_{u=0}$$

$$= \frac{1}{(2\pi)^N} \iint W_f(x', u') W_{h_D}(x - x', -u') dx' du'. \tag{2-81}$$

Therefore, each resolution level of a DOG power representation is a smoothed version of the Wigner distribution, where the smoothing kernel $W_{h_D}(x)$ is the Wigner distribution of the DOG filter. Given the definition of the DOG filter in Eq. (2-79), its Wigner distribution is given by

$$W_{h_D}(x, u) = WD \text{ of } \{h_1(x) - h_2(y)\}$$

$$= W_{h_1}(x, u) + W_{h_2}(x, u) - 2 \operatorname{Re}\{W_{h_1 h_2}(x, u)\}, \quad (2\text{-}82)$$

where $W_{h_1}(x, u)$ and $W_{h_2}(x, u)$ are the (auto-) Wigner distributions of $h_1(x)$ and $h_2(x)$, respectively, and $W_{h_1 h_2}(x, u)$ is the cross-WD of $h_1(x)$ and $h_2(x)$. The auto-WD of an N-dimensional Gaussian already appeared in Eq. (2-78) of the last section; it is a $2N$-dimensional Gaussian. The cross-WD of two N-D Gaussians is a new result. The cross-WD of two one-dimensional Gaussians is derived in Jacobson and Wechsler (1988). Together with the separability of the N-D Gaussians $h_1(x)$ and $h_2(x)$, we obtain the cross-WD $W_{h_1 h_2}(x, u)$ as follows:

$$W_{h_1 h_2}(x, u) = AB\left(\prod_{i=1}^{N} \frac{2\sqrt{\pi}}{\sqrt{a_i + b_i}}\right) \exp\left\{-\sum_{i=1}^{N} (a_i + b_i)x_i^2\right\}$$

$$\times \exp\left\{\sum_{i=1}^{N} \frac{(a_i - b_i)^2 x_i^2 - u_i^2}{a_i + b_i}\right\} \exp\left\{j2 \sum_{i=1}^{N} \frac{a_i - b_i}{a_i + b_i} x_i u_i\right\}.$$

$$(2\text{-}83)$$

The real part of this cross-WD, as required in Eq. (2-82), is then given by

$$\operatorname{Re}\{W_{h_1 h_2}(x, u)\} = AB\left(\prod_{i=1}^{N} \frac{2\sqrt{\pi}}{\sqrt{a_i + b_i}}\right) \exp\left\{-\sum_{i=1}^{N} (a_i + b_i)x_i^2\right\}$$

$$\times \exp\left\{\sum_{i=1}^{N} \frac{(a_i - b_i)^2 x_i^2 - u_i^2}{a_i + b_i}\right\} \cos\left(2 \sum_{i=1}^{N} \frac{a_i - b_i}{a_i + b_i} x_i u_i\right).$$

$$(2\text{-}84)$$

Using the last result and the expression for the (auto-) WD of a Gaussian together in Eq. (2-84), the kernel $W_{h_D}(x, u)$ is given by

$$W_{h_D}(x, u) = A^2 \left(\prod_{i=1}^{N} \sqrt{\frac{2\pi}{a_i}} \right) \exp\left\{ -2 \sum_{i=1}^{N} a_i x_i^2 \right\} \exp\left\{ -\frac{1}{2} \sum_{i=1}^{N} \frac{u_i^2}{a_i} \right\}$$

$$+ B^2 \left(\prod_{i=1}^{N} \sqrt{\frac{2\pi}{b_i}} \right) \exp\left\{ -\sum_{i=1}^{N} b_i x_i^2 \right\} \exp\left\{ -\frac{1}{2} \sum_{i=1}^{N} \frac{u_i^2}{b_i} \right\}$$

$$- 2AB \left(\prod_{i=1}^{N} \frac{2\pi}{\sqrt{a_i + b_i}} \right) \exp\left\{ -\sum_{i=1}^{N} (a_1 + b_i)x_i^2 \right\}$$

$$\times \exp\left\{ \sum_{i=1}^{N} \frac{(a_i - b_i)^2 x_i^2 - u_i^2}{a_i + b_i} \right\} \cos\left(2 \sum_{i=1}^{N} \frac{(a_i - b_i)}{a_i + b_i} x_i u_i \right).$$

$$(2\text{-}85)$$

Each resolution level of a difference of Gaussian power representation is equivalent to a smoothed Wigner distribution. Accordingly, each resolution level of a DOG power representation is a member of Cohen's class of cojoint representations, albeit one that depends only on the spatial position x. A complete DOG representation is, of course, made up of many such functions, each defining a different resolution level determined by the constants a_i and b_i in the definition of the DOG filter [Eq. (2-79)].

2.3.9. Pseudo-Wigner Distribution

2.3.9.1. Background

We have seen that the positive definite Gabor and DOG *spectral* representations can be obtained by smoothing the WD with representation-specific kernels. The implication, although not yet formally stated is that the WD offers higher cojoint resolution than either the Gabor or the DOG representation. This suggests the possibility of the Wigner distribution itself as a signal representation. Since infinite bounds of integration appear in its definition, however, the Wigner distribution is generally not computable. Accordingly, Claasen and Mecklenbrauker (1980) introduced a computable approximation to the WD, as applied to one-dimensional signals, that they called the *pseudo-Wigner distribution* (PWD). Simply stated, the PWD involves using finite bounds of integration (i.e., sliding analysis window) in Eq. (2-49) of the WD. This results in a function that, relative to the true WD, is smoothed with respect to the frequency domain only.

Jacobson and Wechsler (1983) investigated alternative approximations to the WD. Specifically, in order to accommodate nonuniformly sampled approximations to the Wigner distribution that avoid aliasing, they sought an approximation that allows one to control the peak resolution of the representation in a space and spatial-frequency variant fashion. Thus, they developed two additional PWD definitions. The first definition is complementary to that previously introduced by Claasen and Mecklenbrauker. It yields a PWD that, relative to the true WD, is smoothed only with respect to the spatial domain. The second new definition, the *generalized PWD*, is a smoothed version, both in space and spatial frequency, of the Wigner distribution. The generalized PWD is defined in terms of successive integral transforms that use finite analysis windows in both the spatial and spatial-frequency domains.

Flandrin (1984) later introduced yet another class of computable approximations to the Wigner distribution, the *smoothed pseudo Wigner-Ville distribution* (smoothed PWVD). As noted by Flandrin, the smoothed PWVD is closely related to the generalized PWD introduced by Jacobson and Wechsler (1983). In particular, Flandrin's smoothed PWVD, like the generalized PWD, permits considerable flexibility in controlling peak cojoint resolution. The smoothed PWVD encompasses all approximations to the WD in which the space- and spatial-frequency resolution of the representation are independently controlled by a separable smoothing kernel $\Phi(\xi, \alpha) = \Phi_1(\xi)\Phi_2(\alpha)$ in Cohen's class definition.

2.3.9.2. The Generalized Pseudo-Wigner Distribution

Since the generalized PWD is, in practice, a "good" approximation to the WD only over a finite volume of the cojoint representation space, numerous generalized PWDs must be computed over different regions of the cojoint space and then concatenated to adequately approximate the WD everywhere. The resulting concatenation of generalized PWDs are called the *composite pseudo-Wigner distribution* (CPWD) (Jacobson and Wechsler, 1983).

The procedure for computing a single generalized PWD is straightforward: (1) Compute a Fourier transform over a single finite (windowed) region of an image; then (2) employ Eq. (2-51) with a sliding window to compute the generalized PWD from the Fourier transform found in step (1). The nature of the generalized PWD can be better understood by analytically formulating the procedure.

Assume arbitrary N-dimensional signal and window functions $f(x)$ and $h_1(x)$, whose respective Fourier transforms and WDs are denoted by $F(u)$ and $H_1(u)$, and $W_f(x, u)$ and $W_{h_1}(x, u)$. Also assume an arbitrary window function $H_2(u)$, whose inverse Fourier transform, $h_2(x) = \mathscr{F}^{-1}\{H_2(u)\}$ has a WD denoted by $W_{h_2}(x, u)$.

We first apply the window $h_1(x)$ centred on an arbitrary point (x_i) in the signal domain yielding the function

$$f_{x_i}(x) = f(x)h(x - x_i). \tag{2-86}$$

Using the shift and convolution theorems of Fourier analysis, the Fourier transform of the previous expression is given by

$$F_{x_i}(u) = F(u) \otimes [H_1(u)e^{-jx \cdot u}]. \tag{2-87}$$

Next, we apply the window $H_2(u)$ centered at u_0 in the frequency domain to get

$$F_{x_i, u_0}(u) = [F(u) \otimes [H_1(u)e^{-jx_i \cdot u}]]H_2(u - u_0). \tag{2-88}$$

Next, this result is inverse Fourier transformed. Using the shift and convolution theorems of Fourier analysis we arrive at

$$f_{x_i, u_0}(x) = [f(x)h_1(x - x_i)] \otimes [h_2(x)e^{ju_0 \cdot x}]. \tag{2-89}$$

Then, the WD of the expression is found by using the modulation and convolution properties ((P4) and (P5), respectively) of the WD. This yields

$$W_{f_{x_i, u_0}}(x, u) = [\text{WD of } \{f(x)h_1(x - x_i)\} \underset{x}{\otimes} W_{h_2}(x, u - u_0), \tag{2-90}$$

which after using the shift and windowing properties ((P3) and (P6), respectively) of the WD,

$$= \frac{1}{(2\pi)^N} [W_f(x, u) \underset{u}{\otimes} W_{h_1}(x - x_i, u)] \underset{x}{\otimes} W_{h_2}(x, u - u_0) \tag{2-91}$$

$$= \frac{1}{(2\pi)^N} \iint W_f(x, u')W_{h_1}(x - x_i, u - u')du'] \underset{x}{\otimes} W_{h_2}(x, u - u_0) \tag{2-92}$$

$$= \frac{1}{(2\pi)^N} \iint W_f(x', u')W_{h_1}(x' - x_i, u - u')W_{h_2}(x - x', u - u_0)dx'du'. \tag{2-93}$$

Now, since the frequency domain window $H_2(u)$ was centered on the spatial frequency u_0, we take $u = u_0$ in the previous expression to obtain the generalized PWD at u_0:

$$\tilde{W}_{f_{x_i}}(x, u_0) = W_{f x_i, u_0}(x, u)|_{u = u_0}$$

$$= \frac{1}{(2\pi)^N} \int\!\!\int W_f(x', u') W_{h_1}(x' - x_i, u_0 - u') W_{h_2}(x - x', 0) dx' du'. \tag{2-94}$$

Finally, since u_0 was arbitrarily chosen, we write the final form of the generalized PWD as

$$\tilde{W}_{f_{x_i}}(x, u) = \frac{1}{(2\pi)^N} \int\!\!\int W_f(x', u') W_{h_1}(x' - x_i, u - u') W_{h_2}(x - x', 0) dx' du'. \tag{2-95}$$

Accordingly, the generalized PWD, $\tilde{W}_{f_{x_i}}(x, u)$, is a cojointly smoothed version of the true Wigner distribution, $W_f(x, u)$, of the signal f. The effective smoothing kernels for several choices of the independent space and spatial-frequency analysis windows are depicted in Fig. 2.8. We initially assumed that $h_1(x)$ and $H_2(u)$ were uniquely chosen analysis windows. In practice, however, these windows vary as a function of space and spatial-frequency, thus greatly increasing one's flexibility to vary the resolution of the CPWD throughout the cojoint space.

2.3.10. Cojoint Resolution

Section 2.3.2. discussed proposed methods for characterizing the cojoint uncertainty of a *signal*. It was argued that general measures of cojoint uncertainty should be based upon a signal's cojoint energy distribution rather than on its marginal energy distributions. An uncertainty measure based upon a signal's cojoint energy distribution is exemplified by the *cojoint entropic uncertainty* of the signal as defined by Eqs. (2-37) and (2-33), which together yield the uncertainty metric

$$\Omega_f = \exp\{H(P_f(x, u))\} = \exp\left\{-\int\!\!\int P_f(x, u)\log P_f(x, u) dx\, du\right\}, \tag{2-96}$$

where $P_f(x, u)$ is the signal's intrinsic cojoint energy distribution.

For this uncertainty measure to be useful, one must establish (for each possible signal $f(x)$) the existence of a unique distribution function that specifies the signal's intrinsic cojoint energy distribution. This distribution function should, as discussed in Section 2.3.2., have the properties of a probability density function and have the signal's

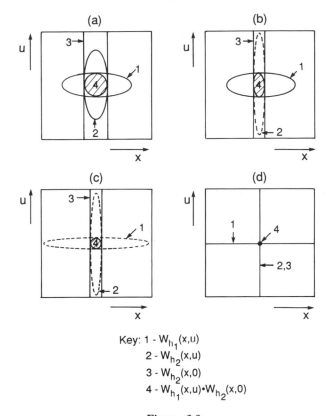

Key: 1 - $W_{h_1}(x,u)$
2 - $W_{h_2}(x,u)$
3 - $W_{h_2}(x,0)$
4 - $W_{h_1}(x,u) \cdot W_{h_2}(x,0)$

Figure 2.8.
Support regions (hatched) of effective smoothing kernels for composite pseudo-Wigner distributions of 1D signal computed at four different resolutions: (a) low spatial, low frequency resolution, (b) high spatial, low frequency resolution, (c) high spatial, high frequency resolution, and (d) full resolution of WD.

energy $|f(x)|^2$ and power spectrum $|F(u)|^2$ as its marginal densities. Appropriately normalized, the Wigner distribution (discussed in Section 2.3.4.) arguably has the previously mentioned properties for some signals, including the class consisting of complex-modulated Gaussians with complex-valued variance.

Though the Wigner distribution of any signal in the minimizing class is nonnegative definite, one may dispute the assertion that the Wigner distribution is *the* intrinsic cojoint energy distribution of a signal. Note that the Wigner distribution is uniquely related, by a Fourier transform, to the raw second-order cross product of a signal with itself —$r_f(x, \alpha)$ as defined by Eq. (2-50). In particular, unlike computable cojoint representations, the Wigner distribution does not involve any free parameters (window sizes, etc.) in its definition. The Wigner distribution of a signal is, therefore, in the previous sense, intrinsic to the signal.

As discussed in Section 2.3.4., many signals possess a Wigner distribution that takes on negative values. It has long been recognized that one cannot reasonably interpret local values of the Wigner distribution as defining local cojoint energy. For this reason, the Wigner distribution and many of its approximations are commonly referred to as pseudo-energy representations. The negative values found in many Wigner-type distributions are especially bothersome if one equates a normalized Wigner distribution with the probability-like function $P_f(x, u)$ in the cojoint entropic uncertainty definition. The question arises whether it makes sense to suppose the existence of a cojoint energy distribution that is *intrinsic* to an arbitrary signal and which might serve, in conjunction with a definition such as Eq. (2-96), to define an arbitrary signal's *intrinsic* cojoint uncertainty.

Given that estimators of energy are generally conceived of as bilinear functionals of a signal, one must consider alternatives to the Wigner distribution for defining a signal's "intrinsic" cojoint energy distribution function. As proved by Wigner (1979), no bilinear function of $f(x)$ has the signal's energy and Fourier power spectrum as its marginals and, at the same time, is nonnegative definite for all (x, u) and all signals. As suggested by Claasen and Mecklenbrauker (1984), one must therefore give up the requirement that a cojoint distribution function give the correct energy marginals if one wishes to obtain a positive definite cojoint representation. Specifically, Flandrin and Escudie (1979) have shown that, among all bilinear signal representations, only the class of spectrograms (and certain linear combinations of spectrograms) are guaranteed to produce nonnegative definite distributions for all signals $f(x)$. Consequently, only the spectrograms appear compatible with the nonnegativity requirement for the energy distribution $P_f(x, u)$ in the measure of entropic uncertainty. Rather than offer a measure of intrinsic *signal* uncertainty, however, such a metric would quantify cojoint *representational* resolution as it applies to some particular combination of signal and spectral analysis window. In conclusion, it remains an open research question as to whether the definition of entropic cojoint uncertainty given by Eq. (2-96) might be modified in some way to yield a metric for the intrinsic cojoint uncertainty of an arbitrary signal.

Note that the use of Eq. (2-96) with normalized signal spectrograms could offer a formal metric for ranking the resolutions of such spectrograms computed from a common signal or ensemble of signals. That spectrogram with the lowest uncertainty would, in effect, define the highest resolution *estimate* of a signal's cojoint energy distribution. The lowest entropy would be associated with the spectrogram whose energy is most concentrated or tightly clumped throughout the joint (x, u) domain of representation.

Though the cojoint resolution metric is applicable to the class of signal spectrograms, the question arises as to how one might specify the cojoint resolution of *linear* representations, such as the complex Gabor representation or the difference of Gaussians representation— as distinguished from their associated power representations. First, note that such linear representations may often serve as complete representations in the sense that the original signal can be fully reconstructed if so desired. Indeed, a signal—simply a complete representation of itself—can be regarded as a cojoint representation, albeit one that does not vary with frequency. An analogous observation also holds for the Fourier transform of a signal. As will become clear shortly, completeness does not necessarily correlate with high cojoint resolution.

Jacobson and Wechsler (1988) claim that linear representations of a signal have cojoint resolution, which is determined by applying some resolution metric to the associated power distribution obtained with the square modulus of a linear representation. Though such a definition is not often stated formally, it conforms with the intuitive notion of cojoint resolution as it pertains to linear cojoint representations of a signal. Note that the square modulus of a linear representation is trivially guaranteed to be nonnegative definite; therefore, one can apply the metric of cojoint entropic uncertainty. Recall that this metric only makes sense for the class of spectrograms, since they exclusively form the class of nonnegative definite bilinear representations. In fact, as discussed in Sections 2.3.7. and 2.3.8., the squared moduli of the Gabor representation and each resolution level of a difference of Gaussians representation are both spectrograms. Having discussed these power representations in earlier sections, we will soon be in a position to comment on how their effective bilinear kernels interact with the signal to determine their cojoint resolution properties that, by definition, they share with their respective linear representations.

We emphasize here that the previous approach to cojoint resolution of a linear or nonnegative definite bilinear signal representation takes into account properties of both the signal and the filter used to obtain its representation. The interaction between signal and filter is evident, for example, in the essential symmetry of a signal and its filter in the definition of the spectrogram. Such signal/filter interactions manifest themselves in the so-called window tradeoff problem, whereby the size of the filter support region determines the signal-domain and spectral-domain resolution tradeoff. Marcelja (1980) provides an in-depth discussion of the implications of this tradeoff for the Gabor image representations. As he notes, the cojoint energy distribution of (a) signal(s) to be represented must be determined (e.g., top-down primed, see also Chapter 6) before a suitable analysis windowing scheme (possibly space- and frequency-variant) can be determined. To assist in this task, Marcelja recommends that the Wigner distribution of a signal be (approximately) computed to give an indication of the intrinsic cojoint distribution of a signal.

Indeed, as discussed in earlier sections, the Gabor power representation, like all spectrograms, is equivalent to a blurred Wigner distribution. Therefore, the window tradeoff problem is readily understood by considering the effective smoothing kernel $\phi(x, u)$ on the Wigner distribution that is associated with any spectrogram. Recall that $\phi(x, u)$ is just the Wigner distribution of the window function used to compute a spectrogram, for example, the kernel $\phi_G(x, u)$ (Table 2.3) associated with the Gabor power spectrum that has the form of a two-dimensional Gaussian. The spatial and spectral variances of this Gaussian are inversely related along any cojoint pair of coordinates. This inverse relation is illustrated in Fig. 2.9., which depicts the $1/e$

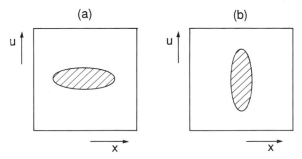

Figure 2.9.
Support regions of the Gabor smoothing kernel $\phi_G(x, u)$ for (a) narrowband, and (b) broadband analysis of a 1D signal. Note the tradeoff in spatial vs. frequency resolution.

contours of the kernels $\phi_G(x, u)$, which are associated with two different window tradeoffs for the Gabor spectral analysis of one-dimensional signals. Exemplary support regions for smoothing kernels associated with these representations in a 4D spatial/spatial-frequency domain are illustrated schematically in Fig. 2.10. Note that, in the case of the DOG and Gabor power spectra, the associated smoothing kernels exhibit a localization tradeoff between space versus spatial frequency. In contract, the CPWD, which approximates the Wigner distribution with arbitrary fidelity, has an associated smoothing kernel whose space and spatial-frequency support can be independently specified. In the limit, one obtains a smoothing kernel that is a four-dimensional delta function (illustrated as a pair of dots in the respective space and spatial-frequency domains in Fig. 2.10.), and the full resolution of the Wigner distribution is obtained.

Note that, by smoothing the WD—a "pseudo" energy representation—the kernels associated with spectrograms act as operators that,

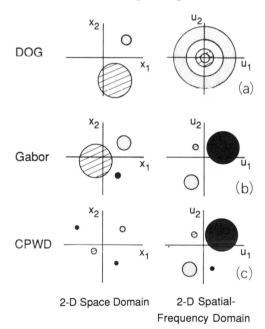

Figure 2.10.
Cojoint energy averaging exhibited by (a) each resolution level of the difference of Gaussians representation, (b) the Gabor representation, and (c) the generalized pseudo-Wigner distribution. The commonly textured regions suggest the 4D spatial/spatial-frequency volume over which the Wigner distribution of a 2D image is averaged by typical filters used to compute these representations.

by averaging over a region in accord with the uncertainty principle, deliver an estimate of the "true" (nonnegative definite) cojoint energy in some region of (x, u)-space. With regard to the application of cojoint representations to visual pattern recognition, it is not apparent why one would require a true energy representation at all. Rather, a likely application of cojoint representations might be the correlation between cojoint prototype representations (previously computed from some image and stored in memory) and a cojoint representation computed from the sensed image. The prototype acts, in effect, as the kernel of an operator that, by normalized correlation, detects specific relevant energy *configurations* in the joint (x, u)-space. The use of a cojoint representation of limited resolution (a spectrogram) for prototype and sensed image representations leads only to blurring of their associated correlation function relative to the result obtained by correlating WDs. This is apparent since the spectrograms of two images are themselves equivalent to blurred WDs, which by associativity of convolution implies a doubly blurred correlation function.

2.3.11. Summary

Cojoint image representations arguably offer improved pattern separability when compared with either pure space- or pure spatial-frequency image representations. Visual tasks such as pattern analysis can, therefore, benefit from the use of cojoint visual representations.

Several cojoint representations of interest to the visual sciences and their associated signal spectral power representations have all been shown to be expressible as *smoothed Wigner distributions*, where smoothing is with respect to a representation-specific kernel. For the spectrogram and the Gabor power spectrum, the associated smoothing kernel is given by the Wigner distribution of the associated analysis window functions. For the DOG representation, the smoothing kernel associated with each resolution level is given by the Wigner distribution of the associated difference of Gaussians filters. These smoothing kernels, by virtue of being Wigner distributions of window (or filter) functions, have a finite cojoint spread that is imposed by the *signal uncertainty principle*. When the spectrogram, Gabor power spectrum, or DOG power spectrum are employed as cojoint representations of some signal's space-varying power spectrum, the uncertainty spread of the analysis window (or filter) becomes confounded with the intrinsic cojoint (pseudo) energy distribution of the analyzed signal. Such is the nature of the well-known resolution tradeoff, when these classical power representations—all being variations of the spectrogram—are interpreted as smoothed Wigner distributions as in Cohen's (1966) general description of linear representations. A possible definition for the *cojoint resolution* of positive-definite power representations (spectrograms) of specified signals was proposed, which uses an entropic measure that would directly be applied to the cojoint power representation of the signal. It was further argued that since cojoint signal uncertainty is always defined in terms of either marginal or joint signal *power* distributions, the cojoint resolution of the *linear* Gabor and DOG representations should similarly be defined in terms of their associated cojoint *power representations*, simply obtained by taking their squared moduli.

Since all cojoint power representations are obtained through bilinear (quadratic) transformation of a signal, there has been a recent emphasis on the explicit construction of bilinear (pseudo) power representations. As has been discussed, such an approach is exemplified by the composite pseudo-Wigner distribution (CPWD) introduced by Jacobson and Wechsler (1983) and the related smoothed pseudo Wigner-Ville distribution (smoothed PWVD) of Flandrin (1984). Whereas the spectrograms involve the specification of the size of a spatial support window in a linear filtering operation whose modulus gives a cojoint power representation, the CPWD and smoothed PWVD involve the specification of separate window supports in both the space- and spatial-frequency domains of an overall bilinear transformation. The added degree of freedom in specifying the bilinear analysis support region eliminates the window tradeoff problem. Cojoint resolution is limited only by the cojoint localization of the signal itself and the computational limitations imposed by incremental increases in the bilinear analysis support region.

2.4.   Texture Analysis

We consider in this section the issue of texture analysis. The representational aspect based on the preceding section on cojoint s/sf representations is discussed here, while the parallel computational aspect is deferred to Chapter 4. The next discussion is based on the work of Reed and Wechsler (1990) and helps in understanding the role cojoint s/sf representations can play in image modeling for tasks like segmentation and/or clustering/grouping.

Texture can be described in a variety of ways. It can be generated by primitives organized by placement rules or as the result of some random process. It can be found in a continuous spectrum from purely deterministic to purely stochastic. Texture is an important surface characteristic and based on it, shape and motion can be estimated. A taxonomy of problems encountered within the context of texture analysis would include classification/discrimination, description, and segmentation—in order of increasing difficulty. Generic segmentation is the partitioning of an image into regions that are "homogenous" with respect to one or more characteristics. A basic issue to be considered is that of cell unit size, i.e., the resolution of the area measured in order to test for homogeneity.

The study of perceptual grouping in the 1920s led to the formulation of what came to be called the Gestalt laws. As suggested by Wertheimer (1958) individual tokens appear to group according to a set of principles. These principles include proximity, similarity, good continuation, symmetry, closure, and common fate. The tendency of seeing elements with similar characteristics as belonging to an approximately homogenous group suggests that the same mechanism might be at work for Gestalt and texture segmentation. While providing a framework for predicting visual grouping effects, the Gestalt laws are not grounded in any specific theory of vision. Our goal is to provide such a theory and test it.

Basic constraints for segmentation and/or clustering are locality, resolution, automaticity, and if possible, consistency with the known findings about the HVS. Locality of analysis was already discussed in Section 2.2.4., and the concept of resolution has been fully explored so far. Any suggested theory and its corresponding implementation should be as automatic as possible, and the parameters used should be such that a wide class of images can be analyzed. The theory is "scientific" if its predictions are met by psychophysical experiments.

2.4.1. Review

The main approaches for texture segmentation are the statistical and spatial-frequency or spatial/spatial-frequency methods, which are related by Eq. (2-11c).

2.4.1.1. Statistical Methods

One of the oldest methods used today in texture segmentation is spatial-gray level dependence (SGLD) (Haralick *et al.*, 1973). The SGLD estimates the (second order) joint gray level distribution for two gray levels α and β displaced by $\mathbf{d}(|d|, \not\prec)$, where d is the distance and $\not\prec$ stands for the orientation of $\mathbf{d}$ with respect to the horizontal. Statistical measurements such as coarseness and entropy can be defined over $SGLD(\mathbf{d})$ and then used to classify/discriminate textures. The first order distribution is called gray-level difference (GLD), and it is the marginal of SGLD. Formally, for any $\mathbf{d} = (\Delta x, \Delta y)$ one estimates $P(g|\mathbf{d})$, where g are the gray levels such that $g \in [0, 2^{n-1}]$, and the density P is defined as

$$P(g|\mathbf{d}) = |f(x, y) - f(x + \Delta x, y + \Delta y)|. \qquad (2\text{-}97)$$

Another popular method is the gray level run length (GLRUM), where a run is defined as a set of linearly adjacent pixels having the same gray-level value. Appropriate statistics can be defined over GLD and/or GLRUM, as was done for SGLD.

A disadvantage is their dependence on the resolution chosen. More recently, of an increasing interest is the Gibbs (Markov) random fields (GRF) (Geman and Geman, 1984; Cohen, 1986; and Goutsias, 1987). As discussed in Section 4.2.3., however, such approaches have yet to consider resolution.

2.4.1.2. Spatial/Spatial-Frequency Methods

Cojoint spatial/spatial-frequency (s/sf) methods discussed in Section 2.3. are based on image representations indicating the frequency content in localized regions in the spatial domain. As such, these methods overcome the shortcomings of traditional Fourier-based techniques. They achieve high resolution in both domains and are consistent with recent HVS theory. Specifically, there is a large and growing body of theory postulating frequency analysis in HVS. Support for a spatial-frequency interpretation of the HVS in predicting object recognition has been reported by Ginsburg (1980). Beck *et al.*, (1987) have shown correlation between the ability of humans to segment tripartite-textured images and the output of a bank of 2D Gabor filters applied to the images.

A key issue in comparing joint spatial/spatial-frequency representations is the resolution that can be attained (simultaneously) in the two domains. Often referred to as uncertainty, this topic has been addressed for image processing applications by Wilson and Grandlund (1984). As mentioned earlier, Daugman (1985) has examined the class of Gabor filters and found that they achieve the lower limit of uncertainty as measured by the product of effective widths corresponding to the spatial and spectral domains, respectively. Specifically, the uncertainty (as it relates to entropy) is measured separately along each dimension (i.e., the s and sf) of the joint s/sf representation. Jacobson and Wechsler (1988) looked further into the resolution/uncertainty issue and came to two conclusions. First, if one talks about cojoint s/sf representations, then the uncertainty should be derived from a joint Cartesian domain ($s \times sf$) rather than from the result of computing it over two independent dimensions (i.e., the s and sf). Furthermore, the spectrogram, DOG, and Gabor representation are smoothed versions of the WD, and they are members of the more general Cohen (1966) class of distributions. The smoothing is a result of convolving the WD with various kernels (shown in Table 2.4 as $W_{h'}$ $W_{h_{D'}}$ and W_{h_G}). As such, these s/sf representations cannot improve on the resolution to be achieved by the WD. These results are summarized (for the N-dimensional case) in Table 2.4.

The relevance of the WD for the HVS is hypothetical at this time. One possible conjecture is that simple cells (characterized by a linear response) implement frequency analysis in the form of the Gabor representation. The next major layer of cells, the complex cells, are known to exhibit a nonlinear response. One possibility is that the second definition of the WD is being implemented. The shifted Fourier transform of the image and its shifted complex conjugate (the outputs of the simple cells) may be multiplied, and the inverse Fourier transform then found, by the complex cells. The net result is an image representation of higher resolution, the WD.

2.4.2. Texture Representation

The generic system for texture segmentation and/or clustering is shown in Fig. 2.11a., and its instantiation, which is used by us, is shown in Fig. 2.11b. The representational components are described next, while the computational relaxation step is deferred to Section 4.2.2. as an example of parallel and distributed computation.

Table 2.4. A summary of results from Jacobson and Wechsler (1988).

Spectrogram:

$$S_f(x, u) = |F_x(u)|^2$$

$$= \frac{1}{(2\pi)^N} \iint W_f(x', u') W_h(x - x', u - u') dx' du'$$

Difference of Gaussians (Single Level):

$$D_f(x) = \frac{1}{(2\pi)^N} \iint W_f(x', u') W_{hD}(x - x', u - u') dx' du'$$

Gabor Power Representation:

$$G_f(x, u) = \frac{1}{(2\pi)^N} \iint W_f(x', u') W_{hG}(x - x', u - u') dx' du'$$

where

$W_f(x, u) =$ the WD of the function of interest,
$W_h(x, u) =$ the WD of the window function,
$W_{hD}(x, u) =$ the WD of $Ae^{-\sum_{i=1}^{N} a_i x_i^2} - Be^{-\sum_{i=1}^{N} b_i x_i^2}$

and

$W_{hG}(x, u) =$ the WD of $e^{-\sum_{i=1}^{N} h_i x_i^2}$

The aliasing effect, well known in digital signal processing, occurs when a continuous signal is sampled at such a rate that the translated versions of its spectrum (created by the sampling process) overlap in the frequency domain. This prevents recovery of the original (continuous) signal from the sampled (discrete) signal. To prevent aliasing, one must sample at twice the highest frequency in the continuous signal (the Nyquist rate). This effect is encountered most often in computer vision in the construction of multiresolution (pyramid) representations. As each level of the pyramid is subsampled, the bandwidth of the image at that level must be reduced by the subsampling ratio in each dimension. The most obvious way to avoid aliasing in the product function characterizing the PWD is to "oversample" the original function by a factor of two in each direction. As this is the same function performed in the resolution reduction for the generation of pyramids, we have used the pyramid generating kernels proposed by Meer *et al.* (1987), which were discussed earlier in Section 2.2.2. The

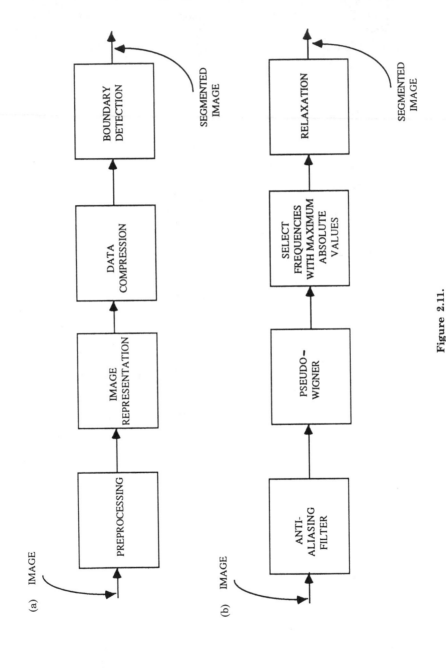

Figure 2.11.

(a) A generic image segmentation system. (b) The implemented segmentation system.

pseudo-Wigner distribution (PWD) is given as

$$\text{PWD}(m, n, p, q) = 4. \sum_{k=-N_2+1}^{N_2-1} \sum_{l=-N_1+1}^{(N_1-1)} h_{N_1,N_2}(k, l)$$

$$\sum_{r=-M_2+1}^{M_2-1} \sum_{s=-M_1+1}^{M_1-1} g_{M_1,M_2}(r, s)f(m+r+k, n+s+l).$$

$$f^*(m+r-k, n+s-l)e^{-j((2\pi kp/P) + (2\pi lq/Q))}, \tag{2-98}$$

where

$$p = 0, \pm 1, \ldots, \pm (N_2 - 1)$$
$$q = 0, \pm 1, \ldots, \pm (N_2 - 1)$$
$$P = 2N_2 - 1$$
$$Q = 2N_1 - 1.$$

m and n are integers, and $h_{N_1,N_2}(k, l)$ and $g_{M_1,M_2}(r, s)$ are windows.

The selection of the window $h_{N_1,N_2}(k, l)$ is governed by the same considerations that hold in any application of the DFT. That is, we wish to select a window that is of a specified size in the spatial domain with "nice" properties in the spatial-frequency domain. The size of the window is dictated by the resolution required in the spatial-frequency domain.

We have chosen to use a 2D extension of the 1D Kaiser window. This parameterized window is an approximation to the family of prolate spheroidal wave functions of order zero, which maximize energy in a given band of frequencies for a specified time duration. That this class of functions is useful in various signal processing applications has long been known and was reaffirmed by Wilson (1987). The window parameter in the Kaiser approximation allows compromise between center lobe width and side-lobe height.

The function of the window $g_{M_1,M_2}(r, s)$ is to allow local averaging. We have chosen to use a normed rectangular window. The larger the values of M_1 and M_2, the more averaging occurs.

The Kaiser window is defined as

$$h_{N_1,N_2}(k, l) = h_{N_1}(k) \cdot h_{N_2}(l). \tag{2-99}$$

where

$$h_{N_1}(k) = \begin{cases} \dfrac{I_0[\alpha_1\sqrt{1 - (k/(N_1 - 1))^2}]}{I_0(\alpha_1)} & \text{if } |k| \le N_1 - 1 \\ 0 & \text{otherwise} \end{cases}$$

$$h_{N_2}(k) = \begin{cases} \dfrac{I_0[\alpha_2\sqrt{1 - (k/(N_2 - 1))^2}]}{I_0(\alpha_2)} & \text{if } |k| \le N_2 - 1, \\ 0 & \text{otherwise} \end{cases}$$

and $I_0(x)$ is the modified Bessel function of the first kind and zero order.
  The normed rectangular is given by

$$gM_1, M_2(r, s) = \begin{cases} \dfrac{1}{(2M_1 - 1)(2M_2 - 1)} & \text{if } |r| \le M_1 - 1 \text{ and } |s| \le M_2 - 1 \\ 0 & \text{otherwise.} \end{cases}$$

$$(2\text{-}100)$$

The result of applying the PWD to an image results in a representation of the frequency content in the image at each pixel. That is, for each pixel in the image, a 2D array representing frequency content is stored. For an $N \times N$ image and the main PWD window size specified by N_1 and N_2, $2N_1 \times 2N_2 \times N \times N$ elements are stored. A viable method to reduce the amount of data retained is to save only the frequency content at frequencies containing the most energy for each pixel. Further compression can be obtained if only the frequency content at frequencies containing the most energy (at each pixel) over all pixels is saved.

  In the system implemented, we consider only the frequency content at the frequency-containing maximum energy for each pixel. The resulting frequencies are ranked for energy (calculated at each pixel over all pixels). The frequency content at a number of the top-ranked frequencies is retained. If F frequencies are retained, then F, $N \times N$, arrays are stored. For most of the examples considered, only the frequency with maximum energy (for all frequencies and pixels) is examined, yielding a single $N \times N$ array, or "frequency plane."

  One might choose to retain more frequencies, particularly in the second step of the compression, because using more frequency planes (and multilevel relaxation) improves discrimination, particularly when more than two textures are involved. The examination of multiple frequency planes can also give insight into psychophysical grouping effects, as will be seen in Section 4.2.2.

2.5. Conclusions

We have made a strong case for the spatial/frequency (s/sf) representation as the ideal candidate for early visual representations. Such representations are distributed and enjoy high resolution. We advanced the Wigner distribution (WD) as the main representative of the class of (s/sf) representations and have shown that the other (s/sf) schemes being used are particular instances of the WD. The mathematics involved are quite complicated but essential to redefine the concept of resolution and to put the issue of low-level representations on a firm foothold. We took the example of texture analysis and developed the underlying s/sf representation. The corresponding parallel and distributed (PDP) algorithm, that of relaxation, to operate on such representations is discussed in Section 4.2.4. Furthermore, we show in the Chapter 3 how such (s/sf) representations can be made invariant to affine transformations.

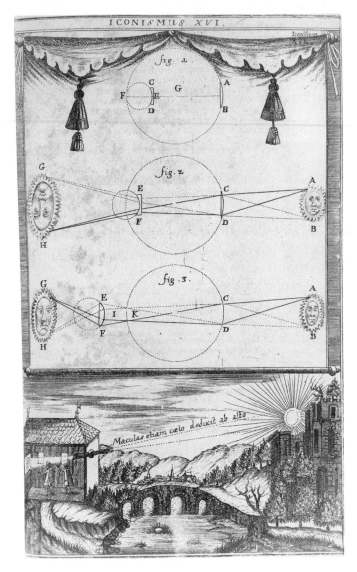

Demonstration of how sunlight travels through convex and concave lenses; demonstration of directing sunlight through camera obscura. (From Zahn, J., Oculus artificialis..., 1685–86.) Reprinted courtesy of the National Library of Medicine.

3

Invariance

Plus ça change plus c'est la même chose.
The more things change, the more they stay the same.

One of the reasons for the high degree of computational complexity of computational vision lies in the inherent variability of the input. Such variability can be the result of noise, incomplete (or missing) data, and geometric or topological changes. Vision systems must not only perceive the identity of objects despite such variability but explicitly characterize the variability, because the variability in the image formation process (particularly that due to geometric distortion and varying incident illumination) inherently carries valuable information about the image.

The variability aspect could also be very harmful, if not appropriately handled, because recognition would then need to deal with a possible infinite number of image occurrences. Since there are many viewpoints, one must deal with the problem of searching a multidimensional parameter space. To deal with viewpoint variability and

stochastic changes in the input, the image invariants must be defined
and captured.

Present research in CV is greatly indebted to James Gibson for his
pioneering work in the psychophysics of perception. Gibson (1950,
1979) emphasized throughout his career the important role of geometric and topological invariants. Gibson's approach has been misunder-
stood and controversial. (Ullman (1980) discusses some of the debated
issues.) Image invariants are crucial, however, and we will discuss how
a visual system could eventually "resonate" through PDP to such
invariants. According to Gibson (1966):

> It can be shown that the easily measured variables of stimulus energy,
> the intensity of light, sound, odor, and touch, for example, vary from
> place to place and from time to time as the individual goes about his
> business in the environment. The stimulation of receptors and the
> presumed sensations, therefore, are variable and changing in the ex-
> treme, unless they are experimentally controlled in a laboratory. The
> unanswered question of sense perception is how an observer, animal or
> human, can obtain constant perceptions in everyday life on the basis of
> these continually changing sensations. For the fact is that animals and
> men do perceive and respond to the permanent properties of the environ-
> ment as well as to the changes in it.
>
> Besides the changes in stimuli from place to place and from time to
> time, it can also be shown that certain higher-order variables of stimulus
> energy—ratios and proportions, for example—do *not* change. They re-
> main invariant with movements of the observer and with changes in the
> intensity of stimulation. And it will be shown that these invariants of the
> energy flux at the receptors of an organism correspond to the permanent
> properties of the environment. They constitute, therefore, information
> about the permanent environment.
>
> The active observer gets invariant perceptions despite varying sensa-
> tions. He perceives a constant object by vision despite changing sensa-
> tions of light; he perceives a constant object by feel despite changing
> sensations of pressure; he perceives the same source of sound despite
> changing sensations of loudness in his ears. The hypothesis is that
> constant perception depends on the ability of the individual to detect the
> invariants, and that he ordinarily pays no attention whatever to the flux
> of changing sensations.

In this chapter, we look for invariants, define them, and show their
usefulness. We also suggest that an invariant representation of the
input might come at an early stage and is coupled with the derivation
of cojoint spatial/spatial-frequency representations. The vision system
could thus make available an invariant cojoint and distributed repre-

sentation useful for recognition and the derivation of intrinsic images, such as optical flow and depth. We consider the geometrical invariance and show how the vision system could yield invariant distributed representations suited for object recognition. We continue by considering the statistical and algebraic/topological invariances, respectively. The role of invariances found in the environment falls under ecological optics, which call an active observer into play, and are discussed in Section 6.2.

3.1. Geometrical Invariance

In this section, we consider ways to achieve invariance to geometrical transformations such as LT (linear transformations) given in terms of scale, rotation, and shift (translation). Pseudoinvariance to perspective distortions (slant and tilt) is dealt with later using reprojections.

3.1.1. Invariance to Linear Transformations

The bread-and-butter approach for dealing with linear transformations is the Fourier transform (FT). We consider 2D images characterized by the picture value function $f(x, y)$. The corresponding (cyclic) FT is given by $F(u, v)$, where

$$F(u, v) = \int\limits_{-\infty}^{\infty}\!\!\!\int f(x, y)\exp[-j2\pi(ux + vy)]dxdy. \qquad (3\text{-}1a)$$

The discrete version of the FT is given by the discrete Fourier transform (DFT), which takes a periodic, discrete spatial image into another discrete and periodic frequency representation. Formally, the 2D DFT is given as

$$F(k, l) = \sum_{n=0}^{N-1}\sum_{m=0}^{M-1} f(n, m)\exp\left[-j2\pi\left(\frac{kn}{N} + \frac{lm}{M}\right)\right], \qquad (3\text{-}1b)$$

where $0 \leq k \leq N - 1$, and $0 \leq l \leq M - 1$.

The magnitude of both the FT and DFT are shift invariant, i.e., invariant to a shift in position due to translation. Specifically, let $F(u, v)$ be the FT of $f(x, y)$; i.e., $F(u, v) = \mathcal{F}\{f(x, y)\}$. Assume that $f(x, y)$

is shifted by $\alpha = (x_0, y_0)$. Then, the FT of $f(x - x_0, y - y_0)$ is given as

$$F_1(u, v) = \mathscr{F}\{f(x - x_0, y - y_0)\}$$

$$= \int\int\limits_{-\infty}^{\infty} f(x - x_0, y - y_0)\exp[-j2\pi(ux + vy)]dxdy. \quad (3\text{-}2a)$$

Substitute $a = x - x_0$ and $b = y - y_0$, and it follows that

$$F_1(u, v) = \exp[-j2\pi(ux_0 + vy_0)] \int\int\limits_{-\infty}^{\infty} f(a, b)\exp[-j2\pi(au + bv)]dadb$$

$$= \exp[-j2\pi(ux_0 + vy_0)]F(u, v).$$

Finally,

$$|F_1(u, v)| = |\exp[-j2\pi(ux_0 + vy_0)]||F(u, v)| = |F(u, v)|. \quad (3\text{-}2b)$$

In a similar way, one can show the same property regarding the DFT. Specifically, let $F(k, l)$ be the DFT of the picture value function $f(n, m)$ and assume that $f(n, m)$ has been shifted by $\alpha = (n_0, m_0)$. The corresponding DFT of $f(n - n_0, m - m_0)$ is given by $F_1(k, l)$ as

$$F_1(k, l) = \sum_{n=0}^{N-1}\sum_{m=0}^{M-1} f(n - n_0, m - m_0)\exp\left[-j2\pi\left(\frac{kn}{N} + \frac{lm}{M}\right)\right]. \quad (3\text{-}3a)$$

Substitute $q = n - n_0$ and $p = m - m_0$. Then,

$$F_1(k, l) = \sum_{q=-n_0}^{N-1-n_0}\sum_{p=-m_0}^{M-1-m_0} f(q, p)\exp\left[-j2\pi\left(\frac{k(q + n_0)}{N} + \frac{l(p + m_0)}{M}\right)\right]$$

$$= \left\{\exp\left[-j2\pi\left(\frac{kn_0}{N} + \frac{lm_0}{M}\right)\right]\right\}\sum_{q=-n_0}^{N-1-n_0}\sum_{p=-m_0}^{M-1-m_0} f(q, p)$$

$$\times \exp\left[-j2\pi\left(\frac{kq}{N} + \frac{lp}{M}\right)\right]$$

$$= \exp\left[-j2\pi\left(\frac{kn_0}{N} + \frac{lm_0}{M}\right)\right]F(k, l).$$

Finally, one obtains

$$|F_1(k, l)| = \left|\exp\left[-j2\pi\left(\frac{kn_0}{N} + \frac{lm_0}{M}\right)\right]\right||F(k, l)| = |F(k, l)|. \quad (3\text{-}3b)$$

An important concept related to the FT is the *similarity theorem*. It states that if $F(u, v) = \mathscr{F}[f(x, y)]$ then,

$$\mathscr{F}\{f(a_1x + b_1y, a_2x + b_2y)\} = (A_1B_2 - A_2B_1)F(A_1u + A_2v, B_1u + B_2v). \quad (3\text{-}4)$$

The theorem can be shown as following from

$$\mathscr{F}\{f(a_1x + b_1y, a_2x + b_2y)\} = \int\!\!\!\int_{-\infty}^{\infty} f(a_1x + b_1y, a_2x + b_2y)$$

$$\times \exp[-j2\pi(ux + vy)]dxdy.$$

Let $\alpha = a_1x + b_1y$ and $\beta = a_2x + b_2y$. Then, $x = A_1\alpha + B_1\beta$ and $y = A_2\alpha + B_2\beta$, where

$$A_1 = \frac{b_2}{a_1b_2 - a_2b_1}; \quad B_1 = \frac{-b_1}{a_1b_2 - a_2b_1}$$

$$A_2 = \frac{-a_2}{a_1b_2 - a_2b_1}; \quad B_2 = \frac{a_1}{a_1b_2 - a_2b_1}.$$

Then one can obtain the formulae (3-4), i.e.,

$$\mathscr{F}\{f(a_1x + b_1y, a_2x + b_2y)\}$$

$$= \int\!\!\!\int_{-\infty}^{\infty} f(\alpha, \beta)\exp\{\ j2\pi[(A_1u + A_2v)\alpha$$

$$+ (B_1u + B_2v)\beta]\}(A_1B_2 - A_2B_1)d\alpha d\beta$$

$$= (A_1B_2 - A_2B_1)F(A_1u + A_2v, B_1u + B_2v).$$

In other words, the similarity theorem relates the FT of a transformed function to its original FT. An interesting application is the *central slice theorem*, which is discussed next. Consider the projection $g_Y(x)$ of the function $f(x, y)$ on the x axis,

$$g_Y(x) = \int_{-\infty}^{\infty} f(x, y)dy. \quad (3\text{-}5a)$$

The corresponding one-dimensional FT of the projection is given by

$$G_Y(u) = \mathscr{F}\{g_Y(x)\} = \int_{-\infty}^{\infty} g_Y(x)\exp(-j2\pi ux)dx$$

$$= \int\!\!\int_{-\infty}^{\infty} f(x, y)\exp(-j2\pi ux)dxdy = F(u, 0). \qquad (3\text{-}5b)$$

From the similarity theorem rotation of $f(x, y)$ through an angle θ yields the original spectrum rotated by the same amount, if we let

$$a_1 = \cos\theta; \quad b_1 = \sin\theta; \quad a_2 = -\sin\theta; \quad b_2 = \cos\theta$$

so that

$$A_1 = \cos\theta; \quad A_2 = \sin\theta; \quad B_1 = -\sin\theta; \quad B_2 = \cos\theta.$$

Finally, $G_Y(u) = F(u, 0)$ combines with the rotation property to imply that the one-dimensional FT of $f(x, y)$ projected onto a line at an angle θ with the x axis is the same as if $F(u, v)$ were evaluated along a line at an angle θ with the u axis. Therefore, one can get a collection of FT evaluated along $\{\theta_i\}$ radial orientations, interpolate along the 2D (u, v) grid, and, if the sampling is satisfactory, reconstruct the original image as

$$\hat{f}(x, y) = \mathscr{F}^{-1}[F(u, v)]. \qquad (3\text{-}5c)$$

The central slice theorem can be recast using the sifting properties of the impulse distribution if the projections $g_Y(x)$, known as the Radon transform (RT), are rewritten as

$$g_\theta(R) = \int\!\!\int f(x, y)\delta(x\cos\theta + y\sin\theta - R)dxdy, \qquad (3\text{-}6a)$$

where $(x\cos\theta + y\sin\theta - R)$ is the parametric equation of a line of slope θ and distance R from the origin. In polar coordinates (r, ϕ), such that $x = r\cos\phi$, $y = r\sin\phi$, and $R = r\cos(\theta - \phi)$, the projection can be rewritten as

$$g_\theta(R) = \int_0^\infty \int_0^{2\pi} f(r, \phi)\delta[r(\cos(\theta - \phi)) - R]rdrd\phi, \qquad (3\text{-}6b)$$

where r is the Jacobian of the transformation. The corresponding FT in polar coordinates is given by

$$F(\rho, \theta) = \int_0^\infty g_\theta(R)\exp(-j2\pi\rho R)dR, \qquad (3\text{-}6c)$$

and the original function $f(x, y)$ can be then recovered by taking the inverse FT, i.e.,

$$f(x, y) = \int_0^{2\pi} d\theta \int_0^\infty F(\rho, \theta)\exp[j2\pi\rho(x \cos \theta + y \sin \theta)]\rho d\rho. \quad (3\text{-}6d)$$

The (central slice) reconstruction method requires evaluating the inverse FT preceded by interpolation. Because central slice reconstruction is computationally expensive, alternative methods have been suggested. *Back projection* smears the projections across the unknown (line) density functions. The back projection, corresponding to a single projection, is called a *laminogram*, and it is given by

$$l_\theta(x, y) = \int g_\theta(R)\delta(x \cos \theta + y \sin \theta - R)dR. \qquad (3\text{-}7a)$$

The reconstructed image $\hat{f}(x, y)$ is obtained by integrating for all angles θ, i.e.,

$$\hat{f}(x, y) = \int_0^\pi l_\theta(x, y)d\theta$$

$$= \int_0^\pi \int_{-\infty}^\infty g_\theta(R)\delta(x \cos \theta + y \sin \theta - R)dRd\theta. \qquad (3\text{-}7b)$$

The "smear" is equivalent to convolution in the spatial domain. The corresponding impulse response and transfer functions can be found as $1/R$, and $1/\rho$, respectively. Then, a suitable correction method would take the FT of $\hat{f}(x, y)$, filter it with ρ, and then take the inverse FT. Formally

$$\hat{f}(x, y) = f(x, y) \otimes \frac{1}{R}$$

and

$$\hat{F}(\rho, \theta) = F(\rho, \theta)\frac{1}{\rho}, \qquad (3\text{-}7c)$$

and the newly reconstructed image is then given as $\mathscr{F}^{-1}[\hat{F}(\rho, \theta)\rho]$. Back projection still suffers from the need for the inverse FT. The suggested alternative, known as the *convolution method*, is widely used in computer tomography (CT) reconstruction. Simply stated, instead of filtering in the frequency domain, one can undo the effects of smearing in the spatial domain, back project, and then get away from the need for expensive 2D transforms. The convolution function $cf(R)$ has to be chosen such that it can correct for the $(1/R)$ blur, and it is given as

$$cf(R) = \mathscr{F}^{-1}\{|\rho|\} \tag{3-7d}$$

and the reconstructed image $\hat{f}(x, y)$ is obtained as

$$\hat{f}(x, y) = \int_0^\pi \int_{-\infty}^\infty [g_0(R) \otimes cf(R)]\delta(x \cos \theta + y \sin \theta - R)dRd\theta. \tag{3-7e}$$

Makovski (1983) discusses different ways for deriving $cf(R)$ and coping with the problem of an undefined transform (because $|\rho|$ is not integrable). Generally, in the vicinity of the origin, $cf(R)$ approaches an impulse distribution as $2/\varepsilon^2$ (for $R^2 \ll \varepsilon^2/4\pi^2$). Otherwise, $cf(R)$ approaches $[-1/(2\pi^2R^2)]$. Further discussion on computer tomography and temporal invariance using the Kalman filter is provided in Section 3.2.3.

Another geometric transformation to be considered is reflection. Based on the similarity theorem,

$$\mathscr{F}\{f(ax, by)\} = \frac{1}{|ab|} F\left(\frac{u}{a}, \frac{v}{b}\right), \tag{3-8}$$

where $F(u, v) = \mathscr{F}\{f(x, y)\}$. If one assumes that $a = b = -1$, then it follows that

$$\mathscr{F}\{f(-x, -y)\} = F(-u, -v), \tag{3-9}$$

i.e., the FT of a function reflected through the origin is given by $F(-u, -v)$. If we were to reflect this image only with respect to the y axis, then $\mathscr{F}\{f(-x, y)\} = F(-u, v)$. Note also that the FT of the FT yields the reflected function. Specifically,

$$\mathscr{F}\{\mathscr{F}\{f(x)\}\} = \int_{-\infty}^\infty F(u)\exp(-j2\pi ux)dx = f(-x). \tag{3-10}$$

Next, we assume two-dimensional picture value functions $f(x, y)$ characterized by circular symmetry, i.e., $f(x, y) = f_R(r)$, where $r^2 = x^2 + y^2$. The corresponding FT is given by (Castleman, 1978)

$$\mathscr{F}\{f(x, y)\} = \int\!\!\!\int_{-\infty}^{\infty} f(x, y)\exp[-j2\pi(ux + vy)]dxdy$$

$$= \int_0^\infty \int_0^{2\pi} f_R(r)\exp[-j2\pi qr \cos(\theta - \phi)]rdrd\theta, \quad \text{(3-11a)}$$

where $x + jy = r[\exp(j\theta)]$ and $u + jv = q[\exp(j\phi)]$. Then

$$\mathscr{F}\{f(x, y)\} = \int_0^\infty f_R(r)\left(\int_0^{2\pi} \exp(-2jnqr \cos \theta)d\theta \right)rdr. \quad \text{(3-11b)}$$

The zero-order Bessel function of the first kind is

$$J_0(\alpha) = \frac{1}{2\pi} \int_0^{2\pi} \exp[-j\alpha(\cos \theta)]d\theta. \quad \text{(3-11c)}$$

Therefore, one obtains by substitution of $J_0(\alpha)$

$$\mathscr{F}\{f(x, y)\} = 2\pi \int_0^\infty f_R(r)J_0(2\pi qr)rdr \quad \text{(3-11d)}$$

and can observe that the FT of a circularly symmetric function is a function of only a single radial frequency variable q, i.e., $F(u, v) = F_R(q)$, where $q^2 = u^2 + v^2$. This pair of transforms is known as the Hankel transform of zero order, i.e., $\langle f_R(r), F_R(q)\rangle$, where

$$F_R(q) = 2\pi \int_0^\infty f_R(r)J_0(2\pi qr)rdr$$

$$\text{(3-11e)}$$

$$f_R(r) = 2\pi \int_0^\infty F_R(q)J_0(2\pi qr)\, qdq.$$

The previous discussion showed that circular symmetry is invariant, i.e., it is preserved by the FT, and that the FT of a circularly symmetric function is a one-dimensional function of a radial variable only.

An additional important property relates the spacing and orientation of the spatial domain to that of the corresponding FT. Specifically, if there is a structure of parallel spatial lines, spaced by Δ and passing through the origin with a slope α, then the Fourier spectrum is spaced $1/\Delta$ and is of slope $-1/\alpha$. Such a relationship can be used for texture classification, discrimination, and segmentation (Bajcsy and Liberman, 1976). Formally, if $f(x, y) = \delta(y - \alpha x)$, i.e., if $f(x, y)$ is a line passing through the origin at a slope α, then by using the sifting property of the impulse distribution, one can derive that the FT of $f(x, y)$ is

$$F(u, v) = \int\limits_{-\infty}^{\infty}\!\!\int \delta(y - \alpha x)\exp[-j2\pi(ux + vy)]dxdy$$

$$= \int_{-\infty}^{\infty} \exp[-j2\pi(ux + \alpha vx)]dx = \delta(u + \alpha v), \qquad (3\text{-}12)$$

i.e., the FT is also a line passing through the origin but at a slope of $-1/\alpha$, and the spatial line and the FT line are perpendicular one to the other since $\alpha(-1/\alpha) = -1$.

Next, assume parallel lines spaced by Δ, i.e., the *comb* given by $f(x, y)$ as

$$f(x, y) = \sum_{-\infty}^{\infty} \delta(x - n\Delta). \qquad (3\text{-}13a)$$

The orientation of the lines is parallel to the y axis, but the result can be easily extended to any orientation α. Because the function $f(x, y)$ is periodic, it can be replaced by its Fourier series representation

$$f(x, y) = \sum_{-\infty}^{\infty} c_n \exp\!\left(\frac{j2\pi nx}{\Delta}\right), \qquad (3\text{-}13b)$$

where the coefficients c_n are given by

$$c_n = \frac{1}{\Delta} \int_0^{\Delta} f(x, y)\exp\!\left(-j\,\frac{2\pi nx}{\Delta}\right)dx$$

$$= \frac{1}{\Delta}. \qquad (3\text{-}13c)$$

The last result is obtained using the sifting property of the impulse distribution when $x = \alpha\Delta$ for some α. The FT corresponding to the *comb*

$f(x, y)$ is given as

$$F(u, v) = \int\int_{-\infty}^{\infty} \sum_{-\infty}^{\infty} \delta(x - n\Delta)\exp[-j2\pi(ux + vy)]dxdy$$

$$= \int\int_{-\infty}^{\infty} \sum_{-\infty}^{\infty} \frac{1}{\Delta} \exp\left(-j\frac{2\pi nx}{\Delta}\right)\exp[-j2\pi(ux + vy)]dxdy$$

$$= \int_{-\infty}^{\infty} \sum_{-\infty}^{\infty} \frac{1}{\Delta} \delta\left(u - \frac{n}{\Delta}\right)\exp(-j2\pi vy)dy$$

$$= \frac{1}{\Delta} \sum_{-\infty}^{\infty} \delta\left(u - \frac{n}{\Delta}\right)\delta(v), \qquad\qquad (3\text{-}13\text{d})$$

where we have again made use of the sifting properties of the impulse distribution and the fact that the FT of $\exp(j2\pi\alpha x)$ is given by $\delta(u - \alpha)$. The previous expression describes a line along the u axis sampled at $1/\Delta$; i.e., the FT of parallel lines spaced at Λ is a sampled line perpendicular to those lines and sampled at the inverse of the distance separating them; i.e., the spacing is given by $1/\Delta$.

3.1.1.1. Mellin Transform

We now introduce the Mellin transform (MT) (Bracewell, 1978), which is either implicitly or explicitly a major component in most invariant pattern recognition applications (Zwicke and Kiss, 1983). Formally, the MT is defined by

$$M(u, v) = \int\int_{0}^{\infty} f(x, y)x^{-ju-1}y^{-jv-1}dxdy, \qquad (3\text{-}14\text{a})$$

where $j^2 = -1$. The discrete version of the MT is given as

$$M(k\Delta u, l\Delta v) = \sum_{n=1}^{N} \sum_{m=1}^{M} f(n\Delta x, m\Delta y)(n\Delta x)^{-jk\Delta u-1}(m\Delta y)^{-jl\Delta v-1}\Delta x\Delta y.$$

$$(3\text{-}14\text{b})$$

The relationship between the FT and MT can be derived if one substitutes $x = Qe^q$ and $y = Pe^p$,

$$M(u, v) = Q^{-ju}P^{-jv} \int\limits_{-\infty}^{\infty}\!\!\!\int f(Qe^q, Pe^p)\exp[-j(qu + pv)]dqdp. \quad (3\text{-}14c)$$

If one notes that $|Q^{-ju}| = |P^{-jv}| = 1$, then the magnitude of the MT is the same as the magnitude of the FT evaluated for an exponentially sampled function. Note that such an exponential sampling is similar to the human retina sampling over a polar grid (high resolution in the "central" fovea and low resolution in the preattentive "periphery").

Invariance properties to distortions such as scaling can be easily proved. Assume, for the continuous case, that $g(x, y)$ is the scaled version of the original image function $f(x, y)$, i.e., $g(x, y) = f(k_1x, k_2y)$. If we denote the MT of $g(x, y)$ and $f(x, y)$ by $G(u, v)$ and $M(u, v)$, respectively, then

$$G(u, v) = Q^{-ju}P^{-jv} \int\limits_{-\infty}^{\infty}\!\!\!\int f(k_1Qe^q, k_2Pe^p)\exp[-j(qu + pv)]dqdp$$

$$= Q^{-ju}P^{-jv} \int\limits_{-\infty}^{\infty}\!\!\!\int f(Qe^{q + \ln k_1}, Pe^{p + \ln k_2})\exp[-j(qu + pv)]dqdp.$$

Substitute $t = q + \ln(k_1)$, and $z = p + \ln(k_2)$

$$G(u, v) = Q^{-ju}P^{-jv} \int\limits_{-\infty}^{\infty}\!\!\!\int f(Qe^t, Pe^z)\exp\{-j[(t - \ln(k_1))^u$$

$$+ (z - \ln(k_2))^v]\}dtdz$$

$$= k_1^{-ju}k_2^{-jv}M(u, v).$$

Therefore, $|G(u, v)| = |M(u, v)|$.

3.1.1.2.   Conformal Mapping

One technique useful in achieving invariance to rotation and scale changes is the complex-logarithmic (CL) transformation generically known as conformal mapping. Specifically, assume that Cartesian plane points are given by $(x, y) = (\text{Re}(z), \text{Im}(z))$, where $z = x + jy$. Thus, one can write $z = r[\exp(j\theta)]$, where $r = |z| = (x^2 + y^2)^{1/2}$, and $\theta = \arg(z) = \arctan(y/x)$. The CL mapping, shown in Fig. 3.1., is simply the conformal mapping of points z onto points w defined by

$$w = \ln(z) = \ln[r(\exp(j\theta))] = \ln(r) + j\theta. \tag{3-15}$$

Therefore, points in the target domain are given by $(\ln(r), \theta) = (\mathrm{Re}(w), \mathrm{Im}(w))$, and logarithmically spaced concentric rings and radials of uniform angular spacing are mapped into uniformly spaced straight lines. If the scale and rotation factors are k and ϕ, respectively, then z_{old} and z_{new} are given by $r(\exp(j\theta))$ and $(kr)\exp[j(\theta + \phi)]$, respectively. After CL mapping, rotation and scaling about the origin in the Cartesian domain become linear shifts in the $\theta(\mathrm{mod}\,2\pi)$ and $\ln(r)$ directions, respectively. (CL mapping has also been conjectured as a cortex function by neurophysiological research (Schwartz, 1977).) Correlation techniques can then be used to detect invariance (to scale and rotation) within a linear shift between objects.

The ideas suggested by the Mellin transform and the conformal mapping are incorporated into an optical system designed by Casasent and Psaltis (1975). Their system recognizes an object invariant to LT (linear transformations) and works as follows:

(1) Evaluate the magnitude of the FT corresponding to the image function $f(x, y)$, i.e., $|F(u, v)| = |\mathscr{F}\{f(x, y)\}|$;
(2) Convert $|F(u, v)|$ to polar coordinates $F_1(r, \theta)$;
(3) Derive $F_2(\zeta, \theta) = F_1(\exp(\zeta), \theta)$, where $\zeta = \ln(r)$;
(4) Conclude that the magnitude of the FT of F_2 yields the PSRI (position, rotation, scale invariant) function $M_1(\omega_p, \omega_o)$.

Step 1 removes translation distortion because the magnitude of the FT is shift invariant. Step 2 changes scale and rotation distortions into

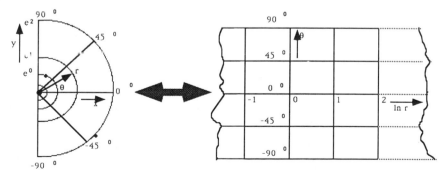

Figure 3.1.
Conformal mapping of the Cartesian half plane.

linear shifts along the $\ln(r) - \theta$ new system of coordinates. Steps 3 and 4 are equivalent to an MT performed with respect to both axes; i.e., linear shifts are recognized as scale (k) and rotation (ϕ) changes.

3.1.2. *Invariant Cojoint Image Representations*

Jacobson and Wechsler (1985) developed FOVEA—a system for invariant image representation. The underlying representations are distributed as cojoint spatial/spatial-frequency (s/sf) representations, while the invariance is with respect to linear transformations (scale and rotation) and perspective distortions. We begin by reviewing the theoretical foundation behind FOVEA and then proceed to describing the system. The actual implementation, including deprojection filters, is deferred to Section 3.1.4.

For the case of an ideally diffusing surface, an image can be modeled as a product of three independent signal components:

$$f(x, y) = i(x, y) \times r(x, y) \times \cos \alpha, \qquad (3\text{-}16a)$$

where $i(x, y)$ is the illumination, $r(x, y)$ is the reflectance (lightness), and α, $0 \leq \alpha \leq \pi/2$, is the angle of incidence of the illuminating light. Assume that any additive noise that is present has small magnitude relative to the three components of the signal. A well-known and effective method, homomorphic filtering, allows one to individually filter out such multiplicative signal components when certain reasonable conditions are met. (See also lightness computation in Section 5.1.1.) This method is briefly reviewed.

Suppose one takes the logarithm of the function $f(x, y)$ defined previously. The so-called "density image" results:

$$\ln(f(x, y)) = \ln(i(x, y)) + \ln(r(x, y)) + \ln \cos \alpha. \qquad (3\text{-}16b)$$

Hence, the product becomes a sum of three density components $\ln(i)$, $\ln(r)$ and $\ln \cos \alpha$. If these three additive density components have Fourier spectra that overlap very little in their regions of significant energy, then linear filtering can be used to extract any one of them. By taking the exponent of the extracted component, one obtains the corresponding multiplicative signal component that appeared in the original expression of the image $f(x, y)$. Homomorphic filtering has been applied to enhance imagery by selectively filtering out the slowly varying illumination component.

The success of homomorphic filtering clearly stems from the fact that the Fourier transform of a density image often has the effect of "representationally" decoupling multiplicative image components. If image pattern recognition (rather than image filtering) could be based on such a Fourier representation, then perhaps the effects of surface reflectance, illumination, and angle of incidence could be decoupled. This in turn could lead to pattern recognition methods that are insensitive to varying illumination conditions. Unfortunately, neither the Fourier transform nor its power spectrum has proved especially useful for image pattern recognition. Alternative representations— namely, joint spatial/spatial-frequency representations such as the Gabor spectrum—can provide decoupling of multiplicative image density components and, at the same time, overcome the classical shortcomings of the Fourier spectrum as an image representation.

The Wigner distribution (WD) of a picture value function is a possible representation for decoupling multiplicative image components. A reasonable, simplified model when imaging planar forms for a distant source of illumination is

$$f(x, y) = i(x, y) \times r(x, y), \tag{3-17a}$$

since the angle of incidence α is constant across any particular planar surface. Taking the logarithm and computing the WD of both sides yields

$$W_{\ln(f)}(x, y, u, v) = W_{\ln(i)}(x, y, u, v) + W_{\ln(r)}(x, y, u, v)$$
$$+ 2\mathrm{Re}\{W_{\ln(i),\ln(r)}(x, y, u, v)\}. \tag{3-17b}$$

The cross-WD term, $W_{\ln(i),\ln(r)}(x, y, u, v)$, results because the WD is a bilinear representation, which can be problematic to the extent that its (s/sf) support overlaps, with the auto-WD term, $W_{\ln(i)}(x, y, u, v)$ of the ln reflectance function. It can be shown that

$$W_{\ln(i),\ln(r)}(x, y, u, v) = 4 \sqrt{S_{\ln(r)}(2x, 2y, 2u, 2v)}, \tag{3-18}$$

where $S_{\ln(r)}(x, y, u, v)$ is the spectrogram of $\ln(r(x, y))$ computed with a sliding window $\ln(i(x, y))$. Therefore, if $\ln(i(x, y))$ varies slowly in comparison to $\ln(r)$, the cross-WD term will resemble a spatially compressed narrowband spectrogram of $\ln(r(x, y))$, which has had all frequencies reduced by one half. Together with the definition of $W_{\ln(f)}$, this suggests that if the band of frequencies in $\ln(r)$, which is to support pattern recognition, is narrower than the difference between the low-frequency side of this band and the high-frequency cut-off of $\ln(i)$, then there will be minimal overlap between the cross-WD and the auto-WD portions of $\ln(r)$ that are relevant to pattern recognition.

Some researchers (Schwartz, 1977, 1981; Weiman and Chaikin, 1979; Schenker *et al.*, 1981) have advocated representations derived by conformally mapping the gray-scale image function. Others (Casasent and Psaltis, 1975; Cavanagh, 1978) have suggested that the conformally mapped Fourier power (or magnitude) spectrum should be used. In fact, neither representation seems entirely adequate. The former representation is indeed invariant, within a linear shift (WALS-invariant), to rotation and scaling of the image about a single image point. It is not, however, invariant to translation of an image, and the effects of illumination and reflectance cannot be decoupled from one another. The second representation mentioned (the conformally mapped power spectrum of an image) is strictly invariant to translation and WALS-invariant to rotation and scaling of an image, but it does not uniquely represent an image since Fourier phase information is discarded.

As alluded to earlier in our discussion of multiplicative signal separability, the loss of phase information makes Fourier spectra especially ill-suited for imagery containing clutter or multiple objects to be recognized. The WALS-invariance properties, which arise exclusively from conformal mapping, could be used to develop considerably more robust image representations (perhaps cojoint spatial/spatial-frequency representations) that are similarly WALS-invariant to rotation and scale changes—as we will soon illustrate.

Assume that we are given an arbitrary image density function $\ln(f(x, y))$ and a function $\ln(g(x, y))$ that is obtained by arbitrarily scaling and rotating $\ln(f(x, y))$ about the origin $(x, y) = (0, 0)$. Then, the corresponding Wigner distributions $W_{\ln(g)}$ and $W_{\ln(f)}$ are related by

$$W_{\ln(g)}(x, y, u, v) = k^2 W_{\ln(f)}\left(\frac{x \cos \phi + y \sin \phi}{k}, \frac{-x \sin \phi + y \cos \phi}{k},\right.$$

$$\left.\frac{u \cos \phi + v \sin \phi}{1/k}, \frac{-u \sin \phi + v \cos \phi}{1/k}\right), \quad (3\text{-}19a)$$

where k and ϕ denote the scale factor and rotation angle, respectively, between $\ln(g)$ and $\ln(f)$. If one conformally maps both the spatial and spatial-frequency domains of $W_{\ln(f)}$ and $W_{\ln(g)}$ with respect to the corresponding domain-specific origins $(x, y, u, v) = (0, 0, u, v)$ and $(x, y, u, v) = (x, y, 0, 0)$, then the corresponding conformally mapped

Wigner distributions $\tilde{W}_{\ln(f)}$ and $\tilde{W}_{\ln(g)}$ are related as follows:

$$\tilde{W}_{\ln(g)}(z', \theta', w', \xi') = k^2 \tilde{W}_{\ln(f)}(z' - \ln(k), \theta' - \phi, w' + \ln(k), \xi' - \phi), \quad \text{(3-19b)}$$

where k and ϕ are as defined previously, and

$$z' = \ln(x^2 + y^2)^{1/2}, \qquad \theta' = \arctan(y/x),$$
$$w' = \ln(u^2 + v^2)^{1/2}, \qquad \xi' = \arctan(v/u).$$

Hence, we see that the 4D conformally mapped WD is WALS-invariant to rotation and scaling of the image $f(x, y)$ about the origin $(x, y) = (0, 0)$. Note that $W_{\ln(g)}$ can only translate with respect to $W_{\ln(f)}$ along a 2D hyperplane $\{(z', \theta', w', \xi') | z' = -w', \theta' = \xi'\}$ in the 4D representational space. The similarity theorem dictates that there are only two degrees of translation freedom in the 4D space—corresponding to the separate effects of rotation and scaling—because the spatial and spatial-frequency domains of the WD undergo reciprocal scaling and equal rotation.

Suppose $f(x, y) = i(x, y) \times r(x, y)$ and $g(x, y)$ corresponds to $r(x, y)$ rotated and scaled about the origin by angle ϕ and factor k, respectively, while $i(x, y)$ remains unchanged. Then if $\tilde{W}_{\ln(i)}$ and $\tilde{W}_{\ln(r)}$ have good separability, both before and after the linear spatial transformation of $r(x, y)$, we obtain

$$\tilde{W}_{\ln(g)}(z', \theta', w', \xi') \simeq k^2 \tilde{W}_{\ln(r)}(z' - \ln(k), \theta' - \phi, w' + \ln(k), \xi' - \phi)$$
$$+ W_{\ln(i)}(z', \theta', w', \xi'). \quad \text{(3-19c)}$$

Therefore, subject to reasonable restrictions, different multiplicative components of an image can undergo independent translations within the 4D rotation and scale WALS-invariant representation.

Recall that the conformally mapped WD is WALS-invariant to rotation and scale changes about a single image point, namely, the spatial domain origin $(x, y) = (0, 0)$. At the cost of adding two additional dimensions to the proposed representation, one could obtain a new representation that provides WALS-invariance with respect to translation as well as rotation and scale changes. Referring to Fig. 3.2., one simply constructs the 6D representation $\tilde{W}_f(z', \theta', w', \xi'; z, \theta)$, by performing spatial conformal mapping about each point $(x, y) = (e^z \cos \theta, e^z \sin \theta)$ in the spatial domain. (As before, spatial-frequency domain conformal mapping is done only with respect to the spatial-frequency origin $(x, y, u, v) = (x, y, 0, 0)$.) Note that multipositional conformal mappings were first proposed for the primary visual cortex model in mammals (Schwartz, 1977, 1981).

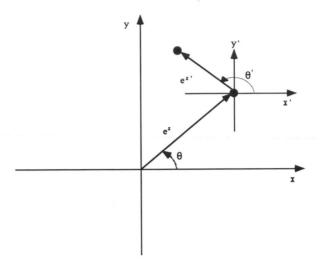

Figure 3.2.
Coordinates used in multipositional conformal mapping of the WD.

Having assumed imagery of planar, nonoccluding forms in some frontoparallel plane, we now relax these conditions to include nonfrontoparallel views and describe approaches for invariance to perspective distortions of gray-scale planar forms. Given the geometry of Fig. 3.3., we assume some arbitrary object plane (σ, τ, z), where σ, τ, and $|z|$ specify the plane's slant, tilt, and line-of-sight distance, respectively, from the viewpoint. Let an arbitrary point in the given object plane, whose object plane coordinates (x_0, y_0) have world coordinates (x_0', y_0', z_0') in the (x', y', z')-space. The geometry of Fig. 3.3. shows that our arbitrary point's object plane and world coordinates are related by:

$$\begin{bmatrix} x_0' \\ y_0' \\ z_0' \end{bmatrix} = \begin{bmatrix} x_0 \cos \tau - y_0 \sin \tau \cos \sigma \\ x_0 \sin \tau + y_0 \cos \tau \cos \sigma \\ z - y_0 \sin \sigma \end{bmatrix}. \tag{3-20}$$

From the familiar formula governing perspective projection onto a plane, a point having world coordinates (x_0', y_0', z_0') projects to a point $(x', y', 0)$ in the image plane given by

$$\begin{bmatrix} x' \\ y' \end{bmatrix} = \frac{f}{f - z_0'} \begin{bmatrix} x_0' \\ y_0' \end{bmatrix}. \tag{3-21}$$

Using Eq. (3-20) to substitute for x_0', y_0' and z_0' in the equation, one obtains the transformation from object plane coordinates (x_0, y_0) to

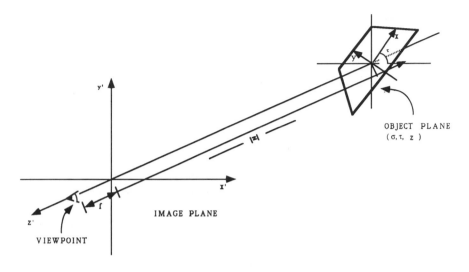

Figure 3.3.
Geometry relating the object and image planes.

image plane coordinates $(x', y', 0)$ as

$$\begin{bmatrix} x' \\ y' \end{bmatrix} = \frac{f}{f - z + y \sin \sigma} \begin{bmatrix} x \cos \tau - y \sin \tau \cos \sigma \\ x \sin \tau + y \cos \tau \cos \sigma \end{bmatrix}, \qquad (3\text{-}22)$$

where we dropped the subscripts of the object plane coordinates since they were abritrarily chosen. Invert the relation and obtain an expression for the object plane coordinates (x, y) in terms of its corresponding image plane coordinates (x', y') as

$$\begin{bmatrix} \cos \tau \left(-\sin \tau \cos \sigma - \dfrac{x' \sin \sigma}{f} \right) \\ \sin \tau \left(\cos \tau \cos \sigma - \dfrac{y' \sin \sigma}{f} \right) \end{bmatrix} \cdot \begin{bmatrix} \dfrac{x}{f - z} \\ \dfrac{y}{f - z} \end{bmatrix} = \begin{bmatrix} \dfrac{x'}{f} \\ \dfrac{y'}{f} \end{bmatrix}. \qquad (3\text{-}23)$$

Finally, using Cramer's rule to solve the system for $x/(f - z)$ and $y/(f - z)$, we get

$$\begin{bmatrix} \dfrac{x}{f - z} \\ \dfrac{y}{f - z} \end{bmatrix} = \frac{1}{\Delta \cdot f} \begin{bmatrix} (x' \cos \tau + y' \sin \tau)\cos \sigma \\ (-x' \sin \tau + y' \cos \tau) \end{bmatrix},$$

where

$$\Delta = \cos \sigma + \frac{\sin \sigma}{f}(x' \sin \tau - y' \cos \tau). \qquad (3\text{-}24)$$

This relation illustrates that the inverse perspective mapping finds the object plane coordinates only to within a scale factor $(f - z)$, which is the distance from the viewpoint to the intersection between the object plane and the line of sight. In fact, the distance-normalized coordinates $[x/(f - z), y/(f - z)]$ characterize *all* object planes having a common value of slant and tilt. This relationship, therefore, defines a family of "reprojective" transformations parametrized on all ordered pairs (σ, τ) of slant and tilt.

Projection invariance can be obtained as follows. Given a density image $\ln(f(x', y'))$ (whose formation is constrained as previously specified), one applies all possible reprojective transformations, i.e., one transformation for each ordered pair (σ, τ), $0 \le \sigma < 90°$, $0 \le \tau < 360°$. This yields a 4D representation, which we denote as $\ln(f^0(x, y; \sigma, \tau))$. Every planar form in the scene that produced the original image will appear in frontoparallel view within one of the reprojected images composing $\ln(f^0(x, y; \sigma, \tau))$. Hence, we say that $\ln(f^0(x, y; \sigma, \tau))$ is WALS-invariant to all possible planar projections.

For each reprojected image obtained, i.e., each $\ln(f^0(x, y; \sigma_0, \tau_0))$ for some $(\sigma, \tau) = (\sigma_0, \tau_0)$, one merely computes the representation $\tilde{W}_{\ln(f)}(z', \theta', w', \xi'; z, \theta)$ that was introduced earlier. This leads to an 8D representation $\hat{\tilde{W}}_{\ln(f)}(z', \theta', w', \xi'; z, \theta, \sigma, \tau)$, which provides WALS-invariance to perspective, to position in 3D space, and also to orientation and size of patterns in the distance-normalized object plane coordinate system.

The dual-domain (spatial and spatial-frequency, respectively) conformal mapping leads to improved efficiency in representational encoding in discretely sampled systems, which must simultaneously support high accuity and a wide field of view. In particular, spatial conformal mapping leads to a "foveated" representation of visual information, whereby high spatial accuity is maintained along the camera line of sight but progressively drops with increasing eccentricity. Next we show how a multidimensional image representation (MIR) can be implemented along the previously discussed concepts. (Section 7.2.3., dedicated to 2D object recognition, implements an MIR instantiation, where the "forested" complex-log representation is subject to a Fourier transform instead of the WD and is recognized by a subsequent pattern recognition subsystem of the distributed associative memory (DAM) type.)

The FOVEA invariant representation system develops canonical representations for interface with visual memory. Early stages of processing in FOVEA, which precede interaction with memory, serve primarily to compute a multidimensional image representation (MIR) that reveals canonical representations consistent with the gray-scale image. FOVEA represents planar shapes and textures that appear in an image with arbitrary perspective, position, size, and orientation.

The MIR derivation, shown in Fig. 3.4., implicitly defines an identification tree whose (leaf) nodes are equivalent to the WALS-invariant $\hat{W}_{\ln(f)}$ representation defined earlier. Note that the sequence of steps used to compute the MIR is quite different from that used to define the analogous 8D representation. (Recall that the analogous sequence discussed earlier is made up from the logarithmic, reprojection, WD, and conformal mappings about multiple fixation points.) These differences are motivated largely by the requirement to minimize the computational burden of the algorithm. Unlike typical image representations, the MIR is based (at an early stage) on a foveated representation similar to that encoded by the retinal ganglion cells in the human retina. The processing stages include

Step 1: Compute the pixel-by-pixel logarithm. This converts a product of image components (e.g., illumination and reflectance) into a linear superposition of their respective logarithms.

Step. 2: Conformally map the image to convert rotation and scaling of the image into simple linear shifts. To avoid aliasing, simultaneously filter the image with a space variant low-pass filter whose pass frequency is constant over a central fixed-accuity disk and drops inversely proportional with radial distance outside of the central disk.

Step 3: Separately reproject the image for numerous combinations of object-plane slant and tilt. Any planar surface in the image will appear in frontoparallel view within only one of the many reprojected images.

Step 4: Conformally map the image for multiple object plane points that differ from the line-of-sight point.

Step	Transformation	Type of Transformation	Effect of Transformation
1	Functional	Logarithm	Changes product of signal components into a sum
2	Functional Geometric	Space-variant low-pass filtering and on-fixation CL conformal mapping	Yields a foveated image representation that is invariant to rotation and scaling about the camera line of sight
3	Geometric	Reprojection	Provides invariance to planar perspective
4	Geometric	Off-fixation conformal mapping	Provides invariance to conformal mapping translation
5	Functional	Computation of patchwork of conformal-mapped finite support Fourier sine and cosine transforms	Intermediate step in computation of the composite pseudo-Wigner distribution (CPWD); provides rotation-and scale-invariant spatial frequency representation; Provides separation of multiplicative signal components in original image
6	Functional	Fourier-domain interpolations, nonlinear correlations, and finite support inverse Fourier transformations	Converts patchwork of Fourier transforms into the CPWD; provides real-valued local 2D spectral representation (s/sf)

Figure 3.4.
FOVEA—a system for deriving invariant distributed spatial/spectral (s/sf) representations.

Step 5: Derive a spatial/spatial-frequency representation by computing a patchwork of Fourier sine and cosine transforms with a Gaussian analysis window sized inversely proportional to spatial frequency. This results in a constant relative bandwidth Gabor representation of the image. Compute the spatial-frequency domain of the representation at sample points of a polar-exponential grid.

Step 6: Compute the modulus of the (complex-valued) Gabor representation (Step 5). Alternatively, one can obtain improved spatial resolution of the spatial/spatial-frequency representation by computing an approximation to the Wigner distribution.

The 4D functions available at the leaf nodes of the MIR are complex-log mapped joint spatial/spatial-frequency representations, where conformal mapping has been performed on spatial and spatial-frequency domains. The resulting 4D representations are invariant (within a linear shift through two degrees of freedom of the underlying 4D space) to rotation and scale changes about a single point in a single family of fixed slant-tilt object planes in the scene. In fact, the 4D representations at the leaf nodes are intimately tied to the canonical patterns that were briefly discussed earlier. The canonical pattern corresponding to each familiar planar form in a scene is strictly invariant to viewing geometry and can be found at a single leaf node of the MIR. As a planar form undergoes changes in attitude and position in a scene, its 4D canonical pattern moves from one MIR leaf node to another; i.e., it translates through the 8D representational space of the MIR. The 4D canonical pattern corresponding to some planar form is, by definition, the conformal mapped space/spatial-frequency representation computed from a frontoparallel view of that planar object, where spatial conformal mapping has been performed for an arbitrary point on the object. In practice, these canonical patterns can be plucked from the appropriate MIR leaf node. If information about a planar pattern's surface slant and tilt is available through stereopsis or kinetic depth cues, then the system could extract an appropriate canonical pattern without human intervention, which brings considerable savings and relief from trying "all" reprojections. Once extracted, canonical patterns can be stored in (learned) memory and used for pattern recognition.

Any pattern recognition system should determine whether familiar canonical patterns are located at any of the leaf nodes in the MIR and, if so, specify the identity of the planar form corresponding to each such canonical pattern. Correlating each 4D canonical prototype pattern stored in memory with the MIR of an image is one method, which for each stored canonical pattern, a 6D correlation function is yielded that can be searched for suprathreshold correlation peaks. Detection of a suprathreshold peak signifies recognition of the corresponding planar form, and the location of the peak uniquely specifies the attitude and position of the recognized form in 3D space.

The FOVEA system (summarized in Fig. 3.4.) implements an invariant low-level distributed representation as conjectured for early vision. Note that the MIR subsystem of FOVEA involves strictly deterministic processing of the sensed image. Low-level segmentation or feature detection is not required, and the canonical representations revealed at the leaf nodes of the MIR are derived entirely from geometrical considerations without making any assumptions about the contents of the image. The intriguing question that arises, however, is whether the FOVEA approach might also be applicable to the recognition of fully three-dimensional objects; i.e., can one use canonical prototypes for 3D objects at the first interface with visual memory?

As currently defined, FOVEA's MIR has six degrees of rigid body freedom. These degrees correspond to the six translational degrees of freedom of 4D canonical patterns in the 8D representation for planar pattern position and attitude changes in the imaged 3D scene. Generalizing to 3D objects would require developing a representation that incorporates information from multiple, locally planar surface patches. The canonical pattern for a 3D object must, in some sense, integrate information over all leaf nodes. To accomplish this, not only must the existing canonical patterns be generalized to 3D objects, but the MIR itself must be substantially redefined as well. We will discuss this issue further in Section 7.4., which is dedicated to 3D object recognition and is where we define visual potentials made up of characteristic views.

3.1.3.   Distributed Filters for Invariance to Linear Transformations

We have shown that scale and rotation invariance can be adequately handled by the conformal mapping mechanism. There is no consensus on how to handle translational invariance, and the pro and cons for different methods are discussed next.

The Mellin transform (MT) is invariant to translation (and rotation/ scale as well), but it does not uniquely represent an image (Oppenheim and Lim, 1981; Cavanagh, 1984) because it discards the (Fourier) phase. Multipositional conformal mappings, implemented in FOVEA, might find support in neurophysical findings (Schwartz, 1977), but they are computationally expensive and derive unnecessary data.

Active perception and ecological optics, to be discussed in Chapter 6, allow an observer actively to look for interesting things and to overcome the translational invariance issue. Large misalignments (i.e., translations) could be accounted for by changing the gaze to bring the object(s) into the center of the field of view (FOV). Deciding what to bring into the center of the FOV is an active function (performed by a preattentive system) and is task dependent. One system that can guide the change of fixation points is the pyramid system developed by Anderson *et al.* (1985). The system analyzes the input image at different temporal and spatial resolution levels. The smart sensor then shifts its fixation point such that interesting parts of the image (i.e., something large and moving) were brought into the central part (fovea) of the FOV for recognition. A similar approach is taken by Wechsler and Zimmerman (1989) for the bin-picking problem (discussed in Section 7.2.3.).

The retinotopic organization of the HVS is such that the relative positioning of the receptive fields (RF) preserves the input's spatial contiguity. Such a neuroarchitecture could be parametrized along several dimensions. We have already discussed (in Section 2.2.4.) the possibility of multiple spatially and spectrally localized RF, implemented in terms of Gabor filters, where the width and orientation dimensions were considered. The parameters corresponding to linear transformations could be the other dimensions encoded. As for FOVEA, a distributed architecture could reduce linear transformations to shifts within the neuroarchitecture.

Zetzsche and Caelli (1989) suggest a scheme for achieving translation (together with rotation and scale) invariance. The scheme registers the actual transformations and lends itself to a distributed implementation through multiple filter image representations. The suggested 4D representation (compare this against the 6D representation employed in FOVEA) is, "encapsulated in a set of filtered images, each containing the original Cartesian coordinates of the image and indexed in terms of the orientation and scale-specific filters used to produce them. The result of this scheme is that a change in position, scale, or orientation of the input pattern has the effect of shifting the distribution of filter outputs within (position) and between (orientation and scale) columns" (Zetzsche and Caelli, 1989). If all transformations are reduced to shifts, then a pattern P given in terms of location $(\bar{x}, \bar{y})$, scale k, orientation θ, and subject to translation (x_0, y_0), scale change k_0 and rotation θ_0 transforms such that

$$T : P(\bar{x}, \bar{y}, k, \theta) \rightarrow P(\bar{x} + x_0, \bar{y} + y_0, k + k_0, \theta + \theta_0). \qquad \text{(3-25a)}$$

where T is defined as

$$\begin{bmatrix} x' \\ y' \end{bmatrix} = a^{k_0} \begin{bmatrix} \cos\theta_0 & \sin\theta_0 \\ -\sin\theta_0 & \cos\theta_0 \end{bmatrix} \begin{bmatrix} x \\ y \end{bmatrix} + \begin{bmatrix} x_0 \\ y_0 \end{bmatrix}. \qquad (3\text{-}25b)$$

To achieve a transformation T as suggested previously, normalized coordinates are introduced as given

$$\begin{bmatrix} \bar{x} \\ \bar{y} \end{bmatrix} = a^{-k} \begin{bmatrix} \cos\theta & \sin\theta \\ \sin\theta & \cos\theta \end{bmatrix} \begin{bmatrix} \bar{x} \\ \bar{y} \end{bmatrix}. \qquad (3\text{-}26)$$

The normalization represents the original image f as a set of image columns with normalized coordinates $(\bar{x}, \bar{y})$, where each column is determined by (k, θ), and a provides a logarithmic scaling. Thus,

$$P(\bar{x}, \bar{y}, k, \theta) = f[a^k(\bar{x}\cos\theta + \bar{y}\sin\theta), a^k(-\bar{x}\sin\theta + \bar{y}\cos\theta)]. \quad (3\text{-}27)$$

Going from 6D to 4D representations has a price in terms of representational redundancy, which is determined by the degree to which the filters are disjoint or uncorrelated. The filtered representation $Q(\bar{x}, \bar{y}, k, \theta)$ derived from the representation P via convolution with the filter kernels $g(x, y)$ is given as

$$Q(\bar{x}, \bar{y}, k, \theta) = P(\bar{x}, \bar{y}, k, \theta) \otimes g(\bar{x}, \bar{y}). \qquad (3\text{-}28)$$

Convolution does not affect linear shift properties as defined in (3-25a) and preserves the shift invariance with respect to (translation, scale, rotation) for the Q representation as well. The normalized $g(\bar{x}, \bar{y})$ filters are identical, but when considered with respect to (x, y) coordinates *their size and orientation specificity is revealed*

$$g(x, y; k, \theta) = g[a^{-k}(x\cos\theta - y\sin\theta), a^{-k}(x\sin\theta + y\cos\theta)]. \quad (3\text{-}29a)$$

This multiple representation scheme (akin to the MIR scheme described in the preceding section) is obtained through filters with specific orientation and scale characteristics. The corresponding FT of such kernels are Gaussian, they are sufficient to reconstruct the input image (Geisler and Hamilton, 1986), and their spectral position and bandwidth define the orientation and scale sensitivity range. This again suggests a distributed image representation, but this time it achieves invariance to linear transformation. (Conceptually related reprojection filters will be discussed in the next section.) The spectral

Gaussian filters are given as

$$G_{ij}(u, v) = \exp\left\{-\pi \cdot \frac{(u - u_{ij})^2 + (v - v_{ij})^2}{((2/3)f_i)^2}\right\}, \qquad (3\text{-}29\text{b})$$

where $u_{ij} = f_i \cos \theta_j$ and $v_{ij} = f_i \sin \theta_j$ correspond to the (spectral) filter center of radial frequency f_i and orientation θ_j. (See the Gabor transform (Daugman, 1988) discussed in Section 4.1.6.) Simulations applied to recognition used $f_i = 4$, 8, 16 picture cycles and the orientations $\theta_j = 0°$, 45°, 90°, and 135°. Shift invariant correlation (or match filtering) searches then for the identity of a (possibly distorted) pattern.

3.1.4. Projection Invariance

We concluded the last section by showing how a distributed system of filters achieves invariance to LT. A conceptually related idea, which applies to projection invariance, is discussed in this section.

The (s/sf) filtered image could be geometrically mapped (or reprojected) back to multiple, distal surface coordinate reference frames. Implemented in FOVEA (see Section 3.1.2.), this concept is supported by McLean-Palmer *et al.* (1985), who have reported that simple cells exhibit a spatial receptive field (RF) skew ranging from 10° to 43° between the main axis of the RF envelope and the axis of RF modulation. Such a finding is consistent with zero-skew Gabor functions seen in perspective view.

The FOVEA system is a logarithmic transformation followed by spatial-variant low-pass filtering and on-fixation conformal mapping. The conformally mapped density function that serves as input to the reprojection step is typically defined over far fewer sample points than the original uniformly sampled image function. Therefore, the convolution underlying the reprojection transformations requires far fewer computations than would be required if directly applying multiple reprojections to the original image.

The previous procedure is applied to all possible reprojective transformations for each slant/tilt pair (see convention in Fig. 3.5.) (σ, τ), $0 < \sigma < 90°$, $0 < \tau < 360°$. Reprojections can be suggested by geometric and spatial reasoning and/or inferences of the generic type "Shape from X," where X might be motion and/or texture. Such transformations usually cause smooth changes and probably only a rough quantization of the reprojection range is needed (Lowe, 1985). Note that every

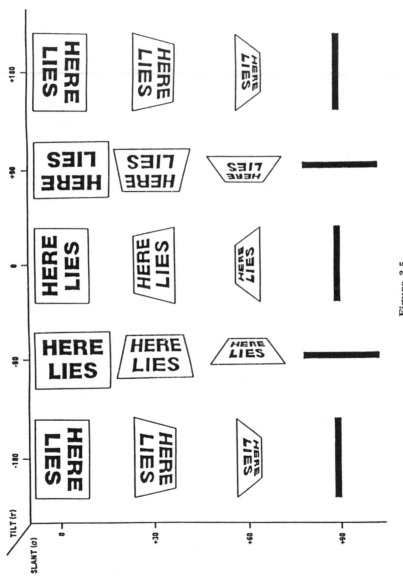

Figure 3.5.
Slant and tilt (σ, τ) conventions.

planar form in the scene that produced the original image will appear in the frontoparallel view within one of the reprojected images. Also, different reprojections can be carried out in parallel.

Specially designed space-variant linear filters are used to compute the complex-log mapped reprojected density images (Jacobson, 1987). The input to this filtering operation is a 2D conformally mapped density image. The output is a 4D function with added dimensions corresponding to object-plane slant and tilt. Next, we derive a general expression for the 4D reprojection filters.

1. Define a space-variant low-pass filtering operation in a Cartesian object plane coordinate system (x, y).

2. Change the integration variables from Cartesian object plane coordinates (x, y) to Cartesian image plane coordinates (x', y').

3. Make a second change of integration variables from Cartesian image plane coordinates to log-polar image plane coordinates (ρ, ϕ) to reflect that a logarithmic conformally mapped function serves as input to the reprojection filtering stage.

4. Convert the continuous integration defining the filter into a discrete summation over the sample points in the function that serves as input to the reprojection filter.

Let $f(x, y)$ be some gray level function defined with respect to object plane coordinates (x, y). This object-plane function is filtered with a space-invariant (low-pass) filter $g(\eta, \xi; x, y)$ to yield a "foveated" object plane image $h(x, y)$; that is

$$h(x, y) = \iint g(\eta - x, \xi - y; x, y) f(\eta, \xi) d\eta d\xi. \tag{3-30}$$

Now, suppose that an object plane's slant and tilt, with respect to an observer's line of sight, is given by (σ, τ). To reexpress the filtering operation for the image plane coordinates of the physical sensor, we must make a change of integration variables from object plane coordinates (η, ξ) to image plane coordinates (η', ξ'). Using Eq. (3-24), the object plane coordinates are expressed for image plane coordinates through the focal-length-dependent transformation

$$\begin{bmatrix} \eta(\eta', \xi') \\ \xi(\eta', \xi') \end{bmatrix} = \frac{1}{\Delta \cdot f} \begin{bmatrix} (\eta' \cos \tau + \xi' \cos \tau)\cos \sigma \\ (-\eta' \sin \tau + \xi' \cos \tau) \end{bmatrix}, \tag{3-31}$$

where

$$\Delta = \cos \sigma + \frac{\sin \sigma}{f} (\eta' \sin \tau - \xi' \cos \tau),$$

and f is the focal length of the system. After making the substitution, we obtain the filtering operation for the image plane coordinates

$$h(x, y) = \iint g[\eta(\eta', \xi') - x, \xi(\eta', \xi') - y; x, y]$$
$$\times f[\eta(\eta', \xi'), \xi(\eta', \xi')]J_1\left(\frac{\eta, \xi}{\eta', \xi'}\right)d\eta'd\xi', \qquad (3\text{-}32a)$$

where J_1, the Jacobian of the deprojective mapping, is given by

$$J_1\left(\frac{\eta, \xi}{\eta', \xi'}\right) = \frac{\cos^2 \sigma}{f^2 \Delta^3}, \qquad (3\text{-}32b)$$

and Δ is the same as given previously.

To re-express the integral for the logarithmic conformally mapped image plane coordinates (ρ', ϕ'), we use the variable transformation

$$\begin{bmatrix} \eta' \\ \xi' \end{bmatrix} = \begin{bmatrix} e^{\rho'} \cos \phi' \\ e^{\rho'} \sin \phi' \end{bmatrix} \qquad (3\text{-}33)$$

to get

$$h(x, y) = \iint g\{\eta[\eta'(\rho', \phi'), \xi'(\rho', \phi')]$$
$$- x, \xi[\eta'(\rho', \phi'), \xi'(\rho', \phi')] - y; x, y\}$$
$$\times f'(\rho', \phi')J_1\left(\frac{\eta, \xi}{\eta'(\rho', \phi'), \xi'(\rho', \phi')}\right)J_2\left(\frac{\eta', \xi'}{\rho', \phi'}\right)d\rho'd\phi', \quad (3\text{-}34a)$$

where

$$f'(\rho', \phi') = f\{\eta[\eta'(\rho', \phi'), \xi'(\rho', \phi')], \xi[\eta'(\rho', \phi'), \xi'(\rho', \phi')]\}, \quad (3\text{-}34b)$$

and the Jacobian of the transformation is given by $J_2 = \exp(2\rho')$.

The FOVEA system uses a discretely sampled version of the conformally mapped image $f'(\rho', \phi')$ as input to the reprojection filters. Therefore, we convert the integral defining $h(x, y)$ to a sum that approximates the integral. To do this, we substitute

$$\begin{bmatrix} \rho' \\ \phi' \end{bmatrix} = \begin{bmatrix} m\Delta_{\rho'} \\ n\Delta_{\phi'} \end{bmatrix} \quad m, n = 0, \pm 1, \pm 2, \ldots . \qquad (3\text{-}35)$$

and obtain

$$h(x, y) \simeq \frac{1}{\Delta_{\rho'}\Delta_{\phi'}} \sum_m \sum_n g\{\eta[\eta'(m\Delta_{\rho'}, n\Delta_{\phi'}), \zeta'(m\Delta_{\rho'}, n\Delta_{\phi'}] - x,$$

$$\zeta[\eta'(m\Delta_{\rho'}, n\Delta_{\phi'}), \zeta'(m\Delta_{\rho'}, n\Delta_{\phi'})] - y; x, y\}$$

$$\times f'(m\Delta_{\rho'}, n\Delta_{\phi'})J_1\left(\frac{\eta, \zeta}{\eta'(m\Delta_{\rho'}), \zeta'(n\Delta_{\phi'})}\right)J_2\left(\frac{\eta', \zeta'}{m\Delta_{\rho'}, n\Delta_{\phi'}}\right)$$

$$= \frac{1}{\Delta_{\rho'}\Delta_{\phi'}} \sum_m \sum_n g'(m\Delta_{\rho'}, n\Delta_{\phi'}; x, y)f'(m\Delta_{\rho'}, n\Delta_{\phi'}), \qquad (3\text{-}36a)$$

where

$$f'(m\Delta_{\rho'}, n\Delta_{\phi'}) = f\{\eta[\eta'(m\Delta_{\rho'}, n\Delta_{\phi'}), \zeta'(m\Delta_{\rho'}, n\Delta_{\phi'})],$$

$$\zeta[\eta'(m\Delta_{\rho'}, n\Delta_{\phi'}), \zeta'(m\Delta_{\rho'}, n\Delta_{\phi'})]\} \qquad (3\text{-}36b)$$

is the discretely sampled (logarithmic) conformally mapped image that is filtered, and

$$g'(m\Delta_{\rho'}, n\Delta_{\phi'}; x, y) = g\{\eta[\eta'(m\Delta_{\rho'}, n\Delta_{\phi'}), \zeta'(m\Delta_{\rho'}, n\Delta_{\phi'})] - x,$$

$$\zeta[\eta'(m\Delta_{\rho'}, n\Delta_{\phi'}), \zeta'(m\Delta_{\rho'}, n\Delta_{\phi'})] - y; x, y\}$$

$$\times J_1\left(\frac{\eta, \zeta}{\eta'(m\Delta_{\rho'}), \zeta'(n\Delta_{\phi'})}\right)J_2\left(\frac{\eta', \zeta'}{m\Delta_{\rho'}, n\Delta_{\phi'}}\right). \qquad (3\text{-}36c)$$

is the space-invariant reprojection filter we set out to derive. The discretely sampled Jacobian functions introduced by the reprojective and logarithmic variable transformations are derived as

$$J_1 = \frac{f\cos^2\sigma}{[f\cos(\sigma) - \exp(m\Delta_{\rho'})\sin(\sigma)\sin(n\Delta_{\phi'} - \tau)]^3}, \quad \text{and} \quad J_2 = \exp(2m\Delta_{\rho'}).$$

To summarize, we defined a space-variant filtering operation on some object plane image using an arbitrary known filter kernel $g(\eta, \zeta; x, y)$, which is specified in terms of the object-plane coordinate system (x, y). We performed two successive variable substitutions under the integral defining the convolution so that an equivalent object-plane filtered output could be obtained through a space-variant filter $g'(m\Delta_{\rho'}, n\Delta_{\phi'}; x, y)$, which operates on a discretely sampled, complex-log mapped image function. Let the object-plane filter, $g(\eta, \zeta; x, y)$, be a space-variant low-pass filter. Assuming that the focal length f of the optical system is

known, and that a particular slant and tilt (σ, τ) are selected, the
equations together define a reprojection filter. Such a filter, if used,
permits one to interpolate a spatial-frequency bandwidth limited version of the object-plane density function emitted in the direction of the
sensor from a surface with slant σ and tilt τ. In the FOVEA system,
such reprojection filters are designed for each of a finite number of
ordered pairs (σ, τ) to provide multiple reprojected versions of the
sensed image.

The filtering operation removes the geometric perspective distortion
that results from imaging a nonfrontoparallel plane—an operation
referred to as image reprojection. Interpolating the object plane image
only at points on a polar-exponential grid results in a complex-log
mapped reprojected image, which is invariant, within a linear shift, to
object plane rotations and scalings about the origin.

3.1.5.   Invariant Cytoarchitecture

We have shown that cojoint s/sf representations can be augmented
to include invariance for linear transformations (LT) and perspective
distortions. The use of LT and/or deprojection filters amounts to the
availability of a cytoarchitecture for achieving geometrical invariance. Specifically, such invariance can be achieved by combining the
specially designed LT and reprojection distributed filters. Such a
distributed architecture is also well suited for fault-tolerant behavior
with respect to noise and occlusion.

The representations derived so far are analogical, as opposed to
symbolic, and could interface directly to the visual memory for recognition or help to derive intrinsic images. There is increasing evidence
that human perception involves the internal manipulation of object
representations, which undergo mental transformations through intermediate states as would a physical object in the external world
(Shepard and Cooper, 1986). Such transformations resemble the shifts
involved in correlation needed for LT invariance (Section 3.1.3.).
According to Jacobson (1987),

> The hypothesized multidimensional representation is viewed as comple-
> mentary to the types of low- to middle-level computations that led to so-
> called intrinsic image representations. In particular, visual processes
> responsible for color constancy, motion sensitivity (optical flow deriva-
> tion), shape-from-shading, etc., can benefit from the explicit representa-

tion of surface in known distal coordinated reference frames. Such computations, together with the transformational and invariance properties of the hypothesized multidimensional image representation, would have to cooperate in a selection process which perceptually activates only those distal surfaces of the representation which correspond to physically plausible percepts. This requirement follows from the fact that, given an arbitrary scene, the hypothesized representation would encode information on multiple distal surface frames of reference, most of which would correspond to surfaces that are not physically present in the imaged scene. Familiarity with imaged objects (learned canonical patterns) could also greatly facilitate this selection process when explicit cues for distal surface attitude are not sufficient to specify scene layout.

Research performed by Hubel and Wiesel (1962) has uncovered a very regular pattern of interconnections for the early stages of visual processing. The pattern, labeled as *cytoarchitecture*, has as its main processing unit a hypercolumn that detects edges within a 3D "hardware" (RF position, orientation, and size). Jacobson and Wechsler (1988) observed that edge detection is a byproduct of frequency analysis and suggested that the previous cytoarchitecture could also be construed as being implemented along spatial position and spatial frequency. The (s) and (sf) dimensions could be given in terms of scale and orientation, such as in Daugman's (1988) Gabor scheme. The possibility of multiple spatially and spectrally localized RF, implemented in terms of Gabor filters, where width and orientation are the dimensions considered, has also been suggested (Koenderink and van Doorn, 1987). The cytoarchitecture of the HVS is characterized by orderly retinotopic mappings, which preserve neighborhood relationships within the visual field. The cytoarchitecture could be parametrized along several dimensions where the parameters corresponding to linear transformations could be some of the dimensions encoded. As for FOVEA, a distributed architecture reduces all the linear transformations to shifts within the cytoarchitecture.

That the visual system takes advantage of geometrical invariances as discussed so far shows that searching for invariant stimulus is a worthwhile pursuit. Invariant stimulus, rather than unstructured and varying input, significantly decreases the computational load. Many theories of machine vision are inadequate because they avoid considering geometric invariance. Because such theories fail to exhibit invariant behavior, they are poor candidates for object recognition applications (for more on object recognition, see Chapter 7).

3.1.6. Moments

The geometric invariance discussed so far has been achieved through the use of a distributed architecture of local spatial filters. The main alternative is to use moments, which are global in nature and not fault tolerant and, consequently, yield disappointing results for the most part in recognition experiments. The difference between theory and practice is due to the very computational aspects that were mentioned in Chapter 1, which include discretization, quantization, noise, and/or occlusion. Nevertheless, we decided to include the method for historical reasons. Assuming a two-dimensional continuous function $f(x, y)$, we define the moment of order $(p + q)$ by the relation

$$m_{pq} = \int\int\limits_{-\infty}^{\infty} x^p y^q f(x, y)dxdy \qquad p, q = 0, 1, 2, \dots . \qquad (3\text{-}37a)$$

The central moments can be expressed as:

$$\mu_{pq} = \int\int\limits_{-\infty}^{\infty} (x - \bar{x})^p (y - \bar{y})^q f(x, y)dxdy, \qquad (3\text{-}37b)$$

where the center of gravity is given by $(\bar{x}, \bar{y},) = (m_{10}/m_{00}, m_{01}/m_{00})$, and m_{00} is the total function mass, i.e.,

$$m_{00} = \int\int\limits_{-\infty}^{\infty} f(x, y)dxdy.$$

The discrete central moments are

$$\mu_{pq} = \sum_x \sum_y (x - \bar{x})^p (y - \bar{y})^q f(x, y). \qquad (3\text{-}37c)$$

Note that for a binary blob, μ_{00} represents its area. A set of normalized central moments can be also defined

$$\eta_{pq} = \frac{\mu_{pq}}{\mu_{00}^\gamma}, \quad \gamma = \frac{p + q}{2} + 1. \qquad (3\text{-}38)$$

The normalized central moments can be used to define a set of seven moments invariant to LT. The set (Hu, 1962) is given

$$M_1 = \mu_{20} + \mu_{02}$$

$$M_2 = (\mu_{20} - \mu_{02})^2 + 4\mu_{11}^2$$

$$M_3 = (\mu_{30} - 3\mu_{12})^2 + (3\mu_{21} - \mu_{03})^2$$

$$M_4 = (\mu_{30} + \mu_{12})^2 + (\mu_{21} + \mu_{03})^2$$

$$M_5 = (\mu_{30} - 3\mu_{12})(\mu_{30} + \mu_{12})[(\mu_{30} + \mu_{12})^2 - 3(\mu_{21} + \mu_{03})^2]$$
$$+ (3\mu_{21} - \mu_{03})(\mu_{21} + \mu_{03})[3(\mu_{30} + \mu_{12})^2 - (\mu_{21} + \mu_{03})^2]$$

$$M_6 = (\mu_{20} - \mu_{02})[(\mu_{30} + \mu_{12})^2 - (\mu_{21} + \mu_{03})^2]$$
$$+ 4\mu_{11}(\mu_{30} + \mu_{12})(\mu_{21} + \mu_{03})$$

$$M_7 = (3\mu_{21} - \mu_{03})(\mu_{30} + \mu_{12})[(\mu_{30} + \mu_{12})^2 - 3(\mu_{21} + \mu_{03})^2]$$
$$- (\mu_{30} - 3\mu_{12})(\mu_{12} + \mu_{03})[3(\mu_{30} + \mu_{12})^2 - (\mu_{21} + \mu_{03})^2].$$

$$(3\text{-}39)$$

The normalized central moments μ_{pq} are also invariant to scale changes. Assuming that α is the scale change, it follows that $x' = \alpha x$, $y' = \alpha y$, and

$$\eta'_{pq} = \frac{(\mu')^{pq}}{(\mu'_{00})^{(p+q)/2+1}} = \frac{\mu_{pq}\alpha^{p+q+2}}{(\alpha^2\mu_{00})^{(p+q)/2+1}} = \frac{\mu_{pq}}{(\mu_{00})^{(p+q)/2+1}} = \eta_{pq},$$

where

$$\mu'_{00} = \int\limits_{-\infty}^{\infty}\!\!\int dx'dy' = \alpha^2 \int\limits_{-\infty}^{\infty}\!\!\int dxdy = \alpha^2\mu_{00}.$$

Note that the discrete nature of the input introduces errors that sometimes mask invariance. Functions M_1 through M_6 are also invariant under reflection, while M_7 changes sign.

Another approach is to use moments in polar coordinates, i.e., radially and angularly. The radial and angular moments are defined (Reddi, 1981) as

$$\psi(k, p, q, g) = \int_0^\infty \int_{-\pi}^{+\pi} r^k g(r, \theta)\cos^p \theta \sin^q \theta d\theta dr. \qquad (3\text{-}40a)$$

One notes that for an image centered at $(\bar{x}, \bar{y})$ the central moments in polar coordinates are

$$\mu_{pq} = \int_{-\infty}^{\infty} \int_{-\infty}^{\infty} f(x, y)x^p y^q dxdy$$

$$= \int_{0}^{\infty} \int_{-\pi}^{\pi} r^{p+q+1} \cos^p \theta \sin^q \theta \, g(r, \theta)d\theta dr$$

$$= \psi(p + q + 1, p, q, g).$$

Since $x = r\cos \theta$, $y = r\sin \theta$ and $dxdy = rdrd\theta$. The seven invariant functions defined before can then be expressed in terms of radial and angular moments as follows, if one defines $\psi_r(k, g)$ as

$$\psi_r(k, g) = \int_{0}^{\infty} r^k g(r, \theta)dr, \quad \psi_\theta(g) = \int_{-\pi}^{\pi} g(r, \theta)d\theta \qquad (3\text{-}40b)$$

$$M_1 = \psi_r(3, \psi_\theta(g))$$
$$M_2 = |\psi_r(3, \psi_\theta(ge^{j2\theta}))|^2$$
$$M_3 = |\psi_r(4, \psi_\theta(ge^{j3\theta}))|^2$$
$$M_4 = |\psi_r(4, \psi_\theta(ge^{j\theta}))|^2 \qquad (3\text{-}40c)$$
$$M_5 = \mathrm{Re}[\psi_r(4, \psi_\theta(ge^{j3\theta}))\psi_r^3(4, \psi_\theta(ge^{-j\theta}))]$$
$$M_6 = \mathrm{Re}[\psi_r(3, \psi_\theta(ge^{j2\theta}))\psi_r^2(4, \psi_\theta(ge^{-j\theta}))]$$
$$M_7 = \mathrm{Im}[\psi_r(r, \psi_\theta(ge^{j3\theta}))\psi_r^3(4, \psi_\theta(ge^{-j\theta}))].$$

Invariance to rotation $(\theta \to \theta + \alpha)$ is shown using M_5 as an example:

$$M_5 = \mathrm{Re}[\psi_r(4, \psi_\theta(ge^{j3(\theta + \alpha)}))\psi_r^3(4, \psi_\theta(ge^{-j(\theta + \alpha)}))]$$
$$= \mathrm{Re}[\exp(j3\alpha)\psi_r(4, \psi_\theta(ge^{j3\theta}))\exp(-j3\alpha)\psi_r^3(4, \psi_\theta(ge^{-j\theta}))]$$
$$= \mathrm{Re}[\psi_r(4, \psi_\theta(ge^{j3\theta}))\psi_r^3(4, \psi_\theta(ge^{-j\theta}))].$$

Changing θ to $-\theta$, for reflection, would leave the first six functions unchanged, but M_7 would change sign.

Advantages of the radial and angular moments are (1) Positive integral powers of r in generating the invariants are unrestricted; and (2) The weighting function is not restricted to r^k but could take the form of exponentials and/or similar functions of r. (Such weighting could be better for target identification with low SNR.) Finally, with proper weighting, the functions could be made scale invariant.

The moments can also be used to define a natural (invariant) system of coordinates determined by the moments of inertia (see the need for frames of references in Section 7.4.). Assume, as before, that the centroid is of coordinates $(\bar{x}, \bar{y})$. Then, the moment of inertia of $f(x, y)$ about the line $y = x \tan \theta$ (i.e., a line through the center of gravity) is given as

$$M_\theta = \sum\sum (x \sin \theta - y \cos \theta)^2 f(x, y). \qquad (3\text{-}41a)$$

Let θ_0 be the angle for which M_θ is minimal. If there is a unique θ_0, then the line $y = x \tan \theta_0$ is a principal axis of inertia with respect to $f(x, y)$. Define

$$M_{20} = \sum\sum x^2 f(x, y)$$
$$M_{11} = \sum\sum xy f(x, y)$$
$$M_{02} = \sum\sum y^2 f(x, y).$$

Then

$$M_\theta = M_{20} \sin^2 \theta - 2M_{11} \sin \theta \cos \theta + M_{02}^2 \cos \theta, \qquad (3\text{-}41b)$$

and the minimum for M_θ is achieved when $\partial M_\theta / \partial \theta = 0$. It follows that θ_0 is a solution to

$$\tan(2\theta_0) = \frac{2M_{11}}{(M_{20} - M_{02})}. \qquad (3\text{-}41c)$$

There are two solutions for θ_0, which are 90° apart, and, therefore, one principal axis is perpendicular to the other. The moment of inertia is maximal for one principal axis and is minimal with respect to the other axis. Define

$$\alpha = \frac{\partial^2 M_\theta}{\partial \theta^2} = 2(M_{20} - M_{02})\cos(2\theta_0) + 4M_{11} \sin(2\theta_0), \qquad (3\text{-}41d)$$

and the moment of inertia will be minimal if $\alpha \geq 0$. Once the principal axes are determined, the x axis of the desired system of coordinates is chosen according to the requirement $M_{20} > M_{02}$, and the positive direction of the x axis is chosen according to whether $M_{30} = \sum x^3 f(x, y)$ is positive or negative. Note that the ratio of principal axes is invariant to rotation and scale.

3.2. Statistical Invariance

Geometrical transformations are not the only source of image variability. Sensory noise also introduces variability, which can be modeled in a statistical or stochastic fashion. Statistical invariance is relevant to signal estimation and perceptual modeling of the environment in terms of learning average prototype image representations. Estimation, another form of achieving invariance, might be characteristic to preprocessing, as discussed next. Prototype image representations and modeling are considered in Chapters 7 and 8, respectively.

Before discussing statistical invariance, we will review some prerequisites. Assuming that $X^T = [x_1, x_2, \ldots, x_n]$ is a random vector, then the covariance matrix of X is defined as $\text{COV}(X) = E[(X - \bar{X})(X - \bar{X})^T]$, where E and $\bar{X}$ stand for the expected value operator and the mean of vector X, respectively. Specifically,

$$\text{COV}(X) = \{\sigma_{ij}^2 | \sigma_{ij}^2 = E[(x_i - \bar{x}_i)(x_j - \bar{x}_j)], \, i, j = 1, 2, \ldots, n\}. \quad (3\text{-}42)$$

It follows that the diagonal terms of the covariance matrix are the variances of $\{x_i\}$, and that the off-diagonal terms are the covariance of x_i and x_j. $\text{COV}(X)$ is obviously symmetric. $\text{COV}(X)$ can be expressed alternatively as $\text{COV}(X) = S - \bar{X}\bar{X}'$, where S is called the scatter or autocorrelation matrix of X, and it is given by $S = E\{XX'\} = [E\{x_i x_j\}]$. Furthermore, if we define the correlation coefficient r_{ij} as $r_{ij} = \sigma_{ij}^2/(\sigma_{ii}\sigma_{jj})$, then $\text{COV}(X) = \Gamma R \Gamma$, where $\Gamma = \text{diagonal}(\sigma_{ii})$, and R is the correlation matrix given by $R = \{r_{ij} | |r_{ij}| < 1, r_{ii} = 1\}$.

The main goal of pattern recognition is to classify an object as belonging to one class among a number of given classes; linear (decision) discriminant functions $d(X)$ are widely used for such a purpose. Given some object with X being its set of features, then $d(X) = W^T X + \omega_0$. This is equivalent to projecting X on the (weight) W direction and comparing the result with the threshold ω_0. As an example one can reduce the dimensionality of X by projecting it onto a line, and then find that orientation for which the projected samples are well separated. The Fisher two class discriminant, which can achieve such a result, is optimal in the minimum squared-error sense. Specifically, one maximizes the ratio of between-class scatter to within-class scatter. Formally, the criterion $J(W)$ is given as

$$J(W) = \frac{(\tilde{m}_1 - \tilde{m}_2)^2}{(\tilde{\sigma}_1^2 + \tilde{\sigma}_2^2)}$$

$$= \frac{W^T S_B W}{(W^T S_W W)}, \tag{3-43a}$$

where $\tilde{m}_i = W^T m_i$, $\tilde{\sigma}_i^2 = W^T \text{COV}_i\, W$, and S_B and S_W are the between- and within-class scatter matrices, respectively. For two classes, the generalized eigenvalue problem $S_B W = \lambda S_W W$ can be solved and yields $W = S_W^{-1}(m_1 - m_2)$ if S_W is not singular. In general $S_W = 1/2(\text{COV}_1 + \text{COV}_2)$. For the case of equal covariances $W = \text{COV}^{-1}(m_1 - m_2)$. The threshold ω_0 is given (Duda and Hart, 1973)

$$\omega_0 = \frac{(m_2 - m_1)^T(\frac{1}{2}\text{COV}_1 + \frac{1}{2}\text{COV}_2)^{-1}(\sigma_1^2 m_2 + \sigma_2^2 m_1)}{\sigma_1^2 + \sigma_2^2}. \tag{3-43b}$$

If the probability distributions are multivariate $N(m_i, 1)$, then the threshold to discriminate signals is the mean of the projected signals.

3.2.1.   Wiener Filter

The generalized Wiener filter assumes a signal $\mathbf{X}$ embedded in (uncorrelated) noise N, where signal and noise are of zero mean, and seeks $\hat{\mathbf{X}}$, the estimate of X. An optimal filter A is found, such that the expected value of the minimum squared-error (MSE) between X and $\hat{\mathbf{X}}$ is minimized, where $\hat{\mathbf{X}} = T^{-1}AT(\mathbf{X} + \mathbf{N})$ for any orthogonal transform T. The optimal filter A is given as

$$A = T\hat{A}T', \tag{3-44a}$$

where the (impulse) response matrix $\hat{A}$ is

$$\hat{A} = \text{COV}(X)(\text{COV}(X) + \text{COV}(N))^{-1}. \tag{3-44b}$$

T is the eigenvector matrix of $\hat{A}$, and T' is the transpose matrix of T. If A is a diagonal matrix, the corresponding transform is called the Karhunen-Loeve transform (KLT) (See also Section 4.1.1.). Note that the covariance matrices $\text{COV}(X)$ and $\text{COV}(N)$ correspond to a stationary process and are fixed.

Feature selection and data compression can be considered within the context of KLT. Specifically, the KLT allows a nonperiodic signal to be expanded in a series of orthogonal functions. In the discrete case, MSE implies that KLT expansion minimizes the approximation error when $m < n$ basis vectors are used, assuming that n is the dimension of the signal. Optimum properties are obtained by choosing as m normalized eigenvectors those corresponding to the largest eigenvalues, i.e., the components of maximum variance, which are the only ones useful for signal discrimination.

Conversely, if we choose $m < n$ normalized vectors associated with the smallest eigenvalues, then entropy is minimized, and the process allows for cluster analysis. Finally, if $Y = TX$, then $T \, \mathrm{COV}(Y) = T \, \mathrm{COV}(X)T' = \mathrm{diag} \, (\lambda_1, \lambda_2 \ldots, \lambda_n)$, i.e., if we subject a signal X to the KLT, the resulting signal is uncorrelated.

The same Wiener filter can be derived using linear system theory and spectral considerations. Assume uncorrelated random variables, $x(t)$ and $n(t)$, corresponding to signal and noise, respectively. The random variables are ergodic (i.e., their time and spatial averages are equal) and are given in terms of their power spectra, $P_X(u)$ and $P_N(u)$, respectively. Assuming that the observed signal $s(t)$ is given as $s(t) = x(t) + n(t)$, one seeks a transfer function G such that $\hat{\mathbf{X}} = GS$ (or alternatively $\hat{\mathbf{x}} = g \otimes s$) and the MSE between the estimated signal $\hat{\mathbf{x}}$ and the original x is minimized. The optimal Wiener estimation for uncorrelated signal and noise is given as

$$\mathbf{G}(u) = \frac{P_X(u)}{P_X(u) + P_N(u)} \tag{3-45}$$

for $u \neq 0$. If either $x(t)$ or $n(t)$ is of zero mean value, $\mathbf{G}(u)$ as given holds for all u. That the optimal Wiener estimators given in terms of statistics [Eq. (3-44)] or in terms of spectral characteristics [Eq. (3-45)] are similar is not surprising if one remembers that the FT of the autocorrelation function is the power spectrum of the same function. [See Eq. (2-12c).]

Smoothing (both spatially and/or temporally) a set of m images increases the SNR (signal-to-noise ratio) by m. (Compare this with the case discussed in Section 2.1.2. which shows that smoothing is equivalent to low-pass filtering, and that the removed high frequency content is equivalent to noise.) Assuming that each image is given as $I_i =$

$X + N_i$, where X is the original signal, N_i is uncorrelated random noise of zero mean value, and the signal-to-noise ratio is $X^2/E(N)^2$, Castleman (1978) shows how smoothing and optimal linear filter (Wiener) can enhance the boundaries of blood vessels for angiography. Specifically, the power spectrum $P_X(u)$ of the signal X is estimated by averaging the scanned lines $f_i(x)$ as

$$P_X(u) = |\mathcal{F}[X(x)]|^2 \cong \left| \mathcal{F}\left[\frac{1}{m} \sum f_i(x) \right] \right|,$$

where the picture value function is scanned along m lines $f_i(x)$. The noise power spectrum is then estimated as

$$P_N(u) \cong \frac{1}{m} \sum |\mathcal{F}[f_i(x) - X(x)]|^2.$$

The Wiener estimator $\mathbf{G}(u)$ is derived according to Eq. (3-45), smoothed into $\hat{\mathbf{G}}(u)$, and the impulse response $\hat{g}(x)$ is then easily obtained. To find accurate (vessel) boundaries one convolves the original scan lines $f_i(x)$ with $\hat{g}'(x)$, the derivative of the impulse response. Clearly, the Wiener filter acts as a low-pass filter. Optimal estimation achieves invariance to noise and recovers the original signal despite its variability.

3.2.2. Matched Filter

We introduced the Wiener filter as the optimal filter for recovering an unknown signal embedded in noise, i.e., approximating it in the MSE sense. In situations where one merely seeks to determine the presence or absence of a given signal (or object) at a given location or moment t_0 in time, i.e., one looks for a known signal in a noisy background, the matched filter (MF) (Castleman, 1978) is appropriate. Its corresponding transfer function $\mathbf{G}(u)$ is given as

$$\mathbf{G}_o(u) = \frac{P_X^*(u)}{P_N(u)} \exp(-j2\pi u t_0) \tag{3-46}$$

where $P_X^*(u)$ is the conjugate power spectrum of the known signal $x(t)$, and $P_N(u)$ is the power spectrum corresponding to the noise. Assuming that the noise is white, i.e., $n(t)$ is Gaussian of zero mean and is uncorrelated, then the impulse response $g(t)$ is merely a reflected and shifted version of the signal itself, i.e.,

$$g_0(t) = \mathscr{F}^{-1}\{G_0(u)\} = \int_{-\infty}^{\infty} P^*(u)\exp(-j2\pi u t_0)\exp(j2\pi u t)du$$

$$= \int_{-\infty}^{\infty} P(-u)\exp[-j2\pi u(t_0 - t)]du = x(t_0 - t). \qquad (3\text{-}47)$$

The MF is implemented as a correlator, and its output is "high" at t_0 when $x(t)$ is present and "small" when it is absent. Specifically, if the (possible corrupted) signal and noise are given by $s(t)$ and $n(t)$, respectively, then

$$s(t) = x(t) + n(t)$$

$$\alpha(t) = x(t)\circledast g_0(t) = \int_{-\infty}^{\infty} x(\tau)x(t_0 - t + \tau)d\tau = R_s(t_0 - t) \qquad (3\text{-}48a)$$

$$\beta(t) = n(t)\circledast g_0(t) = \int_{-\infty}^{\infty} n(\tau)x(t_0 - t + \tau)d\tau = R_{ns}(t_0 - t).$$

The output of the MF,

$$y(t) = \alpha(t) + \beta(t) = R_s(t_0 - t) + R_{ns}(t_0 - t) \qquad (3\text{-}48b)$$

has an autocorrelation component only when the signal is present, and it always has a correlation component. (Assume correlation between noise and signal x is small, i.e., $R_{ns} \simeq 0$ and that $R_s(a)$ peaks at $a = 0$ so the MF output is large at $t = t_0$ as desired.) Both the SDF (synthetic discriminant function) and the GMF (generalized matched filter), to be presented next, try to capture an object's structure by using the MF concept. The MF to be used is obtained by averaging "representative" images defining an object and/or its stable states (i.e., object appearances that are more or less the same and are stable over a given range of viewing conditions—see also aspects in Section 7.4.).

3.2.2.1. Synthetic Discriminant Functions

Hester and Casasent (1981) use the conceptual model of the MF for developing synthetic discriminant functions (SDF), which provide reasonable recognition capability regardless of intensity and geometrical distortions. Assume one object is given by its training set as $\{f_n(x)\}$, and that each generic member $f(x)$ is expanded in terms of orthonormal basis functions $\phi_m(x)$ as

$$f(x) = \sum_{m=1}^{M} a_m \phi_m(x). \tag{3-49a}$$

Then each input $f(x)$ can be expressed as a vector $\mathbf{f} = \langle a_1, a_2, \ldots, a_M \rangle$. Using geometrical considerations the energies E_f and E_{fg} (correlation) are given as

$$E_f = \int f^2(x)dx = \sum_m a_m^2 = \mathbf{f} \cdot \mathbf{f} = R_s$$

$$E_{fg} = \mathbf{f} \cdot \mathbf{g} = \sum_m a_m b_m = f \circledast g|_{\tau=0} = R_{fg}(0) = R_p \tag{3-49b}$$

where R_s and R_p stand for a hypersphere of radius s and a hyperplane (g is fixed), respectively. To perform classification, one can require finding a fixed reference function $\mathbf{g}$, such that objects belong to class i if and only if $\mathbf{f} \cdot \mathbf{g}_i \leq R_{p_i}$. This is equivalent with determining an MSF (matched spatial filter). Alternatively, one can require that objects belong to class i if $\mathbf{f} \cdot \mathbf{f} \leq R_{s_i}$, i.e., be within an equienergy surface sphere of radius s_i. The orthonormal set $\{\phi_m\}$ and the desired filter are yet to be determined. One can write for each member of the training set

$$\mathbf{f}_n = \sum_m a_{nm} \phi_m, \qquad \mathbf{g} = \sum_m b_m \phi_m, \tag{3-49c}$$

and the constraint needed for classification is then

$$\mathbf{f}_n \cdot \mathbf{g} = R_{f_n g}(0) = \sum_m a_{nm} b_m = R_p = \text{constant.} \tag{3-49d}$$

Finding the optimal $\mathbf{g}$ requires finding both b_m and ϕ_m. The algorithm to do just that is:

1. Form the autocorrelation matrix $R_{f_i f_j}(0) = \mathbf{f}_i \cdot \mathbf{f}_j$;
2. Diagonalize [Gram-Schmidt (GS) or KLT] R and determine ϕ_m;
3. Determine a_{nm} coefficients from $a_{nm} = \mathbf{f}_n \cdot \phi_m$;
4. Use some R_p as a constant, possibly $R_p = 1$, and from $\mathbf{f}_n \cdot \mathbf{g} = R_p$ determine b_m.

The algorithm presented provides the optimum filter $\mathbf{g}$. How can such an approach deal with additive (zero-mean) noise? The noisy input is $\mathbf{f}' = \mathbf{f} + \mathbf{n}$. Since the expected value of the noise is zero, its average energy E_n is given by

$$E_n = \langle \mathbf{n} \cdot \mathbf{n} \rangle = \sum_n \sigma_{nn}^2$$

and can be bound by the hypersphere centered at the origin, and of radius $\rho = k \langle \mathbf{n} \cdot \mathbf{n} \rangle^{1/2}$, where k is a function of $\{\sigma_{nn}^2\}$. Then, the correlation threshold criterion is modified as

$$\mathbf{f}_n' \cdot \mathbf{g} = R_p \pm \rho,$$

i.e., $\mathbf{f}'$ is bound to lie within a sphere of radius ρ. This method can be implemented such that the training set is given in terms of aspects (see Section 7.4.) or stable states like f_1, right side (RS); f_2, front side (FS); and f_3, rear (R). In other words, SDF attempts to create a match filter that is an average over all possible object appearance, both the stable states and their translated versions (which are given as part of the training set as $\{f_1^r, f_2^j, f_3^k\}$).

3.2.2.2.   Generalized Matched Filters

   Caulfield and Weinberg (1982) follow a line of thought similar to that of Hester and Casasent's SDF (1981), suggesting generalized matched filters (GMF). They view the $M \times N$ samples of the Fourier transform of the input as an MN-component feature vector and use the Fisher discriminant [Eq. (3-43)] to find the linear combination that best separates in the MSE sense the objects of interest from other patterns and/or noise. Their contribution to tackling computational complexity is in solving large eigenvalue problems. Specifically, an assumption is made that $\mathbf{f}$, the vector derived from the FT (using an arbitrary raster scan like left-to-right, top-down), is such that its individual components have no correlation, and that B (between-class scatter matrix) and W (within-class scatter matrix) are diagonal. As with SDF, the GMF achieves invariance by clustering an object and its different appearances (according to viewpoints) into just one class, which then has to be discriminated/separated from other classes of objects, (white) noise being one among several classes. Note that recognition achieves as a by-product what a matched filter is supposed to, i.e., an indication regarding the presence/absence of a given object.

3.2.3. *Kalman Filter*

The Kalman filter (KF) is the optimal procedure for a continuously changing environment when trying to predict future events. It allows for nonstationary processes and continuously updates the covariance matrices. The KF is a state approach that yields recursive definitions, which describe the current state estimate as a function only of the previous estimate and the new data sample. As such it eliminates the need for expensive data storage.

Assume that $\mathbf{x}(t + 1)$ is the signal state to be estimated at time $(t + 1)$. Then

$$\mathbf{x}(t + 1) = \phi(t + 1, t)\mathbf{x}(t) + \mathbf{W}(t)$$
$$= \phi(t + 1)\mathbf{x}(t) + \mathbf{W}(t), \tag{3-50a}$$

where $\phi(t + 1, t) = \phi(t + 1)$ is a $N \times N$ state transition matrix and $\mathbf{W}(t)$ is zero mean, uncorrelated noise, i.e.,

$$\text{COV}(W) = \text{COV}(\mathbf{W}(t_1), \mathbf{W}(t_2)) = Q(t_1)\delta_{t_1, t_2}. \tag{3-50b}$$

The measurement (sensory) process continuously provides observations

$$\mathbf{y}(t) = H(t)\mathbf{x}(t) + \mathbf{n}(t), \tag{3-50c}$$

where $H(t)$ is a known matrix, and $\mathbf{n}(t)$ is white, zero-mean, uncorrelated (to signal $\mathbf{x}$) noise, i.e.,

$$\text{COV}(n) = \text{COV}(\mathbf{n}(t_1), \mathbf{n}(t_2)) = R(t_1)\delta_{t_1, t_2}. \tag{3-50d}$$

The sensor and process noise are assumed to be uncorrelated, i.e.,

$$\text{COV}[\mathbf{W}(t_1), \mathbf{n}(t_2)] = 0. \tag{3-50e}$$

The a priori assumptions include the initial state of the system $\mathbf{x}(0)$ and its covariance $P(0)$. The diagram of the whole process is shown in Fig. 3.6.

The KF operates in two stages, prediction (extrapolation) and correction (update). The previous state estimate $\mathbf{x}(t)$ is used to predict the current state $\hat{\mathbf{x}}(t + 1)$, i.e.,

$$\hat{\mathbf{x}}(t + 1) = \phi(t)\mathbf{x}(t), \tag{3-51a}$$

and the previous state covariance $P(t)$ is extrapolated to yield $\hat{\mathbf{P}}(t + 1)$

$$\hat{\mathbf{P}}(t + 1) = \phi(t)P(t)\phi^T(t) + Q(t). \tag{3-51b}$$

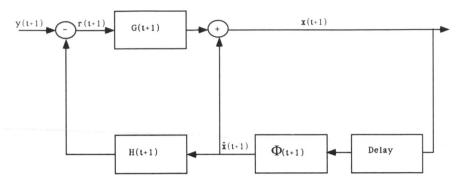

Figure 3.6.
Kalman filter.

The update stage computes the new (time-varying) gain matrix $G(t + 1)$ as

$$G(t + 1) = \hat{\mathbf{P}}(t + 1)H^T(t + 1)[H(t + 1)\hat{\mathbf{P}}(t + 1)H(t + 1) + R(t + 1)]^{-1}.$$
$$(3\text{-}52a)$$

The updated covariance matrix $P(t + 1)$ is given as

$$P(t + 1) = [I - G(t + 1)H(t + 1)]\hat{\mathbf{P}}(t + 1),$$

and the measurement residual

$$\mathbf{r}(t + 1) = [\mathbf{y}(t + 1) - H(t + 1)\hat{\mathbf{x}}(t + 1)] \qquad (3\text{-}52b)$$

is weighted by the gain matrix $G(t + 1)$ and added to the prediction $\hat{\mathbf{x}}(t + 1)$ to yield the updated state $\mathbf{x}(t + 1)$ as

$$\mathbf{x}(t + 1) = \hat{\mathbf{x}}(t + 1) + G(t + 1)\mathbf{r}(t + 1). \qquad (3\text{-}53)$$

Acharya *et al.* (1987) describe how KF can improve the quality of computer tomography (CT). CT scanners require a finite time to collect projections around the object to be reconstructed. In order to obtain good temporal resolution for moving images, reconstruction from relatively few views must be carried out, since the extended time required to collect a large number of views causes blurring in the reconstruction process. One decreases the number of views by providing the reconstruction algorithm with additional unmeasured projections, which are estimated in the time domain with the aid of KF.

Both simulated results (Fig. 3.7.) using a mathematical phantom and

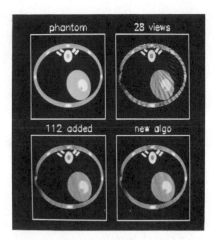

Figure 3.7.
Results from the mathematical phantom. (a) Actual phantom corresponding to the systolic phase. (b) Reconstruction with 28 views using the convolution and back-projection algorithm. Notice the heavy ray-shaped artifacts. (c) Reconstruction obtained by adding four successive time points ($4 \times 28 = 112$ views). Notice the blurring at the ventricular boundaries. (d) Reconstruction obtained from the new temporal synchronization algorithm, showing reduced artifacts and blurring, using 112 views estimated for the same time interval in which the 28 views used for the reconstruction in panel (b) were obtained.

a dog's heart (Fig. 3.8.) scanned live using the Dynamic Spatial Reconstructor (DSR)—a high temporal resolution, synchronous volume (3D) CT scanner—showed a clear improvement over conventional techniques. Artifacts were eliminated and no physiological stationarity, as with gated CT scanning, was assumed.

3.2.4.   Markov Chains

Another way to perform invariant pattern recognition is to look for some internal structure. To capture (internal) relationships, one can use statistical tools such as the Markov chains (see also Section 4.2.3.). Markov chains are characterized by a sequence of trials whose outcomes, say $0_1, 0_2, \ldots$ satisfy the following two properties: (1) Each outcome belongs to a finite set $\{0_1, 0_2, \ldots, 0_m\}$ called the state space; if the outcome of the nth trial is 0_i, then we say that the system is in state 0_i at time n. (2) The outcome of any trial depends at most on the outcome(s) of the n immediately preceding trials (finite memory).

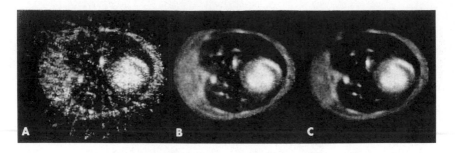

Figure 3.8.
Results from a scan of a dog in the DSR. (a) The 14-view reconstruction (projection data recorded in 1/60 s) of the dog's chest obtained by the conventional convolution back-projection algorithm. (b) The reconstruction obtained by gating four cycles ($4 \times 14 = 56$ views). (c) The reconstruction obtained by use of the temporal synchronization algorithm (112 estimated views for 1/60 s interval).

A transition matrix, P, characterizes the Markov chain, and its entries are given by p_{ij}, corresponding to the transitions $0_i \to 0_j$. One further defines $p_{ij}^{(m)}$ as $0_i \to 0_{k_1} \to \cdots \to 0_{k_{m-1}} \to 0_j$, i.e., going from state 0_i into state 0_j in exactly m steps. The square matrix $P = (p_{ij})$ is called a stochastic matrix, and each of its rows is a probability vector, i.e., $\forall_{i,j}$ $p_{ij} \geq 0$, and $\forall_i$, $\sum_j p_{ij} = 1$. If A and B are stochastic, so is AB. A stochastic matrix A is said to be regular if all entries of some power P^m are positive, i.e., $p_{ij}^m > 0$. If P is the transition matrix, then the n-step transition matrix $P^{(n)} = P^n$. Assuming that $\mathbf{p}^{(0)} = (p_1^{(0)}, \ldots, p_m^{(0)})$ denotes the initial probability distribution, i.e., the likelihood of starting in one of the states $0_1, 0_2, \ldots, 0_m$, then $P^{(n)}$, i.e., the transition matrix after "enough" n steps, approaches a limit matrix T. Informally, this means that the effect of the original state wears off as the number of steps increases; therefore, one captures the internal structure of the image. Formally, if P is a regular stochastic matrix, then

1. P has a unique fixed probability vector $\mathbf{t}$ (i.e., $\mathbf{t}P = \mathbf{t}$), and the components of $\mathbf{t}$ are all positive;

2. $\lim_{n \to \infty} P^n = T$, where $T = \{\mathbf{t}, \mathbf{t}, \ldots, \mathbf{t}\}$;

3. if $\mathbf{p}$ is any probability vector, then the sequence of vectors $\{pP, pP^2, \ldots\} \to \mathbf{t}$. The vector $\mathbf{t}$ stands for the steady (final) state distribution and is independent of the initial-state distribution. One should note that if P has 1 as one of the diagonal entries, then P is not regular; i.e., there are forbidden transitions.

Optical character recognition (OCR) is a practical application, which often requires that garbled text be corrected. Single character recognition is prone to error, so contextual postprocessing helps. Markov chains, as described, can encode (state) transitions between characters for a particular dictionary domain. Such statistics are gathered, and then local processing (single character recognition) is combined with diagrams (probability of a transition between two consecutive characters), which yields a global measure for the fit. The Viterbi algorithm (Forney, 1973) is a recursive optimal solution for estimating the state (character) sequence of a discrete-time finite-state (alphabet) Markov chain embedded in memoryless noise. The algorithm is "analog" to FDP (forward dynamic programming), which is described next.

Assume that $\alpha = (\alpha_1, \alpha_2, \ldots, \alpha_n)$ is the (local) measurement vector corresponding to (letter) position i (within a word of length $(n + 1)$ and of known first letter x_0), and that $P(\alpha) = (p(\alpha_1), \ldots, p(\alpha_M))$ is the a priori letter-class probability vector, i.e., the probability that a given observation α_i could be induced by some letter class $l \in \{1, 2, \ldots, M\}$, where M is the size of the alphabet. The local letter-class assignments can be combined based on known probabilities for Markov-type transitions of a (subset) of the (English) dictionary. Specifically, one then looks for the maximum a posteriori probability (MAP) estimation, i.e., the best global solution. The Viterbi algorithm (VA), like the FDP, transforms the MAP estimation into seeking a shortest-path solution. One can write

$$P(X, \alpha) = P(X|\alpha)P(\alpha) = P(X)P(\alpha|X).$$

$$= \prod_{k=0}^{n-1} P(x_{k+1}|x_k) \prod_{k=1}^{n} P(\alpha_k|x_k), \qquad (3\text{-}54a)$$

where X is the letter-class assignment vector, $X = (x_0, x_1, \ldots, x_n)$, and $P(\alpha)$ is the a priori probability vector. To maximize $P(X, \alpha)$, one needs to find the shortest path through a tree built as follows: (1) Each node corresponds to a letter class at step/position k through the word (letter sequence). (2) Branches β_k correspond to transitions between letter states x_k and x_{k+1}.

The tree starts with $k = 0$ (first letter, x_0) and ends with $k = n$ (last letter, x_n). Then, if the length of branch β_k is defined as

$$\lambda(\beta_k) = -\ln P(x_{k+1}|x_k) - \ln P(\alpha_{k+1}|x_{k+1}), \qquad (3\text{-}54b)$$

the total length for traversing the tree through some path is given as

$$\sum_{k=0}^{N-1} \lambda(\beta_k) = -\ln P(X, \alpha), \tag{3-54c}$$

and it is obvious that MAP is equivalent to finding the shortest path. The shortest-path problem is efficiently solved if for any $k > 0$ and x_k, there is just one best minimal path, which is called a "survivor." The shortest complete path must always begin with one of these survivors, and recursively one extends, step by step, each survivor at step k into a new survivor at step $(k + 1)$.

3.2.5. Eigenfilter Method

The KLT transform, despite being a very expensive computational tool, suggests several ways to perform invariant pattern recognition. First, we recall that the choice of basis vectors aims at the direction of maximum variance, as given by the eigenvectors of the covariance matrix. If an object is rotated by an angle θ, such that $y = Ax$, where

$$A = \begin{bmatrix} \cos\theta & \sin\theta \\ -\sin\theta & \cos\theta \end{bmatrix},$$

then the eigenvectors are rotated by the same angle θ, and registration/ alignment within a standard coordinate system is possible (see Section 7.4. for the need of a coordinate system). If x is normalized and becomes of zero mean, then invariance to translation is achieved as well, while scaling by a factor of k is accounted for by considering the factor of k^2 introduced into the covariance matrix. The invariant aspects of KLT, and MSE approximation, if truncation (for data compression) is considered, were employed by Marinovic and Eichman (1985) using the Wigner distribution (WD) for invariant shape analysis. Specifically, feature extraction and pattern classification are performed in a joint space/spatial-frequency (s/sf) domain and, therefore, enjoys the benefits of simultaneous s/sf representations. Assuming 2D (shape) patterns given in terms of their 1D (one-dimensional) parametric boundary representation, i.e., a sequence of $r(t)$ radial distances measured from the centroid to each of N equidistant points on the boundary (if the shape is not convex, unwrapping is necessary), the corresponding WD is defined as

$$W_r(t, m) = \frac{1}{T} \int_{-T/2}^{T/2} r\left(t + \frac{\alpha}{2}\right) r^*\left(t - \frac{\alpha}{2}\right) \exp\left(-j\frac{2\pi m\alpha}{T}\right) d\alpha, \quad (3\text{-}55)$$

where T is the spatial period, and $0 \leq t < T$. If generalized singular value decomposition (SVD) is performed, the resulting diagonal eigenvalue matrix is given as $D = \text{diag}(\lambda_1, \lambda_2, \ldots, \lambda_N)$. If all but k singular values are close to zero, truncation to k terms is feasible. The corresponding KLT is implemented using a sampled approximation $W(n, m)$ of the WD, where n, $m = 1, \ldots, N$. (Remember that the WD is a 2D representation (s/sf) of a 1D signal $r(t)$.) Then the generalized eigenvalue analysis problem is stated as

$$W(n, m) = UDV' = \sum_{i=1}^{N} \lambda_i u_i(n) v_i'(m),$$

$$\|W\|^2 = \sum_{i=1}^{N} \lambda_i, \quad (3\text{-}56)$$

where $\{\lambda_i\}_{i=1}^{N}$ are the positive square roots of the square matrix WW'. Columns of U (or V) are orthonormal eigenvectors u_i (or v_i) of WW' (or $W'W$). Finally, a set of k descriptors obtained as described previously could discriminate among shapes (such as letters or airplane contours). Note that the invariance mentioned earlier is a kind of pseudoinvariance due to the nonlinear nature of selecting the first k eigenvalues among N.

3.2.6. *Integral Geometry*

Statistical invariance can be achieved not only through digital processing but also through optical pattern recognition. One example is the generalized chord transform (GCT) implemented by Casasent and Chang (1983). The GCT is a nice combination of concepts such as (1) Statistics on the chords of a pattern, a concept coming from integral geometry (Moore, 1972) and (2) polar representation followed by wedge and ring detectors, which are insensitive to scale and rotation.

Specifically, if the boundary of an object is given by $b(x, y) = 1$, the chord distribution (of length r and angle θ) is given by $h(r, \theta)$. A chord exists between two points (x, y) and $(x + r \cos \theta, y + r \sin \theta)$, if

$$g(x, y, r, \theta) = b(x, y)b(x + r \cos \theta, y + r \sin \theta) = 1. \quad (3\text{-}57a)$$

The statistical distribution $h(r, \theta)$ can be defined as

$$h(r, \theta) = \iint g(x, y, r, \theta)dxdy$$

$$= \iint b(x, y)b(x + r \cos \theta,$$

$$y + r \cos \theta)dxdy. \qquad (3\text{-}57b)$$

Substitute (ξ, η) for $(r \cos \theta, r \sin \theta)$, and then one obtains

$$h(\xi, \eta) = \iint b(x, y)b(x + \xi, y + \eta)dxdy = b(x, y) \circledast b(x, y), \quad (3\text{-}57c)$$

i.e., the distribution $h(\xi, \eta)$ is the autocorrelation function of the boundary. If, instead of restricting ourselves to the boundary, one substitutes any picture value function f for the binary valued function b, one obtains the GCT as

$$h_G(\xi, \eta) = \iint f(x, y)f(x + \xi, y + \eta)dxdy. \qquad (3\text{-}57d)$$

Both for computational reasons (data compression) and invariance, ring and wedge detectors, respectively, are defined as

$$h_G(r) = \int h_G(r \cos \theta, r \sin \theta)rd\theta$$

$$\qquad (3\text{-}57e)$$

$$h_G(\theta) = \int h_G(r \cos \theta, r \sin \theta)rdr.$$

One sees that the ring and wedge detection features are independent of the orientation and scale changes in the chord distribution.

3.3.   Algebraic Invariance

Much of CV results could be "expressed concisely in terms of advanced mathematics. The merit of advanced mathematics is not limited to compact expression. It gives us a guideline to proceed, sometimes leading us to previously unexpected new applications. If we know that two apparently different problems have the same underlying mathematical structure, a mathematical tool developed for one problem can be transformed to solve the other" (Kanatani, 1986). As an example, regularization techniques, to be discussed in the next chapter, deal with recovering 3D structure from 2D geometrical projections, while geometrical considerations play a major role in implementing specific minimization designs. Geometry "played a key role in the development of modern mathematics, giving birth to many branches of mathematics such as topology, manifold, and differential geometry" (Kanatani, 1986). In this section, we consider the algebraic aspect in search for image invariants and discuss differential geometry in the context of object recognition (Section 7.3.2.).

Many people seek to model the environment in some abstract sense. While some do so in terms of symbols and logic (see Chapter 8), they can also use mathematical models known as groups and/or algebras. Models are given in terms of specific objects and the corresponding relationships (operations) connecting them. The major benefits derived from abstract mathematical modeling are twofold. First, one can transform the problem of searching for the identity of a (possibly transformed) object into a problem of model equivalence through isomorphism or homeomorphism techniques. Second, models provide constructive rules to generate new elements belonging to the same model. The rules characteristic of geometrical reasoning could be used to check if a specific object instance belongs to a given model or not. The Lie transformation group, to be considered in Section 3.3.2., provides the rules of closure, associativity, and the identity and inverse elements.

3.3.1. Shape Invariance

Quadric surfaces are generally represented as

$$a_{11}x^2 + a_{22}y^2 + a_{33}z^2 + 2a_{12}xy + 2a_{13}xz + 2a_{23}yz$$

$$+ 2a_{14}x + 2a_{24}y + 2a_{34}z + a_{44} = 0. \tag{3-58}$$

Corresponding to the equation, the sets (I, K, D, A) and (sign $\langle A' \rangle$, sign $\langle A'' \rangle$, sign $\langle A''' \rangle$) are invariant with respect to translation and rotation transformations, and thus they define quadric surface properties that are independent of position. The quantities mentioned are defined as

$$I = a_{11} + a_{22} + a_{33}$$

$$K = \begin{vmatrix} a_{11} & a_{12} \\ a_{21} & a_{22} \end{vmatrix} + \begin{vmatrix} a_{22} & a_{23} \\ a_{32} & a_{33} \end{vmatrix} + \begin{vmatrix} a_{33} & a_{31} \\ a_{13} & a_{11} \end{vmatrix}$$

$$D = A_{44} = \begin{vmatrix} a_{11} & a_{12} & a_{13} \\ a_{21} & a_{22} & a_{23} \\ a_{31} & a_{32} & a_{33} \end{vmatrix}$$

$$A = \det|a_{ik}|.$$

where A_{ik} is the cofactor of a_{ik} for $A = \det|a_{ik}|$

$$A' = A_{11} + A_{22} + A_{33} + A_{44}$$

$$A'' = \begin{vmatrix} a_{11} & a_{12} \\ a_{21} & a_{22} \end{vmatrix} + \begin{vmatrix} a_{11} & a_{13} \\ a_{31} & a_{33} \end{vmatrix} + \begin{vmatrix} a_{11} & a_{14} \\ a_{41} & a_{44} \end{vmatrix}$$

$$+ \begin{vmatrix} a_{22} & a_{23} \\ a_{32} & a_{33} \end{vmatrix} + \begin{vmatrix} a_{22} & a_{24} \\ a_{42} & a_{44} \end{vmatrix} + \begin{vmatrix} a_{11} & a_{33} \\ a_{43} & a_{22} \end{vmatrix}$$

$$A''' = a_{11} + a_{22} + a_{33} + a_{44}.$$

Examples of quadric surfaces are the ellipsoid given by $x^2/a^2 + y^2/b^2 + z^2/c^2 = 1$ (for which $A < 0$, DI and K are both greater than 0) and the hyperbolic paraboloid given by $x^2/a^2 - y^2/b^2 = z$ (for which $D = 0$, $K < 0$, and $A > 0$).

With regard to the invariant characterization of shape, we follow the approach suggested by Kanatani (1986) and the previous rational for generalization. (Compare the following derivation to the one concerning moments of inertia presented in Section 3.1.6.)

Assume that 3D points (X, Y, Z) are centrally projected onto points (x, y) in the image plane such that $(x, y) = (fX/Z, fY/Z)$, where f is the camera focal length. Camera rotation around its focus is equivalent to rotation around the origin and is specified by an orthogonal matrix $R = (r_{ij})$. Two points (x, y) and $(x', y') = (fX'/Z', fY'/Z')$ are consistent under the transformation R if the mapping T as given next holds.

$$x' = f\frac{r_{11}x + r_{21}y + r_{31}f}{r_{13}x + r_{23}y + r_{33}f} \text{ and } y' = f\frac{r_{12}x + r_{22}y + r_{32}f}{r_{13}x + r_{23}y + r_{33}f}. \qquad (3\text{-}59)$$

Next, we define the concepts of scalar, point, and line in the context determined by the transformation R. If an image characteristic c does not change its value under transformation R, i.e., $c' \equiv c$, then we call c

a scalar. If a pair (a, b) of numbers is transformed according to T, then it is called a point and indicates a position in the scene. A line given as $Ax + By + C = 0$ is characterized by its ratio $A : B : C$, and under R, transforms into $A'x' + B'y' + C' = 0$, where the test ratio $A' : B' : C'$ is now given by

$$\frac{r_{11}A + r_{21}B + r_{31}C}{f} : \frac{r_{12}A + r_{22}B + r_{32}C}{f} : (f(r_{13}A + r_{23}B) + r_{33}C).$$

Let us now consider a shape S defined by its characteristic function $f(x, y)$ as given by

$$f(x, y) = \begin{cases} 1 & (x, y) \in S \\ 0 & \text{otherwise} \end{cases}$$

and assume that camera rotation R keeps the region S within the image plane. The area A corresponding to S is trivially given as $A = \int_S dxdy$, but it changes with camera orientation. However,

$$c = f^3 \int_S \frac{dxdy}{\sqrt{(x^2 + y^2 + f^2)^3}}$$

does not change, and thus it is a scalar according to our definition. c describes the solid angle the object makes with respect to the viewer.

Next, we consider the center of gravity $(\bar{x}, \bar{y}) = (\int_S xdxdy/\int_S dxdy, \int_S ydxdy/\int_S dxdy)$. Should S be rotated according to R the new center of gravity is given by $(\bar{x}', \bar{y}')$. Note that $(\bar{x}, \bar{y})$ is not mapped into $(\bar{x}', \bar{y}')$ under R but into $\mathbf{a}$ given as

$$\mathbf{a} = (a_1, a_2, a_3) = \left(f \int_S \frac{xdxdy}{(x^2 + y^2 + f^2)^2}, \quad f \int_S \frac{ydxdy}{(x^2 + y^2 + f^2)^2}, \right.$$

$$\left. f^2 \int_S \frac{dxdy}{(x^2 + y^2 + f^2)^2} \right),$$

and thus it is transformed as a vector. Furthermore, $(fa_1/a_3, fa_2/a_3)$, which stands for the invariant center of gravity of S and corresponds to the center of the solid angle the shape S makes, with respect to the viewer, is transformed as a point.

The moments tensor M_{ij} $(i, j = 1, 2)$ defined in Section 3.1.6. as

$$M_{11} = \int_S (x - \bar{x})dxdy, \qquad M_{12} = M_{21} = \int_S (x - \bar{x})(y - \bar{y})dxdy,$$

$$M_{22} = \int_S (y - \bar{y})^2 dxdy$$

does not enjoy invariance properties. The principal values of (M_{ij}) are not scalars, and its principal axes cannot indicate lines on the image plane. The set $B = (b_{ij})$ given next, however, is transformed as a (symmetric) tensor

$$B_{11} = f\int_S \frac{x^2}{\alpha}dxdy; \qquad B_{12} = f\int_S \frac{xy}{\alpha}dxdy; \qquad B_{13} = f^2\int \frac{x}{\alpha}dxdy;$$

$$B_{12} = f\int_S \frac{xy}{\alpha}dxdy; \qquad B_{22} = f\int_S \frac{y^2}{\alpha}dxdy; \qquad B_{23} = f^2\int_S \frac{y}{\alpha}dxdy;$$

$$B_{31} = f^2\int \frac{x}{\alpha}dxdy; \qquad B_{32} = f^2\int_S \frac{y}{\alpha}dxdy; \qquad B_{33} = f^3\int_S \frac{1}{\alpha}dxdy;$$

where $\alpha = \sqrt{(x^2 + y^2 + f^2)^5}$.

The tensor (B_{ij}) is positive definite and its principal values σ_1, σ_2, and σ_3 are positive. The corresponding unit vectors are $\mathbf{e}_1$, $\mathbf{e}_2$, and $\mathbf{e}_3$, where $\mathbf{e}_3$ corresponds to the maximum principal value and points to (x_1, x_2). Lines L_1 and L_2 are defined by (x_1, x_2) and the points corresponding to $\mathbf{e}_1$ and $\mathbf{e}_2$, respectively. Clearly, scalars (σ_1, σ_2), points (x_1, x_2), and lines (L_1, L_2) are invariant quantities standing for invariants (principal values, center of inertia, principal axes).

Since $(c, \mathbf{a}, B) = (\text{scalar, vector, tensor})$, the equivalence of two (shape) images S_1 and S_2 can be tested by comparing the corresponding sets s_1 and s_2 given by

$$s = (c, \mathbf{a}^T\mathbf{a}, \text{Tr}(B), \text{Tr}(B^2), \text{Tr}(B^3), \mathbf{a}^TB\mathbf{a}, \mathbf{a}^TB^2\mathbf{a}), \qquad (3\text{-}60)$$

where Tr stands for the Trace operator. If s_1 is equivalent to s_2 then one can find the corresponding rotation R_{ij} such that

$$S_1 = R_{12}S_2 \text{ and } S_2 = R_{21}S_1.$$

Camera orientation registration benefits from the previous procedure because no point-to-point correspondence is required.

3.3.2. Lie Transformation Groups

Another nice example of applying algebraic consideration to derive invariants is the work by Hoffman (1977) and Dodwell (1983). Their research, consistent with both ecological optics as formulated by Gibson (1979) and active perception, provided an early mathematical account of how invariants could be derived and integrated in visual processing.

Specifically, (Dodwell, 1983) suggested the Lie transformation group (LTG) model "to represent and explain how the locally smooth processes observed in the visual field, and their integration into the global field of visual phenomena, are consequences of special properties of the underlying neuronal complex. The LTG model seeks to relate microgenetic processes in the visual field, which are reflected in strictly localized activities within the nervous system (activity of individual neurons) to more macroscopic aspects of both the visual scene and the neural activities which underlie pattern processing" (Hoffmann, 1977). The LTG model could establish "resonance" relationships between a given operator, L, and a trajectory/orbit f, which it generates. Conversely, if the result of applying operator L to the function f is identically zero, then the operator is the one that generates that function. For example, if the pattern to be generated/or resonated to is a circle, $f = x^2 + y^2 - r^2 = 0$, then the corresponding operator is $L = (-y\partial/\partial x + x\partial/\partial y)$ since

$$L(f) = \left(-y\frac{\partial}{\partial x} + x\frac{\partial}{\partial y}\right)(x^2 + y^2 - r^2) = -2yx + 2xy \equiv 0.$$

The most general form of such operators is

$$L = \left[f(x, y, t)\frac{\partial}{\partial x} + f(x, y, t)\frac{\partial}{\partial y} + f(x, y, t)\frac{\partial}{\partial t}\right] \tag{3-61a}$$

and they can be derived from the linear group for $R^2 \times T$ (plane × time) given as

$$LG = \begin{vmatrix} \dfrac{\partial}{\partial x} & x\dfrac{\partial}{\partial x} & y\dfrac{\partial}{\partial x} & t\dfrac{\partial}{\partial x} \\ \dfrac{\partial}{\partial y} & x\dfrac{\partial}{\partial y} & y\dfrac{\partial}{\partial y} & t\dfrac{\partial}{\partial y} \\ \dfrac{\partial}{\partial t} & x\dfrac{\partial}{\partial t} & y\dfrac{\partial}{\partial t} & t\dfrac{\partial}{\partial t} \end{vmatrix}. \tag{3-61b}$$

Next consider the case of landing on a surface (see Fig. 3.9.), which corresponds to a optical flow (see Section 5.1.2.) whose pattern is that of dilation, with the corresponding operator given as $L = [x\partial/\partial x + y\partial/\partial y]$. Other examples include Lie operators such as $(L_x, L_y, L_t) = (\partial/\partial x, \partial/\partial y, \partial/\partial t)$ whose orbits are (horizontal, vertical, straight lines parellel to the t-axis in the x-y-t space) and who achieve shape constancy under translation or time. Visual invariance to rotation is achieved by $L = -y\partial/\partial x + x\partial/\partial y$ whose orbits are concentric circles. If for any L_1, L_2 Lie operators, $L_1 L_2 - L_2 L_1 \in L$, then the Lie group enjoys the property of closure under commutation, and thus it constitutes a Lie albebra.

Operators such as L could account for a fast response in critical situations, such as collision avoidance, which could be explained by the sensitivity of a given operator to a specific invariant. Sensitivity to invariants could motivate a given system to move and to start or stop "resonating" to a given characteristic. Such behavior benefits the organism by securing desirable conditions and/or avoiding harmful surroundings (such as those of impeding collision). Furthermore, resonance-like behavior would be fault tolerant to noisy and incomplete input patterns due to the holistic aspect of resonance, i.e., $L(f) < \varepsilon$, for some ε. (See also the connection between resonance and

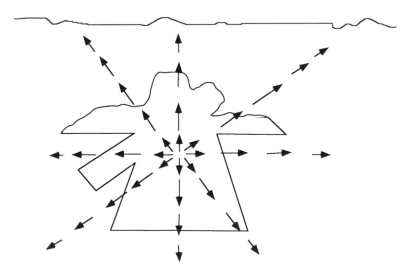

Figure 3.9.
Dilation optical flow corresponding to landing and/or collision.

holistic Gestalt recognition in the next chapter.) Consequently, an organism is motivated to actively look around and to attend to "ecological information." This is why perception should be tightly linked to locomotion/acting.

Papathomas and Julesz (1988) survey the LTG model in neurophysiology (LTG/NP), and the psychophysical evidence, for the most part, related to known phenomena of constancy. They show that an LTG model might provide a unifying theory to identify images that obey affine transformations. Finally, the computation is largely local and could easily be implemented using SIMD (see Section 9.3.) parallel architectures.

3.3.3. Projective Transformations

The retinal image undergoes continuous deformations whenever the head moves, but the visual world remains stable. The only objective perception is the subject's own locomotion. Changes in retinal images are modeled as projective transformations, where point-to-point and line-to-line correspondence is preserved but the shape changes. Gibson (1950), who was well aware of the relevance of such transformation, said

> Transformations are usually represented on a plane, however, whereas the retinal image is a projection on a curved surface. As a matter of fact, the actual retinal image on a curved surface is related to the hypothetical image on a picture-plane only by such a nonrigid transformation. The geometry of transformations is therefore of considerable importance for vision, and it is conceivable that the clue to the whole problem of pattern perception might be found here. The transformation of a given pattern, mathematically defined, does not simply destroy the pattern as one might at first suppose. A transformation is a regular and lawful event which leaves certain properties of the pattern invariant. Moreover, a series of transformations can be endlessly and gradually applied to a pattern without affecting its invariant properties. The features that are preserved may be the mediators of a stable world and the features not preserved the mediators of the visual impression of motion.

Projective transformations are such that lengths and ratios of lengths are altered, but the ratio of two ratios of length (the cross-ratio) is invariant. Clearly, under projective transformations many-to-one mappings prevent one from inferring that two given projections correspond to identical objects. One can only derive necessary rather than sufficient invariant conditions for assessing if two images belong to the same object. Cross-ratio, which allows one to define such necessary conditions, has been discussed by Duda and Hart (1973) in the context of the geometry shown in Fig. 3.10.

$\{D_i\}_{i=1}^4$ is a set of lines having (a, b) as its center. $(x_i, 0)$ and $(0, y_i)$ are corresponding ranges of points lying on lines X and Y, respectively, where the points are specific image features used for object identification. From projective geometry, we know that any two ranges of points such as X and Y can be connected by a chain of two perspective correspondences at most, which are then said to be in projective correspondence. Hence, two images consisting of ranges X and Y show the same object if X and Y are in projective correspondence. For the geometry depicted, where (a, b) is the center of a skewed Cartesian coordinate system (along X and Y), one can write for each line D_i

$$\frac{a}{x_i} + \frac{b}{y_i} = 1, \qquad i = 1, 2, 3, 4. \tag{3-62a}$$

Given two lines (i, j) and their stated corresponding conditions, one can subtract the jth one from the ith one and obtain

$$a\left(\frac{x_j - x_i}{x_i x_j}\right) = -b\left(\frac{y_j - y_i}{y_j y_i}\right). \tag{3-62b}$$

Divide the product of equations (D_3, D_1) and (D_2, D_4) by the product of

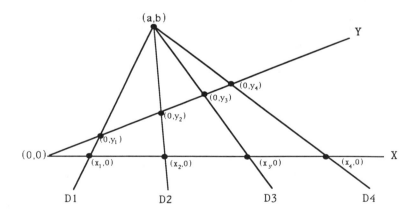

Figure 3.10.
Cross-ratio and projective geometry.

(D_2, D_1) and (D_3, D_4) and then obtain

$$\text{CR} = \frac{(x_3 - x_1)(x_2 - x_4)}{(x_2 - x_1)(x_3 - x_4)} = \frac{(y_3 - y_1)(y_2 - y_4)}{(y_2 - y_1)(y_3 - y_4)}, \tag{3-62c}$$

where the corresponding cross-ratios, $\text{CR}(x_1, x_2, x_3, x_4)$ and CR (y_1, y_2, y_3, y_4) are the same and independent of (a, b). Therefore, the cross-ratio is a projective invariant and remains unchanged under central projection. The 3D case is treated as well by Duda and Hart (1973), who assume that the object plane contains at least five distinguished points, while four of them form the base of a projective coordinate system. Furthermore, such a method requires establishing point correspondence. The resulting necessary condition is that the projective coordinates of the fifth point with respect to the base established previously are identical. Note that the cross-ratio (CR) defined in Eq. (3-62c) is just one of 24 different CRs, only six of which are numerically different. The CRs are invariant under rotation, displacement of line D_1, change of position of line D_2, or movement of the observer. The range of the CR is zero to infinity. A canonical CR of range zero to unity will be defined and derived in Section 6.1.3.

Both the complex-log mapping (Section 3.1.1.) and the projective transformation are examples of conformal mappings of the type $w = f(z)$. At each point z such that $f(z)$ is analytic and $f'(z) \neq 0$, the mapping is conformal, i.e., the angle between two curves through such a point remains invariant in magnitude and sense. Furthermore, the projective transformation is a specific case of a bilinear transformation given as

$$w = \frac{az + b}{cz + d} \qquad (bc - ad \neq 0). \tag{3-63}$$

Straight lines and circles in the z-plane correspond to straight lines or circles in the w-plane and, conversely, the cross-ratio defines the unique bilinear transformation mapping three given points (x_1, x_2, x_3), respectively, into three given points (y_1, y_2, y_3) (Korn and Korn, 1968). Take a permutation on the indexes, such that $3 \to 1$, $2 \to 3$, $4 \to 2$ and keep 1 fixed (as z and w, respectively). The resulting Moebius transformation is then given by

$$\frac{(z_1 - z)(z_3 - z_2)}{(z_3 - z)(z_1 - z_2)} = \frac{(w_1 - w)(w_3 - w_2)}{(w_3 - w)(w_1 - w_2)}. \tag{3-64}$$

The linear transformation $w = Az + B$, where A and B are complex numbers, corresponds to a rotation, through the angle arg(A) together with a stretching by a factor $|A|$, followed by a translation through vector B. The transformation is the most general conformal mapping that preserves similarity of geometrical figures. Finally, it should be noted that bilinear transformations form a group, and that inverses and products of bilinear transformations are bilinear transformations.

Every bilinear transformation may be then expressed as the product of three successive simpler bilinear transformations

$$z' = z + \frac{d}{c} \qquad \rightarrow \text{(translation)}$$

$$z'' = \frac{1}{z'} \qquad \rightarrow \text{(inversion and reflection)}$$

$$w = \frac{bc - ad}{c^2} z'' + \frac{a}{c} \quad \rightarrow \quad \begin{array}{l}\text{(rotation and stretching,} \\ \text{followed by translation).}\end{array}$$

3.3.4. Differential Invariants

Weiss (1988) uses the theory of differential and algebraic invariants to derive projective invariants of shapes. A curve is represented parametrically as $x(t) = x_i(t)$, where $i = 1, 2, 3, 4$ for space curves and $i = 1, 2, 3$ for plane curves, assuming homogeneous coordinates. One then seeks a differential equation whose solution is the curve $x_i(t)$. The advantage of differential equations (de) is that some constants are eliminated. For example, $x_i'' = 0$ represents all straight lines; the invariance achieved is with respect to slope and intersect, and what stays invariant under projection is the straightness of the line. Any curve in nD homogeneous coordinates satisfies the linear de (lde)

$$x^{(n)} + \binom{n}{1}p_1 x^{(n-1)} + \binom{n}{2}p_2 x^{(n-2)} + \cdots + p_n x = 0, \qquad (3\text{-}65)$$

which is projectively invariant, and the solution of the equation is a curve in $(n-1)$ dimensional space, determined up to a projective transformation. Equation (3-65) is representative of the class $\Phi(a_k, x_i) = 0$, where Φ is a lde, a_k are parameters, and x_i are space coordinates of the points on the shape. A projective invariant is then a functions of the parameters a_k, or if we consider Φ as given by Eq. (3-65), in terms of the p_i, which are scalar functions of t. The (vector) equation for plane curves corresponding to the general Eq. (3-65) is given as

$$x''' + 3p_1 x'' + 3p_2 x' + p_3 x = 0. \tag{3-66}$$

Define the semi-invariant (no invariance under change of parameter t) quantities

$$P_2 = p_2 - p_1^2 - p_1'$$
$$P_3 = p_3 - 3p_1 p_2 + 2p_1^3 - p_1''. \tag{3-67a}$$

The quantities θ_3 and θ_8, to be defined next, are invariant under projective transformations, multiplication of the coordinates by a factor $\lambda(t)$ and a change of parameter t.

$$\theta_3 = P_3 - \frac{3}{2}P_2'$$

$$\theta_8 = 6\theta_3 \theta_3'' - 7(\theta_3')^2 - 27P_2 \theta_3^2. \tag{3-67b}$$

Then, the invariants θ_3 and θ_8 uniquely determine a plane curve up to a projective transformation. The invariants are relative, i.e., they contain the Jacobian of the transformation. Absolute invariants are a combination of the relative ones. For plane curves, an example of an absolute invariant is θ_3^8/θ_8^4.

For space curves, the *lde* is given as

$$x'''' + 4p_1 x''' + 6p_2 x'' + 4p_3 x' + p_4 x = 0. \tag{3-68}$$

Then the semi-invariants are given by

$$P_2 = p_2 - p_1' - p_1^2$$
$$P_3 = p_3 - p_1' - 3p_1 p_2 + 2p_1^3$$
$$P_4 = p_4 - 4p_1 p_3 - 3p_2^2 + 12p_1^2 p^2 - 6p_1^4 - p_1''', \tag{3-69a}$$

and the invariants are

$$\theta_3 = P_3 - \frac{3}{2}P_2'$$

$$\theta_4 = P_4 - 2P_3' + \frac{6}{5}P_2'' - \frac{6}{25}P_2^2$$

$$\theta_8 = 6\theta_3 \theta_3'' - 7\theta_3'^2 - \frac{108}{5}P_2 \theta_3^2. \tag{3-69b}$$

Implementation of the previous theory (which can be extended to surfaces as well and involves differentials) is likely to be prone to errors. Algebraic invariants are simpler to derive because they do not require finding derivatives. We defined quadric surfaces in Section 3.3.1. and Eq. (3-58). The quantity $A = \det |a_{ik}|$ is an invariant of order two, i.e., under projection of coordinates x_i, the coefficients a_{ik} change, and the new determinant $\bar{A}$ is related as $\bar{A} = AJ^2$, where J is the Jacobian of the projective transformation. (Note that the general form of the quadric as given in Eq. (3-58) stays unchanged.) The Jacobian is, however, the same for all conic sections, and it can thus be eliminated by choosing one of the conics as the standard A^*, and dividing the "invariant" A of all the others by A^*.

3.4. Conclusions

We took our attempt to elucidate early (low-level) vision one step further and showed the relevance of invariant representations. Cojoint spatial/spatial-frequency representations invariant to linear transformations and projective distortion are feasible. They can be realized through the use of an appropriate distributed cytoarchitecture of filters.

Invariants play a crucial role in human perception, and they correspond to the HVS's fixed principles of causality, space, and time. Such principles, consistent within our own ecological niche rather than learning (i.e., nurturing), impose organization and structure on visual perception. Furthermore, as will be discussed later, such an approach is also consistent with Gestalt psychology, which asserts that the HVS captures at once characteristics of the whole and that perception of components is determined by their relationship to the whole. Witkin and Tenenbaum (1983) subscribe to a similar approach when they argue that

> perceptual organization is a primitive level of inference, the basis for which lies in the relation between structural and causal unit: the appearance of spatiotemporal coherence or regularity is so unlikely to arise by the chance interaction of independent entities that such regular structure when observed, almost certainly denotes some underlying unified cause or process. This will be shown to have broad implications for computational theories of vision, providing a unifying framework for

many current techniques of early and intermediate vision, and enabling
a style of interpretation more in keeping with the qualitative and holistic
character of human vision.

Consequently, the distributed and invariant architecture developed so
far represents a strong case against reductionist (atomistic) approaches, which suggest symbolic representations for early vision.

Our ever-changing environment and the need for safe navigation
suggest an invariant approach as well. Our perceptions remain constant despite continuous changes taking place around us, a phenomena
known to psychophysicists as perceptual constancy, which includes
size, shape, and lightness (reflectance) constancies. Structural invariants, related to high-level, cognitive processes will be considered
within the framework of intelligent systems and discussed in
Chapter 8.

The discussion here reconciles the Gibson, Gestalt, and information
processing schools. Once the need for invariance has been elucidated
and explained, one can comfortably accept that the organism must
somehow computationally resonate to invariants, or as Gibson labeled
them, affordances. The Gibsonians and Gestaltists are split concerning
where ordering takes place. The Gibsonians argue for (sensation)
retinal ordering that reflects natural regularities while the Gestaltists
talk about mental ordering. Both are correct if we first ask the
Gibsonians to *compute* and pick up (bottom-up) image invariants, while
the Gestaltists implement holistic and distributed processing ordering.
Holistic and distributed processing is the very topic of the next
chapter.

Liber.X.Trac.II.de potentijs

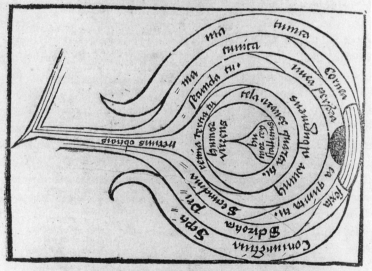

De medio videndi z noctulucis vnde lumen habeant/z quare in die non videntur in lumine proprio.   Capitulũ. X.   Di.

Medium visus. **M**Edium videndi qd est? **D**i. Corpus pspicuũ z puũ ſaltem in vltima pre qua colori coniunctũ est actu dyaphonũ. Necesse est eñ colorem a lũie excitari vt viſum moueat:pp materialitatẽ nãq̃ suam viũſ ſe mĩime ingerere poteſt.qignie ergo de ſe ſpecies a maſia depuratas: qs in lumine z imagine lucis t unꝗ in ꝑncipio ſuo octo i mittit. Ex tñ viſibilia q̃ **noctiluca.** noctiluca dicunf: cũ intrinſecũ z cõnatũ lumẽ habeãt:etiã in nocturnis tenebꝝ videri poſſunt: vt ſunt q̃dam ligna putrida vermiculi:ſquame piſciũ z ſimilia

Front view of eye and cross section of eye. (From Reisch, G., Margarita philosophica . . . , 1508.) Reprinted courtesy of the National Library of Medicine.

4

Parallel and Distributed Processing

Divide et impera.

Having discussed computational constraints and the need for distributed and invariant image representations, we now go one step further and introduce parallel and distributed processing (PDP) as a major paradigm of computation. PDP is also the preferred computational model for fault tolerant and robust behavior. The strong analogy between PDP models of computation and physical systems and their relevance to specific visual tasks is discussed in Section 4.2.

Mead (1989) notes that "the visual system of a single human being does more image processing than the entire world's supply of supercomputers. The difference lies in the analog nature of neural computation; animals are thereby able to accept imprecise inputs and to solve ill-defined, real-world problems. Digital computers, by contrast, can provide answers only to precisely stated questions." There is a fundamental need to handle imprecise inputs and to exhibit fault tolerant and robust behavior. Mead also suggests the neural paradigm as the appropriate computational model when he says that "nature knew nothing of bits, or Boolean algebra. But evolution had access to a vast array of physical phenomena that implemented important functions. It is evident that the resulting computational metaphor has a range of capabilities that exceeds by many order of magnitude the capabilities of the most powerful digital computers."

4.1. Artificial Neural Systems

In this section, we consider artificial neural systems (ANS), also known as neural networks (NN), as the suitable model for PDP computation. We show how ANS, as the result of competitive processes, yield the solution to specific visual tasks, which are appropriately described as minimization problems. The question of how ANS could evolve, i.e., the issue of learning, is deferred to Chapter 8.

Neural networks were originally suggested to account for biological memory systems. PDP (McClelland and Rumelhart, 1986) models correspond to distributed computation, where a large number of highly interconnected "simple" processing elements (PE) operate in parallel. Neural networks are concerned with the study of dynamic systems that implement information processing tasks, such as constrained optimization, through their internal (fixed-point or attractor) state responses to continuous input. Through the NN's collective dynamics, the emergent behavior, which results from competition and cooperation between neighboring PEs, yields the optimal solution subject to contextual constraints implicitly embedded in the net of interconnections. Such computation cannot always continuously enforce those constraints, and there might be a temporary violation of such constraints. (See the basic differential multiplier method in Section 4.1.2.) The PDP strives to reach some optimal solution within a limited set of resources.

The general model of computation used throughout this section is shown in Fig. 4.1. The ANS is made up of nonlinear cells and includes competitive feedback provided by excitatory and inhibitory links. The behavior of the cells, i.e. their state, is governed by some differential equation. Assuming that the state of the cell i is given by x_i, the dynamics of the state changes are usually defined (Cohen and Gross-

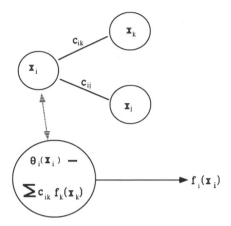

Figure 4.1.
Artificial neural system (ANS) model.

berg, 1983) as

$$\frac{dx_i}{dt} = a_i(x_i)\left[\theta_i(x_i) - \sum_{k=1}^{n} c_{ik}f_k(x_k)\right],$$ (4-1)

where θ_i is the threshold corresponding to cell i, and c_{ik} are the synaptic weights of the links (synapses) to cell i that carry competitive feedback from the cells k according to their firing rate $f_k(x_k)$. The ANS model defined by Eq. (4-1) is quite general and includes the additive (Section 4.1.4.) and shunting models. The shunting model results when $a_i(x_i)$ are not constant and $\theta_i(x_i)$ can be sigmoidal rather than linear. The model describes on-center/off-surround interactions as those existing among cells that obey membrane equations. Specifically, cell states are restricted to bounded intervals $[D_i, B_i]$, and automatic gain control, instantiated by multiplicative shunting terms, interacts with balanced positive and negative feedback signals and inputs, to maintain the sensitivity of each state within the interval.

4.1.1.   What Are Neurons Good For?

Resistance to neural network (PDP models of computation) is largely due to a simplistic understanding of neurons. The McCulloch–Pitts neurons, back in the 1940s, were binary devices characterized by fixed thresholds (Anderson and Rosenfeld, 1988). Excitatory synapses have identical weights, and the inhibitory synapses, if active, prevent the neuron from firing. Finally, during a synaptic delay, weighted integration of inputs takes place.

Nilsson (1974) stated that "knowledge about the structure and function of the neuron or any other basic component of the brain—is irrelevant to the kind of understanding of intelligence that we are seeking. So long as these components can perform very simple logical operations, then it doesn't really matter whether they are neurons, relays, vacuum-tubes, transistors, or whatever." Contrast Nilsson's view, however, to that espoused earlier by Barlow (1972).

> The cumulative effect of all the changes I have tried to outline has been to make us realize that each single neuron can perform a much more complex and subtle task than had previously been thought. Neurons do not loosely and unreliably remap the luminous intensities of the visual image into our sensorium, but instead they detect pattern elements, discriminate the depth of objects, ignore irrelevant causes of variation and are arranged in an intriguing hierarchy. Furthermore, there is evidence that they give prominence to what is informationally important, can respond with great reliability, and can have their pattern selectivity permanently modified by early visual experience. This amounts to a revolution in our outlook. It is now quite inappropriate to regard unit activity as a noisy indication of more basic and reliable processes involved in mental operations: instead, we must regard single neurons, and we should not use phrases like "unit activity reflects, reveals, or monitors thought processes," because the activities of neurons, quite simply, are thought processes. This revolution stemmed from physiological work and makes us realize that the activity of each single neuron may play a significant role in perception.

The traditional view of neuronal function is that the stimulus available in the receptive field (RF) matches the neuron's characteristics. The strength of the match (the firing rate) indicates how fit the neuron is to "resonate" to the input stimulation (Hartline, 1940; Barlow, 1972). Still, the information available in the input can be understood only by looking at the distribution of neuronal activity over large neural areas. The strength of the response clearly fails as an adequate coding tool, because the neuron responds to more than one intrinsic characteristic of the input (such as shape and motion). The traditional and accepted view that all that is relevant about the neuron's response lies in its strength (number of spikes) has been challenged by Richmond *et al.* (1988). The hypothesis they advance is that the spike train has a temporal dimension as well, and that the very temporal distribution allows individual neurons to encode local stimulus features. From information theory and physiological studies, they found that the efficiency of encoding visual information while using the new temporal-modulation hypothesis is about twice as much as when compared against the traditional mean-firing rate. The new conjecture, coined the multiplex-filter hypothesis, further assumes that each neuron can be viewed "as a small number ($\simeq 4$) of simultaneously active spatial-to-temporal (SST) filters." One can think of these filters as forming an orthogonal basis, and if so "multiple messages are carried simultaneously by or are multiplexed onto the spike train, and that information about multiple features is not confounded, as is assumed by the response strength/receptive field match hypothesis." Such a hypothesis still requires some decoding mechanism to unscramble the stimulus features that have been multiplexed together, and such decoding has yet to be found. Even at this early stage, however, it is clear that the idea of temporal modulation is justified. Then, the complexity of processing accomplished by each neuron on its own is upgraded. The same hypothesis has additional implications. To quote again from Richmond *et al.* (1988)

> In the current view, individual neurons cannot convey useful information about the stimulus by themselves. Instead, the information is encoded across a large group of such neurons. In contrast, our results suggest that individual neurons encode information about local stimulus features, conveyed by multiple messages using a temporally modulated code. Consolidation of local messages to determine global properties of images may then be accomplished through compilation of many temporally encoded messages. Processing of information in visual areas may not consist of altering the distribution of active neurons so much as in the transformation of temporally modulated messages. The current view of the many visual areas of the brain is that they represent a hierarchical organization for the successive abstraction of visual features. In contrast, our results suggest that these areas provide stages of spatial-to-temporal filtering that change the emphasis of the visual features, but never confound or ignore information.

Some arguments against neural networks derive from the lack of meaningful decomposition into elementary constituents. Because the network's internal behavior is transparent to the user, expressing its operation in terms of predefined, specific elements as with the more traditional symbolic approach might be difficult. Although the case for a symbolic approach for (at least) early low-level vision is a hard one to make, one could still bridge the decomposition issue by defining appropriate mapping rules between the symbolic and distributed methods. Furthermore, the PDP type of computation, modeled as multilayer networks, was shown by Kolmogorov (1957) to implement needed mappings (transformations) between any given pair of ⟨input, output⟩. The distributed computation discussed so far makes neural networks the equivalent of level two (representation/algorithm) and level three (implementation) within the framework suggested by Marr (1982). PDP models should be understood as computation models that provide a good fit between level one (optimization tasks) and levels two and three, respectively.

The gulf between early (low-level) processing and higher levels of cognition (Klopf, 1982) does not justify the sudden computational jump to the higher level. To quote Rusell (1921): "In attempting to understand the elements out of which mental phenomena are compounded, it is of greatest importance to remember that from the protozoa to man there is nowhere a very wide gap either in structure or in behavior. From this fact it is a highly probable inference that there is also nowhere a very wide mental gap." Combining task levels two and three as suggested can help bridge such a gap within a multistage computational vision theory.

4.1.2. *Regularization*

Visual processing, such as recovering 3D structure from 2D projections, has largely been shown to be mathematically underconstrained, or to quote Poggio (1985), "ill-posed in the sense of Hadamard." Searching for a specific solution requires imposing viable constraints, which could already be known and embedded in the processing mechanism. In evolution, such phylogenetic constraints know about nature's regularities and prevent the visual system from starting with a "tabula rasa"—devoid of knowledge and having to learn everything from scratch. The regularities referred to relate to nonaccidentalness characteristics (Witkin and Tenenbaum, 1983; Lowe, 1986) of nature.

These ill-posed, inverse problems yield a solution through minimization techniques. If the algorithm performing the task is tractable, then perhaps efficiency is only an implementation detail. If the task is an intractable one, however, as vision seems to be, complexity satisfaction is not simply a detail to contend with during implementation. Complexity satisfaction provides a major source of constraints on the problem solution (Tsotsos, 1987). To deal with such problems, Ullman (1979a) suggested the use of distributive computation for constrained optimization problems by networks of locally interconnected simple processors. Such networks are the equivalent of today's parallel distributed processing (PDP) and neural networks (NN) models.

Methods that deal with ill-posed problems are called regularization techniques and usually use the Lagrange multiplier method. We will be able to compare regularization and PDP models in deriving the optical flow of an image or solving the stereo problem in Chapter 5, which is dedicated to the recovery of intrinsic images.

Regularization concerns itself with the optimization of a function f of n variables, x_1 through x_n. The necessary condition for f to reach an extremum assuming the n variables being independent is given as $\partial f / \partial x_i = 0$ for $i = 1, \ldots, n$. Should some of the n variables be dependent and their dependency given as $\psi_\alpha(x_1, x_2, \ldots, x_n)$ $\alpha = 1, 2, \ldots, m < n$, then optimizing f amounts to optimizing

$$\hat{f} = f(x_1, x_2, \ldots, x_n) + \sum_{\alpha=1}^{m} \beta_\alpha \psi_\alpha(x_1, x_2, \ldots, x_n), \qquad (4\text{-}2)$$

where β_α are the Lagrange multipliers associated with the m dependency constraints given previously.

An interesting application of this technique to data compression, is discussed by Ahmed and Rao (1975). The application is related to the Karhunen–Loeve transform (KLT) and to the variance criterion. (Such issues, as we will see later in Chapter 8, are essential to neural models of learning based on statistical analysis (Anderson *et al.* 1977).) The problem is to search for an orthogonal transform that yields efficient signal representation and is also optimal in the minimum squared-error (MSE) sense. Assume T to be an orthogonal transform, that its transpose is $T' = [\phi_1, \phi_2, \ldots, \phi_N]$, and that the base vectors are such that $\phi_i' \phi_j = \delta_{ij}$ (Kroenecker symbol). For each vector X one can obtain the corresponding Y as $Y = TX$. Clearly $T'T = I$, so $X = T'Y$ or $X = \sum_{i=1}^{N} y_i \phi_i$. The goal is that of retaining a subset $\{y_i\}_{i=1}^{M}$ and yet obtain a good estimate of X. One could neglect the $(N - M)$ terms by choosing preselected constants b_i in some optimal way where the corresponding error $\varepsilon(M) = \sum_{i=M+1}^{N} E\{(y_i - b_i)^2\}$ is minimized, and E is the expectation operator. The optimal b_i are obtained from

$\frac{\partial}{\partial b_i}E\{(y_i - b_i)^2\} = 0$ as $b_i = E\{y_i\} = \phi_i'E\{X\}$, and the error that results is $\varepsilon(M) = \Sigma_{i=M+1}^{N}\phi_i'\,\mathrm{COV}(X)\phi_i$, where $\mathrm{COV}(X)$ is the covariance matrix of X. To obtain the optimum transform T, one must not only minimize $\varepsilon(M)$ with respect to ϕ_i but also obey the constraint $\phi_i'\phi_i = 1$. Using Lagrange multipliers, the new function to be minimized is given as $\hat{\varepsilon}(M) = \varepsilon(M) + \sum_{i=M+1}^{N}\beta_i[\phi_i'\phi_i - 1]$. The solution to the restated optimization problem yields ϕ_i as the eigenvectors of the covariance matrix $\mathrm{COV}(X)$ and β_i as the corresponding eigenvalues λ_i. The minimum mean-square error is then given as the sum of the eigenvalues λ_i for $i = M+1, \ldots, N$, and the optimal transform T is known as the Karhunen–Loeve transform. KLT is related to factor or principal component analysis, and it is theoptimal transform for signal representation with respect to the MSE criterion. Furthermore, it is easy to show that the transform vector components y_i are uncorrelated. Another example of data compression, yielding the coefficients of the 2D Gabor transform, is treated according to the PDP paradigm in Section 4.1.6. Contrast the two approaches and notice the possible advantages of distributed computation for representation and processing.

The PDP model takes a similar approach to the regularization problem. The approach uses constrained differential optimization, and the constraints restrict the space of feasible solutions. Such constraints act as penalty factors and are crucial to optimization—violating them is costly. Applications of such concepts are shown in Section 4.1.4., while here we discuss their mathematical foundations. As we remarked earlier, optimization methods of the PDP penalty type are not necessarily guaranteed to continuously satisfy the constraints precisely. Platt and Barr (1988) suggest the basic differential multiplier method (BDMM), which satisfies the constraints by using neurons that estimate the Lagrange multipliers.

The Lagrange multiplier method (LMM), discussed earlier, is conceptually equivalent to the PDP penalty approach. The PDP penalty approach needs to set the constraints' strengths or relevance. If the strengths are too small, then the system finds a deep local minimum, but does not fulfill all of the constraints. If the strengths are too large, then the system quickly fulfills the constraints, but gets stuck in a poor local minimum. The Lagrange multiplier method, on the other hand, has a natural way of finding those strengths that correspond to the β_α in Eq. (4-2). Furthermore, the generic LMM seeks to optimize $E(\mathbf{x}) = f(\mathbf{x}) + \lambda g(\mathbf{x})$, where λ are the Lagrange multipliers, $g(\mathbf{x})$ a subspace of the solution state space, and the corresponding unconstrained extremum problem yields $\nabla E(\mathbf{x}) = \nabla f(\mathbf{x}) + \lambda \nabla g(\mathbf{x})$, i.e., $-\lambda$ is the constant of proportionality between ∇f and ∇g, which are also collinear. Note that λ is a variable, and that one takes the derivative of $E(\mathbf{x})$ with respect to λ and sets it equal to zero, i.e., one obtains an auxiliary equation, $\dot{\lambda} = -g(\mathbf{x})$. Gradient descent techniques are inappropriate for the LMM, because there are critical points that can take the form of saddle points. The energy $E(\mathbf{x})$ can be simply decreased by setting $\lambda = \pm\infty$, while keeping $\mathbf{x}$ unchanged. Thus, critical points are not necessarily attractors, and gradient descent does not necessarily converge.

The BDMM performs gradient ascent rather than descent by using an auxiliary equation of the form $\dot{\lambda} = +g(\mathbf{x})$. Platt and Barr (1988) show how and why the BDMM gradually fulfills the constraints. BDMM always converges for a special class of constrained optimization, for whom $f(\mathbf{x})$ is quadratic, $(\partial^2 f)/(\partial x_i \partial x_j)$ is positive definite for all $\mathbf{x}$ and λ, $g(\mathbf{x})$ is piecewise linear continuous, and $(\partial^2 g)/(\partial x_i \partial x_j) = 0$. .

The equations corresponding to gradient ascent can be written as

$$\dot{x}_i = -\frac{\partial f}{\partial x_i} - \lambda \frac{\partial g}{\partial x_i}$$
$$\dot{\lambda} = +g(x),$$
(4-3)

where "$\dot{x}_i$" is the first derivative of x_i. Using second-order differentials, the equation can be rewritten as

$$\ddot{x}_i + \sum_j \left(\frac{\partial^2 f}{\partial x_i \partial x_j} + \lambda \frac{\partial^2 g}{\partial x_i \partial x_j} \right) \dot{x}_j + g \frac{\partial g}{\partial x_i} = 0.$$
(4-4)

The last equation however, describes a damped system with damping matrix A_{ij}, where

$$A_{ii} = \frac{\partial^2 f}{\partial x_i \partial x_j} + \lambda \frac{\partial^2 g}{\partial x_i \partial x_j}$$
(4-5a)

and an internal force $g(\partial g)/(\partial x_i)$, which is the derivative of the internal energy

$$U = \tfrac{1}{2}(g(\mathbf{x}))^2.$$
(4-5b)

The total energy of the system is the sum of kinetic and potential energies, i.e.,

$$E = T + U = \sum_i \tfrac{1}{2}(\dot{x}_i)^2 + \tfrac{1}{2}(g(\mathbf{x}))^2.$$
(4-6)

If the total energy is decreasing with time and the state remains bounded, the system gradually loses energy and settles down into a constrained extremum of the original optimization problem. The time derivative of the energy E is

$$\dot{E} = \frac{\partial E}{\partial t} = -\sum_{i,j} \dot{x}_i A_{ij} \dot{x}_j. \tag{4-7}$$

The BDMM approach for quadratic programming assumes a positive definite damping matrix A_{ij}. Consequently, $\dot{E} < 0$, and the system eventually converges. The BDMM can easily accommodate penalty terms $E_p = (c/2)(g(\mathbf{x}))^2$, and then a corresponding change in the first equation of the system given by Eq. (4-3) yields $\dot{x}_i = -\partial f/\partial x_i - \lambda(\partial g/\partial x_i) - cg(\partial g/\partial x_i)$. We conceptually refer back to the BDMM physical analogy in the next section and also in Section 6.1. when we discuss snakes and thin plates.

4.1.3.   Cohen–Grossberg Stability Model for Competitive Neural Networks

We began by assuming that an ANS, as shown in Fig. 4.1., has symmetric synaptic interconnections $C = (c_{ik})$, and that its dynamics are governed by the differential Eq. (4-1). An energy function $E(\mathbf{X})$, where $\mathbf{X} = (x_1, x_2, \ldots, x_n)$ can then be defined as:

$$E(\mathbf{X}) = -\sum_{i=1}^{n} \int_0^{x_i} \theta_i(\xi_i) f_i'(\xi_i) d\xi_i + \frac{1}{2} \sum_{j,k=1}^{n} c_{jk} f_j(x_j) f_k(x_k). \tag{4-8}$$

This energy function is of the Lyapunov type, because if $a_i(x_i) \geq 0$, and $f_i(x_i)$ is strictly monotonic non-decreasing, i.e., $f_i'(x_i) \geq 0$, then $dE/dt \leq 0$. Specifically,

$$\frac{dE(\mathbf{X})}{dt} = -\sum_{i=1}^{n} \theta_i(x_i) f_i'(x_i) \frac{dx_i}{dt}$$

$$+ \sum_{i,k=1}^{n} c_{ik} f_i(x_i) f_k'(x_k) \frac{dx_i}{dt}$$

$$= -\sum_{i=1}^{n} a_i(x_i) f_i'(x_i) \left[\theta_i(x_i) - \sum_{k=1}^{n} c_{ik} f_k(x_k) \right]^2. \tag{4-9}$$

Cohen and Grossberg (1983) have shown that the points $\mathbf{X}$ characterized by $dE(\mathbf{X})/dt = 0$ are local minima and that the equilibrium is reached for

$$\theta_i(x_i) = \sum_{k=1}^{n} c_{ik} f_k(x_k) \tag{4-10}$$

when C is positive definite. Basically, the case considered is that of nonconvex minimization characteristic to n nonlinear equations in n variables.

4.1.4. Additive Model

The differential equations governing the dynamics of a given physical system were considered by Cohen and Grossberg (1983) and described by Hopfield (1984) as

$$\frac{du_i(t)}{dt} = -\frac{1}{R_i C_i} u_i(t) + \sum_{j=1}^{n} T_{ij} f_j(u_j) - I_i, \tag{4-11}$$

where $V_j = f_j(u_j)$ is the gain in the input-output (sigmoid) relation, and the energy function E is given by

$$E = -\frac{1}{2} \sum_{i,j=1}^{n} T_{ij} V_i V_j$$
$$+ \sum \left(\frac{1}{R_i C_i}\right) \int_0^{V_i} f_i^{-1}(V) dV + \sum I_i V_i \tag{4-12}$$

for $1/R_i = 1/\rho_i + \sum_{j=1}^{n} |T_{ij}|$. Note that $du_i(t)/dt = -\partial E/\partial V_i$, and that the previous equations characterize an RC network; neurons are modeled as analog amplifiers, the weights T_{ij} correspond to the conductance between amplifier i and j, C_i is the capacitance, and ρ_i is the resistance connected to the input of the ith neuron. For steep sigmoid functions, the energy E given in Eq (4-12) can be approximated by an energy function E_1 given as

$$E_1 = -\frac{1}{2} \sum_{i,j-1} T_{ij} V_i V_j + \sum I_i V_i, \tag{4-13}$$

where n is the number of neurons in the network, V_i is the activity or firing rate of neuron i, and T_{ij} are the interactions between neurons i and j. I_i, known as the bias, is the activity threshold for neuron i. The elements of the connection matrix $T_{ij} = T_{ji} = -\partial^2 E_1/(\partial V_i \partial V_j)$ are completely determined from E_1. The term dropped in approximating E of

Eq. (4-12) with E_1 of Eq. (4-13) corresponds to a leak or self-decay. The network's energy is again Lyapunov, and its minima are the network's steady states. The state of the network, i.e., the firing rates or activities of neurons, through interaction with each other, change with time but eventually the network settles into a steady state where neuronal activities cease to change.

Note that the additive model is a particular instantiation of the Cohen and Grossberg model presented in Section 4.1.3. Specifically, Carpenter *et al.* (1987) showed how the particular instantiation can be derived. The additive model in Eq. (4-11) results from the more general case given in Eq. (4-1), repeated next for easy reference as

$$\frac{du_i}{dt} = a_i(u_i)\left[\theta_i(u_i) - \sum_{k=1}^{n} c_{ik} f_k(u_k)\right]$$

when $a_i(u_i) = 1$ or $a_i(u_i) = C_i^{-1}$ if we substitute $\tau_i = R_i C_i$ in Eq. (4-11) for R_i and adjust the other terms accordingly as $\theta_i(u_i) = -(1/R_i)u_i + I_i$ and $c_{ij} = -T_{ij}$. Note that for $I_i = 0$ the additive model reduces to the Brain-State-in-a-Box model (BSM) (Anderson *et al.*, 1977), which is discussed in Section 8.4.5. The Boltzmann machine (Ackley *et al.*, 1985) is also an additive model with symmetric coefficients and is regulated by simulated annealing (see Section 4.1.5.).

There are two distinct additive models, the analog type and the digital type which correspond to Eqs. (4-11) and (4-13), respectively. For the digital case, $V_i = 1$ or 0, while for the analog case, $0 \le V_i \le 1$. The digital model allows 2^n states for n-neuron networks, and the corresponding computations proceed by randomly jumping (searching) among the corners of the unit hypercube in the n-dimensional space. The analog model, however, operates within the entire volume of the hypercube and thus performs a more deliberate search. The dynamics of the networks are given as

(a) *Digital/Discrete Network*

Input to neuron i at time $(t + 1)$ is

$$u_i(t + 1) = \sum_{j=1}^{n} T_{ij} V_j(t) - I_i, \qquad (4\text{-}14a)$$

and the output of neuron i is

$$V_i(t) = \begin{cases} 0 & \text{if } u_i(t) \leq 0 \\ 1 & \text{if } u_i(t) > 0. \end{cases} \tag{4-14b}$$

The corresponding BSB model implements the binary McCulloch–Pitts model and is described by Hopfield (1982). Content-addressable memories (CAM), implemented assuming $I_i = 0$ and using the energy function given in Eq. (4-13), are known also as crossbar associative networks (CAN).

(b) Analog Networks

Input to neuron i is determined by

$$\frac{du_i}{dt} = -\frac{u_i}{\tau} + \sum_{j=1}^{n} T_{ij} V_j - I_i, \tag{4-15a}$$

where τ is the characteristic decay time of neurons and may be set to 1. The output is then given as (a sigmoid-like function)

$$V_i = \frac{1}{2}\left(1 + \tanh\frac{u_i}{u_0}\right) = \frac{1}{1 + e^{-2u_i/u_o}}, \tag{4-15b}$$

where u_o determines the steepness of the gain function.

The analog network has two advantages. First, the function E is Lyapunov, while for a digital network it is not. Second, deeper minima of energy have generally larger basins of attraction. A randomly selected starting state has a higher probability of falling inside the basin of attraction of a deeper minimum. Kamgar-Parsi and Kamgar-Parsi (1987a) showed how one could take advantage of the nice properties of analog networks and still not have to solve differential equations. Define a network with the following dynamics (D):

$$u_i = \sum_{j=1}^{n} T_{ij} V_j(t) - I_i$$

$$\Delta V_i = g(u_i) = \alpha \tanh\frac{u_i}{u_o} \quad \text{with} \quad |\Delta V_i| < 2\left|\frac{u_i}{T_{ii}}\right| \quad \text{if} \quad T_{ii} < 0$$

$$V_i(t+1) = V_i(t) + \Delta V_i \quad \text{with} \quad 0 \leq V_i \leq 1, \tag{4-16}$$

where α and u_o are parameters. The function $g(u_i)$ may be chosen to be any (monotonically increasing) odd function, to ensure that the energy function is Lyapunov. Since the dynamics are asynchronous, all the neurons are updated only once in every iteration according to a randomly generated scheme.

It might be helpful to grasp first the analogy between finding a solution to a given optimization problem and setting an appropriate Lyapunov function corresponding to the additive model. The traveling salesman problem (TSP) is a classic example of an NP-*complete* problem, which was introduced in Chapter 1. The solution to an n-city TSP problem consists of an ordered visiting list of n cities. Mapping the TSP into the corresponding crossbar associative network (Hopfield and Tank, 1986) requires a $n \times n$ square array C, where entry $c_{ij} = 1$ *iff* (if and only if) city j is the ith in order to be visited. Clearly, if $c_{i,j_1} = 1$ and $c_{i-1,j_2} = 1$ then the added cost to the tour is the weight (distance) corresponding to traveling at instance i from cities j_2 to j_1. There are $(n-1)!/2n$ distinct alternatives for closed TSP paths, and one attempts to find the optimal one. Finding that part of the (Lyapunov) energy function corresponding to minimizing the cost of the tour is easy; it is the sum of the distances between the cities on one of their *legal* permutations, where the *legal* part is enforced through the added use of penalty-like constraints. The energy E_{II} corresponding to the legal aspect can be built as

$$E_{II} = \frac{\alpha}{2} \sum_{i_2 \neq i_1} \sum_{i_1} \sum_{j} c_{i_1,j} c_{i_2,j} + \frac{\beta}{2} \sum_{j_2 \neq j_1} \sum_{j_1} \sum_{i} c_{i,j_1} c_{i,j_2} + \frac{\gamma}{2} \left(\sum_{j} \sum_{i} c_{ij} - n \right)^2.$$

The first sum is zero *iff* each city column j contains no more than one "1" (i.e., is visited no more than once), the second sum is zero *iff* each row instance in tour contains no more than one "1" (i.e., no two cities are visited at the same time), and the last sum is zero *iff* there are n entries of "1" (i.e., n cities are being visited). The three constraints enforced via E_{II} allow only those paths that are legal for the TSP. The energy E_I corresponding to the attempt to minimize the length of the tour is given as

$$E_T = \frac{\delta}{2} \sum_{j_1 \neq j_2} \sum_{j_2} \sum_{i} d_{j_1 j_2} c_{i,j_1} (c_{i+1,j_2} + c_{i-1,j_2}).$$

The appropriate energy function for the TSP network is then given as $E = E_I + E_{II}$. The shortest path corresponds to the lowest energy state. The optimization process is implemented by performing nonconvex minimization, and one can get stuck in a local minima.

The previous method of solving the TSP does not guarantee finding the global minima. The ways to escape local minima using simulated annealing are discussed in Section 4.1.5. There are many fine details concerning the Hopfield and Tank (HT) solution to the TSP, which Wilson and Pawley (1988) discuss. They deal with algorithm stability, the possibility of finding illegal minima tours, and that a particular simulation of the system dynamics might not converge. Finally, the BDMM optimization approach discussed in Section 4.1.2. applies as well to the TSP and is conceptually modeled by Platt and Barr (1988) as elastic snakes (see Section 6.1. on snakes). Specifically, the snake minimizes the length $\Sigma_i \ [(x_{i+1} - x_i)^2 + (y_{i+1} - y_i)^2]$ subject to the constraint that the snake must lie on the cities, i.e., $k(x^* - x_c) = k(y^* - y_c) = 0$, where (x^*, y^*) are city coordinates, (x_c, y_c) is the closest snake point to the city, and k is the constraint strength. The minimization is quadratic, the damping positive definite, and the physical system can converge to a state where the constraints are fulfilled, and a legal tour is obtained. The tour lengths are only slightly longer than those obtained using simulated annealing, and the time needed scales as $n^{1.6}$, where n is the number of cities.

Clustering, another significant problem for computational vision can be solved using a refined Hopfield and Tank model (Kamgar-Parsi et al. (1990)). They address the fitting of (hyper) planes to high dimensional data points, a problem of exponential complexity. The fitting of planes is relevant when one needs to reconstruct a planar world from range data or to recognize point patterns in an image. The problem can be cast as an optimization problem, and PDP models yield good but not necessarily optimal solutions. The dynamics, given in Eq. (4-16), enjoy the advantages of analog networks. Such an approach finds the best K lines that fit a set of n points (where best means to minimize the sum of the residuals). As was the case for the TSP, the energy function has two parts: one enforces the constraint that a point belongs to one and only one line (or cluster); and the other is a cost term that we want to minimize. Assume a matrix of entries (neurons) $V_{i\alpha}$, where indexes i and α refer to line and point, respectively. Then, the energy function E is given as the sum of the constraint part E_{II} and the cost term E_I

$$E_{II} = \frac{A}{2} \sum_{\alpha=1}^{N} \sum_{i=1}^{K} \sum_{j \neq i}^{K} V_{i\alpha} V_{j\alpha} + \frac{B}{2} \sum_{\alpha=1}^{N} \left(\sum_{i=1}^{K} V_{i\alpha} - 1 \right)^2$$

$$E_I = \frac{C}{2} \sum_{i=1}^{K} \sum_{\alpha=1}^{N} R_{i\alpha} V_{i\alpha}^2,$$

where the residual $R_{i\alpha}$ is given as

$$R_{i\alpha} = \frac{(y_\alpha - b_i - s_i x_\alpha)^2}{1 + s_i^2}.$$

for (x_α, y_α) the coordinates of point α, and b_i and s_i are the y-intercept and the slope of the line, respectively. Note that the syntax of a line enforced through the E_{II} term implies that each point lies on exactly one line. The collective signal received by each neuron $V_{i\alpha}$ is

$$u_{i\alpha} = -A \sum_{j \neq i}^{K} V_{j\alpha} - B \left(\sum_{j=1}^{K} V_{j\alpha} - 1 \right) - CR_{i\alpha} V_{i\alpha} - I_{i\alpha}.$$

where $I_{i\alpha}$ is the threshold (bias) of the neurons. The elements of the CAN are given by

$$T_{i\alpha, j\beta} = -\frac{\partial^2 E}{\partial V_{i\alpha} \partial V_{j\beta}} = \frac{\partial u_{i\alpha}}{\partial V_{j\beta}}$$

$$= -A(1 - \delta_{ij})\delta_{\alpha\beta} - B\delta_{\alpha\beta} - CR_{i\alpha}\delta_{ij}\delta_{\alpha\beta}.$$

Since all the neurons are identical we take $I_{i\alpha} = I$. Assuming that the solution is a stable state (fixed-point) of the network given by the dynamics D (Eqs. 4-15 and 4-16) i.e., $V_{i\alpha}(t + 1) = V_{i\alpha}(t)$ $\forall i, \alpha$, then one can show that

$$CR_{i\alpha} \leq -I \leq A \qquad \forall i, \alpha,$$

and the parameter set (A, B, C) could be given as $(1, 1, 1/\bar{R})$, where the C-term provides the force that drives the network toward deeper minima and $\bar{R}$ is the average residual. One can further determine the number of lines (clusters) that fit the data by comparing the average residual against some given tolerance for each K. Simulations show that the method described is fault-tolerant because it is less sensitive to noise than traditional methods, such as K-means. Some neural networks methods are merely fault-tolerant distributive computations implementing conceptually similar (non)parametric estimation methods. Furthermore, one could extend the method by including curves of different shapes through appropriate energy components (C-terms).

To decide whether the additive model of neural networks is suitable for a computationally difficult problem, one has to consider how good the solutions are and what the success rate for finding valid solutions is (Kamgar-Parsi *et al.*, 1989). When the analog network finds a solution, the quality is generally very good. A specific solution, of course, depends on the chosen initial state. Overwhelming empirical evidence (based on Monte Carlo estimates) suggests that deeper energy function minima of an analog network have generally larger basins of attraction. Consequently, a randomly selected initial state has a higher chance of falling inside a deep minimum basin and finding a good solution. Additive models are complicated dynamical systems, and there are still no theoretical estimates for the sizes of their basins. Although one cannot give the probability of finding the best solution, analog networks greatly enhance the search characteristic of TSP and clustering problems.

The success rate appears to be largely problem dependent, and TSP yields a rather modest success rate. The solution scales poorly as the problem size increases, which suggests that neural nets may be unsuitable for solving TSP. For clustering, however, the success rate is very high, nearly 100%, and the solution scales smoothly as the number of points increases.

Matrix representations of these two problems drive the higher success rate of clustering problems in finding valid solutions. A given tour in the n-city TSP problem is represented by a $n \times n$ permutation matrix with constraints on the rows and columns, which is a difficult syntax to satisfy. A given partitioning of n points among K clusters in the clustering problem is represented by a $K \times n$ matrix with constraints on only the columns, which is a much easier syntax to satisfy. Furthermore, the scaling of success rate with problem size may be related to the density of "on" neurons in the syntax matrix, i.e., the ratio of 1s to the total number of elements in the syntax matrix. For TSP this ratio is $n/n^2 = 1/n$, which scales inversely with the size of the problem n. For the clustering problem, this ratio is $n/Kn = 1/K$, which scales inversely with the number of clusters K. Since in most problems K is small and much less than n, our results confirmed that increasing the number of points should not adversely affect scaling.

Based on these observations, it appears that problems suitable for an additive neural net approach are those whose valid solutions can be represented with few constraints, i.e., easy syntax.

4.1.5. *Nonconvex Optimization and Simulated Annealing*

The additive model implements nonconvex minimization problems through minimizing Lyapunov functions. Gradient-descent-like techniques only ensure that a local rather than a global minimum is reached. To avoid entrapment in a local minimum, one needs methods to escape from a given basin of attraction by jumping over the wall separating the two minima. Such methods, based on the work of Metropolis *et al.* (1953), were reintroduced by Kirkpatrick *et al.* (1983). Basically, simulated annealing methods generate perturbations on the system state and decide if the newly generated state is acceptable as the next point in the search space. The Metropolis method implements a Monte-Carlo approach to simulate the equilibrium state for a given collection of particles (neurons) at any given temperature T by analogy to the corresponding Boltzmann distribution. The relative probability of any configuration of particles is given as $P = \exp[-E(s)/(k_B T)]$, where $E(s)$ is the energy associated with configuration s, and k_B is the Boltzmann constant. The lower the temperature T, the more likely the corresponding configuration s. At each step, a state s_1 is chosen at random and subjected to a random perturbation Δs resulting in a new configuration s_2. If $\Delta E = E(s_2) - E(s_1) < 0$, then the new configuration becomes the starting point for the next adjustment.

Unlike standard iterative search techniques for the Lyapunov-type functions, the Metropolis method can accept higher energy state configurations with an escape probability $P_e^B = \exp[-\Delta E/(k_B T)]$, where P_e^B is compared to a randomly drawn number in the range $[0, 1]$. The acceptance of higher energy states allows the system to escape entrapment in local minima and also allows a chance for neighboring basins of attraction to be visited.

Simulated annealing schedules vary as a function of the cooling process. The Metropolis method reduces to standard gradient descent-like methods as $T \to O$ and $P_e^B \to 0$. In analogy to physical systems, low energy states (configurations) dominate as the system cools. An exponentially decreasing temperature schedule as a function of time (i.e., the temperature drops quickly in the beginning and then is lowered gradually) was suggested by Kirkpatrick *et al.* (1983). Geman and Geman (1984) proved that if the cooling schedule is logarithmic, then global minimum is reached as time goes to infinity. Note that cheaper, Monte Carlo (heuristic interchange) perturbations methods for finding global minimum can still be more efficient (Nahar *et al.*, 1984) than simulated annealing.

Simulated annealing methods are characterized by two functions: temperature schedule to cool off; and probability distributions to extricate the system from local minimum. The model discussed so far, labeled as classical simulated annealing (CSA), is characterized by a cooling schedule inversely proportional to the logarithmic function of time, i.e., $T(t)/T_o = 1/\log(1 + t)$ and a corresponding Boltmann distribution, where T_o is a sufficiently high initial temperature. In an alternate model of fast simulated annealing (FSA), suggested by Szu and Hartley (1987), the cooling schedule of the FSA algorithm is inversely linear in time, i.e., $T(t)/T_o = 1/(1 + t)$, and is much faster than CSA. The corresponding FSA probability distribution is the n-dimensional Cauchy distribution given as $P_e^c = T(t)/[T(t)^2 + \Delta E]^{(n + 1)/2}$. The Cauchy distribution allows for occasional long jumps to speedup state generation and to extricate the system from entrapment in local minima. As such, FSA proved superior to CSA when the initial state was far from global minimum. Note that long jumps characteristic of far away connections are observed in biological networks and proved to be useful artifacts for processes such as relaxation, which we discuss in Section 4.2.2.

Many visual tasks are related to visual reconstruction and can be cast as optimization problems. The optimization is of some energy function related to the quality of the reconstruction. Regularization (Section 4.1.2.), the Cohen and Grossberg (Section 4.1.3.) and additive (Section 4.1.4.) models, Markov random field (MRF) (Section 4.2.3.) lead naturally to optimization. As we have seen so far, the optimization involved is of the nonconvex type and traditional gradient descent techniques can be trapped in finding local rather than global optimal solutions. Dynamic programming techniques, such as those discussed in Section 3.2.4., could in principle seek global optimal solutions and find exact solutions, but their time complexity is exponential. Simulated annealing, as was introduced in this section, is characteristic of stochastic algorithms targeted to work on nonconvex optimization. Such algorithms find global optimal solutions using random perturbations. Graduated nonconvexity (GNC) algorithms (Blake and Zisserman, 1987) are deterministic and offer an alternative to the stochastic algorithms mentioned earlier. GNC algorithms work by approximating the (nonconvex) energy function as a convex function using continuation methods.

Blake (1989) reports on a comparison of the efficiency of deterministic and stochastic algorithms for 1D visual reconstruction tasks. The comparison between the deterministic GNC algorithm and three stochastic algorithms (including those of Metropolis and Geman and Geman) showed the latter three to be outstripped both in computational efficiency and in their problem-solving power for dealing with moderately high levels of noise. The comparison led Blake to argue that "the potential of MRFs as a general vehicle for specifying and integrating visual tasks must be regarded as highly questionable." The results reported were for 1D visual reconstruction, but arguments were made for the conjecture that they will apply to 2D visual reconstruction as well.

4.1.6.   Gabor Transform through PDP Minimization

Demonstrating the interrelationship between distributed representations, covered in Chapter 2, and distributed processing, introduced in this chapter, is one of the major thrusts of this book. We repeatedly emphasize the misguided approach of considering only the algorithmic aspect of the computational process, while neglecting the representational aspect. A case in point was the huge effort (mis)spent on syntactical pattern recognition methods at the raw pixel level.

In this section, we review Daugman's (1988a) work on deriving the coefficients of the 2D Gabor transform (cojoint s/sf representations introduced in Sections 2.2.4. and 2.3.7. and suggested as being characteristic of early low-level image representations) using minimization techniques similar to those introduced in the preceding sections. The work described next can be cast within a data compression framework to capture the image structure in the original picture and to allow for its later reconstruction. Earlier attempts at low-level image representation based on zero crossings failed to capture the true image structure and the reconstruction aspect (Daugman, 1985). (See Section 2.2.4.) The cojoint 2D Gabor transform extracts "locally windowed 2D spectral information," and if implemented using PDP models could prove the strong interaction between distributed representations and processing at very early stages of visual processing.

The 2D Gabor transform is given in terms of a projection (elementary) vector base that is not orthogonal. As such, computing the coefficients of the transform, i.e., performing data compression, is quite difficult and expensive. A neural network architecture could prove its usefulness here by attempting to find the optimal coefficients as an optimization problem.

Assume that a 2D image function is given as $f(x, y)$ and that we seek expansion coefficients $\{a_i\}$ in terms of projection. vectors $\{g_i(x, y)\}$, which might be linearly dependent. If, as it is in our case, $\{g_i(x, y)\}$ do not form either a complete or orthogonal set, then the reconstructed image $R(x, y)$ is in general inexact, and the desired set of coefficients $\{a_i\}$ must be determined by an optimization criterion, such as the MSE of the difference vector

$$E = ||f(x, y) - R(x, y)||^2 = \sum_{x,y} ([f(x, y) - R(x, y)])^2, \qquad (4\text{-}17)$$

where $R(x, y) = \sum_i^n a_i g_i(x, y)$. The norm E will be minimized iff ($\leftrightarrow$) $\forall\, i$, $i = 1, \ldots, n$, $\partial E/\partial a_i = 0$, where

$$\frac{\partial E}{\partial a_i} = -2 \sum_{x,y} (f(x, y) g_i(x, y)) + \sum_{x,y} \left[2\left(\sum_k a_k g_k(x, y) \right) \cdot g_i(x, y) \right] = 0,$$

or, equivalently,

$$\sum_{x,y} f(x, y) g_i(x, y) = \sum_{x,y} \left[\left(\sum_k a_k g_k(x, y) \right) g_i(x, y) \right]. \qquad (4\text{-}18a)$$

Even for the nonorthogonal case, the previous system of n equations could still be solved; however, the general matrix solution would require numerous floating-point multiplications. The suggested solution yields the desired set $\{a_i\}$ by gradient descent along the $E(a_i)$ energy function surface, i.e., by implementing an iterative optimization scheme. A three-layered neural network for finding the optimal coefficients is suggested by Daugman (1988a). Equilibrium is reached for those stable states that minimize E and yield the transform coefficients in the middle layer. The weight adjustment for the coefficient layer is the difference between the feed forward of the first layer and feedback of the third layer and is given as

$$\Delta_i = \sum_{x,y} g_i(x, y) f(x, y) - \sum_{x,y} \left[\left(\sum_k a_k g_k(x, y) \right) g_i(x, y) \right]. \qquad \text{(4-18b)}$$

and the iterative rule for adjusting the coefficients is $a_i \leftarrow a_i + \Delta_i$, for $\Delta_i = -\frac{1}{2} \partial E / \partial a_i$. Note that the minus sign implies that the weight adjustment is always downhill on $E(a_i)$ and that consequently, $\forall_i$, $\Delta_i = 0 \leftrightarrow \partial E / \partial a_i = 0$. As in the Lyapunov functions case studied in the additive model context, the stable states yield the optimal set of coefficients.

The 2D Gabor transform can be specified in terms of the space-domain impulse response function $g(x, y)$ and its corresponding 2D Fourier transform $G(u, v)$

$$g(x, y) = \exp(-\pi[(x - x_o)^2 \alpha^2 + (y - y_o)^2 \beta^2])$$

$$\times \exp(-j2\pi[u_o(x - x_o) + v_o(y - y_o)]) \qquad \text{(4-19a)}$$

$$G(u, v) = \exp\left(-\pi \left[\frac{(u - u_o)^2}{\alpha^2} + \frac{(v - v_o)^2}{\beta_2} \right] \right)$$

$$\times \exp(-j2\pi[x_o(u - u_o) + y_o(v - v_o)]). \qquad \text{(4-19b)}$$

For Daugman's network, the 2D joint Gabor elementary functions (e.g., projection vectors) are given as

$$g_{mnrs}(x, y) = \exp(-\pi \alpha^2[(x - mM)^2 + (y - nN)^2])$$

$$\times \exp\left(-j2\pi \left[r\frac{x}{M} + s\frac{y}{N} \right] \right),$$

where invariant Gaussian windows are positioned on (fully overlapping) Cartesian lattice locations $\{x_m, y_n\} = \{mM, nN\}$ and are modulated by corresponding complex exponentials sampled over a lattice of 2D spatial frequencies $\{u_r, v_s\}$ given as $\{u_r, v_s\} = \{r/M, s/N\}$ when integer increments of r and s span $\{-(M - 1)/2, (M - 1)/2\}$ and $\{-(N - 1)/2, (N - 1)/2\}$, respectively. The desired set of coefficients is then obtained as $\{a_{mnrs}\}$ with conjugate symmetry due to the input being real. Note also the uncertainty principle as expressed in the sets $\{x_m, y_m\}$ and $\{u_r, v_s\}$, respectively. As expected, the fewer the samples in one domain, the larger the number in the dual domain; in other words, good (representational) fidelity in one domain comes at the expense of that corresponding to the cojoint domain. Implementing the neural network corresponding to the Gabor transform allows a significant rate of data compression (an entropy of 2.55 bits for a $256 \times 256 \times 8$ bit image) and almost exact original image reconstruction. Furthermore, the same Gabor coefficients could be used for segmentation tasks.

The 2D Gabor transform derivation as shown demonstrates the utility of using PDP computation models to derive cojoint spatial/spatial-frequency representations, characteristic of early vision. Interaction between distributed representation and processing is the cornerstone of early vision and its implementation.

4.2. PDP Models and Physical Systems

Analogies between specific PDP models and corresponding physical systems illustrate distributed computations.

4.2.1. Distributed Associative Memory (DAM) and Holography

DAMs described by Kohonen (1987) are analogous to holograms. The underlying behavioral model is Hebbian (classical conditioning) learning through synaptic connections. Specifically, one attempts to relate stimuli $\mathbf{s}_i$ to responses $\mathbf{r}_i$ via a projection matrix M such that

$$M\mathbf{s}_i = \mathbf{r}_i \quad \forall_i. \tag{4-20}$$

Assuming n relationships are to be learned and that the stimulus and response matrices are given by

$$S = [\mathbf{s}_1, \mathbf{s}_2, \ldots, \mathbf{s}_n] \text{ and } R = [\mathbf{r}_1, \mathbf{r}_2, \ldots, \mathbf{r}_n],$$

the Eq. (4-20) can then be rewritten as

$$MS = R, \tag{4-21}$$

and the MSE solution is found by minimizing $\|MS - R\|^2$ and yields

$$M = RS^+, \tag{4-22}$$

where S^+ is the pseudoinverse matrix of S. If $R = S$, the memory is autoassociative; otherwise, M is said to be heteroassociative. Once the optimal memory M has been found, the retrieval (recall) phase, such as that corresponding to holography, becomes operational. Given an unknown stimulus $\mathbf{s}_\alpha$, the recall is obtained as

$$\mathbf{r}_\alpha = M\mathbf{s}_\alpha. \tag{4-23}$$

Recall accuracy depends on several factors, including the orthogonality and dissimilarity of the original training set. If the memorized stimulus vectors are independent, and the unknown stimulus vector $\mathbf{s}$ is one of the memorized vectors $\mathbf{s}_k$, the recalled vector will then be the associated response vector $\mathbf{r}_k$. If the memorized stimulus vectors are dependent, then the vector recalled by one of the memorized stimulus vectors will contain the associated response vector and some *crosstalk* from the other stored response vectors.

Viewed as the weighted sum of the response vectors, recall begins by assigning weights according to how well the unknown stimulus vector matches the memorized stimulus vectors using a least squares classifier. Response vectors are multiplied by the weights and then summed together to build the recalled response vector. The recalled respsonse vector is usually dominated by the memorized response vector closest to the unknown stimulus vector.

The fault-tolerant aspect of the DAM can be described in quantitative and qualitative terms. Assume that there are n associations in the memory and that each of the associated stimulus and response vectors have m elements. This means that the memory matrix has m^2 elements. Also assume that the noise added to each element of a memorized stimulus vector is independent, of zero mean, and variance σ_i^2. The recall from the memory is then

$$\mathbf{r} = M(\mathbf{s}_k + \mathbf{v}_i) = M\mathbf{s}_k + M\mathbf{v}_i = \mathbf{r}_k + \mathbf{v}_o, \tag{4-24}$$

where $\mathbf{v}_i$ and $\mathbf{v}_o$ are the input and output noise vectors, respectively. The ratio of the average output noise variance to the average input noise variance is

$$\frac{\sigma_o^2}{\sigma_i^2} = \frac{1}{n} \, \text{Trace} \, [MM^T]. \tag{4-25a}$$

For the autoassociative case this simplifies to

$$\frac{\sigma_o^2}{\sigma_i^2} = \frac{n}{m}. \tag{4-25b}$$

Thus, when a noisy version of a memorized input vector is applied to memory, recall is improved by a factor corresponding to the ratio of the number of memorized vectors to the number of elements in the vectors. For the heteroassociative memory matrix, a similar formula holds as long as n is less then m (Stiles and Denq, 1985)

$$\frac{\sigma_o^2}{\sigma_i^2} = \frac{1}{m}\text{Tr}[RR^T]\text{Tr}[(S^TS)^{-1}]. \tag{4-25c}$$

Another error-correcting aspect of fault-tolerant behavior is that the memory matrix is the orthogonal projection matrix for the set of stimulus vectors. The noise vector in this m-dimensional space will be projected unto the space spanned by the n memorized vectors. The parts of the noise vectors that are orthogonal to the n-memorized stimulus vectors will be lost, which accounts for the noise reduction in the output recall vector.

Fault tolerance is a by-product of the distributed nature and error-correcting DAM capabilities. By distributing the information, no single memory cell carries a significant portion of the information critical to overall memory performance. Fault tolerance refers to a system's ability to cope with faulty (noisy or possibly variable) inputs. Another relevant constraint is the visual system itself, which is prone to damage over its life span. (Brain cells die all the time, not to mention major lesions.) Fault-tolerant behavior is then facilitated by the distributed aspect of both memory representation and processing as embedded in PDP models. We show in Section 7.2.3. that one can randomly destroy up to 75% of a DAM, and that the system's performance is only slightly affected.

We apply the DAM concept first to derive optimal template edge operators (Meer *et al.*, 1989) and then to obtain the DOG (difference of Gaussians) as adaptive invariant novelty filters (Szu and Messner, 1986).

4.2.1.1. Edge Detection

Edge detection, which seeks local, usually context-free changes in an image's gray level values, is a much researched computer vision problem. Edge detection is known to be an ill-posed problem—small input changes may lead to large output deviations. The problem's ill-posedness, as well as the data's discretness, make arriving at a general solution difficult. A vast amount of literature has been dedicated to edge detection.

A local operator detects edges at every pixel of an image. These operators can be loosely classified into two classes: discrete approximations of differential operators and templates. In what follows, we refer to only template edge operators.

Consider a 3×3 neighborhood on a square lattice. Eight different ideal step edges passing through the central pixel and oriented at multiples of 45 degrees can be defined as

$$s_1 = \begin{bmatrix} 1 & 1 & 1 \\ 1 & 1 & 1 \\ 0 & 0 & 0 \end{bmatrix} \qquad s_2 = \begin{bmatrix} 1 & 1 & 1 \\ 1 & 1 & 0 \\ 1 & 0 & 0 \end{bmatrix} \qquad s_3 = \begin{bmatrix} 1 & 1 & 0 \\ 1 & 1 & 0 \\ 1 & 1 & 0 \end{bmatrix}$$

$$s_4 = \begin{bmatrix} 1 & 0 & 0 \\ 1 & 1 & 0 \\ 1 & 1 & 1 \end{bmatrix} \qquad s_5 = \begin{bmatrix} 0 & 0 & 0 \\ 1 & 1 & 1 \\ 1 & 1 & 1 \end{bmatrix} \qquad s_6 = \begin{bmatrix} 0 & 0 & 1 \\ 0 & 1 & 1 \\ 1 & 1 & 1 \end{bmatrix} \qquad \text{(4-26a)}$$

$$s_7 = \begin{bmatrix} 0 & 1 & 1 \\ 0 & 1 & 1 \\ 0 & 1 & 1 \end{bmatrix} \qquad s_8 = \begin{bmatrix} 1 & 1 & 1 \\ 0 & 1 & 1 \\ 0 & 0 & 1 \end{bmatrix}.$$

In template matching a 3×3 mask corresponds to each one of the stimuli, and the corresponding eight masks are applied at every pixel of the image. The pixel is considered to belong to an edge whenever the response that corresponds to that edge exceeds a prespecified threshold.

The four most widely employed template matching operators associated with $0°$ and $45°$ orientations are given next

$$\text{—"3-level"} \quad \frac{1}{3} \begin{bmatrix} 1 & 1 & 1 \\ 0 & 0 & 0 \\ -1 & -1 & -1 \end{bmatrix} \qquad \frac{1}{3} \begin{bmatrix} 1 & 1 & 0 \\ 1 & 0 & -1 \\ 0 & -1 & -1 \end{bmatrix}$$

$$\text{—"5-level"} \quad \frac{1}{4} \begin{bmatrix} 1 & 2 & 1 \\ 0 & 0 & 0 \\ -1 & -2 & -1 \end{bmatrix} \qquad \frac{1}{4} \begin{bmatrix} 2 & 1 & 0 \\ 1 & 0 & -1 \\ 0 & -1 & -2 \end{bmatrix}$$

$$\text{—"compass"} \quad \frac{1}{5} \begin{bmatrix} 1 & 1 & 1 \\ 1 & -2 & 1 \\ -1 & -1 & -1 \end{bmatrix} \qquad \frac{1}{5} \begin{bmatrix} 1 & 1 & 1 \\ 1 & -2 & -1 \\ 1 & -1 & -1 \end{bmatrix}$$

$$\text{—"Kirsch"} \quad \frac{1}{15} \begin{bmatrix} 5 & 5 & 5 \\ -3 & 0 & -3 \\ -3 & -3 & -3 \end{bmatrix} \qquad \frac{1}{15} \begin{bmatrix} 5 & 5 & -3 \\ 5 & 0 & -3 \\ -3 & -3 & -3 \end{bmatrix}.$$

The complete set of masks for each operator type can be derived by rotating the pairs shown by 90°, 180°, and 270° around the center of the neighborhood. All the preceding masks respond to more than one ideal edge. For example, the "five-level" masks return

$$1, .75, 0, -.75, -1, -.75, 0, .75$$

when applied to the stimulus s_1. While the largest response is correctly given by the first mask, the maximum is indistinct from some of the other responses.

The responses of the masks are proportional to the amplitudes of gray level discontinuities in the 3×3 neighborhood. Establishing a predefined threshold introduces uncertainty. Some weak edges may have a good match but large discontinuities in gray level may poorly resemble an edge. To reduce the artifacts, adaptive thresholding methods are required, which significantly increases computation. To avoid such inconveniences, one can develop a new procedure in which: (a) each mask responds only to one ideal stimulus, and (b) the decision about the pixel is made based on a *confidence measure* of the match to the ideal stimulus. Mathematical results in optimal linear associative mapping can be employed as described earlier.

Consider the 3×3 neighborhood as a nine-dimensional linear vector space. Left-to-right scanning of the neighborhood transforms the two-dimensional patterns into nine-dimensional vectors. Thus,

$$\mathbf{r}_j = \begin{bmatrix} r_j(1) & r_j(2) & r_j(3) \\ r_j(4) & r_j(5) & r_j(6) \\ r_j(7) & r_j(8) & r_j(9) \end{bmatrix}.$$

becomes the column vector

$$\mathbf{r}_j = [r_j(1)\, r_j(2)\, r_j(3)\, r_j(4)\, r_j(5)\, r_j(6)\, r_j(7)\, r_j(8)\, r_j(9)]^T.$$

The Euclidean inner product of two vectors is defined as

$$\langle \mathbf{r}_j, \mathbf{r}_k \rangle = \sum_{i=1}^{9} r_j(i) \cdot r_k(i),$$

and the norm of a vector is given as

$$\|\mathbf{r}_j\| = \left(\sum_{i=1}^{9} [r_j(i)]^2 \right)^{1/2}.$$

The vectors s_j are linearly independent, i.e., for any pair the inner product divided by the norms is less than 1 in absolute value. Their linear independence allows us to associate each $\mathbf{s}_j$ with a unit vector $\mathbf{r}_j$:

$$\mathbf{r}_j = \{\delta_{ij}; i = 1, \ldots, 8\} \quad \text{for } j = 1, \ldots, 8.$$

These unit vectors form an orthonormal basis. To have a complete basis for the nine-dimensional space, the unit vector $\mathbf{r}_9 = [0\,0\,0\,0\,0\,0\,0\,0\,1]^T$ must be added. It will be associated with stimulus $\mathbf{s}_9$, the uniform pattern

$$\mathbf{s}_9 = \begin{bmatrix} 1 & 1 & 1 \\ 1 & 1 & 1 \\ 1 & 1 & 1 \end{bmatrix}. \tag{4-26b}$$

The stimuli $\mathbf{s}_j$, $j = 1, 2, \ldots, 9$, written as column vectors, define the 9×9 stimulus matrix $S = [\mathbf{s}_1\,\mathbf{s}_2\,\mathbf{s}_3\,\mathbf{s}_4\,\mathbf{s}_5\,\mathbf{s}_6\,\mathbf{s}_7\,\mathbf{s}_8\,\mathbf{s}_9]$. Similarly, we define the response matrix $R = [\mathbf{r}_1\,\mathbf{r}_2\,\mathbf{r}_3\,\mathbf{r}_4\,\mathbf{r}_5\,\mathbf{r}_6\,\mathbf{r}_7\,\mathbf{r}_8\,\mathbf{r}_9]$. Note that R is the identity matrix I, but we keep the notation R for the purpose of generality. Association of the stimulus with the response vectors is equivalent to the mapping performed by the 9×9 solution matrix M of the linear equation $R = MS$. Because S has maximum rank, the exact solution is $M = RS^{-1}$ where S^{-1} is the inverse of S. The associative mapping between a specific stimulus $\mathbf{s}_j$ and the desired response $\mathbf{r}_j$ is performed by the row vector $\mathbf{m}_j$ of M, where $M = \{\mathbf{m}_j; j = 1, \ldots, 9\}$. The resulting patterns are the sought after template masks. The following nine masks are obtained when computing M:

$$m_1 = \frac{1}{3}\begin{bmatrix} 1 & -2 & 1 \\ 1 & 1 & 1 \\ -2 & 1 & -2 \end{bmatrix} \quad m_2 = \frac{1}{3}\begin{bmatrix} -2 & 1 & 1 \\ 1 & 1 & -2 \\ 1 & -2 & 1 \end{bmatrix}$$

$$m_3 = \frac{1}{3}\begin{bmatrix} 1 & 1 & -2 \\ -2 & 1 & 1 \\ 1 & 1 & -2 \end{bmatrix} \quad m_4 = \frac{1}{3}\begin{bmatrix} 1 & -2 & 1 \\ 1 & 1 & -2 \\ -2 & 1 & 1 \end{bmatrix}$$

$$m_5 = \frac{1}{3}\begin{bmatrix} -2 & 1 & -2 \\ 1 & 1 & 1 \\ 1 & -2 & 1 \end{bmatrix} \quad m_6 = \frac{1}{3}\begin{bmatrix} 1 & -2 & 1 \\ -2 & 1 & 1 \\ 1 & 1 & -2 \end{bmatrix} \tag{4-27}$$

$$m_7 = \frac{1}{3}\begin{bmatrix} -2 & 1 & 1 \\ 1 & 1 & -2 \\ -2 & 1 & 1 \end{bmatrix} \quad m_8 = \frac{1}{3}\begin{bmatrix} 1 & 1 & -2 \\ -2 & 1 & 1 \\ 1 & -2 & 1 \end{bmatrix}$$

$$m_9 = \frac{1}{3}\begin{bmatrix} 1 & 1 & 1 \\ 1 & -5 & 1 \\ 1 & 1 & 1 \end{bmatrix}$$

As with the template masks mentioned earlier, the first eight masks are permutations of the same pattern around the center of the neighborhood. When applied to an ideal stimulus, however, these masks respond to only their associated stimulus.

Our specific stimulus-response mapping method has several computational features that assist in edge detection. The stimuli employed in computing the masks are built from zeros and ones. In an image, however, an ideal step edge s corresponding to the stimulus s_j has the expression $s = hs_j + bs_9$ where h, the height of the edge, and b, the gray level of the background, are scalar constants. After mapping, the response vector r is $r = Ms = hr_j + br_9$. Note that all components except the ninth component of r_9 are zero, and we have

$$r(i) = \begin{cases} 0 & i \neq j \\ h & i = j. \\ b & i = 9 \end{cases}$$

For any ideal step edge, only two response vector components are not zero. The component conveying the edge's orientation gives the edge's amplitude. The last component provides the gray level value of the background. We shall refer to the subspace spanned by the first eight response vector components as an eight-dimensional *edge subspace* of the nine-dimensional vector space.

When applied to an image, the set of template masks shifts along the pixels during scanning. The response of the masks to an ideal step edge that is not passing through their center is also significant. The particular choice of s_9, however, preserves the property of only one nonzero response vector component in the edge subspace. For example, the pattern

$$s = \begin{bmatrix} 0 & 0 & 0 \\ 0 & 0 & 0 \\ 1 & 1 & 1 \end{bmatrix}$$

obtained when the 3×3 neighborhood is centered at one pixel offset from a $180°$ edge s_5 can be written as $s = s_9 - s_1$ and after the mapping yields $r = r_9 - r_1 = [-1\,0\,0\,0\,0\,0\,0\,0\,1]^T$. In the edge subspace the response vector has its components zero at all but the first position. The nonzero component has negative sign and corresponds to the edge lying on the border of the neighborhood rotated by 180°.

For an arbitrary 3×3 pattern of pixels, the response vector $\mathbf{r}$ has more than two nonzero components. The tilt of the vector $\mathbf{r}$ from the coordinate axes of the nine-dimensional space (the vectors $\mathbf{r}_j$) indicates how close the 3×3 pattern is to the ideal stimuli $\mathbf{s}_j$. The cosines of these angles are the most convenient *confidence measures*.

We have two different ways to compute confidence measures. In the edge subspace, the inner product is restricted to eight dimensions

$$P_j = \frac{\langle \mathbf{r}, \mathbf{r}_j \rangle_8}{\|\mathbf{r}\|_8 \cdot \|\mathbf{r}_j\|_8} = \frac{r(j)}{(\sum_{i=1}^{8} [r(i)]^2)^{1/2}} \qquad j = 1, 2, \ldots, 8.$$

The confidence measure P_j can take any value between -1 and $+1$. It is equal to one if and only if $\mathbf{r}$ coincides with $\mathbf{r}_j$ in the edge subspace, and it is zero if $\mathbf{r}$ is the response to any other stimulus $\mathbf{s}_j$. Note that the background amplitude returned by $r(9)$ is absent in the equation. High confidence values are thus obtained for weak but well-fitting edges present on large background levels. Note the simplicity of confidence measure expression in the stimulus-response mapping employed here.

A serious drawback to most edge detectors is their sensitivity to stimuli other than edges such as thin lines and isolated points. In our case, however, these stimuli will elicit responses at several components of $\mathbf{r}$, pushing the values of P_j down. Thus, the edge detection procedure can be made insensitive to artifacts by employing the confidence measure in a decision criterion.

A second confidence measure, computed across the entire space, can also be defined

$$Q = \frac{\langle \mathbf{r}, \mathbf{r}_9 \rangle}{\|\mathbf{r}\| \cdot \|\mathbf{r}_9\|} = \frac{r(9)}{(\sum_{i=1}^{9} [r(i)]^2)^{1/2}}.$$

The value of Q indicates how far the 3×3 pattern is from a uniform patch $\mathbf{s}_9$ and is useful for rejecting weak edges.

The edge subspace dimension can be reduced by associating more than one stimulus $\mathbf{s}_j$ with one response vector. The dimension is reduced to four when the stimuli $\mathbf{s}_j$ are associated pairwise with response vectors. The corresponding response matrix

$$R = \begin{bmatrix} 1 & 1 & 0 & 0 & 0 & 0 & 0 & 0 & 0 \\ 0 & 0 & 1 & 1 & 0 & 0 & 0 & 0 & 0 \\ 0 & 0 & 0 & 0 & 1 & 1 & 0 & 0 & 0 \\ 0 & 0 & 0 & 0 & 0 & 0 & 1 & 1 & 0 \\ 0 & 0 & 0 & 0 & 0 & 0 & 0 & 0 & 0 \\ 0 & 0 & 0 & 0 & 0 & 0 & 0 & 0 & 0 \\ 0 & 0 & 0 & 0 & 0 & 0 & 0 & 0 & 0 \\ 0 & 0 & 0 & 0 & 0 & 0 & 0 & 0 & 0 \\ 0 & 0 & 0 & 0 & 0 & 0 & 0 & 0 & 1 \end{bmatrix}$$

is no longer the identity matrix. Five masks are now obtained, and it follows that the four masks operating in the edge subspace are given by the pairwise sums of the masks corresponding to the two stimuli that were mapped together. We have

$$\mathbf{m}_{41} = \mathbf{m}_1 + \mathbf{m}_2 = \frac{1}{3} \begin{bmatrix} -1 & -1 & 2 \\ 2 & 2 & -1 \\ -1 & -1 & -1 \end{bmatrix}$$

$$\mathbf{m}_{42} = \mathbf{m}_3 + \mathbf{m}_4 = \frac{1}{3} \begin{bmatrix} 2 & -1 & -1 \\ -1 & 2 & -1 \\ -1 & 2 & -1 \end{bmatrix}$$

$$\mathbf{m}_{43} = \mathbf{m}_5 + \mathbf{m}_6 = \frac{1}{3} \begin{bmatrix} -1 & -1 & -1 \\ -1 & 2 & 2 \\ 2 & -1 & -1 \end{bmatrix} \qquad \text{(4-28a)}$$

$$\mathbf{m}_{44} = \mathbf{m}_7 + \mathbf{m}_8 = \frac{1}{3} \begin{bmatrix} -1 & 2 & -1 \\ -1 & 2 & -1 \\ -1 & -1 & 2 \end{bmatrix}$$

The fifth mask is $\mathbf{m}_{45} = \mathbf{m}_9$.

In a similar way, we can reduce the dimension of the edge subspace to two by employing the masks

$$\mathbf{m}_{21} = \sum_{i=1}^{4} \mathbf{m}_i = \frac{1}{3} \begin{bmatrix} 1 & -2 & 1 \\ 1 & 4 & -2 \\ -2 & 1 & -2 \end{bmatrix}$$

$$\mathbf{m}_{22} = \sum_{i=5}^{8} \mathbf{m}_i = \frac{1}{3} \begin{bmatrix} -2 & 1 & -2 \\ -1 & 4 & 1 \\ 1 & -2 & 1 \end{bmatrix} \qquad \text{(4-28b)}$$

If the dimension of the edge subspace is one the resulting mask is

$$\mathbf{m}_{11} = \sum_{i=1}^{8} m_i = \frac{1}{3} \begin{bmatrix} -1 & -1 & -1 \\ -1 & 8 & -1 \\ -1 & -1 & -1 \end{bmatrix}. \qquad (4\text{-}28c)$$

We observe that $\mathbf{m}_{11}$ corresponds to the Laplacian, a second order differential mask derived earlier as Eq. (2-18). Thus, our sets of masks for the edge subspace can be regarded as different linear decompositions of a Laplacian-type operator. As the number of terms in the decomposition increases, discriminating edge orientations are more accurate. An optimal tradeoff between resolution and computation can be determined for each specific task.

4.2.1.2.   DOG Derivation

We showed in Section 4.1.6. how minimization techniques implemented as PDP models can obtain a particular cojoint s/sf representation—the 2D Gabor transform. In this subsection we derive the DOG representation using DAM modeling (Szu and Messner, 1986).

The optimal DAM memory M can be found as $M = RS^{+}$. If the stimuli vectors $\mathbf{s}_i$ are orthonormal, then M reduces to a correlation matrix given as

$$M = RS^{+} = R(S^{T}S)^{-1}S^{T} = RS^{T} = \sum_{k=1}^{m} r_k s_k^{T}. \qquad (4\text{-}29)$$

because $S^{T}S = I$. According to Von Neumann (Kohonen, 1987), the matrix M can also be expressed as

$$M = \alpha R \sum_{k=0}^{\infty} (I - \alpha S^{T}S)^{k} S^{T}, \qquad (4\text{-}30)$$

where $0 < \alpha < 2c$, c is the largest eigenvalue of $S^{T}S$, and the first term in the expression (4-30) is $M_0 = \alpha RS^{T}$.

A novelty filter serves as the remainder operator $(I - P)$, where $P^2 = P$ is a projection operator, but differs from the latter by allowing an incomplete base set and gradually building up to the complete subspace. Given M in terms of its Von Neumann expression, a family of novelty filters $(I - \alpha SS^{T})^{k}$ can be defined. Assuming that the connecting network (relating input intensity vectors $\mathbf{s}_i$ to yet to be determined (early) feature vectors r_i) can be described in terms of the coupling matrix G as a negative-feedback system (Hall and Hall, 1979), i.e., $R = S - GR$, one could derive the modulation transfer function (MTF) of the human visual system (HVS) from

$$R = S - GR$$
$$= S - GS + G^2S - \ldots$$
$$= (I + G)^{-1}S, \tag{4-31a}$$

i.e., the MTF or memory M is given as

$$M = (I + G)^{-1}$$
$$= (I - G)(I + G^2 + G^4 + \ldots). \tag{4-31b}$$

The multichannel system corresponding to the DOG can be derived if one models G as a Gaussian matrix and approximates I by a narrow-width Gaussian G_o. The Fourier transform of a Gaussian is still a Gaussian, and the filtering operations corresponding to Eq. (4-31b) rewritten as

$$M = (G_o - G) + (G_o - G)G^2 + (G_o - G)G^4 + \ldots \tag{4-31c}$$

are then implemented as the difference of two Gaussian convolutions of widths (σ_i) and heights (h_i). The MTF given in Eq. (4-31c) corresponds to the Von Neumann expression from Eq. (4-30) and the DAM mechanism thus accounts for deriving the DOG representation. Thus, one can show how a specific early low-level visual representation, that of the DOG, could also be obtained as a result of PDP computation, as was the case with the Gabor transform in Section 4.1.6.

4.2.2. Gestalt Clustering and Relaxation

Relaxation, a basic staple of computational vision, has a long history and is analogous to physical systems that implement large optimization problems. Ballard *et al.* (1983) describe relaxation as follows:

> The goal of many problems in vision is to find the optimal interpreta-
> tion of an image consistent with known optical and geometrical con-
> straints. The difficulty is that an enormous number of constraints must be
> simultaneously satisfied. One way to solve such large optimization
> problems is to implement the constraints as excitatory and inhibitory
> links between processing units and to allow the units, whose values

represent the physical properties of objects, to iteratively approach a self-consistent solution.

The classical problem of finding the shape of a soap film bounded by a nonplanar wire hoop illustrates how a global solution can be achieved through successive local interactions. Imagine that the soap film is viewed from above and that its height is represented by a number in each cell of a two-dimensional array. The wire hoop fixes the heights around the edge, but an interior height is only constrained by being equal to the average of its neighbors. Regardless of the initial assignment of the interior heights, the correct shape of the soap film can be computed by iteratively assigning to each interior cell the average of its neighbours, a process called relaxation.

Just as the balance of forces acting on a piece of soap film act as physical constraints on the solution, so the optical constraints of image formation and the general nature of the surfaces causing the image restrict the plausible solutions that are consistent with the intensity array. The boundary condition for the computation is the raw input —that is, the whole intensity array—just as the height of the wire hoop is the boundary condition for the soap-film problem.

The corresponding discrete and stochastic/probabilistic relaxation models according to Rosenfeld *et al.* (1976), Hummel and Zucker (1983) and Levine (1986) follow.

Discrete relaxation, related to genetic algorithms, assumes a set of nodes $S = \{\alpha_1, \alpha_2, \ldots, \alpha_m\}$ and a set of legal labels $\Lambda = \{\lambda_1, \lambda_2, \ldots, \lambda_n\}$ for S, where, as an example, labels could be the edge/nonedge type. The node consistency for α_k is expressed as $\Lambda_k \subseteq \Lambda$, while the compatibility constraint between two nodes α_k and α_j is given a $\Lambda_{kj} \subseteq \Lambda_k \times \Lambda_j$. $L = \{L_1, L_2, \ldots, L_m\}$ defines a labeling of the nodes in which a subset of m labels $L_k \subseteq \Lambda$ is assigned to each node α_k. The set L is consistent if $(\{\lambda\} \times L_j) \cap \Lambda_{kj} \neq \varnothing$ for $\forall \lambda \in L_k$. $L^{\max}$ is the greatest consistent labeling, and a labeling is called unambiguous if it is consistent and if each node has only one label assigned to it. The relaxation algorithm is iterative, and since the number of both nodes and labels is finite, an unambiguous labeling can be found if one exists. The algorithm starts by deriving $L^{\max}$ and then looks for an unambiguous labeling by exploring different (tree) alternatives, which result after discarding "extra" elements from those sets L_j whose cardinality is greater than one.

The more interesting and realistic relaxation model, related to simulated annealing, is stochastic, and assigns to all nodes $\alpha_k \in S$ a corresponding label probability vector $\mathbf{p}_k(\lambda)$ such that

$$0 \leq p_k(\lambda) \leq 1 \text{ and } \sum_{\beta = 1}^{n} p_k(\lambda_\beta) = 1. \qquad (4\text{-}32a)$$

The compatibility function is $r_{kj}: \Lambda_k \times \Lambda_j \to [-1, 1]$. The step (K) update or change in confidence $\Delta^{(K)} p_k(\lambda)$ is

$$\Delta^{(K)} p_k(\lambda) = \sum_j d_{kj} \left[\sum_{\lambda'} r_{kj}(\lambda, \lambda') p_j^{(K)}(\lambda') \right], \qquad (4\text{-}32b)$$

where d_{kj} weights the contributions from the neighbors of α_k and $\Sigma_j d_{kj} = 1$. The update step measures the degree of support in the neighborhood of α_k for label λ. The new, updated probability vector is then given as

$$p_k^{(K+1)}(\lambda) = \frac{p_k^{(K)}(\lambda)[1 + \Delta^{(K)} p_k(\lambda)]}{\Sigma_{\lambda'} p_k^{(K)}(\lambda')[1 + \Delta^{(K)} p_k(\lambda')]}. \qquad (4\text{-}32c)$$

Many generic relaxation schemes have been used to derive characteristic intrinsic images, such as optical flow and stereo/disparity images, which will be discussed in Chapter 5.

Relaxation has also been widely used for solving optimization problems, such as those encountered in early vision. The corresponding algorithm is local and iterative, and thus lends itself to parallel implementations (see Section 9.3.). Early vision is characterized by large amounts of data, however, and, consequently, relaxation algorithms converge slowly toward the desired solution. Terzopoulos (1986) suggests that multigrid relaxation methods can overcome the problem by dealing with imagery at different resolution levels. Scale-space and cojoint representations (Chapter 2) and pyramidal hardware structures (Chapter 9) are the appropriate levels of representation and hardware implementation for the multigrid relaxation algorithms.

We will now return to the texture analysis problem. We began in Section 2.4. by looking at underlying s/sf representations and eventually derived an s/sf representation based on the Wigner distribution (WD). We continue our original texture segmentation task and Gestalt clustering by implementing the last step of the system outlined in Fig. 2.11., that of relaxation (Reed and Wechsler, 1990).

Texture segmentation requires the identification of boundaries between homogenous regions. Relaxation, which amounts to self-organization, is applied to each of the 2D arrays selected in the data compression phase, and was proposed by Caelli (1985). It consists of two steps

(1) Averaging

$$A^{(i)}(x, y) = \sum_{m=-L}^{L} \sum_{n=-L}^{L} \frac{1}{(2L + 1)^2} I^{(i-1)}(x + m, y + n)$$

(2)   Transformation

$$I^{(i)}(x, y) = L(A^{(i)}(x, y)) = \frac{1 - \exp\left(\dfrac{A^{(i)}(x, y) - \alpha}{\beta}\right)}{1 + \exp\left(\dfrac{-A^{(i)}(x, y) - \alpha}{\beta}\right)},$$

where $I^{(i)}(x, y)$ is the 2D array at the ith iteration.

The $L(\bullet)$ transformation is a sigmoidal function that facilitates a labeling decision by pushing the corresponding estimate towards $+1$ (class A) or -1 (class B).

The averaging and transformation steps are repeated as required. Termination could occur at the point where the 2D array changes by only a small amount, or after a specified number of iterations. Region labels are assigned on a pixel-by-pixel basis. When the normalized inner product between the feature vectors (after relaxation) for adjacent pixels is above a threshold, the pixels are grouped together.

Beck (1983) and Beck *et al.* (1987) have examined a number of textured images in order to characterize spontaneous segregation in the human visual system, and they have found evidence relating spontaneous segregation to the frequency content of image regions. Here, we examine textured images, which are patterned after those used in the previous experiments and provide evidence that cojoint spatial/spatial-frequency representations result in spontaneous texture segmentation.

Figs. 4.2a. and 4.2b. show an image of textures found by Beck to be spontaneously discriminable, before and after anti-aliasing filtering, respectively. The primary function plane of the Pseudo-Wigner distribution (PWD) is shown in Fig. 4.2c. Relaxation, as shown in Fig. 4.2d., clearly separates the two regions.

Another texture segmentation example is shown in Fig. 4.3. The texture on the left is D102 from Brodatz (1966), a shadowgraph of cane. The second texture is D103, a shadowgraph of loose burlap. The 64×64 pixel versions of the image, before and after anti-aliasing filtering, are shown in Figs. 4.3c and 4.3d. The primary frequency plane of the PWD is shown in Fig. 4.3d., and the result of relaxation is shown in Fig. 4.3e.

We now present experimental results using s/sf representations to explain and predict Gestalt-type phenomena. The results illustrate different aspects of Gestalt organization. The proximity law, which specifies that elements group together based on nearness, is reflected in

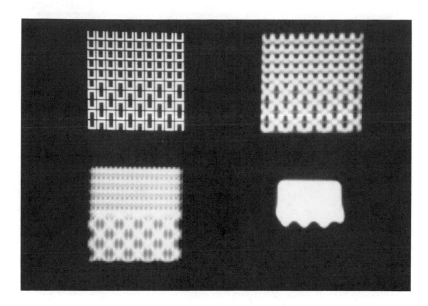

Figure 4.2.

Beck-type texture segregation. (a) The input image. (b) The image after anti-aliasing filtering. (c) The primary frequency plane of the PWD. (d) The result of the relaxation step.

Fig. 4.4a. Following the established procedure, Fig. 4.4b. shows the image after anti-aliasing filtering. The primary frequency plane of the PWD is shown in Fig. 4.4c. Finally, the result of relaxation is shown in Fig. 4.4d. The grouping is clear and consistent with visual experience and proximity law prediction.

Similarity law specifies that elements group together based on the similarity of their appearances (i.e., features). In the following examples, we address grouping based on size and shape similarity.

Similarity in size: Figure 4.5a. is an example of grouping according to size. The image after anti-alias filtering is shown in Fig. 4.5b. The primary frequency plane of the PWD, before and after relaxation, is shown in Figs. 4.5b. and 4.5c., respectively. The grouping is consistent with human perception and Gestalt laws.

Similarity in shape: Figure 4.6a. shows an example of grouping based on shape. After anti-aliasing filtering, Fig. 4.6b. results. Figure 4.6c. shows the primary PWD plane. After relaxation, the grouping

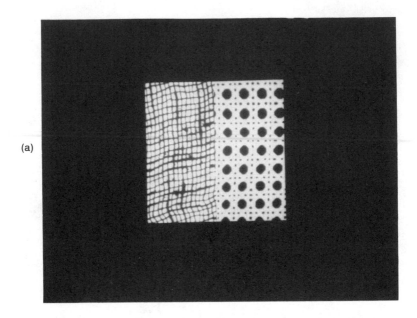

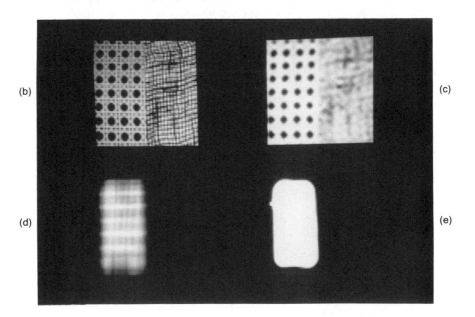

Figure 4.3.
Brodatz-type texture segmentation. (a) The original 256 × 256 image. (b) The resolution reduced (64 × 64) image. (c) The 64 × 64 image after anti-aliasing filtering. (d) The primary frequency plane of the PWD. (e) The result of the relaxation step.

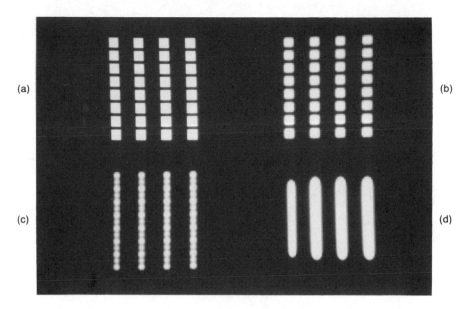

Figure 4.4.
Gestalt (proximity) clustering. (a) The input image. (b) The image after anti-aliasing filtering. (c) The primary frequency plane of the PWD. (d) The result of the relaxation step.

shown in Fig. 4.6d. occurs. Examination of the original image would lead one to expect groupings smaller than those shown and perhaps more of them (although to group more than two types of objects, one typically needs more than one PWD frequency plane). Examining Fig. 4.6a. more closely, we see how this grouping results. With low-image resolution, the tops of the inverted triangles (rows three and six) are identical to the tops of the squares (rows two, five, and eight) and are grouped with the squares into the white regions of Fig. 4.6d. The bottoms of the inverted triangles are identical to the bottoms of the circles (rows one, four, and seven) and are grouped with the circles into the black regions. The use of a higher resolution image and/or multiple frequency planes of the PWD might allow better grouping of such images.

The good continuation law predicts that elements group in such a way as to minimize discontinuities. The closure law (the Gestalt law that predicts preference for closed groups) could be considered a corollary of the continuation law.

Figure 4.7a. shows a constellation of dots. The image after anti-aliasing filtering is shown in Fig. 4.7b., while Figs. 4.7c. and 4.7d. show

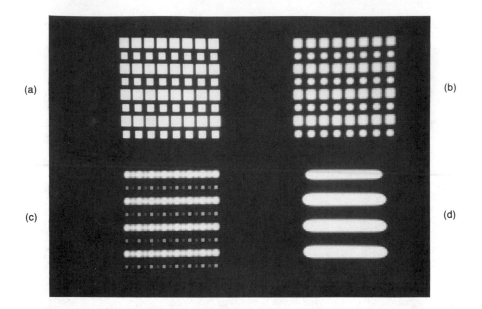

Figure 4.5.
Gestalt (size) clustering. (a) The input image. (b) The image after anti-aliasing filtering.
(c) The primary frequency plane of the PWD. (d) The result of the relaxation step.

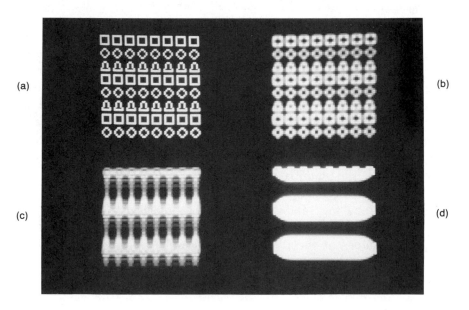

Figure 4.6.
Gestalt (shape) clustering. (a) The input image. (b) The image after anti-aliasing
filtering. (c) The primary frequency plane of the PWD. (d) The result of the relaxation step.

the primary PWD plane and its absolute value. After relaxation, the grouping in Fig. 4.7e. results. This grouping is predicted by good continuation and closure.

In addition to its consistency with Gestalt laws, the grouping obtained in the previous example is the same that would be achieved by linking the dots in a Euclidean minimal spanning tree (EMST). (The EMST is the tree of minimum total length whose vertices are the given points.) In the examples considered so far, we have been concerned with only the most (visually) obvious groupings and their calculation using the primary frequency plane of the PWD. Less obvious groupings of objects, still perceived by humans and explainable by Gestalt laws, may also exist in an image. A less obvious clustering of objects is the columnar, proximity-based one. By examining the second frequency plane in the PWD (the plane corresponding to the second highest energy content, Fig. 4.8a.), evidence for such a grouping can be seen, mixed with the (still dominant) row grouping. The third plane in the PWD (Fig. 4.8b.) shows the columnar clustering clearly. Comparing the

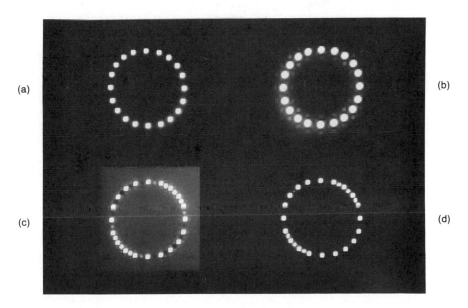

(a)

(b)

(c)

(d)

Figure 4.7.
Gestalt (good continuation and closure) clustering. (a) The input image. (b) The image after anti-aliasing filtering. (c) The primary frequency plane of the PWD. (d) The absolute value of the primary plane. (e) The result of the relaxation step.

energy levels of the second and third planes with the first (Figs. 4.8c. and 4.8d.), the second plane is nearly as strong as the first, while the third is much lower. The strength of a grouping, as perceived visually, is apparently related to the energy level of the PWD plane displaying that grouping.

The ability of humans to discriminate between regions based on multiple cues has been studied by Treisman (1985). Briefly stated, human subjects can quickly differentiate between regions when only one feature is examined (e.g., if the elements in one region are circles and all others are squares). If, however, two or more features are examined, (e.g., red circles and blue squares in one region and blue circles and red squares in the other) discrimination takes longer. The activity required by humans to differentiate between regions with these characteristics was termed "parallel" (preattentive) and "serial" (attentive), respectively. To test this conjecture, Reed and Wechsler (1990) experimented with an image where one region consists of circles and another of squares. The circles and squares were of two intensities

Figure 4.8.
Alternative groupings: (a) the second frequency plane of the PWD; (b) the third frequency plane of the PWD; (c) a comparison of the first and second PWD planes; (d) a comparison of the first and third PWD planes.

(white and gray), which were selected randomly. Although the circles and squares vary in brightness, only the shape needs to be determined in order to discriminate between these regions. Thus, this image is an example of one requiring "parallel" processing in humans, and the resulting segmentation was as expected. Next, we considered a case where the first region consists of white circles and gray squares, while the second is made up of white squares and gray circles. This image requires "serial" processing in humans, and indeed the result of segmentation indicates that the system does not differentiate between the regions we have specified. The last two examples supply evidence that the system we have implemented might perform like the human visual system for cases when humans segregate regions preattentively, and priming does not occur. Cases requiring human subjects to focus their attention in order to segregate regions apparently cannot be segmented by the system in its current form.

Our experiments show good agreement between groupings predicted by Gestalt laws and those produced using our s/sf approach. Grouping is often described as a low frequency phenomenon, achievable in machine vision simply through the application of low-pass filters. Janez (1984) has shown, however, that grouping can also occur using high frequencies. It may be necessary, therefore, to consider the entire frequency spectrum in order to achieve grouping in the general case. We have demonstrated a system capable of texture segmentation (which cannot be accomplished with low-pass filters), which also produces object groupings consistent with those produced by the HVS.

Further work could be performed along the qualitative and quantitative dimensions. The qualitative aspect could provide additional insights into the human visual system's ability to group. The quantitative aspect could reveal sensitivity to perceptual parameters, such as distance. Since the frequencies perceived vary with distance, and since we have shown that different groupings can be seen in different frequency planes, one would expect different groupings to be perceived at different distances.

4.2.3. Markov Random Fields (MRF) and Spin Systems

Two-dimensional lattices of pixels can be viewed as Ising spin systems. Assume that a particle with a magnetic spin (± 1) is placed at each pixel and that the interactions between neighbors are translation-invariant and isotropic. The energy of such a (CAN) system can then be written as

$$E = -\frac{J}{2} \sum C_{P_1, P_2} \mu_{P_1} \mu_{P_2}.$$ (4-33)

where $C_{i,j}$ is the connection matrix for four-neighborhood connectivity among pixels (P_1, P_2), J is a positive constant, and μ_P is the spin at a given location P. The system, in order to minimize its energy, clearly tends to align its spins in the same direction. An additional feature could be the presence of an external magnetic field γ that tends to align the spins in the direction prescribed by its sign. Then, the updated energy function E is given as

$$E = -\frac{1}{2} \sum_P \gamma_P \mu_P - \frac{J}{2} \sum_{P_1, P_2} C_{P_1, P_2} \mu_{P_1} \mu_{P_2}.$$ (4-34)

Each spin is, therefore, under the influence of two (possibly competing) forces—the external field and local interactions, i.e., one had to trade global vs. local constraints.

Carnevali *et al.* (1985) drew the analogy between Ising spin systems and the image restoration task. The corresponding energy function for an Ising system yields in its minimum energy state a ground state configuration corresponding to the restored image, subject to given constraints. To speed up the minimization process, Carnevali *et al.* used simulated annealing techniques (see Section 4.1.5.) for parameter estimation. For example, assume a binary image given by a noiseless configuration of 50 rectangles. Image degradation results by adding random noise, where pixels have their binary value changed according to some probability p. The task is to estimate the location of the original, noise-free rectangles. The energy E, which is to be minimized, is made up of two components: (1) The difference in the number of pixels between the original and estimated image, f and $\hat{f}$, respectively. (Note that the image model is known and that the parameters to be estimated are the number of rectangles and their locations.) (2) The Lagrange-type multiplier k is linked to α, the number of rectangles, where four parameters describe each rectangle. The energy E to be minimized is given as $E = \beta + k\alpha$, β is the discrepancy between the original and estimated image, and k is a penalty-like price to be paid for too many rectangles. (When k is made large, small rectangles are considered noise and discarded if the corresponding ΔE change in energy, which results from adding such rectangles, is positive.) The perturbations considered for this problem were add/delete rectangle, stretch, and/or split rectangle. The results, even for low signal-to-noise ratio (SNR), were fairly good but required expensive computation.

Another Ising-type problem that is solved using simulated annealing is image smoothing. The energy E to be minimized is a function of β, the discrepancy between the smoothed and the original image and R, the roughness of the smoothed image. Specifically, $E = \beta + kR$, where k, the Lagrange-type multiplier, trades off roughness and fidelity of restoration. β can be estimated using either the L^1 or the L^2 distance between the original "a" image and its smoothed version "b" [for binary a and b, $|a - b| = (a - b)^2$], while R is estimated using the Laplacian. Minimizing E is then equivalent to finding the ground state of an Ising ferromagnetic system embedded in an external field. The external field clearly is analogous to fidelity (β), while neighborhood alignment is analogous to roughness (R). The experiments showed that for low SNR, simulated annealing yields better results than standard smoothing algorithms. The main drawback of simulated annealing is still its large computational requirements.

Geman and Geman (1984) exploit the equivalence between Markov random fields (MRF) and Gibbs distributions for image restoration. The MRF is a 2D lattice-like array of pixels where the conditional probability that a pixel has a given value is a function of a finite ("memory") neighborhood (see also Section 3.2.4.). The probability that a MRF is in a given state configuration $\mathbf{x} = \{X_i\}$ of pixel values can be given by a Gibbs (Boltzmann—see Section 4.1.5.) distribution given as

$$P(\mathbf{X}) = \frac{1}{Z} \exp \frac{-E(\mathbf{X})}{T}. \tag{4-35}$$

Similar to simulated annealing, gradual temperature T reductions yield low energy states $E(\mathbf{X})$ and thus lead to most probable states under the Gibbs distribution. The equivalence between MRF and Gibbs implies that such states $\mathbf{X}$ are the very ones sought as the maximum a posteriori (MAP) estimate of the original image given the degraded data. The whole process is characteristic of highly parallel relaxation, to which we refer to in Section 9.2.3. for showing SIMD (single instruction multiple data) architecture and efficient Geman and Geman algorithm implementation. We will briefly review the restoration algorithm using specific implementation details of the SIMD architecture described by Murray *et al.* (1986). Assume that D stands for the degraded image being observed and that one looks for the MAP estimate $\mathbf{X}$. Clearly max $P(\mathbf{X}|D)$ is the MAP estimate sought and is given as

$$\max_{\mathbf{X} \in \Omega} P(X/D) = \max_{\mathbf{X} \in \Omega} [P(D|\mathbf{X})P(\mathbf{X})P(D)]. \tag{4-36}$$

We estimate only $P(D|\mathbf{X})$ and $P(\mathbf{X})$, because if one assumes an ergodic process, $P(D)$ is fixed and can be discarded. First, we handle $P(\mathbf{X})$ by remembering that it can assume the Gibbs distribution given in Eq. (4-35), and it depends on $E(\mathbf{X})$. The energy term, like a penalty cost function, has to be minimized in terms of explicit discontinuities introduced by edge elements e_{ij} between pixels i and j and by the continuity (homogeneity) of a restored image.

Assume that the domain of image definition is a square lattice $\Omega = \{(i, j), i = 1, \ldots, m; j = 1, 2, \ldots, m\}$ and define the neighborhood of type c for pixel (i, j) as

$$N_c = \{(k, l) \in \Omega; 0 < (k - 1)^2 + (l - j)^2 \le c\}.$$

Clearly, the type (order) c determines the processes resolution and computational complexity. The four-nearest neighbor connectivity (or dependence using the MRF concept) is obtained for $c = 1$. Furthermore, a subset $C \subseteq \Omega$ is a "clique" if every pair of distinct pixels in C are neighbors; $\mathscr{C}$ denotes the set of cliques. Again for $c = 1$, and omitting the singleton clique (i, j), one obtains $\mathscr{C} = \langle\{(i, j), (i, j + 1)\} \{(i, j), (i + 1, j)\}\rangle$. The number of clique types grows fast as c increases; we restrict ourselves (as Geman and Geman did) to $c = 1$. One energy term component provides for nonhomogeneities within Ω by introducing edge elements (e_{ij}) and thus allows a break in spatial coherence of the MRF. There are six basic possible edge configurations; the others are rotations of these while keeping the potentials rotationally invariant. The six configurations are

$$V_0 = \begin{pmatrix} x & x \\ x & x \end{pmatrix} \text{ (off—no edge)};$$

$$V_1 = \begin{pmatrix} x & x \\ \overline{x} & \overline{x} \end{pmatrix}; \qquad V_{2_a} = \begin{pmatrix} x & x \\ \overline{x} & x \end{pmatrix}; \qquad V_{2_b} = \begin{pmatrix} x & x \\ x| & x \end{pmatrix};$$

$$V_3 = \begin{pmatrix} x| & x \\ \overline{x} & x \end{pmatrix}; \qquad V_4 = \begin{pmatrix} x| & x \\ x| & x \end{pmatrix},$$

and the corresponding potentials (Marroquin, 1984) used for the SIMD implementation are $\langle V_0, V_1, V_{2_a}, V_{2_b}, V_3, V_4 \rangle = \langle 0.0, 2.2, 0.4, 1.2, 1.4, 2.2 \rangle$. The penalty component is then

$$T_1(\mathbf{X}) = \sum_k V_k(c),$$

where the summation is over all the cliques. The penalty component corresponding to lack of homogeneity is given as

$$T_2(\mathbf{X}) = \sum\sum U(i, j, e_{ij}),$$

where

$$U(i, j, e_{ij}) = \begin{cases} \hat{U}(i, j) & \text{if } e_{ij} \text{ is off} \\ 0 & \text{if } e_{ij} \text{ is on} \end{cases}$$

and

$$\hat{U}(i, j) = \begin{cases} -1 & \text{if } x_i = x_j \\ +1 & \text{otherwise.} \end{cases}$$

A third term to be considered in Eq. (4-36) is $P(D|\mathbf{X})$. It corresponds to the discrepancy between the degraded and restored image; the corresponding $T_3(\mathbf{X})$ component will thus penalize lack of fidelity. Assuming local, uncorrelated Gaussian white noise of zero mean and standard deviation σ, the discrepancy term is given as

$$T_3(\mathbf{X}) = P(D|\mathbf{X})$$

$$= \exp\left\{ -\sum_j (D_j - X_j)^2 / 2\sigma^2 \right\} / (2\pi\sigma^2)^{(m/2)},$$

where m is the number of points. By taking logarithms in Eq. (4-36) and by discarding constant terms such as $P(D)$ and those depending on z (see Eq. (4-35)) and σ (see the $T_3(\mathbf{X})$ component), the MAP estimate is found by minimizing the cost function $E(\mathbf{X})$ given as

$$E(X) = \sum_k V_k(c) + \sum_i \sum_{j=1}^4 U(i, j, e_{ij}) + \sum_j \frac{(D_j - X_j)^2}{2\sigma^2}. \tag{4-37}$$

The function $E(\mathbf{X})$ can be minimized by simulated annealing, and state configurations were drawn from a Gibbs sampler (Geman and Geman, 1984). Geman and Geman have shown that if the temperature is lowered as $T = T_0 \ln(a)/[\log_2(K + 1)]$ (where T_o is the starting temperature and K is the iteration step), the algorithm converges to the minimal energy. The theoretical value found for T_0 was too high to allow convergence in a "reasonable" amount of time; however, empirically found lower values proved to work quite well. (See, however, the comparison with GNC algorithms in Section 4.1.5.)

We conclude with two observations. First, the use of MRF for image restoration has yet to address the resolution problem (it arbitrarily chooses the neighborhood size), and there still remains a complexity issue related to choosing the starting temperature T_0. Second, by including edge discontinuities in the image restoration process, one can fit (reconstruct) surfaces while preserving discontinuities (Marroquin, 1984). The same breakdown in spatial coherence could then be used for texture segmentation, a problem that we already addressed in Section 4.2.2.

Toffoli and Margolus (1987) consider Cellular Automata Machines the analogy of a general class of deterministic physical systems whose local interactions are well suited to model image processing tasks. Such systems are related to diffusion (and relaxation), fluid dynamics, Ising systems, and ballistic computation. Only imagination and current knowledge of physics would limit this fruitful line of research, where PDP models, by using the analogy to specific physical systems, can describe collective visual phenomena.

4.2.4. Connectionism, Cytoarchitecture, and Visual Maps

The human visual system (HVS) is a "physical" system that has provided much inspiration for computational vision. We briefly consider in this section the internal HVS architecture and its implication for the development of connectionist models.

Research by Hubel and Wiesel (1962) has uncovered a regular interconnection pattern for early visual processing stages. The 2D pattern, or cytoarchitecture, has as its main processing unit a hypercolumn that detects edges (as originally envisioned by Hubel and Wiesel) within a 3D "hardware" (receptive field position, orientation, size). Following our discussion on s/sf representations and the observation that edge detection could be considered a byproduct of frequency analysis, early visual processing could be construed as being performed along spatial position and frequency (scale and orientation), as in Daugman's (1988a) Gabor scheme. The human visual system architecture is further characterized by orderly retinotopic mappings that preserve neighborhood relationships of the visual field. Van Essen (1979, 1985) showed the existence of quite a large number of well-defined subdivisions, which he called visual maps. Such visual maps could correspond to the intrinsic images (to be discussed in Chapter 5) and could be conceptually thought of as feature maps. Evidence suggests that there exist at least two major functional streams, one related to motion analysis and the other related to form and color analysis. Information from such visual maps has to be integrated at some point using the accrued evidence. We will discuss data fusion methods appropriate to such a task in Chapter 8 and restrict ourselves for now to how knowledge of the HVS might have led to connectionism. In fact, it did not lead straight there, but through another relay research station known as the Hough transform, which will be discussed next.

The Hough transform was originally suggested to detect lines in noisy point patterns by looking into a parameterized (feature) space where evidence can accumulate to indicate the presence of specific line patterns. The axes of the evidence space, known as accumulator arrays, correspond to the features characterizing the event one wants to detect. Line detection can be easily handled if one remembers the parametric equation of a line, given in terms of (angle, distance to origin) $= (\theta, \rho)$, as $\rho = x_i \cos \theta + y_i \sin \theta$. Clearly, $f(x, y) = 1$ for those points obeying the relationship $\rho_o = x \cos \theta_o + y \sin \theta_o$, i.e., the accumulator array A is parametrized as $A(\theta, \rho)$, and the search is for a high-valued cell $A(\theta_o, \rho_o)$ that would draw its strength from the many points "ready" to lie on the line given by the pair (θ_o, ρ_o). The same approach could, in principle, be extended to the detection of parametric curves of arbitrary shapes. An interesting application of this concept is that of vertex detection, which could be implemented as a two-stage Hough transform. Assuming that the dot pattern is characteristic of a number of lines, those would be detected as high-valued cells (θ_o, ρ_o) in some accumulator array. If those high-valued cells are considered as $x - y$ locations, their collinearity, indicated by a second stage Hough transform, will indicate that those lines go through a common point, i.e., there is a joint vertex.

The Hough transform is conceptually nothing more than a match filter (see Section 3.2.2.) (Sklansky, 1978), which can be derived from the Radon transform (Deans, 1981) and operated by accruing evidence pointing to some specific interpretation. The sources (features) to be fused can originate from different visual (intrinsic) maps, and their specific mix is like a template seeking its interpretation. PDP models based on this analogy came to be known as connectionist models. Ballard (1986) and Fahlman and Hinton (1987) provide a good review on connectionist models for distributed problem solving. The pattern of connections and their strengths determine for a given input the best possible interpretation. Learning the weighted network of connections is deferred to Chapter 8 when we introduce the concept of back propagation. Connectionist models are also conceptually akin to marker-passing architectures such as NETL (Fahlman, 1979) and semantic networks, and they have also inspsired new parallel architectures such as the Connection Machine (Hillis, 1985). To summarize, the system knowledge is held in synaptic-like connections rather than memory cells, which led to such models being labeled connectionist.

An example of computational capability of connectionist models is given by Feldman *et al.* (1988). The example chosen is related to a classical illusion known as the Necker cube. (Illusions might be harmful in real life, but they offer many useful clues on how to decipher the mysteries of the HVS.) Here, the illusion is described by Feldman *et al.*:

> Most people initially see the cube with the vertex *B* closer to them, but it also can be seen as a cube with a vertex *H* closest to the observer. If you focus on vertex *H* and imagine it coming out of the paper toward you, the picture will flip to the *H*-closer cube. The flip takes less than a second. The Necker cube is interesting to psychologists because it will flip spontaneously between the two views if you keep looking at it.

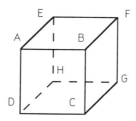

Before seeing how connectionism explains the illusion, let us first try to find a generic "who done it," using the PDP modeling discussed so far. The illusion is clearly one instantiation of the so-called ill-posed problem, and in our case, it is due to our attempt to derive a 3D interpretation from an impoverished 2D image. We already mentioned that regularization problems, such as the one under consideration, can

be solved through constraint optimization, where the added constraints help to choose among several possible alternatives. Clearly,
the constraints require that the (reconstructed) perceived 3D volume is
such that solid angles are straight, and surfaces are of equal size and
shape, and so on. The resulting ambiguity is a clear indication that the
resulting energy function alternates between two minima, i.e., stable
states, and is possibly moving from one basin of attraction to another
one.

Let us now see how the connectionist approach tries to model the
same problem. A Necker cube network consisting of 26 units and 138
links has been set up. Units 0–7 represent the N1-closer cube's vertex
nodes, ABCDEFGH in order. Units 8–11 represent the depth level,
$A < E$, $B < F$, $C < G$, and $D < H$ in order. Unit 12 is the decision (the
N1-closer) unit. Units 13–25 are encoded similarly for the N2-closer
network. Excitatory and inhibitory links are set between nodes within
a single view network and between the two views' networks, respectively. The connections can take a range of weights and asynchronous
simulation, which is useful to break symmetry conditions, yields the
same ambiguity exierenced by humans. (A similar approach has been
applied to language parsing by Waltz and Pollack (1985) for finding the
best among competing alternative sentence interpretations.)

4.3. Conclusions

We showed in this chapter that many problems encountered in computer vision are ill-posed, i.e., underconstrained. Optimization techniques work by introducing additional constraints for the sought-after
solution. Implementing any optimization technique is quite expensive,
and distributed processing in the form of a neural network is a possible
solution. So far, we have advanced the ideas of distributed representations and processing for overcoming computational constraints. Furthermore, the same distributed aspect of computation, as embedded in
PDP models, enables the visual system to display fault-tolerant and
robust behavior. (See also Section 7.2.3. on the use of DAMs for
invariant and fault-tolerant object recognition.)

(34)

tural Figure. In *Fig.* 1. the Rays falling nearly Parallel on the Eye, are by the Cryſtalline A B refracted, ſo as their *Focus*, or Point of Union F falls exactly on the *Retina*. But if

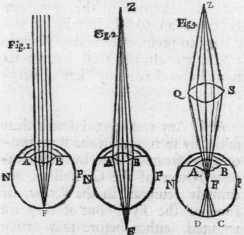

the Rays fall ſenſibly diverging on the Eye, as in *Fig.* 2. then their *Focus* falls beyond the *Retina*: Or if the Rays are made to converge by the *Lens* Q S, before they come at the Eye, as in *Fig.* 3. their *Focus* F will fall before the *Retina*. In which two laſt Caſes,

An essay towards a new theory of vision. (From Berkeley, G., 1709.) Reprinted courtesy of the National Library of Medicine.

5

Intrinsic Representations

By substance *I understand what is in itself and is conceived through itself,
i.e., that whose concept does not require the concept of another thing, from
which it must be formed. Necessarily constitutes the* essence *of a thing . . . ,
what the thing can neither be nor be conceived without, and vice versa,
what can neither be nor be conceived without the thing.*

From *The Ethics* by Baruch Spinoza

Modeling and recognizing 3D objects is one of the goals of any visual
system. Gibson (1950) suggests that 3D modeling revolves around
surfaces. Thus, perception of visual surfaces is crucial for 3D object
recognition. Barrow and Tenenbaum (1978) suggest that

An appropriate role of early visual processing is to describe a scene in
terms of intrinsic (veridical) characteristics—such as range, orientation,
reflectance, and incident illumination—of the surface element visible at
each point of the image. Support for this idea comes from three sources:
the obvious utility of intrinsic characteristics for higher-level scene
analysis; the apparent ability of humans to determine these characteristics, regardless of viewing conditions or familiarity with the scene; and a
theoretical argument that such a description is obtainable, by a noncog-
nitive and nonpurposive process, at least, for simple scene domains. The
central problem in recovering intrinsic scene characteristics is that the
information is confounded in the original light-intensity image: a single

213

intensity value encodes all the characteristics of the corresponding scene point. Recovery depends on exploiting constraints, derived from assumptions about the nature of the scene and the physics of the imaging process.

Visual surfaces could then be identified in terms of their intrinsic characteristics.

Psychophysical and physiological experiments suggest biological vision is modular in design. Random-dot stereograms (Julesz, 1975) indicate that humans have the ability to interpret images in 3D using only cues such as depth, stereopsis, and texture, and that stereopsis precedes object recognition. Psychophysical tests involving human infants suggest a modularity of sensitivity to different (intrinsic) cues as well. According to Yonas *et al.* (1987), the development of an infant's sensitivity follows a definite pattern. Retinal size and motion parallax are early cues used by infants to discriminate depth. These are followed by binocular parallax and then pictorial depth. Staggered development of the ability to use different depth cues indicates that different processing occurs in parallel on the same visual stimulus. Van Essen and Maunsell (1983) showed that there are a large number of well-defined subdivisions in the visual cortex. Evidence also suggests the existence of at least two major functional streams, one related to the analysis of motion and the other to that of form and color. These functional streams seem to be independent in many respects, and the information processing is done continuously and in parallel.

The question that immediately comes to mind is that if these processes are independent and modular, how are they integrated to maintain a continuously, complete, coherent perception of the world. We shall argue later that the derivation of intrinsic image representations is an active process, and that the integration of such representations belongs to middle-level vision under the guidance and control of a (high-level) memory system. The integration process (considered in Section 6.1.) assumes a hierarchical organization where modules are ordered in time. For example, experiments done by Frisby and Mayhew (1979) showed that texture differences, which are clearly visible monocularly, disappear when the stereo pair is fused.

A detailed taxonomy of intrinsic representations and specific methods for deriving them is presented throughout the chapter. One class, the psychophysical one, includes lightness (reflectance), color, optical flow, and depth. Then, there are artificial (man-induced) intrinsic images, such as those encountered in robotics and target tracking and recognition, and they include Doppler shift, ultrasound, and infrared.

Biological perception of space is not limited to visual cues but can also be driven by integration of entirely different modalities. An example of the influence of extravisual sources in spatial orientation tasks can be found in the barn owl. In barn owls the auditory system plays a crucial role in prey capture; since it hunts at night, visual cues are of limited value. As a consequence, owls rely heavily on their sense of hearing to localize prey and to guide their attack. The owl has cells in its optic tectum (the main visual center) that have topographic representations of visual space, which are bimodally sensitive to auditory and visual stimulation (Knudsen, 1982). The topographic representations of visual space depend on point-to-point projections from the retina. The map of auditory space is an emergent property of higher-order processing. The auditory system must derive its map from the relative patterns of auditory input arriving at each ear (Knudsen and Konishi, 1978). Despite different ways of deriving the spatial information, the visual and auditory maps were found to be remarkably similar and closely aligned. Other studies of bimodally sensitive cells have found that interactions between modalities can be nonlinear and complex (Hartline *et al.*, 1978). Thus, we include a class of biologically driven intrinsic images. We call this class ethological because ethology—a relatively new science—seeks to explain behavior through biological neural machinery, which includes genetic programming.

The computational methods used to derive the psychophysical intrinsic images are usually of the PDC (parallel distributed computation) type and include both PDR and PDP, i.e., parallel distributed representations and processing. Furthermore, as it is the case for optical flow, the PDC is well grounded in the conjoint space/spatial-frequency representations discussed in Section 2.3. The invariance aspect discussed in Chapter 3 is again considered here under the psychophysical heading of constancy and is addressed in the context of lightness (reflectance) derivation.

5.1. Psychophysical Images

We consider in this section the derivation of lightness and color, optical flow, and depth intrinsic images.

5.1.1.   *Lightness and Color*

One of the most relevant intrinsic characteristics is that of reflectance (R) or lightness, if one were to use instead the corresponding psychophysical term. However, the image formation process yields the luminance (L) rather than the sought after lightness. The reflectance is embedded in the (luminance) brightness image along with the intensity (I) of the illumination source according to the following equation:

$$L = R \times I. \tag{5-1}$$

We now derive the reflectance and account for its constancy, i.e., the known fact that the lightness of an object is largely independent of its illumination. The first algorithm to consider is based on the Land–McCann (1971) retinex spatial theory, which was developed by Horn (1974). Its basic assumption is that variations in illumination are relatively gradual, whereas variations in reflectance are rather sudden. The algorithm proceeds in two rather simple steps. First, it detects edges in the luminance profile and continues by building up the required lightness profile by reconstructing between edges. Specifically, the algorithm is given as

(a) **Edge detection**   It is implemented via convolution, (using center surround (on-off units)) through lateral inhibition, i.e.,

$$f_1(\mathbf{x}) = f(\mathbf{x}) - \tfrac{1}{6} \sum f(N(\mathbf{x})), \tag{5-2a}$$

where $N(\mathbf{x})$ are the neighbors of point $\mathbf{x}$ on a hexagonal grid, and $f(\mathbf{x})$ is the corresponding picture value function. The hexagonal grid is preferred to the standard rectangular one, where the adjacent pixels are closer than the ones on the diagonal, since all neighbors are equidistant from the central pixel. The hexagonal grid was discovered long ago by nature and has been incorporated into the honeycomb grid.

(b) **Thresholding**

$$f_2(\mathbf{x}) = \theta(f_1(x)). \tag{5-2b}$$

for some threshold function θ

(c) **Reconstruction**   It is implemented via deconvolution through lateral facilitation, i.e.,

$$f_3(\mathbf{x}) = f_2(\mathbf{x}) + \tfrac{1}{6} \sum f_2(N(\mathbf{x})). \tag{5-2c}$$

The algorithm consists of two main steps, those of spatial differentiation and integration, separated by a thresholding step. There are
several variations of the generic algorithm, and they were discussed by
Hurlbert (1986). The algorithm has been used quite successfully on
Mondrian paintings made up of patches of uniform reflectance. Several
objections against the retinex theory were raised by Marr (1982), such
as the question of how to choose the threshold θ.

The retinex theory has been extended to color vision as well. It is
well known that any color can be approximately matched by a vector-
ially additive combination of three primaries (blue at 440 nm, yellow/
green at 540 nm, and red at 575 nm). The retinex theory suggests that
the algorithm is applied independently for each one of the three
primary channels. The assumption used is that the average surface
reflectance for each channel is the same gray, the average of the
lightest and darkest naturally occurring surface-reflectance values
(Hurlbert, 1986). Objections have been raised against such a theory,
most notably by Marr (1982). Specifically, there is much evidence that
the color channels are not processed independently, and that most
color-sensitive cells have an opponent color organization. The opponent theory of color vision suggests that there are two color channels,
those of (red − green) and (yellow − blue), and one intensity channel.
Interestingly, the existence of such channels has been predicted as well
by factor analysis implemented by the KLT. One should keep in mind,
however, that the opponent theory is a point not a spatial theory, as is
the retinex. As such, there are difficulties accommodating the phenomena of simultaneous contrast, i.e., those phenomena occurring when
different stimuli, whose luminance is in the same proportion to their
surroundings, have equal lightness.

Hurlbert and Poggio (1988) have recently suggested an improved
lightness computation algorithm that does not need the nonlinear
thresholding step [Eq. (5-2b)] to eliminate smooth gradients of illumination. The algorithm, discussed next, is based on a PDP model, that of
DAM (Section 4.2.1.). Similar to Land's most recent retinex theory, it
divides the image irradiance at each pixel by a weighted average of the
irradiance (luminance) at all pixels in a large surround and takes the
logarithm of that result to yield lightness. The lightness algorithm
separates reflectance from illumination or in the authors' terminology,
synthesizes it automatically from a set of examples. The image irradiance, or luminance L is given as in Eq. (5-1) for each pixel (x, y) on the

image surface. Taking the logarithm of Eq. (5-1) then yields $L(x, y) = R(x, y) + I(x, y)$. (Homomorphic filtering, discussed in Section 3.1.2. is used in signal processing and proceeds similarly by filtering out low frequencies due to slow gradients of illumination and then exponentiates the result to yield the original reflectance.) The DAM learning amounts to determining a projection memory M, such that $ML = R$, where L and R are the stimulus and response matrices made up of pairs of luminance and reflectance vectors, respectively. As we showed in Section 4.2.1., the memory matrix M is given as $M = RL^+$. A second memory operator could recover the illumination I as well.

The derived lightness memory operator M is, in its central part, a space-invariant filter with a narrow positive peak and a broad, shallow, negative surround. The corresponding Fourier transform of M approximates a band-pass filter that cuts out low frequencies due to slow gradients of illumination and keeps the other frequencies corresponding to steep changes in reflectance. The memory operator that recovers the illumination yields a low-pass filter. The approach discussed shows both the strong connection between vision and traditional signal-processing techniques, and the increasing relevance of PDP modeling for deriving improved vision algorithms. Last but not least, the lightness algorithm is one further example of how ill-posed problems can be handled via PDP modeling.

5.1.2.   Optical Flow

It is well known that shape and depth information can be readily perceived by humans when monocularly viewing appropriately moving random dot fields in the absence of nonmotion cues. This applies both to movement of individual objects in a random dot scene and to movement of one's head when viewing a random dot scene under conditions that simulate visual movement parallax. To understand how one might computationally extract and utilize such motion information, researchers have proposed that time-varying imagery is first processed to obtain an intrinsic optical flow function. The optical flow intrinsic image is a vector flow field that records, pointwise, the instantaneous velocity of gray-scale pattern displacements within the plane of image formation. It is widely recognized that the optical flow function must be subjected to further processing in order to recover surfaces and/or shapes whose identity, location, size, attitude (slant and tilt), rotation, and translation parameters are concealed within the spatiotemporal patterns of the time-varying optical flow field.

At least three criteria are generally judged to be important for rating the performance of optical flow methods. First, optical flow methods should lead to flow fields that have high resolution in both space and time. The optical flow vector obtained at each spatiotemporal point should accurately represent the velocity within a small volume $dV = dxdydt$ of image (space, time) rather than representing the averaged velocity over a more extensive volume of space and time. Second, flow derivation methods should be sufficiently general so as to be applicable to a wide range of natural imagery rather than require strong restrictions on the image formation process and scene content. Third, optical flow methods should not be overly sensitive to noise, which is introduced, for example, by sensor electronics.

We will now review the basics of several methods that have been proposed for computing optical flow of time-varying imagery. These methods include those based on (a) the correspondence of local conspicuous features, (b) spatiotemporal gradients, and (c) spatiotemporal frequency (STF).

5.1.2.1. Feature Correspondence Approach

Given a pair of successive images in time, the first step in this method is to detect local conspicuous gray-scale features (tokens) in each image. These might, for example, be ad hoc features (e.g., corners, arc and line segments, etc.) that are detected by correlation of the image function and specific feature templates. Conversely, a more general scheme might be to detect arbitrary features that are simultaneously well localized and of high contrast without regard to their specific shape or identity. In either case, the feature-detection stage will produce for each image a list of spatial coordinates giving the location of each detected feature along with corresponding information (e.g., parametric, symbolic, or gray-scale information) describing it.

Having obtained a pair of feature lists, a pair-wise matching between corresponding features in the two lists must be obtained that minimizes some cost function based on (a) pair-wise feature affinities, as measured by the quality of the match between pairs of the previously obtained feature descriptions; (b) information about reasonable constraints on maximum interframe feature displacement, readily determined by multiplying the maximum expected flow velocity by the interframe time interval; (c) the uniqueness constraint on feature pairs, which is valid only if flow field "sinks" and "sources" are always absent; and (d) various other global correspondence consistency requirements. One must also deal with additional complications introduced by features that appear only in one of the feature lists. From the correspondence list, a sparse optical flow field is obtained. To refine this general approach, multiscale image representations help to overcome handling variability difficulties in flow velocity and feature size.

The viability of the whole approach is challenged, however, by Jenkin and Kolers (1986). They argue that the notion of correspondence seems ill suited to the task of accounting for how an object is positioned in time and space. Computer algorithms based on simple token matching have not been too successful and, as a consequence, increasingly sophisticated and elaborate token models and selection criteria are needed to accommodate visual information. Such comments argue against symbolic ("tokens") discrete processing.

5.1.2.2. Spatiotemporal Gradient Approach

This approach for computing optical flow employs first-order spatial and temporal gradients of a time-varying image and, at each image point, estimates the component of motion in the direction of maximally increasing gray-scale intensity. Underconstrained, the gradient-based local velocity estimation procedure characterizing this approach cannot unambiguously estimate the velocity on a local basis. This difficulty, often referred to as the aperture problem, requires that one employ some type of global relaxation method to obtain an unambiguous estimate of the true optical flow field. We will first discuss the initial underconstrained local velocity estimation procedure.

Given a continuous, differentiable time-varying image, $f(x, y, t)$, one can form the Taylor series expansion of this function as

$$f(x + dx, y + dy, t + dt) = f(x, y, t) + f_x\, dx + f_y\, dy + f_t\, dt$$
$$+ \text{ higher order terms}, \tag{5-3a}$$

where $f_x = \partial f / \partial x$. Assuming that the time-varying image is band limited, we disregard the second- and higher-order terms for sufficiently small interframe displacements (dx, dy, dt). One can make the further reasonable assumption that, during the time interval $(t, t + dt)$, the gray-scale function defined over each small image domain patch undergoes pure translation over some short distance (dx, dy). That is, for each (x, y, t), there exists some triple (dx, dy, dt) such that

$$f(x + dx, y + dt, t + dt) = f(x, y, t). \tag{5-3b}$$

Combining Eqs. (5-3a) and (5-3b), one obtains

$$f_x\,dx + f_y\,dy + f_t\,dt = 0$$

or equivalently,

$$f_x\frac{dx}{dt} + f_y\frac{dy}{dt} + f_t = 0. \tag{5-3c}$$

Noting that the velocity components v_x and v_y are $v_x = dx/dt$ and $v_y = dy/dt$, the last equation becomes

$$f_x v_x + f_y v_y = -f_t, \tag{5-4}$$

which one can rewrite using the inner product as

$$(f_x, f_y) \cdot (v_x, v_y) = -f_t. \tag{5-5}$$

Fig. 5.1, depicts the locus of possible solutions (v_x, v_y) to the gradient underconstrained equation in terms of the measurable quantities f_x, f_y, and f_t. Having superposed the velocity plane on the image plane centered at (x_0, y_0), one can see that any of the depicted vectors could correspond to the true flow vector at (x_0, y_0). This ambiguity is known as the aperture problem. The particular component of velocity $v^\parallel$ that is parallel to the direction of the spatial gradient $\nabla f = (f_x, f_y)$ at each point (see Fig. 5.1.) is given by

$$v^\parallel = -\frac{f_t}{|\nabla f|}\frac{\nabla f}{|\nabla f|} = -\frac{f_t}{|\nabla f|^2}\nabla f = -\frac{f_t}{f_x^2 + f_y^2}\begin{bmatrix} f_x \\ f_y \end{bmatrix}. \tag{5-6}$$

The previous relation is of course valid for each point (x, y, t), where the time-varying image f is differentiable. Eq. (5-4) or Eq. (5-5) does not completely specify the velocity vector at each point, but rather specifies the particular component of velocity that is parallel to the local spatial gray-scale gradient (i.e., $v^\parallel$ as defined). Therefore, some sort of global relaxation method must be subsequently employed to estimate the true velocity at each image point.

Disambiguation methods needed to overcome the aperture problem gather supporting evidence for a unique solution and are usually some instantiation of a specific PDP model. The Fennema and Thompson (1979) method, based on the Hough algorithm (see Section 4.2.4.), is a

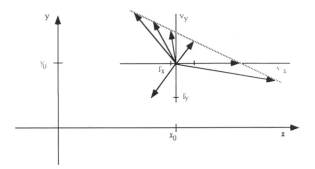

Figure 5.1.
Aperture problem.

variation on the connectionist model. Assume that the basic motion equation is given as $ds = vdt$, and that the gradient G is given in terms of (magnitude, phase) as $G = (|G|, \psi_G)$. Further, assume that the velocity v is given as $v = (|v|, \psi_v)$ and that the camera is fixed while the scene moves. Then, one can write

$$ds = \frac{ds}{df}df \simeq -\frac{df}{\dfrac{df}{ds}} = -\frac{df}{G}, \tag{5-7a}$$

where f and s stand for image intensity and (distance) space, respectively. Then,

$$df = -Gds = -Gvdt$$

$$df = -|G||v|\cos(\psi_G - \psi_v)dt \tag{5-7b}$$

$$|v| = \frac{-df}{|G|\cos(\psi_G - \psi_v)}.$$

For each image point, df and G trace a curve through the $(|v|, \psi_v)$ space corresponding to possible velocities. The operating assumptions are

(a) $||G(t_1)| - |G(t_2)|| < a_1$

$$|\psi_G(t_1) - \psi_G(t_2)| < a_2$$

or

$$2\pi - |\psi_G(t_1) - \psi_G(t_2)| < a_2,$$

i.e., the gradient at a point remains constant over time

(b) $$|\Delta f| > a_3,$$

i.e., eliminate from further consideration points where the intensity change is due primarily to scanner noise.

(c) Low-pass filter the image before deriving the optical flow in order to increase the signal-to-noise (SNR) ratio and to blur discontinuities due to change in the velocity. The corresponding blur or, equivalently, the point-spread function (PSF) is a function of expected velocity.

Horn and Schunk (1981) suggest an alternative method for disambiguation based on relaxation. The spatiotemporal gradient method yields $\nabla f \cdot \mathbf{v} = -f_t$, where $\nabla f = (f_x, f_y)$ and $\mathbf{v} = (v_x, v_y)$. The ambiguity problem is solved subject to a smoothness constraint given in terms of measured Laplacians, $\nabla^2(v_x)$ and $\nabla^2(v_y)$. (Recall from Section 2.1. that smoothness amounts to the Laplacian being almost nil.) The Laplacians can be approximated as

$$\nabla^2(v_x) \cong v_x - \bar{v}_x \quad \text{and} \quad \nabla^2(v_y) \cong v_y - \bar{v}_y,$$

where $\bar{v}_x$ and $\bar{v}_y$ are the average velocity components for some image neighborhood. The derivation of an unique solution results from the minimization of a specific energy function given as

$$E^2(x, y) = (\nabla f \cdot \mathbf{v} + f_t)^2 + \lambda^2 [(\nabla^2(v_x))^2 + (\nabla^2(v_y))^2]. \qquad (5\text{-}8)$$

The first term ensures that the solution is a close approximation to the spatiotemporal equation, while the second term introduced by a Lagrange multiplier ensures vector field solution smoothness. To minimize the energy function E, one has to set the derivative pair $(\partial E^2/\partial v_x, \partial E^2/\partial v_y)$ equal to zero. This yields the following equation:

$$(\lambda^2 + f_x^2)v_x + f_x f_y v_y = \lambda^2 \bar{v}_x - f_x f_t$$

$$f_x f_y v_x + (\lambda^2 + f_y^2)v_y = \lambda^2 \bar{v}_y - f_y f_t.$$

Assuming that $\partial^2[\nabla^2(v_x)]/\partial v_x \equiv 1$, the solution to the system of equations is

$$v_x = \bar{v}_x - f_x \frac{P}{D} \quad \text{and} \quad v_y = \bar{v}_y - f_y \frac{P}{D}, \qquad (5\text{-}9)$$

where

$$P = f_x \bar{v}_x + f_y \bar{v}_y + f_t$$
$$D = \lambda^2 + f_x^2 + f_y^2.$$

The iterative (relaxation) algorithm for $(n = 2)$ frames is

$k = 0$

Initialize v_x^k and v_y^k to zero

Until some error measure is satisfied *do*

$$v_x^k = \bar{v}_x^{k-1} - f_x \frac{P}{D}$$

$$v_y^k = \bar{v}_y^{k-1} - f_y \frac{P}{D}.$$

(5-10a)

This procedure could be extended to multiple frames $(n > 2)$ and yield a better estimate of the optical vector field. Specifically,

$t = 0$

Initialize $v_x(x, y, 0)$, $v_y(x, y, 0)$

for $\dot{t} = 1$ *until* maxframes *do*

$$v_x(x, y, t) = \bar{v}_x(x, y, t - 1) - f_x \frac{P}{D}$$

$$v_y(x, y, t) = \bar{v}_y(x, y, t - 1) - f_y \frac{P}{D}.$$

(5-10b)

Many assumptions are underlying both of the disambiguation algorithms discussed. Furthermore, as in all methods that derive raw data from some type of differentiation operator, these methods are also very sensitive to noise. Integral, rather than differential methods, could improve the solution by gathering appropriate supporting evidence, which brings us to the spatiotemporal frequency (STF) method.

5.1.2.3.   Spatiotemporal-Frequency (STF) Approach

The spatiotemporal frequency (STF) approach to optical flow derivation encompasses all methods that are based upon some underlying spatiotemporal frequency image representation. Mammalian vision literature reveals one advantage for using STF image representation to compute optical flow. Recent investigations have demonstrated that many neurons in various visual cortical areas of the brain behave as spatiotemporal frequency bandpass filters. Studies in human vision of stroboscopic apparent motion have also produced results that appear explicable in terms of STF bandpass channels (Watson and Ahumada, 1983).

Gafni and Zeevi (1977, 1979) have proposed an STF method for computing optical flow such that, given a time-varying image, $f(x, y, t)$, one first computes its Fourier transform denoted $F(w_x, w_y, w_t)$. An arbitrary point $(w_{0_x}, w_{0_y}, w_{0_t})$ of the Fourier transform corresponds to a sinusoidal grating of spatial frequency $w_0^T = (w_{0_x}, w_{0_y})$ moving past a fixed image point with temporal frequency w_{0_t}. The velocity component of such a grating in a direction orthogonal to its axis of constant luminance is given by

$$v^{\perp} = -\frac{w_{0_t}}{|w_0|}\frac{w_0}{|w_0|} = -\frac{w_{0_t}}{|w_0|^2}w_0 = \frac{-w_{0_t}}{w_{0_x}^2 + w_{0_y}^2}\begin{bmatrix} w_x \\ w_y \end{bmatrix}. \qquad (5\text{-}11)$$

The Fourier transform $F(w_x, w_y, w_t)$ provides a set of coefficients that, using the inverse Fourier transform, allows one to uniquely represent the original time-varying image as a weighted superposition of moving sinusoidal luminance gratings.

Equation (5-11) defines a unique orthogonal velocity $v^{\perp}$ for each spatiotemporal frequency (w_x, w_y, w_t). Conversely, considering all spatiotemporal frequencies that satisfy Eq. (5-11) for some arbitrary fixed orthogonal velocity $v_0^{\perp}$, one obtains an infinite set $S_{v_0}^{\perp}$ of spatiotemporal frequencies given by

$$S_{v_0}^{\perp} = \left\{ (w_x, w_y, w_t) : \frac{w_{0_t}}{w_{0_x}^2 + w_{0_y}^2}\begin{bmatrix} w_x \\ w_y \end{bmatrix} + v_0^{\perp} = 0 \right\}. \qquad (5\text{-}12)$$

The locus of points $S_{v_0^{\perp}}$ defined by Eq. (5-12) is just a single line (or pencil) passing through the origin of the spatiotemporal frequency space. In fact, since $v_0^{\perp}$ was arbitrarily chosen, we observe that in general, for each orthogonal velocity $v^{\perp}$, there corresponds a unique pencil $S_{v^{\perp}}$ of consistent spatiotemporal frequencies. Exploiting this observation, Gafni and Zeevi (1977, 1979) have proposed forming velocity channels by separately integrating the magnitude of the Fourier transform $|F(w_x, w_y, w_t)|$ over the corresponding members of the family of infinite (1D) pencils passing through the origin of the spatiotemporal frequency space. Their definition of velocity channel can be formalized by defining a velocity polling function, labeled as $C_f(v^\perp)$, as

$$C_f(v^\perp) = \iiint_{S_v^\perp} |F(w_x, w_y, w_t)| dw_x dw_y dw_t, \qquad (5\text{-}13)$$

where $F(w_x, w_y, w_t)$ is the Fourier transform of the time-varying image, and $S_{v^\perp}$ is as defined in Eq. (5-12).

Finally, Gafni and Zeevi suggest that the particular velocity where $C_f(v^\perp)$ obtains its global maximum should be chosen as the velocity of object motion in the corresponding time-varying image. This scheme for flow derivation does not, however, provide for optical flow function generation, but yields a single velocity characterizing the entire image for all time. To compute a time-varying optical flow function, Gafni and Zeevi have suggested that numerous regionally computed (3D) Fourier transforms should be employed instead of just one global transform. A regional velocity could then be determined for each of the regional Fourier transforms in the previously described manner.

Two limitations of this method for motion analysis require attention. First, with regard to Gafni and Zeevi's suggestion that time-varying optical flow of an image sequence be computed by employing the moduli of locally computed Fourier transforms as the underlying STF representation, the local Fourier analysis suffers from severe tradeoffs in spatiotemporal vs. spatiotemporal frequency resolution because of the varied (3D) analysis window size. This tradeoff induces a corresponding tradeoff in spatiotemporal resolution vs. the optical flow function reliability. An approximation to the Wigner distribution—a truly STF image representation rather than a simple patchwork of Fourier power spectra—can minimize this limitation.

A second serious limitation is Gafni and Zeevi's decision to estimate $v^\perp$, the velocity orthogonal to the isoluminance bars of moving gratings that comprise the image. They overlook the highly ambiguous nature of true velocity corresponding to a particular spatiotemporal frequency, which one must estimate to derive an image's optical flow. Velocity ambiguity associated with moving sinusoidal gratings and its significance for motion coherence in human perception is illustrated in Figs. 5.2a–c. Specifically, as seen in Fig. 5.2a., a given spatiotemporal frequency $(w_{0_x}, w_{0_y}, w_{0_t})$ corresponds to an infinite number of possible

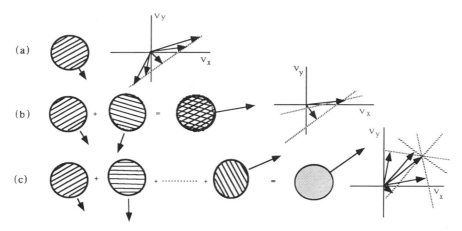

Figure 5.2.
Velocity ambiguity associated with viewing a moving grating through a circular aperture.

velocities, v, satisfying the relationship

$$v^T \cdot w_0 + w_{0_t} = 0, \tag{5-14}$$

where $v^T = (v_x, v_y)$ and $w_0^T = (w_{0_x}, w_{0_y})$. Although the true velocity of a single moving grating (Fig. 5.2a.) can never be uniquely known, an image composed of multiple gratings at various orientations (Figs. 5.2b. and 5.2c.) has a unique velocity determined by the intersection of the loci of those velocities consistent with the separate spatiotemporal frequency components of the image.

Having briefly discussed the true versus the orthogonal velocity estimation using the STF approach for optical flow derivation, we will now consider the obvious similarities between the Eqs. (5-6) and (5-11), which are associated with the spatiotemporal gradient and Gafni and Zeevi's spatiotemporal frequency methods, respectively. Earlier, we defined the parallel velocity $v^{\parallel}$ as the component of the true velocity in the direction of the image's local spatial gradient. The local gradient, of course, defines the direction of maximum increase in the spatial luminance function, which in turn is orthogonal to the local isoluminance axis. Therefore, $v^{\parallel}$ in Eq. (5-6) and $v^{\perp}$ of Eq. (5-11) represent essentially equivalent velocities, which correspond to projections of the true velocity onto axes perpendicular to the local isoluminance orientation. Accordingly, neither the spatiotemporal gradient approach nor Gafni and Zeevi's STF method allows one to estimate directly the true local velocity of motion as informally defined in the case of superposed moving gratings. The spatiotemporal gradient approach requires what often amounts to ad hoc postprocessing of the initial flow field estimate using, for example, relaxation methods to force global consistency on the flow field. Gafni and Zeevis' STF method includes no explicit provision for estimating the true flow field although postprocessing methods similar to those used to disambiguate gradient-based flow fields would be equally applicable to their method. Using the Wigner distribution in conjunction with an improved definition of the velocity polling function, our modified STF method (Jacobson and Wechsler, 1987) proves to be an explicit analytic procedure for directly estimating true optical flow without requiring any further postprocessing.

Velocity ambiguity associated with a single moving grating arises naturally when considering the Fourier spectrum of an image in which all points are uniformly translated with a single constant velocity (Eq. 5-14). In this special case, it suffices to use a single global Fourier power spectrum as an STF image representation.

Suppose we are given a time-varying image $f(x, y, t)$ that is uniformly translating at some constant velocity (v_x, v_y). That is, we have

$$f(x, y, t) = f(x - v_x t, y - v_y t)$$
$$= f(x, y) \otimes \delta(x - v_x t, y - v_y t), \qquad (5\text{-}15)$$

where $f(x, y) = f(x, y, 0)$, $\delta(x, y)$ is the impulse distribution, and "$\otimes$" denotes convolution. The Fourier transform of the translating image $f(x, y, t)$ is given by

$$F(w_x, w_y, w_t) = F(w_x, w_y)\delta(v_x w_x + v_y w_y + w_t) \qquad (5\text{-}16)$$

yielding a very interesting result: a uniformly translating image with velocity (v_{0_x}, v_{0_y}) has a Fourier spectrum that is zero everywhere in the 3D spatiotemporal frequency space (w_x, w_y, w_t) except on the single plane defined by

$$\{(w_x, w_y, w_t) : v_{0_x} w_x + v_{0_y} w_y + w_t = 0\}. \qquad (5\text{-}17)$$

In general, for each velocity $v = (v_x, v_y)$, Eq. (5-17) allows one to define a unique planar locus of spatiotemporal frequencies that are consistent with that velocity. Therefore, rather than forming a velocity polling function $C_f(v^\perp)$ by integrating over pencils in the STF space, as has been previously suggested, one should integrate over planar regions determined for each v, by Eq. (5-17), in order to obtain a velocity polling function $C_f(v)$.

An improved STF method is based on the Wigner distribution (WD). Given a time-varying image $f(x, y, t)$, its Wigner distribution is a 6D function defined as

$$W_f(x, y, t, w_x, w_y, w_t) = \int\!\!\!\int\!\!\!\int_{-\infty}^{\infty} R_f(x, y, t, \alpha, \beta, \tau)e^{-j(\alpha w_x + \beta w_y + \tau w_t)}d\alpha d\beta d\tau,$$

$$(5\text{-}18a)$$

where

$$R_f(x, y, t, \alpha, \beta, \tau) = f\left(x + \frac{\alpha}{2}, y + \frac{\beta}{2}, t + \frac{\tau}{2}\right)$$
$$\times f^*\left(x - \frac{\alpha}{2}, y - \frac{\beta}{2}, t - \frac{\tau}{2}\right),$$

and "*" denotes complex conjugation. Similarly, the WD is defined in terms of the Fourier transform of the time-varying image as

$$W_f(x, y, t, w_x, w_y, w_t) = \frac{1}{8\pi^3} \int\!\!\!\int\!\!\!\int_{-\infty}^{\infty} S_f(\eta, \xi, \gamma, w_x, w_y, w_t)$$
$$\times e^{j(\eta x + \xi y + \gamma t)}d\eta d\xi d\gamma, \qquad (5\text{-}18b)$$

where

$$S_f(\eta, \xi, \gamma, w_x, w_y, w_t) = F\left(w_x + \frac{\eta}{2}, w_y + \frac{\xi}{2}, w_t + \frac{\gamma}{2}\right)$$
$$\times F^*\left(w_x - \frac{\eta}{2}, w_y - \frac{\xi}{2}, w_t - \frac{\gamma}{2}\right).$$

The WD of a time-varying image, as defined, is therefore a simultaneous spatiotemporal/spatiotemporal-frequency representation.

We derived earlier a general expression for the Fourier transform of a uniformly translating image. While highlighting similarities and differences between the Fourier transform and the Wigner distribution as an STF representation, we derive next a general expression for the WD of a uniformly translating image. Recall from Eq. (5-15) that an image $f(x, y, t)$ undergoing uniform translation may be expressed as the convolution between a static image and a translating delta function. By the convolution and windowing properties of the Wigner distribution, the WD of $f(x, y, t)$, denoted $W_f(x, y, t, w_x, w_y, w_t)$, is given by

$$W_f(x, y, t, w_x, w_y, w_t)$$
$$= [W_f(x, y, w_x, w_y)\delta(w_t)] \underset{x,y,w_t}{\otimes} W_\delta(x, y, t, w_x, w_y, w_t), \quad (5\text{-}19)$$

where $W_f(x, y, w_x, w_y)$ is the WD of $f(x, y, 0)$, W_δ is the WD of $\delta(x - v_x t, y - v_y t)$ and "$\underset{x,y,w_t}{\otimes}$" denotes convolution with respect to the spatial variables, x, y, and the temporal-frequency variable w_t. With W_δ as defined, one can derive

$$W_\delta(x, y, t, w_x, w_y, w_t) = \delta(x - v_x t, y - v_y t)\delta(v_x w_x + v_y w_y + w_t). \quad (5\text{-}20)$$

Substituting Eq. (5-20) into Eq. (5-19), one obtains

$$W_f(x, y, t, w_x, w_y, w_t) = [W_f(x, y, w_x, w_y)\delta(w_t)]$$

$$\underset{x,y,w_t}{\otimes} [\delta(x - v_x t, y - v_y t)\delta(v_x w_x + v_y w_y + w_t)]$$

$$= [\delta(w_t) \underset{w_t}{\otimes} \delta(v_x w_x + v_y w_y + w_t)]$$

$$\times [W_f(x, y, w_x, w_y) \underset{x,y}{\otimes} \delta(x - v_x t, y - v_y t)]$$

$$= \delta(v_x w_x + v_y w_y + w_t)W_f(x - v_x t, y - v_y t, w_x, w_y).$$
$$(5\text{-}21)$$

We find, therefore, that the WD of a linearly translating image with velocity $v^T = (v_x, v_y)$ is everywhere zero except in the 2D linear subspace defined by

$$\{(x, y, t, w_x, w_y, w_t) : v_x w_x + v_y w_y + w_t = 0\}$$

for fixed v_x, v_y. Equivalently, we observe that, for arbitrary x, y, t, each local STF spectrum of the WD, denoted $W_{f_{x,y,t}}(w_x, w_y, w_t)$, is zero everywhere except on the plane R_v defined as

$$R_v = \{(w_x, w_y, w_t) : v_x w_x + v_y w_y + w_t = 0\} \quad (5\text{-}22)$$

for fixed v_x, v_y.

Note that Eq. (5-22) defines the same region as Eq. (5-17) when equivalent velocities of translation are considered. Therefore, for a linearly translating image, the local STF spectra $W_{f_{x,y,t}}(w_x, w_y, w_t)$, like the global Fourier transform $F(w_x, w_y, w_t)$, contains energy on only a single plane in the STF space, which is uniquely determined by the image's translation velocity. Unlike the Fourier transform, however, the WD (a spatiotemporal/spatiotemporal-frequency representation) assigns a 3D STF spectrum to each point (x, y, t). Therefore, given a procedure for estimating velocity associated with a given STF spectrum, the WD allows separate estimations at each point (x, y, t) in order to get a time and space-varying optical flow function.

A modified STF approach overcomes the two limitations (resolution vs. reliability tradeoff and inappropriate definition of the velocity polling function—VPF) of the STF flow derivation method proposed by Gafni and Zeevi. Rather than using the moduli of regionally computed Fourier transforms as the underlying STF representation, we suggest that one employs (an approximation to) the Wigner distribution of the time-varying image instead. In particular, for each point (x, y, t), the WD defines a high resolution (3D) spatiotemporal-frequency function $W_{f_{x,y,t}}(w_x, w_y, w_t)$ from which a local velocity polling function (VPF) $C_{f_{x,y,t}}(v_x, v_y)$ is computed. Second, one must reformulate the VPF definition in accordance with the spatiotemporal frequency constraint given in Eq. (5-17). Recall that all spatiotemporal frequencies (w_x, w_y, w_t) satisfying the relationship $v_{0_x} w_x + v_{0_y} w_y + w_t = 0$ are consistent with the velocity $v_0^T = (v_{0_x}, v_{0_y})$. As such, we suggested earlier that to obtain the VPF for some arbitrary velocity v_0^T, one should integrate the STF representation over the planar region R_{v_0} defined according to Eq. (5-22). Combining the definition of the VPF with the use of the WD as an underlying STF representation, we explicitly define the velocity polling function, denoted by $C_f(v_x, v_y; x, y, t)$, as

$$
C_f(v_x, v_y; x, y, t) = \iiint\limits_{-\infty}^{\infty} |W_f(x, y, t, w_x, w_y, w_t)|^n
$$

$$
\times \, \delta(v_x w_x + v_y w_y + w_t) d\alpha d\beta d\tau. \qquad (5\text{-}23)
$$

Finally, in terms of the VPF $C_f(v'_x, v'_y; y, t)$ of an image $f(x, y, t)$, the optical flow $v_f(x, y, t)$ can be defined as

$$
\mathbf{v}_f(x, y, t) = \left\{ \begin{bmatrix} v_x(x, y, t) \\ v_y(x, y, t) \end{bmatrix} : \begin{array}{c} C_f(v_x, v_y; x, y, t) > C_f(v'_x, v'_y; x, y, t) \\ \forall (v'_x, v'_y) \neq (v_x, v_y) \end{array} \right\}. \qquad (5\text{-}24)
$$

We conclude by emphasizing that the STF approach handles

arbitrary time-varying imagery and not just uniform translation. We developed certain STF aspects using the analytically tractable model of uniform image translation. Due to the high cost of computation, the experiments performed (Jacobson and Wechsler, 1987) also employ time-varying images characterized by simple uniform translation. Our goal was to use simple examples to graphically illustrate how the STF approach directly estimates the true velocity vector at an arbitrary image point, so long as the image is locally composed of multiple, nonidentically oriented spatial-frequency components. Finally, the uniformly translating images employed in our experiments were patterned after those used in psychophysical experiments studying motion coherence phenomena in humans.

The STF approach suggests the use of STF filters corresponding to spatiotemporal receptive fields (RF). The (invariant) cytoarchitecture discussed earlier (Section 3.1.5.) can also be extended along the temporal axis. More recently, Heeger (1987) has suggested a similar STF approach using Gabor filters for underlying 3D (space-time) RF, while Fleet and Jepson (1989) have suggested to compute normal velocity from local phase information. Waxman *et al.* (1988) suggest the use of Gaussian activation profiles in a spatiotemporal neighborhood of specified scale around detected features. The profiles are then convected with the feature's motion and thus avoid the assumption of convected intensity. Such an assumption [see Eq. (5-3b)] carries the intensity of a pixel with the local flow to its new image location and is quite restrictive. Actual implementation employs a spatiotemporal RF family of varying scale. All the STF architectures include a selection step, which is characteristic of the velocity polling function (VPF). The VPF can be interpreted as competition among motion-sensitive cells. One could account for motion transparency if multiple motion interpretations to a single location are allowed, instead of winner take all, as in Eq. (5-24). Koenderink (1988) considers some consequences of the STF organization with regard to scale-time and scale space-time, specifically, the need to conserve causality in the resolution domain at any given moment in time. Causality means that "blurring only destroys structure but is not allowed to generate it. Linkages between the different levels of resolution are only permitted to bifurcate towards the direction of increased resolution."

Adequately assessing the relative strengths and weaknesses of the different solutions for estimating optical flow fields is difficult. Each has many published variations, continues to evolve, and has only been superficially evaluated. With this in mind, we compare these methods according to a number of criteria. Owing to the great popularity and relative success of the spatiotemporal-gradient approach, comparisons between this method and the modified STF approach dominate our discussion.

Among the methods reviewed, the STF approach stands for its unique flow estimate at a particular point (x_0, y_0, t_0) based upon the integration of information over the entire space-time extent of the image $f(x, y, t)$. Recall that each local velocity estimate is obtained by choosing the velocity for which the VPF attains its maximum. But each locally defined VPF is computed from the local 3D spatiotemporal-frequency (STF) spectrum of the WD—a function computed by integrating information over the entire extent of the time-varying image. It follows that the velocity estimate at each point is also dependent upon information from the entire spatiotemporal extent of the image. Use of a conventional STF representation, e.g., the power spectrum, as a basis for computing the VPF, leads to tradeoffs between information integration and the spatial resolution of the flow field as the size of the power spectral analysis window varies. Only by exploiting the unique properties of a cojoint representation such as the WD can one fully integrate spatiotemporal information while, at the same time, retain uncompromised spatiotemporal resolution of the estimated optical flow field.

Many researchers have discussed the importance of integrating information over extended regions of space and time when making velocity estimates. Such integration of information over spatial neighborhoods and multiple ($n > 2$) time frames is generally envisioned as a postprocessing stage, which follows the initial estimation of some local velocity component at each point. Recall, for example, the postprocessing of the initial underconstrained velocity flow-field estimates is the final processing step in the spatiotemporal-gradient approach. Use of postprocessing to achieve temporal and spatial integration of information begs the question, however, of the size of the region such postprocessing should cover. Naturally, one would like this region to be very large so as to maximize information integration before arriving at a final flow field estimate. Most approaches to postprocessing do little more, however, than assure smoothness of the field of velocity estimates. Relatively restricted regions of space and time must, therefore, be employed in these conventional postprocessing stages to avoid spatiotemporal blurring of important information in the true flow field. In contrast to most common flow-field estimation approaches, the modified STF method does not require a postprocessing stage to disambiguate an initial estimate of velocity components. Instead, direct true velocity estimates for each point and all local velocity estimates are computed without considering estimates of neighboring points in space and time.

As discussed, the Wigner distribution, as an intermediate representation in the modified STF approach, corrects these deficiencies to optical flow computation. Each local spatiotemporal-frequency spectrum of the WD is computed from globally integrated information. Hence, the local velocity estimate subsequently made at each point is inherently based on global information. Contrast this with the spatiotemporal gradient approach, which is based on a local representation (the spatiotemporal gradient function) that is computed at each point from purely local information. Since each initial velocity estimate in this approach is based solely on the corresponding local spatiotemporal gradient, any global integration of information must of necessity be incorporated as a postprocessing step.

We will now discuss an important distinction between the constraint Eqs. (5-5) and (5-14). The spatiotemporal gradient constraint relation Eq. (5-5) provides a single equation in two variables for each point (x, y, t). Assuming that the measurable differentials, defining the known quantities in this equation, are accurately known (which is often not the case), one is still faced with the problem that the velocity estimate is underconstrained. Additional ad hoc constraints must, therefore, be introduced if one wishes to compute an estimate of the true flow velocity at each image point. Unlike the spatiotemporal gradient relation, the STF constraint relation Eq. (5-14) defines an infinite number of constraint equations at each image point (x, y, t). Namely, for each image point, one STF constraint equation corresponds to each spatiotemporal frequency (w_x, w_y, w_t) in the local spectrum of the WD at that image point. Since typical imagery contains many spatiotemporal frequency components, the STF constraint relation greatly overconstrains the velocity estimate at each image point. Therefore, one does not need to devise further (ad hoc) constraints in order to estimate the true velocity field.

Continuing our comparison of the spatiotemporal gradient and modified STF approaches, we point out that, as an intermediate step, the former requires the computation of derivatives whereas the latter involves only integral transformations having large spatiotemporal support. Others have elaborated on the many error sources in estimates of spatiotemporal derivatives along with their combined effect on inaccuracies in the resulting velocity estimates. Complex procedures are often used to deal with such difficulties. Compared to differential methods, integral transformations of the modified STF approach are inherently much less likely to lead to serious velocity estimation errors. A source of major error will probably be real-world computational constraints that dictate using finite support for all integral transformations. Such errors can in theory, however, be reduced to an arbitrary degree as computational capabilities expand. Future work will nonetheless be required to quantify error sources and magnitudes in the modified STF approach so that a more quantitative comparison with other approaches can be made. Much of this error assessment will undoubtably involve discussions of the uncertainty principle as it applies to the conflict between simultaneous energy localization of spatiotemporal vs. spatiotemporal-frequency WD domains of a time-varying image, and how this uncertainty relation affects the accuracy of velocity flow estimates. Such uncertainty based estimation errors are intrinsic to the signal and will ultimately limit attainable resolution in future flow analysis systems.

5.1.3. Depth and Stereopsis

The topic of this section is to derive depth maps via stereopsis. Two eyes or cameras receive slightly different views. (Place a finger in front of your nose and alternate closing each of your eyes.) The relative difference in the observed layout of the two images is called binocular disparity and by using triangulation, depth can be derived. There are additional ways or cues for inferring depth. Monocular cues, as studied in psychophysics, include aerial perspective, texture gradients, perspective, shading, masking, and position in the field of view. Bringing an image into sharp focus and vergence of the two eyes are examples of active perception (see Chapter 6) helping to derive a depth map. Vergence keeps the disparities small, and furthermore, facilitates the make-up of object-centered coordinate systems (Ballard, 1988).

The geometry of stereopsis, the process of fusing two disparate

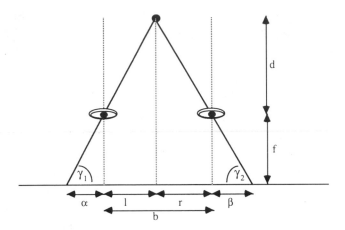

Figure 5.3.
The geometry for stereopsis.

images, is given in Fig. 5.3. Crucial to stereopsis is the correspondence process aimed at establishing a linkage between similar areas in the left and right image, respectively. Given the focal length of the camera f and the base (separation) b, what remains to be established for deriving the depth (distance) d is the displacement pair (α, β).

The geometry for triangulation is then quite easy to use.

$$\frac{d}{l} = \frac{d+f}{l+\alpha}; \quad \frac{d}{r} = \frac{d+f}{r+\beta}; \quad b = l + r$$

$$\therefore \quad d = \frac{fb}{(\alpha + \beta)}.$$

(5-25)

The correspondence process is more difficult and raises concerns similar to those we voiced in the case of optical flow derivation. The methods here are iterative in nature and involve some type of PDP computation, which aims at minimizing some (energy cost) function.

The Barnard and Thompson (1980) algorithm matches discrete tokens so that the resulting depth map is smooth. Interest operators choose those (discrete) pixels (i.e., edges) that enjoy high gray-level variance in all directions. All but the local maxima are suppressed. The correspondence process then takes over by looking for similar (token) pixels in terms of their neighborhoods. The algorithm constructs search neighborhoods in the right image of size $d_{\max} \times d_{\max}$, for candidate pixels (x_i, y_i) selected as interesting in the left image, where $d_{\max}$ is a function of the expected maximum disparity. For each possible (matching) pixel (x_m, y_m) a disparity $\mathbf{d} = (x_i - x_m, y_i - y_m)$ is established and a probability $P_m^0(d)$ is assigned to it according to the following process. Calculate the similarity $S_m(d)$ as the square difference between the windows surrounding (x_m, y_m) and (x_i, y_i), respectively. Then

$$W_m(d) = \frac{1}{1 + C \times S_m(d)} \qquad (5\text{-}26)$$

such that C is a constant and $W_m(d) \in [0, 1]$. Assume d^* as the null disparity and the probability that no match is feasible is

$$P_m^{(0)}(d^*) = 1 - \max_d [W_m(d)], \qquad (5\text{-}27\text{a})$$

where the superscript "0" stands for the first iteration, and the assumption is that the largest value of $W_m(d)$ probably corresponds to the true match. For all disparities d, such that $d \neq d^*$

$$P_m^{(0)}(d) = P_m(d|\alpha)[1 - P_m^{(0)}(d^*)], \qquad (5\text{-}27\text{b})$$

where $P_m(d|\alpha)$ is the probability that the disparity is d assuming that pixel (x_m, y_m) has a match. The probability for such disparities is

$$P_m(d|\alpha) = \frac{W_m(d)}{\sum_{d' \neq d^*} W_m(d')}$$

$$\sum P_m^{(0)}(d) + P_m^{(0)}(d^*) = 1. \qquad (5\text{-}28)$$

The consistency constraint ensures that the disparity varies smoothly and can be implemented through relaxation updating. One can then increase the probabilities of those disparities, which occur often in a region, and decrease those that do not. Disparities d and d' are consistent if

$$\max\{|d_x - d'_x|, |d_y - d'_y|\} \leq 1.$$

The probabilities $P_m^{(k)}(d)$ are updated at iteration k into new probabilities $P_m^{(k+1)}(d)$ by checking the disparities of all the left image candidate points in an $R \times R$ neighborhood of (x_m, y_m).

$$Q_m^{(k)}(d) = \sum_{\substack{n \in R \times R \\ n \neq m}} P_n^{(k)}(d)$$

$$\hat{P}_m^{(k+1)}(d) = \{A + B \times Q_m^{(k)}(d)\} P_m^{(k)}(d) \qquad \text{for } d \neq d^*$$

$$\hat{P}_m^{(k+1)}(d^*) = P_m^{(k)}(d^*). \qquad (5\text{-}29\text{a})$$

The normalized updated probabilities are given as

$$P_m^{(k+1)}(d) = \frac{P_m^{(k+1)}(d)}{\Sigma P_m^{(k+1)}(d)}. \tag{5-29b}$$

The constants (A, B) used in the scheme influence the convergence rate. "A" allows information to flow in from distant points, while B determines how fast the consistency measure affects the disparity probability. For efficiency, if $P_m^{(k)}(d) < 0.1$, "d" is purged. If this were to happen for all "d", then $P_m(d^*) = 1$. The process iterates until the steady state is reached, or if $P_m^{(k)}(d) > 0.7$.

Zhou and Chellappa (1988) have suggested a stereo method based on the additive model. The method aims at overcoming the usual problems associated with stereo-like amplitude bias, edge sparsity, and noise distortion. To overcome noise, a polynomial is first fit to the picture value function. The amplitude bias is handled next by applying differential operators that yield $\dot{g}_L(\bullet)$ and $\dot{g}_R(\bullet)$, the first order intensity derivatives of the left and right images, respectively. The epipolar constraint is then used in order to reduce the number of potential matches and thus to enhance the efficiency of the method. A neural network approach modeled after the CAN model is used to determine disparity values between the two images. There are $N_r \times N_c \times D$ neurons, where D is the maximum disparity, and the image is the size of $N_r \times N_c$. Let $V = \{v_{i,j,k}, 1 \leq i \leq N_r, 1 \leq j \leq N_c, 0 \leq k \leq D\}$ be the discrete neural network with $v_{i,j,k}$ standing for the state of the (i, j, k)th neuron. When $v_{i,j,k}$ is 1, the disparity value at (i, j) is k. Due to the uniqueness constraint, the $(D + 1)$ neurons corresponding to the (i, j) location are mutually exclusive and are updated at each step simultaneously. The interconnections between neurons (i, j, k) and (l, m, n) are symmetric and are given by $T_{i,j,k;l,m,n}$. Each neuron randomly and asynchronously receives input $u_{i,j,k}$ given as

$$u_{i,j,k} = \sum_{l=1}^{N_r} \sum_{m=1}^{N_c} \sum_{n=0}^{D} T_{i,j,k;l,m,n} v_{l,m,n} + I_{i,j,k}. \tag{5-30a}$$

where $I_{i,j,k}$ is the usual bias. The threshold function $\theta(x_{i,j,k})$ is defined as

$$\theta(x_{i,j,k}) = \begin{cases} 1 & \text{if } x_{i,j,k} = \max(x_{i,j,l}; \ l = 0, 1, \dots, D) \\ 0 & \text{otherwise} \end{cases} \tag{5-30b}$$

and $v_{i,j,k} = \theta(u_{i,j,k})$.

The energy function E_A corresponding to the CAN additive model is given as

$$E_A = -\frac{1}{2} \sum_{i=1}^{N_r} \sum_{l=1}^{N_r} \sum_{j=1}^{N_c} \sum_{m=1}^{N_c} \sum_{k=0}^{D} \sum_{n=0}^{D} T_{i,j,k;l,m,n} v_{i,j,k} v_{l,m,n}$$

$$- \sum_{i=1}^{N_r} \sum_{j=1}^{N_c} \sum_{k=0}^{D} I_{i,j,k} v_{i,j,k}. \tag{5-31a}$$

The centers of projection of the two cameras together with any 3D point define an epipolar plane, whose intersection with the image plane determines an epipolar line. The epipolar constraint restricts then the 2D search for corresponding matches to linear 1D searches along the corresponding epipolar lines. Under the epipolar line assumption, the energy can be rewritten as

$$E_B = \sum_{i=1}^{N_r} \sum_{j=1}^{N_c} \sum_{k=0}^{D} (\dot{g}_L(i,j) - \dot{g}_R(i,j \oplus k))^2 v_{i,j,k}$$

$$+ \frac{\lambda}{2} \sum_{i=1}^{N_r} \sum_{j=1}^{N_c} \sum_{k=0}^{D} \sum_{s \in S} (v_{i,j,k} - v_{i,j \oplus s,k})^2, \tag{5-31b}$$

where S is an index set for all the neighbors in a 5×5 window centered at point (i,j), λ is the Lagrange multiplier, and $\oplus$ is defined as

$$f_{a \oplus b} = \begin{cases} f_{a+b} & \text{if } 0 \le a + b \le N_c, N_r \\ 0 & \text{otherwise.} \end{cases}$$

Clearly, E_B is of the form $E_B = E_1 + E_2$, where the first term seeks disparity values, such that the match is as close as possible in a least square sense, while the second term is the smoothness constraint on the solution. As usual, one matches E_A against E_B and determines the synaptic weights and bias as

$$T_{i,j,k;l,m,n} = -48\lambda \delta_{i,l}\delta_{j,m}\delta_{k,n} + 2\lambda \sum_{s \in S} \delta_{i,l}\delta_{j,m \oplus s}\delta_{k,n}$$

and

$$I_{i,j,k} = -(\dot{g}_L(i,j) - \dot{g}_R(i,j \oplus k))^2, \tag{5-31c}$$

where $\delta_{a,b}$ is the (Dirac) impulse distribution, and λ determines the relative relevance of the two terms (E_1, E_2) appearing in the energy E_B for achieving good solutions.

Stereo matching can now proceed following the CAN model, which is applied to the previously defined network. The initial state of the neurons is set to

$$v_{i,j,k} = \begin{cases} 1 & \text{if } I_{i,j,k} = \max(I_{i,j,l}; \, l = 0, 1, \ldots, D) \\ 0 & \text{otherwise.} \end{cases} \qquad \text{(5-31d)}$$

The networks allow for feedback, i.e., $T_{i,j,k;i,j,k} \neq 0$. Consequently, there are perturbations for which the corresponding ΔE is positive and no state change is implemented.

Jepson and Jenkin (1988) performed considerable work on different aspects related to visual stereoscopic computation. Their conclusions cast great doubts on the main tenets of classical stereopsis algorithms. They make a compelling case "against the extraction of complex monocular features such as zero-crossings, against the use of inhibiting surface constraints and against the use of global stereopsis to produce a single-valued retinotopic depth map." Such comments are closely related to the ones espoused earlier by Jenkin and Kolers (1986) in the optical flow derivation context. Their significance lies in the development of a new method that relates disparity to the local phase difference between band-pass versions of the input images. No complex token extraction or correspondence is needed, and the spatial and temporal properties of the disparity measurement allow the construction of filter detectors that are tuned to surfaces with particular 3D orientations and of specific 3D trajectories. The approach is conceptually similar to the STF (see preceding Section 5.1.2.) and reaffirms that intrinsic images can be derived from cojoint (distributed) s/sf representations. The analogy to a massive cytoarchitecture of appropriate filters along several spatial and temporal dimensions again comes to mind.

The disparity-as-phase difference method suggested by Jepson and Jenkin (1989) works as follows. Sine and cosine Gabor filters can extract local band pass approximations of a 1D signal perceived by the left and right retinas of a binocular system as:

$$R_{\sin}(x) = A_r \sin(\omega_r x + \phi_r)$$

$$R_{\cos}(x) = A_r \cos(\omega_r x + \phi_r)$$

$$L_{\sin}(x) = A_l \sin(\omega_l x + \phi_l)$$

$$L_{\cos}(x) = A_l \cos(\omega_l x + \phi_l)$$

where A_r and A_l are constants dependent upon the response of the Gabor filter, ω_r and ω_l are nearly the peak pass frequency of the response, while ϕ_r and ϕ_l are the relative phase at the bandpass frequency. If the Gabor filter for the right and left signal is the same, $\omega_r \approx \omega_l$, and the local phase difference is

$$\phi(x) = [(\omega_l - \omega_r)x + (\phi_l - \phi_r)] \in [-\pi, \pi),$$

the local disparity $d(x)$ can then be derived as

$$d(x) = \phi(x)/\bar{\omega}, \text{ and } \bar{\omega} = \tfrac{1}{2}(\omega_r + \omega_l).$$

The same massively visual cytoarchitecture together with knowledge about ocular dominance columns have motivated Yeshurun and Schwartz (1989) to suggest a fast parallel algorithm for deriving a sparse disparity map. Again, there is no need for solving the correspondence problem, and iterative solutions are thus avoided.

The mathematics behind the algorithm are straightforward and follow Yeshurun and Schwartz (1989). Assume an interlaced image $f(x, y)$ to consist of a single columnar (stereo) pair made up of the corresponding left and right images and that the columnar size, the same as the magnitude of the disparity shift, is of size D. The left and right images are identical for the case where there is no disparity and can be represented by $s(x, y)$. One can then write that

$$f(x, y) = s(x, y) \otimes \{\delta(x, y) + \delta(x - D, y)\},$$

and the corresponding Fourier Transform is found as

$$F(u, v) = S(u, v)\{1 + \exp[-j\pi(D - u)]\}.$$

Finally, one can take the logarithm of $F(u, v)$ and obtain that

$$\log F(u, v) = \log S(u, v) + \log\{1 + \exp[-j\pi(D - u)]\}.$$

The spectrum corresponding to $\log F(u, v)$ known as the *cepstrum*, will exhibit a significant peak, because the spectrum of the second term in the expression consists of a series of delta functions whose weights are decreasing as

$$F[\log\{1 + \exp[-j\pi(D - u)]\}] = \sum_1^\infty (-1)^{n+1} \frac{\delta(X - nd)}{n}.$$

The peak referred to corresponds to the horizontal shift $(D, 0)$. The spectrum, known as *cepstrum* is the power spectrum of the log of the power spectrum of the interlaced input and can be also used to measure auditory "echo." The full stereo algorithm proceeds in three steps. First, the left and right stereo images are patchwise interlaced. The size of the patch (window) is determined by the size of the ocular dominance columns and corresponds to 5 to 10 minutes of arc. The second step applies the cepstral filter, in parallel, across all windows. The last step seeks a peak in the cepstrum limited to the interval $[D/2, 3D/2]$, assuming that disparities lie in the Panum's area, i.e., assuming that they are less than D.

5.2.   Artificial Images

Intrinsic representations, characteristic to a given object, are not limited to those of the psychophysical type discussed so far. Both robotics and target tracking tasks led to the development of specific sensing devices based on sound physical principles. The resulting intrinsic representations, which can be labeled artificial or man in- duced, have found wide use in machine vision applications and are characteristic of active perception (to be discussed in Chapter 6). The derivation of such representations involves special sensing devices operating in specific portions of the electromagnetic spectrum. Signifi- cant amounts of signal processing are usually needed to interpret the derivation of a particular intrinsic characteristic. Our treatment is brief and only aims at illustrating the concept of artificial representa- tions, which are essential for (mobile) robotics and target tracking.

There is much interest in determining the position (distance) and velocity of surrounding objects. One usually uses sonar techniques to determine the position based on time of flight methods. Many mobile robots use ultrasound. One example of such a robot is HILARE, which beams ultrasound pulses at 36 kHz. To compensate for the small ($30°$) sensitive area of each emitter, the robot has fourteen of them placed at different sites.

Velocity information can be derived using the Doppler-shift effect. Assume the configurations shown in Fig. 5.4. for purposes of illustra- tion. The target is moving at a constant speed v toward the radar, which is stationary and emits waves of frequency $\omega = 1/T$. The time Δt necessary for some point A on the wave to travel from the radar to a given target at distance d_o is given as

$$\Delta t = \frac{d_o - v\Delta t}{c} = \frac{d_o}{c + v}. \tag{5-32}$$

where c is the speed of the light. (For different wavelengths, λ, the $c = \omega\lambda$ relationship holds.) The total round trip is $2\Delta t$, and point A returns to the radar at time

$$t_1 = t_0 + \frac{2d_0}{c + v}. \tag{5-33a}$$

Similarly, for a point B on the same wave, the return time is

$$t_2 = t_0 + T + \frac{2d_1}{c + v}. \tag{5-33b}$$

The period of the received wave, T', is given as

$$T' = t_2 - t_1 = T - \frac{2(d_0 - d_1)}{c + v} < T. \tag{5-33c}$$

and as a consequence, the period (frequency) for an approaching target decreases (increases). Since $vT = d_0 - d_1$, one can write

$$T' = T\frac{c - v}{c + v} = T\frac{1 - \dfrac{v}{c}}{1 + \dfrac{v}{c}}, \tag{5-33d}$$

or, in terms of the received frequency, $\omega' = 1/T'$

$$\omega' = \omega\,\frac{1 + \dfrac{v}{c}}{1 - \dfrac{v}{c}} \tag{5-34a}$$

$$\omega' = \omega\left(1 + \frac{v}{c}\right)\left(1 + \frac{v}{c} + \frac{v^2}{c^2} + \cdots\right)$$

$$= \omega\left(1 + \frac{2v}{c} + \frac{2v^2}{c^2} + \cdots\right). \tag{5-34b}$$

For most cases of interest, $v/c \ll 1$ and then

$$w' \cong \omega\left(1 + \frac{2v}{c}\right) = \omega + \frac{2v}{\lambda}. \tag{5-34c}$$

Therefore, the received wave has been shifted by $\omega_D = 2v/\lambda$. As an example, assume a wave pulsating at 1 GHz, which peaks at $\omega_D = 10$ Hz. First, because $c = \omega\lambda$, where $\omega = 1$ GHz, one can easily obtain $\lambda = 0.3$ m. Then, because $\omega_D = 2v/\lambda$, the velocity of the moving target is $v = \lambda\omega_D/2 = 1.5$ m/sec.

There are additional artificial intrinsic representations. The availability of contact sensors able to derive tactile information is very important for robotics. The sensors need to measure touch, proximity, and slip, and to provide information for grasping. Finally, the infrared ($\approx 10^{12} - 10^{14}$ Hz) portion of the spectrum allows the derivation of "heat" maps. These intrinsic signatures help with object identification and are useful for night vision applications. Clearly, one has to have a good understanding of the physical process behind the derivation of artificial intrinsic images before using them for image/scene understanding. Fig. 5.5. considers scene interpretation (Kanade, 1980), and the right-hand side of the loop suggests that any image interpretation system should include the image formation process in order to map appropriately the picture domain cues ("raw signal") into scene domain cues ("features"). Otherwise, a stick seen through water will look bent and lead to the detection of two line segments.

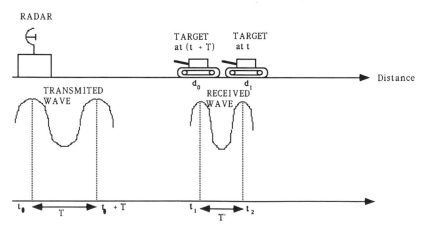

Figure 5.4.
Doppler shift for velocity estimation.

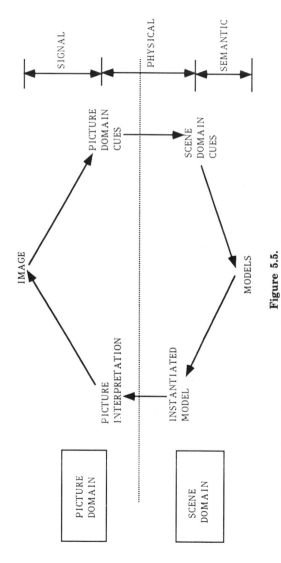

Figure 5.5.

Kanade's image interpretation scheme. Reprinted with permission from Kanade, T. (1980). Survey: region segmentation signal vs. semantics. *CGIP* **13**, 279–297. © Academic Press.

5.3.   Ethological Images

Intrinsic representations are useful for object recognition and safe navigation. Nature, long before people, had to deal with similar problems and come up with clever solutions. Next, we present designs that made navigation possible, by drawing from Gould's (1982) fascinating book on ethology.

The auditory system is a rich source of information. Localization of sounds whose wavelengths are small compared to the head can be accomplished by being attuned to intensity differences between the two (ears) receptors. Echolocation, as used by bats, is a sonar system using the Doppler shift for determining the relative motion of the target. Since the wavelength at which objects become acoustically transparent depends on size, the bat can also determine the size of the prey. Examples of acoustic releasers in song recognition suggest that the auditory system is quite complex, and that recognition was one of its main original tasks.

In Chapter 6, we will see that there is ecological information available in our natural surroundings that helps humans with orientation and navigation (Gibson, 1966). In contrast, there seem to be three major systems that provide orientation information for the animal kingdom. Solar orientation is by far the most common, and the sun appears to be moving east to west at 15° per hour. In the sun's absence, its position can still be determined from the pattern of polarized light in the sky. The earth's magnetic field is another orientation source. The sea is a world of its own, where salmon homing is accomplished through olfactory cues. Navigation over long distances usually uses a celestial compass—the sun, the zenith polarization pattern, or the stars—and the earth's magnetic field. The strategy, interesting to note, makes use of a long-distance navigation system followed by a finely tuned homing system.

5.4.   Conclusions

In this chapter we have introduced intrinsic images and showed their ability to describe specific characteristics of the material environment surrounding us. In discussing psychophysical images such as lightness (reflectance), optical flow, and depth, we conclude that parallel and distributed computation (PDC) plays a major role in deriving such images. We emphasize, however, that the PDC aspect relates to both representation and processing, rather than processing only. The concept of a massively visual cytoarchitecture, in terms of distributed and multiscale representations, gained additional support. The cytoarchitecture is pervasive in its functionality and facilitates the derivation of intrinsic images such as optical flow and disparity. As we are going to see in the next chapter, the same cyctoarchitecture, at a higher level of functionality, can also facilitate the interpretation of 3D motion and the recovery of shape from shading.

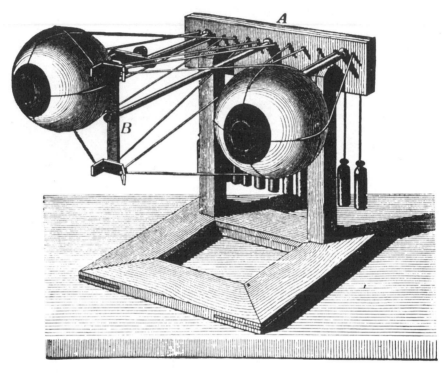

Mechanism derived to show eye movements and muscles. (From Helmholtz, H., Treatise on Physiological Optics, 1858.) Reprinted courtesy of the National Library of Medicine.

6

Active Perception

Seek and ye shall find.

In this chapter we suggest that active perception is essential for decreasing the computational load on the visual system. Active perception leads naturally to exploration and mobility. It is this very mobility that provided people with complex stimulations and demands, which eventually lead to human evolution.

Moravec (1983a), writing on utility of mobility, says that "instinctive skills are much better developed in humans than high level thinking, and are thus the difficult part of the human emulation problem. Instinctive skills are overwhelmingly the difficult part of human intelligence. All animals that evolved perceptual and behavioral competence comparable to that of humans first adopted a mobile way of life. A mobile way of life favors general solutions that tend towards intelligence, while nonmotion favors deep specializations." Moving out of a simple world made up only of blocks is thus a prerequisite for developing intelligence.

Neisser (1976) anticipated much of the present interest in active perception by suggesting that perception is a matter of discovering *quality of sensory input which bears a direct rel. to the interp.*

cognitive maps.

what the environment is like and adapting to it. Discovering what the
environment is like is worth doing only if the individual can pick up
appropriate information, labeled by Gibson (1966) as *affordances*.
Affordances, the ecological link between an observer and his environment, allow us to move and manipulate safely. Neisser suggests that
both spatial and temporal continuity of the environmental representations are facilitated through the use of cognitive maps, which he coins
"schematas." People interact with their environment according to
preestablished schematas and as a result we can see only what we
know how to seek. Perception is a constructive but controlled process
and active perception can help us fill in missing information. The role
of schematas in actively seeking specific affordances leads naturally to
the concept of directed perception, which will be discussed in Section
10.2. The sensory information acquired is assimilated within existing
schematas to update the world models held in memory. Provision has to
be made also for accommodating new information about the world, i.e.,
allowing the organism to adapt to its environment. Perception depends
further on the skill and experience of the perceiver—on what he knows
in advance. Anticipation of what the environment is likely to look like
in the form of mental imagery (see Section 8.2.1.) is related to the
suggested schematas and enhances our perceptual capabilities and
behavioral performance.

I would like to suggest that there is more to active perception than
just exploration, and that the word *active* could be rewritten as ACTive
to emphasize the role ACTIVITY plays in our interactions with the
environment. We are more than simple observers and our perceptual
activities are task dependent. A captivating book by Winograd and
Flores (1987) is quite provocative, and rightly so, in the way it looks at
what computers are and how they can be best used. The book, which
deals with hermeneutics (interpretational issues) and Heidegger's
philosophy, is mostly on language but one can easily map its contents
to perception as well. The book argues that much of present computer
usage is predicated on our rationalist assumptions and our "prejudice"
towards the mind and body dualism. The dichotomy referred to assumes that there is an objective world of physical reality and that there
is also a subjective mental world. The interpreted and the interpreter,
however, do not exist independently: existence is interpretation and
interpretation is existence. The observer can not be objective in his/her
interpretation of the world and by default has to bring his whole
background to the very act of interpretation. Active perception can

then be redefined as *throwness*, i.e., our condition of understanding in which our actions find some resonance or effectiveness in the world. We usually become aware of objects and their properties when they break down. The process of things or activities breaking down is thus essential to make them concrete to us and suggests that perceptual activities are task dependent and that their functionality is related to us acting on our environment. One of the conclusions suggested is that "practical understanding is more fundamental than detached theoretical understandings. Detached contemplation can be illuminating, but it also obscures the phenomena themselves by isolating and categorizing them." (Winograd and Flores, 1987). Ideal types (schemata) as generalized models of real situations and emphasis on action are also a cornerstone of the sociology of Max Weber.

The role of perception is to know WHAT we perceive and WHERE it is. To achieve this, perception is most likely implemented along two distinct perceptual systems, those of direct perception and recognition, respectively (Neisser, 1989). According to this line of thought, "such a proposal is nothing like the traditional chain in which perception" (or segmentation) "comes first and recognition later. On the contrary, the two systems are roughly parallel. Direct perception enables us to perceive our immediate ecological situation: where we are, where objects are, and what affordances (physical actions) those objects afford. The recognition subsystems are not at all concerned with the layout of the environment: they mediate identification and classification by accumulating evidence in relation to stored representations." Following the preceding dichotomy, Chapter 6 is mostly concerned with WHERE things are and the layout of the environment, Chapter 7 considers the recognition aspect, i.e., WHAT is there, while a wrap up discussion is deferred to Section 10.2.

6.1. Dynamic Vision

We must cope with much variability in our environment. Over the long term, we adapt by learning to prototype the world surrounding us. On a day to day basis, we adapt, however, by looking and probing around us. To quote from Bajcsy (1988): "However, it should be axiomatic that perception is not passive, but active. Perceptual activity is exploratory, probing, searching; percepts do not simply fall onto sensors as rain falls onto ground. We do not just see, we look. And in the course, our pupils adjust to the level of illumination, our eyes bring the world into sharp focus, our eyes converge or diverge, we move our heads or change our position to get a better view of something, and sometimes we even put on spectacles. This adaptiveness is crucial for survival in an uncertain and generally unfriendly world!"

Clearly, the whole sensing and perceptual process is actively driven by cognitive processes incorporating a priori knowledge. The observer receives permanent feedback and actively follows up by seeking novel information. Such exploratory behavior underlines the symbiotic relationship between the sensory and motor systems. According to Hering (1868): "It is obvious that the motor apparatus of the visual organ has to fit the sensory apparatus as the shell does an egg. For, whether one assumes that they were set up according to a wise plan, or that they developed with each other and through each other in an inevitable way as the evolutionary series is traversed, in any case: the capabilities of the one have to correspond to the needs of the other."

The exploratory behavior alluded to is one of the characteristics of dynamic vision. The actual correspondence between sensory inputs and exploratory and performatory motor actions could be encoded as the links of the aspect graphs to be discussed in Section 7.4., and/or through specific stimulus-response pairs encoded through DAM mechanisms. Furthermore, dynamic vision is also characterized by flexible perception, whereby top-down modeling can prime the operation and integration of modular processes toward actual recognition. We label this last aspect of dynamic vision as functional (or regulatory) perception.

Dynamic vision is synonimous with active perception—acquiring and processing time-varying imagery for scene interpretation and successful behavior. The first section of this chapter considers exploratory and functional aspects of active perception and suggests a theoretical framework for understanding and implementing visual processes.

6.1.1. Orienting and Perceptual Sensory Systems

This section is based on Gibson's (1966) book, which treats the senses as perceptual systems. The book provides an informative discussion on the specific role of stimulation in perception. As active observers, we investigate stimulations and their interrelationships.

Stimulation can be external to the organism, if it originates in the environment, or internal to the organism, if its origin lies within the organism itself. The corresponding two types of perception are extero-ception and proprioception. The afferent inputs provided by exterocep-tion (sense organs) and proprioception (passive organs or limbs) are (mostly) exafferent and reafferent, respectively, and can be either imposed or obtained. Proprioception, or autostimulation, is self-speci-fying, visual information concerned among other things with posture, equilibrium, and self-locomotion. It is intrinsic to the flow of the organism's activity, corresponds to internal feedback, is contingent upon efferent output, and does not depend upon external affordances. The organism, modulated by its own efferent output, obtains external information through performatory motor behavior or exploratory sensory activity.

The perceptual systems most involved in visual processing are orienting and sensory systems of the auditory and visual type. The information content of different stimulations are in terms of orienta-tion and locomotion. The orienting system provides feedback that guides behavior and helps the body maintain equilibrium. Awareness of movement, or kinesthesis, originates within the orienting system, which is anatomically located within the vestibular organ. Attuned to gravity and acceleration, the orienting system provides the organism with a stable framework of the environment. Steering posture and locomotion, it also provides relevant information regarding the ground-support level and the upright sense of body posture. It enables the organism to sense its own locomotion and to home in over extended stretches of time and space. The auditory sensory system, anatomically located in the cochlear organs, orients the organism toward vibratory events. The visual sensory system, located within the ocular mechan-ism, guides accommodation, fixation and vergence, and exploratory activity. These three systems are interdependent. The eye-head system, for example, uses exteroception and proprioception, enabling an ob-server to move his or her head and perceive a stable world (except when the retinal image swings as a result of our pushing the eyelids). According to ocular-drift cancellation theory, the vestibular-occular reflex "knows" that the eye rotation rate is equal to the head rotation rate, but of opposite sign, and "can" cancel the illusory movement and thus perceive a stable world.

The orienting system attempts to align the body's frame to the gravitational frame, using feedback in the form of reflex compensation. The acceleration we are sensitive to is both linear and rotary (turn) displacement. As is the case with any system, however, the vestibular apparatus can fail due to illusions of either passive transpositions or head rotation. Specifically, the orienting system cannot distinguish between a state of rest and uniform motion in a straight line (as would be the case for smooth airline travel), or between a state of rest and spin. We are thus sensitive only to transitions between steady states, indicated by starts and stops and/or pushes and pulls.

Additional visual information is required to disambiguate such illusions. The auditory system has as its main exteroceptive function to pick up the direction of an event, orient toward the event, and possibly identify it. The proprioceptive function is to listen to one's own speaking voice and modulate it as appropriate. The angle at which the auditory wavefront meets the frontal plane of the head determines priority of onset and disparity of intensity. This differential stimulus asymmetry is picked up by the organism to orient and localize events of interest. Examples of sensory visual tasks involving feedback are discussed in later sections and include kinetic depth, vergence, and depth from focus.

6.1.2. *Exploratory Perception*

Exploratory perception (EP) has two main goals—locating regions of interest and selectively processing them. Visual systems require these capabilities to cope with the computational complexity of visual tasks. Burt (1988) appropriately describes exploratory perception as "smart sensing." Locating regions of interest for further processing is done over space and time through peripheral alerting and subsequent foveation-like processes. Selectivity of processing, under high-level control guidance, chooses how to process the region of interest. Thus, the basic algorithm of smart sensing involves locating, selectivity, and high-level control guiding the first two modules. Locating is a major navigation component, which involves alerting and tracking, and consequently, is a major component in surveillance and object recognition systems. According to Bajcsy (1988), selectivity will amount, among other things, to control of the sensor and/or the low-level vision modules. Sensor control includes setting up and operating the sensor to adjust optimally the camera and bring the scene into focus. Clearly, one can see a need to define some error measure, which drives and controls the sensor. Control of low-level vision can be thought of as using multiscale to determine the appropriate level of resolution. Recall the discussion on cojoint s/sf representations regarding the setup of appropriate parameters to compute image representations, such as the pseudo-Wigner distribution (PWD).

Ballard (1988) suggests additional advantages of exploratory perception, compared to the usually underconstrained or ill-posed problems one has to solve when using fixed camera vision. The constraints needed to make the problems well posed flow naturally from behavioral states, and the processing is *local* and thus computationally feasible. Furthermore, gaze control leads naturally to object-centered coordinate systems, which are invariant to observer motion and correspond to viewer-oriented rather than viewer-centered coordinates.

Burt (1988) describes a surveillance system that integrates spatial and temporal information to locate moving targets. Temporal Laplacian pyramids can be integrated and facilitate homing or foveation processes. Qualitative statements, such as something big is moving quite slow, can be derived from the integrated domain as well. The process of integrating different information sources is deferred to Chapter 8 under multisensory data integration. Finally, the same system referred to previously again operates under the assumption that the visual system is not invariant to translation and that foveation precedes recognition. The functional split between locating (WHERE) and identifying (WHAT) inputs is also confirmed by neurophysiological findings (Mishkin *et al.*, 1983), showing spatial relationshisps in the parietal cortex and recognition in the infero-temporal cortex.

Smart-sensing hardware has been built around the pyramid and pipeline architectural concepts. PIPE built by NBS (Kent, 1985) and the Pyramid Vision Machine (PVM) (Burt, 1988) are two exploratory perceptual visual systems.

6.1.3. *Functional Perception*

Functional (or regulatory) perception (FP), another major component of dynamic vision, is primed by top-down processes triggered by exploratory perception and bottom-up processing. Functional perception provides for the integration of modular processes, characterizes middle-level vision, and allows for flexible interpretation.

Some snakes are poisonous, some such as the python can even

strangle a victim to death, but Kaas' *et al.* snakes (1987) have been tamed and illustrate how "energy-minimizing splines guided by external constraint forces and influenced by image forces" can lock on specific image features. (See Section 4.2.3. for the analogies to Ising spin systems in terms of external magnetic field and local spin interactions, respectively.) Snakes have active contours that attempt to match malleable models to a given image through energy minimization. The approach, quite general in its applicability to vision problems, allows for active (and/or interactive) interpretation. The energy function E corresponding to the spline-like snakes comprises smoothness constraints, internal image forces—looking for the salient image features provided by early vision, and external forces—pushing the spline near the desired local minimum. The smoothness, internal image, and external forces contribute three energy components E_s, E_{int}, and E_{ext}, respectively. The energy E is given by $E = E_s + E_f = E_s + (E_{int} + E_{ext})$, where E_f corresponds to the fidelity energy component. Assuming a parametric spline representation given as $s(t) = (x(t), y(t))$, the bending energy E_s can be written as $E_s = (\alpha(t)|\dot{s}(t)|^2 + \beta(t)|\ddot{s}(t)|^2)$. Adjusting the weights $\alpha(t)$ and $\beta(t)$ "controls the relative importance of the membrane and thin plate, respectively. Setting $\beta(t)$ to zero at a point allows the snake to become second order discontinuous and develop a corner." (Kass *et al.*, 1987). The bending energy, similar to a Lagrange multiplier, must be minimized to ensure smoothness of surface/contour fitting. Illustrating an internal image force component, one can write the corresponding energy term E_{int} as $E_{int} = \Sigma \omega_i E_i$, where the feature i is weighted by ω_i. If one is looking for both lines and edges, then $E_{int} = \omega_{line} E_{line} + \omega_{edge} E_{edge}$, where $E_{line} = f(x, y)$, $E_{edge} = -|\nabla f(x, y)|^2$, and the weights ω are set via high-level control. The snake is thus attracted to either light or dark lines and to contours exhibiting large image gradients. (Interaction between E_s and E_{line} ensures minimum bending energy and thus yields straight lines.) Finally, the external force E_{ext} can assist in semiautomatic image interpretation by providing anchor points—the equivalent of knots for splines. Anchoring a spring at a knot x_1 (desired fixed location) and at a knot x_2 (moving along the snake) adds $-k(x_1 - x_2)$ to E_{ext}.

The snakes are also useful tools for deriving intrinsic representations where raw data provided by (early) low-level vision is eventually interpreted under higher-level processes. Eventually, such intrinsic representations have to be integrated and fused toward 3D recognition.

Active 3D shape models, coined as thin plates, which can be warped and are conceptually similar to the snakes, have been used for 3D object reconstruction (Terzopoulos *et al.*, 1987). The active aspect of the models, changing their shape according to variational principles, contrasts passive parametrized geometrical shapes of the CAD/CAM type. The 3D reconstruction process is not different from matching or recognizing 3D objects. While the memory provides the internal and generic models, the intrinsic representations of the data itself provides the external constraints. Thus, higher-level mechanisms generate external criteria of smoothness and fidelity for intrinsic representations, which in turn drive the matching process against memory models. The loop repeats itself until the object is recognized.

The snake approach derives its generality from the visual system's ability to set the appropriate energy terms. The approach is characteristic of active vision and top-down guidance of low-level processing. Specifically, the model actively allows external forces to select among alternative local minima under the guidance of higher-level processes. Furthermore, such an approach enables early vision to operate in a least-commitment mode, deferring the final choice of a global minima to later stages of visual processing, which act as the final referee in choosing interpretation rules by appropriately setting E_{ext}.

6.1.4. Multilevel Visual Architecture

Exploratory and functional perception are essential for the visual system to operate. The dynamics of perception save on computation and link a multistage visual architecture as suggested next.

Marr's (1982) model for vision is a bottom-up approach that does not provide for top-down priming. Intensity representations (raw images, primal sketch, intrinsic images) are eventually transformed into visible surface representations (the 2.5D sketch). The Barrow and Tenenbaum (1978) model is a loop consisting of data-driven and goal-driven processes. The data-driven (domain-independent) process is iconic. It starts with intensity images, seeks image features (edges, regions) and ends by deriving intrinsic images. The goal-driven (domain specific) process is symbolic. Top-down, it starts with some interpretation that requires specific segmentation. The interface between iconic and symbolic (low and high) level processing lies at the boundary between intrinsic images and segmentation.

We question the need for symbolic processing and segmentation in Chapters 7 and 8, where we suggest an analog and holistic process of interpretation. For both Barrow and Tenebaum, and Marr, however, low- and high-level processing is a dichotomy. The human vision system, however, seems to point in a different direction. It suggests a continuation of processing (Russell, 1921) and a design that is modular and allows for intrinsic representations to cooperate. Terzopoulos (1988) suggests that the visible-surface reconstruction process belongs to another stage of processing, which he labels as intermediate-level vision. Such a stage could then bridge the low- and high-level processing gap. To quote from Terzopoulos (1988),

> the visible-surface reconstruction process proposed for generating and dynamically updating visible-surface representations, a generalization of the distributed, data-driven algorithms, unifies the following four computational goals: (1) Integration: The visible-surface reconstruction process integrates local surface shape constraints from multiple sources and it fuses this information across multiple scales of resolution. Moreover, it coordinates two categories of low-level shape estimation processes, those of 'correspondence' and 'shape-from.' (2) Interpolation. The visible-surface reconstruction process continuously propagates the integrated shape information into regions lacking shape constraints. (3) Discontinuities. The refined discontinuity maps provide (dynamic) boundary conditions, which limit the interpolation of shape constraints. (4) Efficiency. Visible surface representations form and evolve in real time. Massive, fine-grained parallelism appears to be the most viable computational architecture for this purpose. Lack of high bandwidth connections can be overcome by employing multiresolution relaxation.

The solution to visible-surface reconstruction (VSR), one instantiation of the class of regularization problems, could be obtained through PDP minimization techniques, as suggested in Chapter 4. The formulation of such problems could take the standard form of an energy function made up of one term to minimize the discrepancy to a given set of constraints with the other terms looking for a smooth solution. Specifically, according to Terzopoulos (1988), "assume a linear space of admissible functions. Let $P(f)$ be a penalty functional on H which provides a measure of the discrepancy between f and the given constraints. Let $S(f)$ be a stabilizing functional which measures the (lack of) smoothness of a function $f \in H$. The regularized visible-surface reconstruction problem is formulated then according to the following variational principle:

VP1: Find $g \in H$ such that

$$\mathscr{E}(g) = \inf_{f \in H} \mathscr{E}(f),$$

where the energy functional $\mathscr{E}(f)$ is

$$\mathscr{E}(f) = S(f) + P(f)." \tag{6-1}$$

We now suggest a model of computational vision that seems compatible with the operational constraints under which it functions. The model (Fig. 6.1.) includes low-, middle-, and high-level visual processing, and active perception of the exploratory (EP) or functional (FP) type connects the three stages. The matching box, analogous to a blackboard (see Section 8.1.2.), is the medium through which "agents" responsible for the derivation of intrinsic images, among other things, incrementally build a solution that can resonate into meaningful interpretation. (See Section 8.3. on data fusion and Van Essen and

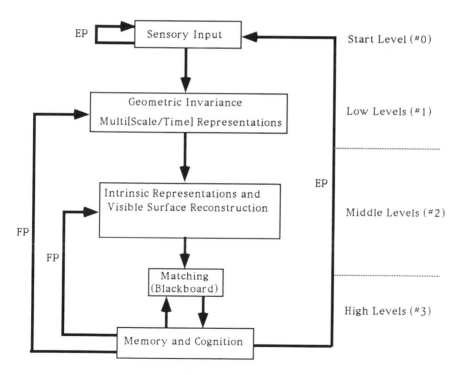

Figure 6.1.
Multilevel visual architecture.

Munsell (1983) on the existence of many well-defined functional subdivisions within the cortex.) The agents complement each other and, characteristic of distributed but "guided" computational strategies (see Section 8.2.4. on planning), attempt to make sense of the scene being interpreted. The memory must be triggered for recognition (see Section 7.5.) and can redirect the interpretation process by resetting the energy functional $\mathscr{E}(f)$ (i.e., by reformulating the VSR problem).

Computational reasons for the existence of the loop shown in Fig. 6.1. are suggested by Tsotsos (1989). The loop referred to consists of a bottom-up part and a top-down task directed part. Tsotsos shows that the bottom-up case is NP-*complete* in the size of the image, while the task directed case has linear time complexity in the number of items in the input. The need for active perception, both exploratory and functional, to implement such a loop is then motivated by computational constraints. To quote from Tsotsos, "the NP-Completeness of the bottom-up case provides the strongest possible evidence for the abandonment of purely bottom-up schemes that address the full generality of vision. It is thus necessary to sacrifice generality in order to re-shape the vision problem and to optimize the resources dedicated to visual information processing so that a tractable problem is addressed."

6.2.   Ecological Optics

Active perception allows the organism to enrich itself with environmental clues about spatial layout. The active observer looks around, probes the surroundings, and actively seeks for interpretation cues. Ecological optics (Gibson, 1979) refers to the human ability to perceive a stable and consistent world by following a systemic approach whose main ingredients are (physical) optics, perspective geometry, and ecological information.

Gibson (1950) advanced the hypothesis that "the basis of the so-called perception of space is the projection of its objects and elements as an image, and the consequent graduate changes of size and density in the image as the objects and elements recede from the observer." The goal is then to "consider, one by one, these various so-called cues for distance perception when they are reformulated as gradients of the retinal image." Since shape can be inferred from depth changes, we start our discussion by introducing "gradients" of depth, not surprisingly called the gradient space.

6.2.1. Gradient Space

The gradient space is easily defined if one assumes that depth z is given as $-z = f(x, y)$. Then the specific gradient is $(p, q) = (\partial f/\partial x, \partial f/\partial y)$ and stands for the rate of depth change in the x and y directions, respectively. Assuming that the equation of a plane is written as $Ax + By + Cz + D = 0$, it follows that

$$-z = \frac{A}{C}x + \frac{B}{C}y + \frac{D}{C} = px + qy + r. \tag{6-2}$$

The direction $\arctan(q/p)$ corresponds to the direction of fastest change of surface depth, while $(p^2 + q^2)^{1/2}$ is the rate of this change. An alternative for representing surface orientation is (slant, tilt) = (σ, τ), which was introduced in Chapter 3 and shown in Fig. 3.5. The relationship between the two alternatives is given as $\tan(\sigma) = (p^2 + q^2)^{1/2}$ and $\tau = \arctan(q/p)$. Gradient space is a handy tool for "shape from X" methods used to reason about surface orientation, where X is a cue-like shade, motion, or texture. If 3D shape is indeed stored as a collection of 2D surfaces, then specific physical constraints among the gradients could disambiguate the recognition process.

An important theorem, which helps us reason about surface orientation, states that if two surfaces meet along a concave or convex edge, their gradients lie along a line in the gradient space that is perpendicular to that edge in the image. Assume that the two surfaces S_1 and S_2 are given in terms of their gradients $G_1(p_1, q_1)$ and $G_2(p_2, q_2)$ respectively. The surfaces could share a vertical hinge, parallel to the y-axis; then $q_1 = q_2$, and the gradients G_1 and G_2 lie along a horizontal line in the (p, q) space. If the hinge is set at an angle, one can rotate the x-y system of coordinates until the hinge becomes vertical and then generalize the previous result. The relationship between G_1 and G_2 determines the property of the edge. Fix G_1 (i.e., the orientation of surface S_1) and rotate surface S_2 around the hinge. It is easy to see that if an edge is convex $(+)$, the gradients of its planes are ordered in the same fashion as the corresponding regions in the image. If the edge is concave $(-)$, their order is changed. Let us consider the cube example shown in Fig. 6.2a. Fix the gradient of S_1 at G_1. AB is a $(+)$ convex edge

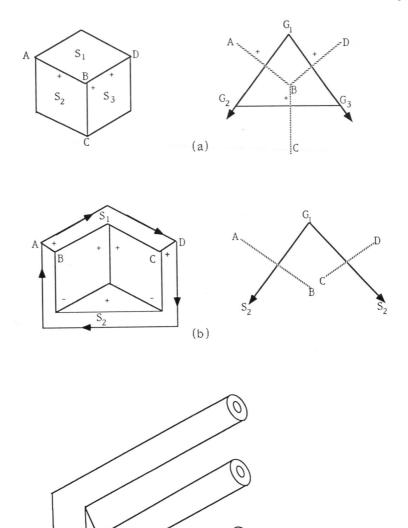

(a)

(b)

(c)

Figure 6.2.
Labeling of 3D shapes in the gradient space.

and thus G_1G_2 is perpendicular ($\perp$) on AB. Fix then the gradient of S_2 at G_2. Then G_3 is uniquely determined by $G_2G_3(\perp BC)$ and $G_1G_3(\perp BD)$. The location and scale of triangle $G_1G_2G_3$ are arbitrary, but its shape and orientation are determined by the original image. Thus, we can say that the labeling of the cube is feasible and that the 3D structure given by its line drawings (or surface representation) is realizable. The example shown in Fig. 6.2b. is characteristic of a structure that is unrealizable. While the gradient of S_1 is fixed, the gradient of S_2 cannot satisfy both constraints imposed by the convex edges AB and CD. How about the structure shown in Fig. 6.2c.?

Early use of gradient space was to derive surface orientation from reflectance models under laboratory conditions. Generically, the approach became known as "shape from shading," and a good overview is provided by Ballard and Brown (1982). The human shape-from-shading process is quite robust and can be manipulated to advantage by changing the surface shading in prespecified ways. Similarly, cosmetics change the reflectivity properties of the skin and creates illusory shapes. While the HVS is robust, shape-from-shade algorithms are quite involved, sensitive to noise, and very dependent on a priori assumptions regarding the scene being analyzed.

Research done by Mingolla and Todd (1986) suggests that "several assumptions for perception of shape from shading are not psychologically valid. The most notable of these assumptions is that the visual system initially assumes that all surfaces have Lambertian reflectance and that illuminant direction must be known before shape detection can proceed. These assumptions are often accompanied by a third assumption that surface orientation is detected locally, and global shape determined by smoothing over local surface orientation estimates." The Lambertian assumption considers those surfaces characterized by both an ideal matte finish and a reflectivity function proportional only to the cosine of the incident angle. Thus, under uniform or collimated illumination, they look equally bright from any direction. The smoothing referred to is generally accomplished through standard relaxation techniques, such as those used for deriving optical flow and shown in Section 5.1.2. Another aproach for shape-from-shading psychophysical modeling contributed by Koenderink and van Dorn (1980) and referred to by Mingolla and Todd (1986) takes a global view and follows the assumption mentioned previously. Specifically, "instead of trying to locally invert the image formation process, it can be shown that the solid structure of an object modulates the structure of light available at station points in globally constrained ways. By structure of light is meant the positions of singularities of luminance (local maxima and minima) and the topology of connection and closure of isophotes, contours of equal luminance. The implication is that perceivers need not bother determining what variables generated a given intensity at all. Instead, the nestings and inflections of possible luminance distributions are shown to be both highly constrained and highly specific to those configurations of hills, valleys, and saddles which make up solid objects." It seems then that "the visual system extracts impressions of these visual properties directly from the input, with no mediating steps that attempt to represent what condition in the world *could* have produced the observed luminance pattern according to some internally embodied theory of image generation. If this is so, the one-to-many mapping problem still needs accounting. For this reason, analyses of global and contextual constraints on visual information and on visual processes are paramount."

Recent research performed by Pentland (1988) and described here shows that people assume a linear reflectance function when interpreting shading information. The solution suggested to recover shape-from-shade discards the usual assumptions about surface smoothness and advances instead the use of distributed and multiscale representations in terms of orientation, spatial frequency, and phase. The mathematics (Kube and Pentland, 1988) assume a linear reflectance function for the imaged surface, such that the normalized image intensity approximates a (constant average) Lambertian reflectance as

$$I(x, y) \approx \cos \sigma + p \cos \tau \sin \sigma + q \sin \tau \sin \sigma,$$

where (σ, τ) is the (slant, tilt) pair related to the single light source, and (p, q) is the gradient space. Under this assumption, the Fourier transform of the image $I(x, y)$ (ignoring the DC term) is

$$F_I(u, v) = 2\pi(\sin \sigma)u \left\{ m_z \exp\left[j\left(\phi_z + \frac{\pi}{2} \right) \right] \right\} [\cos v \cos \tau + \sin v \sin \tau],$$

where (m_z, ϕ_z) is the (magnitude, phase) spectrum of the surface $z = z(x, y)$, i.e., the image intensity is a linear function of the height surface $z(x, y)$. Assuming that the spectrum of the image intensity is given by (m_I, ϕ_I) and that the source of light given in terms of slant and tilt is known one obtains that the Fourier transform of the surface $z(x, y)$, except for a scale factor, is given by

$$F_z(u, v) = m_I \exp\left[j\left(\phi_I - \frac{\pi}{2} \right) \right] / \{2\pi(\sin \sigma)u[\cos v \cos \tau + \sin v \sin \tau]\}.$$

The distributed and multiscale representations can now facilitate the recovery of surface shape from the (shading) information $I(x, y)$. Spectral information is derived using Gabor filters and yields localized measurements of sine and cosine phase frequency contents (the equivalent of $F_I(u, v)$). An inverse transformation according to the last equation then yields the desired surface shape $z(x, y)$. The usefulness of distributed and multiscale representations for recovering shape from shading again shows the functional versatility of a massively visual cytoarchitecture.

Our discussion so far has emphasized the relevance of global luminance patterns to determine shape from shade. Learning and recognition could thus play a major role in recovering 3D shape. The learning is akin to that used by Hurlbert and Poggio (1988) to recover lightness (see Section 5.1.1.) and to holistic (Gestalt) recognition, to be covered in Chapter 7. Finally, the constraints could be environmental, also called ecological. Much work remains to be done, but again, PDP models could prove beneficial to the task of recovering shape information.

6.2.2. Shape from Texture

Much of the work on shape from texture assumes that texture is made up of regular unit primitives. Changes in the form (size, shape) and density of such units (see Fig. 6.3a.) are then directly related to the layout of the surface. One then measures the gradients of those changes to estimate surface orientation or distance from the observer.

Assume the geometry of Fig. 6.3b., where the optical slant refers to the angular relation with which a surface, at a given point, is intersected by the line of sight (which makes an angle α with the horizontal) (Sedgwick, 1983). Then, a texture element is given in terms of angular measures such as radial (γ_1) and tangential (γ_2) solid angles, texture size $S = \gamma_1\gamma_2$, the solid angular area of a unit of texture, density $D = 1/S$, compression $C = \gamma_1/\gamma_2$, and linear perspective in terms of

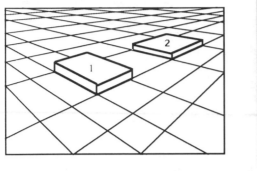

(a)

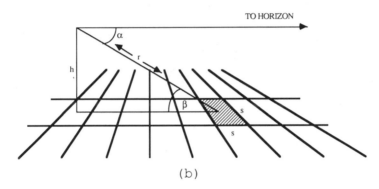

(b)

Figure 6.3.
Texture gradients.

angular width $L = \gamma_2$. The corresponding gradients can then be approximated as

$$G_S = (dS/d\alpha)/S \simeq \frac{3}{\tan \beta}$$

$$G_D = (dD/d\alpha)/D \simeq \frac{-3}{\tan \beta}$$

$$G_C = (dC/d\alpha)/C \simeq \frac{1}{\tan \beta} \qquad\qquad (6\text{-}3)$$

$$G_L = (dL/d\alpha)/L \simeq \frac{1}{\tan \beta},$$

where the line of sight, whose direction α is measured relative to the horizontal, from the observer O, meets the texture element slanted at an angle β. If the size s and/or the range r are sought, one can use the following relationships

$$\frac{s}{r} = 2 \tan \frac{\gamma_2}{2}$$

$$\frac{r_1}{r_2} \simeq \frac{\gamma_2}{\gamma_1}.$$

$$(6\text{-}4)$$

Kanatani (1986) considers the texture density to recover shape from texture on curved surfaces. Assume a stationary planar surface with a unit normal vector $\mathbf{n} = (n_1, n_2, n_3)$ to the plane. If α, β, and γ are the angles $\mathbf{n}$ makes with the x, y, and z axes, respectively, then $n_1 = \cos \alpha$, $n_2 = \cos \beta$, $n_3 = \cos \gamma$. If the planar surface is given as $z = px + qy + r$, where $(p, q) = (\partial z/\partial x, \partial z/\partial y)$ then

$$\cos \alpha = \frac{p}{\sqrt{p^2 + q^2 + 1}},$$

$$\cos \beta = \frac{q}{\sqrt{p^2 + q^2 + 1}},$$

$$\cos \gamma = \frac{1}{\sqrt{p^2 + q^2 + 1}}$$

$$(6\text{-}5)$$

because at (p, q) the surface normal is $(p, q, 1)$. If the plane is projected ortographically onto the x-y plane, it is subject to a $\cos \gamma$ magnification factor. Hence, the area projected onto the infinitesimal square defined by $\{(x, y), (x + dx, y), (x, y + dy), (x + dx, y + dy)\}$ on the image plane is given by $\sqrt{(\partial z/\partial x)^2 + (\partial z/\partial y)^2 + 1} \, dxdy$. Similar arguments apply to measured texture density $\Gamma(x, y)$ in terms of texture elements per unit area. Then one can write

$$\Gamma(x, y) = \rho\sqrt{(\partial z/\partial x)^2 + (\partial z/\partial y)^2 + 1}, \qquad (6\text{-}6)$$

where ρ is the true texture density of the surface. One could devise ad hoc procedures based on a priori knowledge of surface texture density, boundary conditions and the ubiquitous relaxation or use the elegant but complicated procedure suggested by Kanatani.

Moving on to how surface orientation can be estimated from texture distortion, we consider the case where orientation is estimated from the distortion of regular known patterns (Ikeuchi, 1980). A distortion measure α can be defined as $\alpha = (\cos \omega)/(1 + \cos^2\omega)$, where ω denotes the angle between the direction of the line of sight and the direction of the surface normal. The derivation of α uses the (texture) gradient space under spherical perspective projection, which is invariant to surface rotation of a texture element and changes in distance between the viewer and textural pattern. The measure α is independent of the viewer's orientation, since in the spherical perspective projection the viewer is at the center of the image sphere. Alternatively, it can be shown that $\alpha = fg \cdot (\sin \gamma)/(f^2 + g^2)$, where f and g are the lengths of the first and second axis vector of the distorted pattern, and γ is the angle between f and g. Hence (f, g, γ) are measurable quantities, and one could then easily obtain ω, i.e., the surface orientation. A regular pattern, such as a circle, yields under distortion an ellipse and can be tested for estimating the ω orientation parameter. Still, who is to say that an ellipse could not be just what is seen rather than a distorted circle? Clearly, much more information is needed to estimate surface orientation from texture/pattern distortion.

Inverse perspective of a road (from a single image) (DeMenthon, 1986), as it applies to mobile robotics and autonomous land vehicles (ALVs), is a specific application where shape must be inferred from a distorted contour. Generically, it is an instance of "shape from regular patterns."

Contours arise in an image from discontinuities in the depth or orientation just derived or from changes in surface reflectance and illumination effects, such as shadows. Contours are two dimensional, yet one sees them in three dimensions (Marr, 1982). Marr goes further to suggest that subjective contours could result from grouping tokens corresponding to line endings in the primal sketch. Barrow and Tenenbaum (1978) offer as an alternative explanation that the abrupt line endings (or discontinuities), locally interpreted in 3D as evidence of occlusion, cause discontinuity edges to be established in the distance image. Consider the "symbolic" drawing of the sun shown in Fig. 6.4. Subjective contours then result from making the distance image consistent with boundary conditions and the assumptions of continuity of surfaces and their boundaries. Subjective contours are primarily contours of distance, rather than intensity. The net result is the interpretation of the image as a disk occluding a set of radiating lines on a more distant surface.

Figure 6.4.
A subjective contour.

6.2.3. Spatial Layout

As we look around us, we gauge the spatial layout of the environment, i.e., objects' location in space and their relationships. Plastic arts reflect artists' knowledge—going back to Renaissance—of layout cues. These cues are largely qualitative in nature and do not require explicit derivation of depth. Known as monocular cues, they include the texture gradients and linear perspectives, already discussed, and also shading, masking, aerial perspective, and position. Objects closer to us mask (occlude) objects further away; distant objects look hazy and are tinged with blue; and the higher an object is on the ground plane, the further away it is, while the opposite holds for objects situated in the ceiling plane (Frisby, 1980).

Perspective and texture gradients, discussed in the preceding section, provide quantitative information regarding the slant of a given surface in a viewer-centered representation. Transforming the problem of perspective from viewer centered to environment centered, the representation becomes much easier to analyze. Specifically, in 3D, lines characterized by the same direction correspond to lines converging to a point called the vanishing point (VP). Conversely, the VP of a set of parallel lines specifies their direction as the line of sight from the viewer. The horizon (H) of a surface is specified by at least two vanishing points. The horizon structure and the vanishing points "taken together constitute the perspective structure of the optic array arising from a given environment. The perspective structure then directly specifies the environment-centered orientation in space of every line segment and every surface in the environment" (Sedgwick, 1983). We will add the ground (G) to the structure later on for obvious reasons, such as support for most objects of interest.

Vanishing points can be detected using accrued evidence from lines characterized by similar direction. Accrued evidence suggests the use of the Hough transform (HT), which was introduced for such needs in Section 4.2.4. Assume a collection of lines $\mathscr{L}$ characterized by a common $VP = (x_v, y_v)$. Then, for each line $L_i \in \mathscr{L}$, one can write $\rho_i = x_v \cos \theta_i + y_v \sin \theta_i$. Also assume that the gradient g at each point in the image is $\mathbf{g} = (\Delta x, \Delta y)$ and that for each $(x, y, \Delta x, \Delta y)$ corresponds a pair (ρ, θ) such that $\rho = x \cos \theta + y \sin \theta$, where

$$\rho = \frac{(x\Delta x + y\Delta y)}{(\Delta^2 x + \Delta^2 y)^{1/2}}$$

$$\theta = \tan^{-1} \frac{\Delta y}{\Delta x}.$$

(6-7)

One can then write for each $\mathbf{x} = (x, y)$ lying on the line described by (ρ, θ)

$$\rho = \|\mathbf{x}\|\cos \gamma = (\mathbf{x} \cdot \boldsymbol{\rho}) = \left(\mathbf{x} \cdot \frac{\mathbf{g}}{\|g\|} \right)$$

$$= (x, y) \frac{\begin{pmatrix} \Delta x \\ \Delta y \end{pmatrix}}{\sqrt{\Delta^2 x + \Delta^2 y}} = \frac{x\Delta x + y\Delta y}{\sqrt{\Delta^2 x + \Delta^2 y}},$$

(6-8)

where $\boldsymbol{\rho}$ is the unit vector and γ is the angle between $\boldsymbol{\rho}$ and the vector pointing to $\mathbf{x} = (x, y)$ as it spans the line described by (ρ, θ). A polar space representation corresponding to the pair (ρ, θ) is given by

$$\mathbf{a} = \begin{pmatrix} a_x \\ a_y \end{pmatrix} = \begin{pmatrix} \rho \cos \theta \\ \rho \sin \theta \end{pmatrix} = \left(\mathbf{x} \cdot \frac{\mathbf{g}}{\|g\|} \right) \begin{pmatrix} \cos \theta \\ \sin \theta \end{pmatrix}.$$

(6-9a)

The corresponding unit vector $\hat{\mathbf{g}}$ is given as

$$\hat{\mathbf{g}} = \frac{\mathbf{g}}{\|g\|} = \begin{pmatrix} \cos \theta \\ \sin \theta \end{pmatrix},$$

(6-9b)

and then one can rewrite the vector $\mathbf{a}$ as

$$\mathbf{a} = \frac{\mathbf{x} \cdot \mathbf{g}}{\|g\|} \cdot \frac{\mathbf{g}}{\|g\|} = \frac{(\mathbf{x} \cdot \mathbf{g})}{\|g\|^2} \cdot \mathbf{g}.$$

(6-9c)

The lines L_i defined by (ρ_i, θ_i) and going through the VP (x_v, y_v) require that the VP lies on each of them, i.e.,

$$\rho_i = x_v \cos \theta_i + y_v \sin \theta_i. \tag{6-10}$$

One can then prove a lemma saying that the Hough transform (HT) takes the set of lines L_i into a circle in the (ρ, θ) space of center $(x_v/2, y_v/2)$ and radius $1/4(x_v^2 + y_v^2)$. To prove the lemma, one has to show that for all lines L_i, after the HT mapping,

$$\left(a_x - \frac{x_v}{2}\right)^2 + \left(a_y - \frac{y_v}{2}\right)^2 = \frac{1}{4}(x_v^2 + y_v^2). \tag{6-11a}$$

The previous expression can be rewritten as

$$\left(\rho \cos \theta - \frac{x_v}{2}\right)^2 + \left(\rho \sin \theta - \frac{y_v}{2}\right)^2 = \frac{1}{4}(x_v^2 + y_v^2) \tag{6-11b}$$

or equivalently,

$$\rho(\rho - x_v \cos \theta - y_v \sin \theta) = 0. \tag{6-11c}$$

But we already know that each line in the set $\mathcal{L} = \{L_i\}$ obeys the relationship stated by Eq. (6-10). The method suffers from the difficulty of detecting circles and from the fact the VP at ∞ project at ∞. We cope with these drawbacks by using an alternative mapping of $(x, y, \Delta x, \Delta y)$ into $(k/\rho, \theta)$ for some k. If we choose $k = 1$, then $(\rho, \theta) \rightarrow (\rho', \theta') = (1/\rho, \theta)$, and the new pair (ρ', θ') corresponding to $\mathcal{L}$, the set of lines L_i, is projected onto a line perpendicular to $\boldsymbol{\rho}_v = (x_v, y_v)$ and lies at $1/\|\boldsymbol{\rho}_v\|$ from the origin. Furthermore, VP at infinity project into the origin. To prove the preceding statements we have to show that

$$(\rho' \cos \theta')\cos \alpha + (\rho' \sin \theta')\sin \alpha = l, \tag{6-12a}$$

where

$$(x_v, y_v) = (\|\rho_v\| \cos \alpha, \|\rho_v\| \sin \alpha) \quad \text{and} \quad l = \frac{1}{\|\rho_v\|}.$$

But, to show Eq. (6-12a), it is equivalent to derive that

$$\rho' \cos \theta' x_v + \rho' \sin \theta' y_v = 1. \tag{6-12b}$$

Remember that $\rho'_v = 1/\rho_v = (x_v \cos \theta + y_v \sin \theta)^{-1}$ and $\theta' = \theta$. Then Eq. (6-12b) reads

$$(x_v \cos \theta + y_v \sin \theta)^{-1}(x_v \cos \theta + y_v \sin \theta) = 1, \tag{6-12c}$$

so the original Eq. (6-12a) holds true. If $V = (x_v, y_v)$ is a VP, the set $\mathscr{L}$ is such that its lines pass through V, and VP at infinity are projected into the origin, such that $l = 0$. There are no VPs at the origin, so none will be projected at infinity. If one uses the HT a second time (in the (ρ', θ') domain) one could indeed detect such lines. For each $\mathbf{a}' \in (\rho', \theta')$ one can write

$$\text{HT}: \mathbf{a}' \rightarrow \rho' = a'_x \cos \theta' + a_y \sin \theta', \tag{6-13}$$

and the VP is detected as

$$\left\langle \rho' = \frac{k}{\| \rho_v \|}; \qquad \theta' = \tan^{-1}\!\left(\frac{y_v}{x_v}\right) \right\rangle. \tag{6-14}$$

The line joining VP provides the orientation of the surface and its vertical position with respect to the z-axis (i.e., intersection of VPs with $x = 0$) determines the plane's tilt.

Much ecological information comes from the ground plane (G), horizon (H), and vanishing points (VP). The ground plane should probably be among the first to be determined because most objects rest on it. For autonomous land vehicles (ALV) the terrestrial horizon separates the ground from the sky. To locate the horizon easily, look for parallel lines whose projections are intersected by zero-texture gradient (regularly spaced texture elements) lines. The intersection of perspective projections determines the VP, and the horizon is the line through the VP and parallel to the zero-texture gradient lines.

The horizon, invariant by remaining unchanged at eye level, might provide a convenient framework for environment-centered representations. Knowing the horizon is useful for estimating, within a scale factor, the size of objects lying on the ground. Sedgwick (1983), using simple trigonometry, defines the horizon-ratio relation as

$$h_1/h = \frac{\tan \alpha + \tan \beta}{\tan \beta}, \tag{6-15a}$$

where h_1 is the height of the object resting on the ground relative to the height h of the observer (the line of sight intercepts the object at a height equal to that of the observer), and α and β are the angles made by the line of sight with the lines to the top and bottom of the object, respectively. When α and β are small, Eq. (6-15a) can be rewritten as

$$h_1 = \frac{h(\alpha + \beta)}{\beta}. \tag{6-15b}$$

Note that the horizon-ratio relation defines size independently of distance. Furthermore, one could easily determine the relative sizes of different objects resting on the ground disregarding the observer's height. Finally, if γ is the angle formed by the line of sight intersecting ground plane at distance d and the vertical corresponding to the height of the observer, the distance d can be estimated as

$$d = \frac{h}{\tan \gamma}. \tag{6-15c}$$

6.3. Motion Interpretation

We are motivated to move around and seek new sensory information in order to derive a consistent 3D structure of the world. The proximal-distal relationship between the retina and the world is one to many. The geometrical structure of 2D retinal projections, however, changes over time in some nonaccidental way. Motion information sifts through incoming streams of sensory inputs, usually leading to unique and consistent interpretations—illusions such as the Necker cube (see Section 4.2.4.) notwithstanding. The observer's motion is one way to achieve invariant perception and to deal with the complex inverse problem of recovering 3D structure from 2D inputs, which is usually ill-posed. To quote from Helmholtz (1878): "I should like, now, to return to the discussion of the most fundamental facts of perception. As we have seen, we not only have changing sense impressions which come to us without our doing anything; we also perceive while we are being active or moving about. Each movement we make by which we alter the appearance of objects should be thought of as an experiment designed to test whether we have understood correctly the invariant relations of the phenomena before us, that is, their existence in definite spatial relations." Helmholtz's view is a major component of expert vision systems discussed in Section 8.1. and relates to the topic of spatial reasoning discussed in Section 8.2.

6.3.1. Egomotion

Egomotion relates to the relative motion of the observer with respect to the environment. Assume that the observer (O) or the target (T) can be either in motion (M) or stationary (S). We consider only those real ecological situations where one sees the target (T) against some background (B). The ocular-drift cancellation theory, mentioned earlier, is one possibility to detect egomotion. We consider three cases: either the target, the observer, or both are in motion. A moving target and a stationary observer interpretation, i.e., M(T) and S(O), results from detecting a stationary background, i.e., S(B), and through the phenomena of background accretion and deletion. Then for a moving target and a moving observer in pursuit of the target, i.e., M(T) and M(O), the background has to be characterized by the same motion, but in opposite direction, i.e., M^{-1}(B). Finally, if the target is stationary and the observer is in motion, i.e., S(T) and M(O), both the target and the background appear to move in opposite directions, i.e., M^{-1}(T) and M^{-1}(B). The analysis can be carried out both qualitatively and quantitatively.

In Section 5.1.2., we defined optical flow $\mathbf{v} = (v_x, v_y)$ and showed different derivation methods. Thompson *et al.* (1984) consider the qualitative interpretation of optical flow for situations when the motion of a target is characterized by specific optical flow patterns. The targets are planar, originally perpendicular to the line of sight, and their motion belongs to one of four classes—translation and rotation at constant depth or in depth along the line of sight.

Translation at constant depth can be easily detected by looking for an above threshold average flow value. Constant horizontal or vertical translation is detected by looking for flow fields such that they are not *nil* and $\partial v_x/\partial x = 0$ or $\partial v_y/\partial y = 0$, respectively. We can analyze translation or rotation in depth using the Lie operators developed in Section 3.3.2. Translation and rotation in depth correspond to shrinking/dilation patterns and concentric circles, respectively, and they can be detected using the operators L_s and L_r, respectively, given next

$$L_s(\mathbf{v}) = x\frac{\partial \mathbf{v}}{\partial x} + y\frac{\partial \mathbf{v}}{\partial y}$$

$$L_r(\mathbf{v}) = -y\frac{\partial \mathbf{v}}{\partial x} + y\frac{\partial \mathbf{v}}{\partial y}.$$

(6-16)

Detection is recorded if $L_i(\mathbf{v}) < \varepsilon$ for some prespecified threshold. The thresholds needed to perform the analysis are noise dependent. The analysis is quite restrictive in terms of allowed motions and target position, even though compound motions could be defined as well. Such compound motions, once detected, could predict impending collisions, and thus provide for collision avoidance. Specifically, a target moving toward an observer, such that it will not pass to one side or another, can be detected by $L_s(\mathbf{v}) < \varepsilon$ (approach of target), and $\partial \mathbf{v} = 0$ (zero average flow indicating constant bearing).

Optical flow patterns also provide approximate information for orientation through a fixed point called the focus of expansion (FOE), from which all motion radiates (see Fig. 3.9.). If the FOE is on the horizon, there is a decreasing velocity gradient toward the horizon where it vanishes, due to the great distance from the observer. If we were to look backward, the optical flow pattern is reversed, and the focus of contraction (FOC) replaces the FOE. For a pilot landing a plane, the FOE is somewhere on the ground rather than at the horizon, and it indicates that point at which landing is aimed. Gibson believed, despite lacking empirical support, that the FOE is the information used for visual guidance. Locating the FOE requires specific psycho-physical representational assumptions regarding the optical flow field. Longuet–Higgins and Prazdny (1980) and Koenderink and van Doorn (1981) suggest that optical flow can be separated into lamellar and solenoidal components, corresponding to exterospecific and proprio-specific stimuli. The exterospecific component depends on movement trajectory. The propriospecific component is the retinal field, which results from eye rotation and depends on where the observer looks. Feedback from eye muscles generally cannot be subtracted from the optical flow pattern to yield the lamellar component responsible for orientation, because there is still another rotational component during curvilinear translation that is exterospecific rather than proprio-specific, as would be the case when driving a car through a curve (Cutting, 1986). Cutting also provides conclusive evidence that none of the schemes offered to calculate the FOE are correct, and he suggests that differential motion parallax is the visual clue used in orientation.

Optical flow patterns provide additional information. Discontinui-ties or breaks in optical flow indicate boundaries between objects moving at different speed. Patterns of accretion and deletion caused by motion are a clue to the layout of surfaces in depth and to what occludes and is occluded. Clearly, accretion indicates a surface that was hidden and gradually becomes visible. Our earlier example of an approaching target of constant bearing illustrates how optical flow patterns can detect impending collision. Flow patterns can be further analyzed to provide information for navigation. One such piece of information, related to free space, might be required for path planning. Imagine a wall and within it a door, possibly ajar. A perceptual sensory system approaching the wall would scan across the obstacle and seek possible differentials in time to collision and their extent. A uniform time to collision indicates impending collision and no free passage.

Finally, we offer another example of egomotion. In particular, we show how a moving observer learns about ground, its planarity (flatness), and slant (orientation).

In Section 3.3.3. we saw how the cross-ratio is a projective invariant. Cutting (1986) presents an alternative proof based on projected angles. The angles referred to are those subtending the segments in the cross-ratio. Because the proof is in terms of angles, the shape of the projection surface is irrelevant and, furthermore, one can easily see that the cross-ratio stays invariant as the observer moves around. Assume that A, B, C, and D are four collinear points, and that the two observation points (corresponding to polar projection) are P_1 and P_2 as seen in Fig. 6.5. Specifically, the canonical cross-ratio CR, whose range is 0 through 1, is given in terms of angles as

$$
\begin{aligned}
CR &= \frac{AD \cdot BC}{AC \cdot BD} = \left(\frac{AD}{AC}\right) \Big/ \left(\frac{BD}{BC}\right) \\[2mm]
&= \left[\frac{\frac{1}{2} \cdot P_1 Q_1 \cdot AD}{\frac{1}{2} \cdot P_1 Q_1 \cdot AC}\right] \Big/ \left[\frac{\frac{1}{2} \cdot P_1 Q_1 \cdot BD}{\frac{1}{2} \cdot P_1 Q_1 \cdot BC}\right] \\[2mm]
&= \left[\frac{\text{Area } (\triangle AP_1 D)}{\text{Area } (\triangle AP_1 C)}\right] \Big/ \left[\frac{\text{Area } (\triangle BP_1 D)}{\text{Area } (\triangle BP_1 C)}\right] \\[2mm]
&= \left[\frac{\frac{1}{2} \cdot AP_1 \cdot DP_1 \cdot \sin(\alpha + \beta + \gamma)}{\frac{1}{2} \cdot AP_1 \cdot CP_1 \cdot \sin(\alpha + \beta)}\right] \Big/ \left[\frac{\frac{1}{2} \cdot BP_1 \cdot DP_1 \cdot \sin(\beta + \gamma)}{\frac{1}{2} \cdot BP_1 \cdot CP_1 \cdot \sin(\beta)}\right] \\[2mm]
&= \left[\frac{\sin(\alpha + \beta + \gamma)}{\sin(\alpha + \beta)}\right] \Big/ \left[\frac{\sin(\beta + \gamma)}{\sin(\beta)}\right] \\[2mm]
&= \left[\frac{\sin(\alpha' + \beta' + \gamma')}{\sin(\alpha' + \beta')}\right] \Big/ \left[\frac{\sin(\beta' + \gamma')}{\sin(\beta')}\right].
\end{aligned}
\tag{6-17}
$$

Cutting (1986) further discusses the utility of the CR to determine the rigid flatness of a moving plane. Specifically, the moving planes include rotating and toppling surfaces (surfaces that, like the ground,

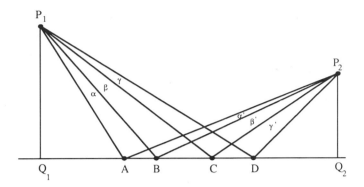

Figure 6.5.
Invariance of cross-ratio under motion.

might support a moving observer) and a circling surface. Observers can indeed pick up the invariant CR, and if faced with nonplanarities caused by the shift of one element only, the CR again can predict the outcome. Basically, the moving planes consist of primary motion, while shifts of single elements constitute secondary motion. If viewers pick up secondary motion as displacement of single elements, then under the condition of equal displacement, one would expect no difference in accuracy across the four positions A through D. If, on the other hand, viewers perceive changes in the stimuli according to changes in the CR, then performance ought to be worse for displacements of A or D than for B or C. In fact, as Cutting (1986) reports, this latter pattern occurs for the subjects in the previous experiment. Still, there are major limitations in the use of the CR, such as those involving collinear points and their fix number (i.e., four) or coplanar parallel lines. As discussed earlier, however, the concepts of nonaccidentalness and analogy might help.

The CR somehow measures the density and distribution of points whose density and spatial distribution might indicate specific relationships and/or properties. Consequently, Cutting (1986) suggests density indexes for situations where the CR is inappropriate. Such density indexes can be easily defined in terms of (Euclidean) distances between detected points of interest. The main question to be asked, however, is the existence and reliability of such points. (See the discussion on motion tokens in Section 5.1.2.)

The CR is not the only invariant the visual system uses to detect

planarity. Pseudo-flow vectors can specify coplanarity. Remember that
the observer sees the horizon from its height h. The line of sight to the
horizon makes an angle α with a line of sight, swapping the ground axis
x as the observer moves along the ground. One can then write

$$\alpha = \arctan\left(\frac{h}{x}\right) \tag{6-18a}$$

or, as the observer moves,

$$\frac{d\alpha}{dx} = \frac{-h}{h^2 + x^2}. \tag{6-18b}$$

One can consider $d\alpha/dx$ as some momentary displacement v and rewrite
the eye height h as

$$h = \frac{-(\tan^2 \alpha)}{v(1 + \tan^2 \alpha)} \tag{6-19a}$$

or for objects near the horizon, h can be approximated as

$$h \approx \frac{-(\tan^2 \alpha)}{v}. \tag{6-19b}$$

Clearly, for a moving observer, the relative ordering of the h values
indicates if the surface is flat (coplanar) or not; it can further indicate
even the slant of a flat surface.

6.3.2. Motion Parallax

Here, we consider the benefits of being able to fixate a point in our 3D
environment. Such a point is a point of reference and points farther
away ("behind") appear to move in the same direction as the viewer,
while points that are nearer ("in front") appear to move away in the
opposite direction. The previous description is that of motion parallax,
and kinetic depth is the sensation one gets when moving the head while
fixating a target.

Ballard (1988) shows how a fixation point facilitates the derivation of
depth by providing an instantaneous origin at $(0, 0, z_0)$, where z_0 is the
distance of the fixation point from the viewer.

Vergence geometry, similar to that shown in Fig. 5.3. for binocular vision, derives the depth z_0 as

$$z_0 = \frac{b \sin \gamma_2}{\sin(\gamma_1 + \gamma_2)}, \tag{6-20}$$

where b is the base line between the two retinas, γ_1 and γ_2 the vergence line of sights from the left and right eye, respectively, and the head is moving right, a distance V_H in the vestibular head system. Using standard projection, and the geometry of Fig. 6.6., any point in the image plane of coordinates (x, y) is given by

$$x = \frac{-fX}{Z}; \qquad y = \frac{-fY}{Z}, \tag{6-21}$$

where (X, Y, Z) are the 3D coordinates, and f is the focal length. Assume that (V_X, V_Y, V_Z) and (v_x, v_y) stand for the 3D (world) and 2D (optical flow retinal) velocities, respectively. Then differentiating Eq. (6-21) and assuming only translational velocity one obtains

$$\begin{aligned}
- Zv_x &= fV_X + xV_Z \\
- Zv_y &= fV_Y + yV_Z.
\end{aligned} \tag{6-22}$$

Assuming that we fixate the target, its foveal position implies that the retinal coordinates are approximately zero. Further assume that V_X and V_Y are comparable or greater than V_Z, and that $f \gg x, y$. Then

$$\frac{v_x}{v_y} \simeq \frac{V_X}{V_Y}. \tag{6-23}$$

If $f(x, y, t)$ stands for the image gray-level function, the spatiotemporal gradient equation (Eq. 5-3) reads

$$f_x v_x + f_y v_y + f_t = 0. \tag{6-24}$$

The last two equations yield the retinal (foveal) optical flow as

$$v_x = \frac{-f_t}{f_x + \dfrac{V_Y}{V_X} f_y}$$

$$v_y = \frac{-f_t}{f_x \dfrac{V_X}{V_Y} + f_y}. \tag{6-25}$$

The 3D velocity $\mathbf{V}$ can be obtained from Eqs. (6-22) and (6-23) as

$$|V| = (V_X^2 + V_Y^2)^{1/2} = \frac{Z}{f}(v_x^2 + v_y^2)^{1/2}. \qquad (6\text{-}26)$$

Then the vestibular command motion V_C is parallel and of opposite sign to $\mathbf{V}$, and if one assumes the V_Z is small and $Z_0 > Z$, from similar triangles one can then write

$$\frac{|V_H|}{|V|} = \frac{Z_0}{Z_0 - Z}, \qquad (6\text{-}27)$$

where Z is the distance of the point displaying velocity $\mathbf{V}$, and Z_0 is the distance from the fixation point.

Finally, from Eqs. (6-26) and (6-27) one readily obtains that

$$Z = Z_0 \left[1 + \frac{Z_0(v_x^2 + v_y^2)}{V_H f} \right]^{-1}. \qquad (6\text{-}28)$$

For $Z_0 < Z$, the sign is negative, as one would expect when moving right and observing targets nearer than the fixation point.

The geometry considered approximates optical flow by translation only. As Cutting (1986) points out, however, both rotational and translational components are involved in motion parallax. To quote from Cutting "when the viewer looks off to the right at some object, maintaining fixation on it while moving forward, the changing optic array is the same as if the ground and objects on it were rotating counterclockwise and expanding around a vertical axis through the

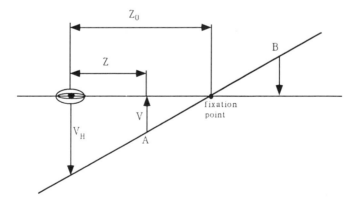

Figure 6.6.
Kinetic depth.

fixation point. Similarly, when the viewer looks left, objects in the array rotate clockwise and expand around the fixated object. Because the world is seen in polar projection, these rotations generate important asymmetries of flow." Assume that the observer's path and line of sight make an angle α, that the gaze is on the left of the path, and that a, b, and c are points along the gaze, where b is the fixation point. Further assume that β and γ are the angles between the planes of sight containing (a, b) and (b, c), respectively. As the observer moves forward, the angle β grows faster than angle γ. Objects closer to the observer shear faster than those farther away, and they shear leftward if the observer's movement with respect to gaze is to the right. Cutting strongly suggests that motion parallax is based on the differential growth of β and γ, and that it is the visual information used for orientation in movement. Finally, such parallax is "contingent on polar perspective" because under parallel projection, the angles β and γ, corresponding to projected motion, are the same. The qualitative discussion presented emphasizes that there might be multiple ways to represent the same phenomena but some might miss the relevant information. Capturing information may, therefore, be a function of active perception under the guidance of cognitive processes and according to specific needs of information contents.

6.3.3. Three-dimensional Motion

Motion parallax and Eq. (6-22) are restrictive in that they assume only translational components. The more general case (Waxman and Duncan, 1986) considers both translational velocity (T) and angular velocity (R). For a 3D pixel, P, the differential motion is defined as

$$\frac{dP}{dt} = -T - R \times P. \tag{6-29a}$$

where $T = (T_x, T_y, T_z)$, $R = (R_x, R_y, R_z)$ and

$$\frac{dX}{dt} = -T_x - R_y Z + R_z Y$$

$$\frac{dY}{dt} = -T_y - R_z X + R_x Z \tag{6-29b}$$

$$\frac{dZ}{dt} = -T_z - R_x Y + R_y X.$$

If one were to project $P(X, Y, Z)$ onto a unit focal length image ($f = 1$), the retinal coordinates are again given as $(x, y) = (X/Z, Y/Z)$, and the optical flow (v_x, v_y) is obtained as

$$
\begin{bmatrix} v_x \\ v_y \end{bmatrix} = \frac{1}{Z} \begin{bmatrix} -1 & 0 & x \\ 0 & -1 & y \end{bmatrix} \begin{bmatrix} T_x \\ T_y \\ T_z \end{bmatrix}
$$

$$
+ \begin{bmatrix} xy & -(1+x^2) & y \\ (1+y^2) & -xy & -x \end{bmatrix} \begin{bmatrix} R_x \\ R_y \\ R_z \end{bmatrix}
$$

(6.30)

The qualitative interpretation of optical flow (Thompson et al., 1984) reported in Section 6.3.1. has been extended by Eagleson (1987) to interpret the general 3D motion (Eq. (6-30)) in terms of perceptual invariants and the corresponding spatiotemporal filters needed to capture them. Three-dimensional motion is factored out into localized 2D horizontal and vertical translations, divergence, curl, and 2D shear optical flows. The flow fields correspond to specific motion deformations and a 6D Lie Group (see Section 3.3.2.), which consists of six orthogonal multiscale and distributed convolution kernels, can then resonate to these specific types of motion. The optical flows can be estimated using the phase shift of spatiotemporal filters (similar to the filters used for disparity estimation in Section 5.1.3.) rather than slope measurements based on the intensity function. The localized deformations and the resonators are related through exponential conformal mappings. The factorization of 3D motion and the corresponding 6D Lie group points out to the strong relationship between distributed and multiscale representations, conformal mappings, perceptual invariants, and resonance.

The 6D Lie Group consists of $\langle L_x, L_y, L_s, L_r, L_b, L_B \rangle$, where the first two operators would resonate to horizontal and vertical translation, the following two (divergence and curl) (see Eq. (6-16)) to dilation and rotation patterns, respectively, and the last two operators resonate to shear. The operators estimate the spatial derivatives of optical flow in terms of filter-based operators. Assuming that projected motion, i.e., optical flow, is

$$
(v_x, v_y) = (u, v) = \left(\frac{\partial x}{\partial t}, \frac{\partial y}{\partial t} \right),
$$

then one defines the spatial derivatives of optical flow as

$$divergence = \mu = u_x + v_y$$
$$curl = \lambda = u_y - v_x$$
$$shear\ (b) = \rho = u_x - v_y$$
$$shear\ (B) = \sigma = u_y + v_x.$$

The resonance to 3D motion is then accomplished by estimating the above 6D Lie group from 2D bandpass spatial frequency filters (Eagleson, 1987).

6.3.4. Depth from Motion

We now consider how incremental and known lateral camera motions can derive a dense disparity map, or equivalently, a dense depth map. The approach uses the Kalman filter's (KF) (see Section 3.2.3.) specific advantages. Because of the uncertainty in deriving disparity, error analysis and ways to reduce the uncertainty are needed. The KF is the optimal linear filter that integrates uncertainty information and iteratively attempts to reduce it. Furthermore, the same KF yields substantial savings in storage because it needs to store only the last update of the system.

Matthies *et al.* (1987) use the KF as defined by Eqs. (3-50) through (3-53) to derive a dense depth map under the assumption of small lateral camera displacements. The measurement process yields some disparity d as the result of correlating two temporally successive image frames. Correlation yields a local new disparity $d(x, y)$ as the minimum of a parabolic error surface $e(d)$ given as $e(d) = ad^2 + bd + c$, where the variance of the measurement is given as $\text{Var}(d) = \sigma_d^2 = 2\sigma^2/a$, where σ^2 is the variance of the image noise process. The matrix H involved in the measurement process is the identity matrix, i.e., $H = I$. The prediction process uses the perspective Eqs. ($x_t = X/Z; y_t = Y/Z; d_t(x, y) = d_t = 1/Z$) at time t (assuming unit focal length) and the premise of known lateral motion T_x to make the predictions $\langle \hat{x}_{t+1} = x_t - T_x \tilde{d}_t; \hat{y}_{t+1} = y_t; \hat{d}_{t+1} = \tilde{d}_t \rangle$, based on the smooth estimate $\tilde{d}_t$ (see next). This prediction yields estimates of disparity between pixels and thus requires resampling. The error and uncertainty are local (pixelwise) and depend on the variance of the local disparity. The covariance prediction yields $\hat{P}(t + 1) = (1 + \varepsilon)P(t)$ for some "inflation" factor ε. The update (integration) process yields new estimates for the covariance, gain, and disparity, at time t, given as

$$P(t) = \frac{1}{\hat{P}(t)} + \frac{1}{\sigma_d^2} = \frac{\hat{P}(t)\sigma_d^2}{\hat{P}(t) + \sigma_d^2}$$

$$G(t) = \frac{P(t)}{\sigma_d^2} = \frac{\hat{P}(t)}{\hat{P}(t) + \sigma_d^2}$$

$$d(t) = \hat{d}(t) + G(t)[d - \hat{d}(t)]$$

$$= \hat{P}(t)\left|\frac{\hat{d}(t)}{\hat{P}(t)} + \frac{d}{\sigma_d^2}\right|, \tag{6-31}$$

where d and σ_d^2 are the actual measured disparity and its variance at time t. The regularization process yields those smooth values d_t to be used by the prediction process on the next cycle. Confidence is the inverse of variance, and the smoothing process is conceptually similar to functional dynamic vision as discussed in Section 6.1.3. (snakes and thin plates) and to the smoothing of optical flow using relaxation (Section 5.1.2.).

Depth derivation can be further enhanced if the motion that controls input acquisition, i.e., gaze control, is optimally controlled. Shmuel and Werman (1989) propose an active (perception) solution, which includes input dependent data acquisition and considers the reliability of the acquired information as well. (See also Abbott and Ahuja (1988) work, described in Section 8.3.2., on dynamic integration of focus, vergence and stereo disparity for surface reconstruction.) After each phase of computation (between pictures), when information is still needed, the camera is placed in a new optimal position for sensing the next input. To optimally place the camera for the next picture, the knowledge of the scene is described by the Gauss–Markov discrete time model. Application of Kalman filtering to this model makes it possible to quantify the information and compute optimal estimations. The process is repeated until sufficient accuracy is achieved. This approach helps in making ill-posed problems become well-posed and improves the ability to perform tasks in a noisy environment. Accuracy and reliability of solutions is improved and the quantity of data being processed, i.e., the computational complexity, is reduced.

6.3.5.   Shape from Motion (SFM)

Structure or shape from motion corresponds to the human capability to interpret changing projections of 3D structures. The term structure

"S" (in SFM) is distinguished from absolute depth; only relative internal depths are recovered. Early explanations that SFM is achieved by recovering distance from velocity disparity turned out to be invalid. Specifically, Ullman (1979b) took the case of two cylinders, a smaller one inside a larger one, and showed that within each cylinder velocity changes in accordance with depth, but points with same velocity belong to different cylinders. Familiarity and recognition were shown to be irrelevant for the interpretation of motion.

Much research related to recovering structure from motion was done using MLD (moving light displays). MLD isolate and present geometric evidence of motion apart from other factors as texture, color, and lighting. The only source of information in an MLD is the position and velocity of its points; and position by itself does not provide sufficient data for MLD interpretation (Rashid, 1980). Note that any single MLD frame shows only a meaningless pattern. If we only glanced once, i.e., if we had access to only one frozen image, we could not succeed in interpreting the MLD. It is only when we can see a whole MLD sequence that a meaningful pattern emerges. This suggests that visual events are perceived and integrated over time. The perceptual cycle (Neisser, 1976; see Section 8.2.4. on planning), which provides spatial and temporal continuity, facilitates the interaction between the observer and the environment.

Several theories advanced to explain SFM were tested on moving light displays. Johansson (1976) argued that motion is interpreted within a hierarchy of coordinate systems set according to vector analysis. The total motion of each point is seen as the composition of a movement relative to its particular coordinate system with the motion of that coordinate system relative to the next one in the hierarchy. The selection of a coordinate system would depend on 2D velocities, where the lowest velocity is interpreted relative to the background. In the context of MLD, however, at certain points of the walker's step, e.g., when the foot is in contact with the floor, the movement of the ankle is actually less than the movement of the hip or knee.

Ullman (1979b) makes the assumption that any set of elements undergoing 2D transformation, which has a unique interpretation as a rigid body moving in space, should be interpreted as such a body in motion. Then, it can be shown that given three distinct orthographic projections of four noncoplanar points in a rigid configuration, the structure and motion compatible with the three views are uniquely determined. Some ambiguity remains because 3D recovery depends on reflection from the frontal plane. The approach cannot cope with a low degree of connectivity, perspective distortion, and its fundamental assumption of rigidity has been challenged. (Remember, for example, that optical flow derivation does not assume any rigidity.)

Rashid (1980) undertook the same well-known cylinder demonstration that was quoted and showed that correspondence (tracking), if done in terms of velocity and position, can yield internal structure. Specifically, minimum spanning trees (MST) are built according to both position (x, y) and velocity (v_x, v_y). To separate objects and determine subparts, one has to split the MST as it is done for taxonomy purposes, i.e., along those edges (links) of relatively high value. Experiments done on MLD corresponding to the two cylinders and animation sketches indeed showed the feasibility of such an approach.

Not only is there disagreement regarding how to solve the SFM problem, but even the basic assumption of a two-step process has recently been challenged. The two-step process involves first determining the optical flow and then proceeding to establish the object's internal structure. Negahdaripour and Horn (1985), however, suggest a SFM technique that bypasses computing optical flow or matching discrete features.

Finally, it is interesting to note that within the context of motion analysis, modularity of processing can lead to ambiguity and/or conflict. Specifically, the SFM module is probably complemented by a motion from structure (MFS) module. Take a trapezoidal window and let it rotate. What one sees is a window appearing to oscillate instead. Conflictive interpretation between SFM and MFS could be resolved by some yet to be specified top-down priming.

6.4.   Navigation

An active observer seeks new information by exploring the environment. Exploration usually involves moving around, tracking objects of interest, planning a safe route to avoid collision, and integrating information for better and more efficient navigation. Such issues are relevant for robots on the factory floor (near the assembly lines and conveyor belts), autonomous land vehicles (ALV) moving around, or someday for exploring remote planetary sites. One could devote an entire book on each. Here, we consider only basic examples to illustrate the concepts of tracking, path planning and collision avoidance, and ALV.

6.4.1. Tracking

Preattentive vision and foveal fixation for object recognition and manipulation require tracking. Tracking implies prediction of future movement and the ability to account for predictive errors and to correct them in future estimates. We consider several tracking methods, characteristic of extrapolation, regression, and the Kalman filter (KF) and assume that we track the center of gravity $\mathbf{x}$ for some object.

The simplest method, that of two-point extrapolation, assumes the trajectory of the moving target to be linear and of constant velocity v. If space location at time t is given as $\mathbf{x}_t$ then the predicted location at time $(t + 1)$ is given as

$$\mathbf{x}_{t+1} = \mathbf{x}_t + v \cdot T = \mathbf{x}_t + (\mathbf{x}_t - \mathbf{x}_{t-1})$$
$$= 2\mathbf{x}_t - \mathbf{x}_{t-1}, \qquad (6\text{-}32)$$

where T is the sampling time interval.

Linear regression predicts the $(t + 1)$ location by finding the best linear fit in the MSE for the given set of locations $\{\mathbf{x}_i\}_{i=1}^t$. Again assume linear trajectory of constant velocity. The ith location for the given set is estimated as $\hat{\mathbf{x}}_i$ and given by

$$\hat{\mathbf{x}}_i = \begin{pmatrix} \hat{x}_i \\ \hat{y}_i \end{pmatrix} = \begin{pmatrix} a_t i + b_t \\ c_t i + d_t \end{pmatrix}, \qquad (6\text{-}33a)$$

while the location at time $(t + n)$ is found as

$$\hat{\mathbf{x}}_{t+n} = \begin{pmatrix} \hat{x}_{t+n} \\ \hat{y}_{t+n} \end{pmatrix} = \begin{pmatrix} a_t(t + n) + b_t \\ c_t(t + n) + d_t \end{pmatrix}. \qquad (6\text{-}33b)$$

The set of coefficients (a_t, b_t, c_t, d_t) vary with the memory span t. The estimates are an approximation of the true locations, and in the process of minimizing the error of approximation E, one derives the set of coefficients and thus defines the regression model. Specifically,

$$x_i = \hat{x}_i + \varepsilon_i = a_t i + b_t + \varepsilon_i$$

$$E = \sum_{i=1}^t \varepsilon_i^2 = \sum_{i=1}^t (x_i - a_t i - b_t)^2$$

$$\frac{\partial E}{\partial a_t} = -2 \sum_{i=1}^t i(x_i - a_t i - b_t) = 0$$

$$\frac{\partial E}{\partial b_t} = -2 \sum_{i=1}^t (x_i - a_t i - b_t) = 0$$

and the coefficients (a_t, b_t) are found as

$$a_t = \frac{\Sigma\, ix_i - 1/t\, \Sigma\, x_i \Sigma\, i}{\Sigma\, i^2 - 1/t(\Sigma\, i)^2}$$

$$b_t = \frac{1}{t(\Sigma\, x_i - a_t\, \Sigma\, i)}.$$

(6-33c)

The coefficients (c_t, d_t) are found in a similar way. The regression model given by Eq. (6-33a) is linear. Higher-order models, computationally more expensive, could be derived. Second-order regression will use a parabolic rather than linear prediction model.

The Kalman filter, as introduced in Section 3.2.3. (and used in Section 6.3.3. to derive depth from motion), is appropriate for tracking tasks if the trajectory results from motion whose transition matrix ϕ is linear or can be approximated by a linear expansion. Assume that some robot is tracking an object rolling over a conveyor belt. The tracking task involves predicting the location (x_t, y_t) and velocity $(\dot{x}_t, \dot{y}_t)$ where the dot stands for time derivative. The state vector $s(t)$ is given as

$$\mathbf{s}(t) = \begin{bmatrix} s_1(t) \\ s_2(t) \\ s_3(t) \\ s_4(t) \end{bmatrix} = \begin{bmatrix} x_t \\ \dot{x}_t \\ y_t \\ \dot{y}_t \end{bmatrix}.$$

(6-34a)

One writes the motion equations as

$$x_{t+1} = x_t + T\dot{x}_t + W_t^1$$

$$\dot{x}_{t+1} = \dot{x}_t$$

$$y_{t+1} = y_t + T\dot{y}_t + W_t^2$$

$$\dot{y}_{t+1} = \dot{y}_t,$$

(6-34b)

where T is the sampling interval, and the velocity is constant (no acceleration). Then the transition matrix ϕ is given as

$$\phi = \begin{bmatrix} 1 & T & 0 & 0 \\ 0 & 1 & 0 & 0 \\ 0 & 0 & 1 & T \\ 0 & 0 & 0 & 1 \end{bmatrix},$$

(6-34c)

and the noise vector $\mathbf{W}_t$ of covariance Q, is defined as

$$
\mathbf{W}_t = \begin{bmatrix} W_t^1 \\ 0 \\ W_t^2 \\ 0 \end{bmatrix}.
\tag{6-34d}
$$

The measurement process senses the new position as

$$
\mathbf{Y}(t) = \mathbf{H}s(t) + \mathbf{n}(t),
\tag{6-34e}
$$

where $H = \left(\begin{smallmatrix} 1 & 0 & 0 & 0 \\ 0 & 0 & 1 & 0 \end{smallmatrix}\right)$ and $\mathbf{n}(t)$ is noise of covariance R. One needs to model the uncertainty in sensing (R) and state transition (Q) and to initialize the (covariance) uncertainty of the system state (P) before the first measurement is taken. Eqs. (3-50) through (3-53) can then be used to predict the motion.

Hunt and Sanderson (1982) applied the predictive methods to robotic tracking of a moving target. The ranking according to timing requirements is obvious, and simple extrapolation is fastest. The two-point extrapolator has a very short memory span and can respond quickly to deviations from linear trajectories. An augmented Kalman filter (which also accounts for acceleration in the state transition matrix) performed only slightly better than the two-point extrapolator, but at a much higher cost in execution time. Clearly, regression models are the worst, because they move away slowly from an established path. The larger the memory span (t), the less flexible regression is, and, as a consequence, the performance degrades.

6.4.2. Path Planning

When we navigate, one key goal is to move safely and quickly between two locations. Basically, one has to plan a path that avoids collision with fixed (static) obstacles (PPP/path planning problem), and at the same time plan the velocity of motion to avoid colliding with obstacles possibly moving across the path (VPP/velocity planning problem). The trajectory planning problem (TPP) is quite complex. Specific assumptions are always made, beginning with the independence of the moving robot's trajectories and the obstacles. Following Kant and Zucker (1984), we consider the decoupling of the original problem TPP into two new problems, those of the PPP and VPP type, and show how they can be solved.

Path planning requires representing the free space available (space void of potential static collisions) and optimally moving a robot through it. The time complexity of algorithms suggested to solve the TPP is usually polynomial. The V-graph algorithm sets as a necessary condition for a minimum length path that it can be composed of straight line segments (SLS) connecting a subset of the vertices of the polygonal obstacles. Graph search techniques then seek the shortest path through the graph defined by the SLS. Here, the obstacles are additionally assumed to be polygonal. Furthermore, one can consider a point-like robot if one compensates by expanding the size of the static obstacles by the size of the robot.

The robot also needs to avoid moving obstacles, which further constrain its movement by denying it additional volumes of space × time (VST). The VST are easily defined according to a moving obstacle's path. Again, we assume that the VST is polygonal or rectangular, so we can use again the V-graph algorithm mentioned earlier. Because there is a tradeoff between accuracy and computation cost, we must often compromise accuracy. The VPP can now be reformulated so that given two locations (those of start (S) and end (E)), find a path in the $s \times t$ space, joining S with E, while avoiding the rectangular obstacles. There are additional constraints that make the VPP different from PPP. Those constraints are:

(a) *Monotonicity.* Time moves forward; the robot cannot retrace its path back through time, and the search graphs is directed.

(b) *Maximum Velocity.* Robot's velocity is bounded, i.e., $|v(t)| < v_{max}$. This constraint can be rewritten as $|ds/dt| < v_{max}$, i.e., the slope of each line segment in the optimal path should not exceed the maximal velocity.

(c) *Finite Acceleration.* Discontinuous paths imply infinite acceleration. As a consequence, the optimal path needs to be smoothly interpolated between the vertices of the rectangular sites while not intersecting any of the forbidden areas.

The optimal path is then found using the V-graph as the one that minimizes the time needed to go from *start* to *end* and does not violate any of the constraints (a)–(c).

The scenario for the TPP is quite simple. Additional constraints, such as the number of moving obstacles and the randomness of their movement might have to be taken into consideration. The global solution approach with respect to both path and velocity might not be feasible. Local or opportunistic planning by default then becomes the only computationally feasible solution.

6.4.3. Mobile Robotics

The technology of mobile robotics is continuously evolving. Several basic concepts, however, will likely endure. First, any mobile robot has to sense its environment using one or more sensory devices. Each of these sensory devices produces intrinsic characteristics of the environment needed for tasks such as location, orientation, and navigation. Intrinsic characteristics, including depth and slant of obstacles (from sonar or stereo), which are resting or in motion, can be interpreted based on sensor properties and appropriate world modeling. The world model is continuously updated using methods characteristic of spatial reasoning (see Section 8.2.) and data fusion (see Section 8.3.). Data fusion reduces the uncertainty of interpretation for safe navigation and behavior. It also keeps likely errors, related to motion and/or manipulation, below acceptable thresholds. Furthermore, the same data fusion approach is also responsible for allocating appropriate resources for sensing and processing and keeping the computational burden manageable. The process of continuously updating the world model also facilitates other intelligent tasks, such as path planning, motor behavior, and learning. The whole process is conceptually akin to AI image interpretation and understanding systems and to the blackboard approach (see Section 8.1.), where different agents cooperate toward solving (interpreting) some jigsaw puzzle by filling in the missing parts one at a time. A major component of any mobile robot is the ability to maintain a dynamic model of the external world and to be able to translate between different frames of reference, specifically those of the observer, the objects, and the environment. Mobile robotics thus requires to develop an intelligent computational capability for visual processing for safe navigation and manipulation.

Many mobile robots "ressemble" to some degree the first homanoids in their upright position, trying to find their way around. Such robots are usually slow, and the world they are aware of is quite limited. Characteristic examples of such robots, which hold promise for future developments, include the Stanford Cart and the CMU Rover

(Moravec, 1983b), Alvin, the Autonomous Land Vehicle (ALV) (Turk *et al.*, 1988), and the Navlab (Thorpe *et al.*, 1988).

The most impressive ALV so far is the one developed by Dickmanns and Graefe (1988a and 1988b). Their success in running an ALV over the autobahn (at speeds exceeding 60 mph but with specific limitations regarding collisions) is due to the ALV being able to maintain a dynamic model of the external world, to incorporate the motion dynamics, and to relate between them using the Kalman filter.

6.5. Structured Light

Structured light is a machine vision approach that takes advantage of the physics of ambient illumination and the image formation process, acts as a kind of "intelligent filter" by discarding irrelevant details and enhancing the signal itself, and increases the SNR where noise means the irrelevant "background." Structure light is usually implemented through active sensing, and it is the human engineer equivalent to active perception. An early system is CONSIGHT-I (Holland *et al.*, 1979). Two light sources project angled lines of light onto the conveyor belt and to some mechanical part lying on it. The two sources are adjusted so that the lines of light are superimposed on one another at the level of the belt. Because of the thickness of the mechanical part, two lines of light would then separate in passing over it. The points where the superimposed lines break into separate lines define the edge of the part. Therefore, the whole process of edge detection needed to obtain the contour or shape of the object is greatly simplified because the search area is much reduced.

6.5.1. Shape from Grid Projections

Acquisition of scene properties, such as depth and surface orientation, is a key problem in computer vision. Structured light methods are often used for quick acquisition of range. These methods take advantage of texture gradients and utilize a projection of a regular light pattern. Establishing correspondence between stripes in the image and the light stripes in the scene is easier than solving the stereo correspondence problem. The pattern of light being projected can be as simple as a point, or it can be either one light bar or several bars in a regular pattern, as can be seen from Fig. 6.7. (Kak, 1983). Once correspondence is established, depth is derived via triangulation.

Qualitative analysis can also be performed based on depth change. As seen in Fig. 6.8., one can easily distinguish among flat surfaces, convex edges, concave edges, and obscuring edges.

Asada and Tsuji (1987) present an elegant method to determine an object's shape by projecting a stripe pattern onto it. The system uses orthographic projection as a camera model and implements a quantitative analysis by locally deriving the surface normals from the slopes

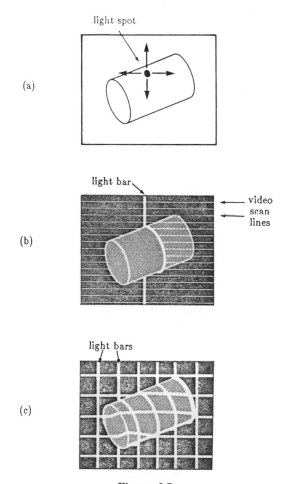

Figure 6.7.
Stereo range mapping using stripe patterns. Reproduced with permission from Kak, A., 1983, Depth perception for robots, TR-EE 83–44, Purdue University.

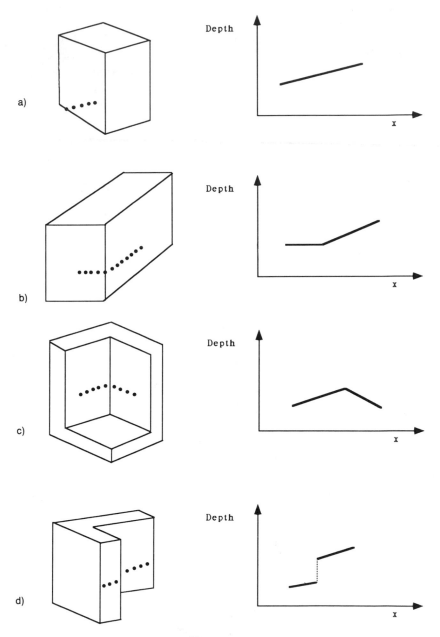

Figure 6.8.
Qualitative analysis of change in depth. (a) Change of depth on a flat surface, (b) change
of depth at a convex edge, (c) change of depth at a concave edge, and (d) change of depth
at an obscuring edge.

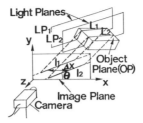

Figure 6.9.
Surface orientation from stripe patterns. Reproduced with permission from Asada, M., and Tsuji, S., 1987. Shape from projecting a stripe pattern, CAR-TR-263, Computer Vision Lab, University of Maryland.

and intervals of the stripes in the image. Assume the geometry of Fig. 6.9. Then, the equations of the parallel light planes (LP_1) and (LP_2) and object plane (OP) on which a stripe pattern is projected are given as

$$(LP_1): P_S X + Q_S Y + Z = D_1$$

$$(LP_2): P_S X + Q_S Y + Z = D_2$$

$$(OP): P_O X + Q_O Y + Z = C, \qquad (6\text{-}35)$$

where $(D_2 - D_1)$ is the width of the stripe with respect to the camera-centered coordinate system.

The object plane (OP) and the light planes $LP_1(LP_2)$ intersect at a line $L_1(L_2)$. The projections of lines L_1 and L_2 onto the image plane (the $x - y$ plane) are labeled as $l_1(l_2)$. The slopes of these lines (θ) and the distance between them (Δx) in the x-axis direction are easily measured in the image using orthographic projection. Specifically,

$$\tan \theta = -\frac{P_S - P_O}{Q_S - Q_O}$$

$$\Delta x = \frac{D_2 - D_1}{P_S - P_O}. \qquad (6\text{-}36)$$

and then,

$$P_O = P_S - \frac{D_2 - D_1}{\Delta x}$$

$$Q_O = \frac{D_2 - D_1}{(\tan \theta) \times \Delta x} + Q_S. \qquad (6\text{-}37)$$

Thus, surface orientation (such as in gradient space) $\langle P_O, Q_O \rangle$ of a point on an object can be locally determined from the slope (θ) and interval (Δx) of the stripe pattern projected onto the image. The accuracy of estimated surface orientation depends on the accuracy of measuring the slopes and spacings of the projected stripes, first detected as edge points.

Range data can also be acquired almost automatically using calibration charts (Albus, 1981). The pixel row and column of any point in the TV image can be converted to (x, y) position in some coordinate system. One example of such a coordinate system would consider the x-axis passing through two robot fingertips and the y-axis pointing into the same direction as the fingers do. The plane of the projected light is the $(x - y)$ plane so that the z coordinate is zero.

6.5.2. Moiré Patterns

Moiré patterns result when the illumination of two gratings interfere on the surface (in depth) of some object. Next we will show how one can derive depth by using the light and dark fringe contours created by Moiré patterns. Each topographic contour stands for a plane of uniform depth. Following Meadows *et al.* (1970), assume a source illuminating a grating at an angle α. The grating has an intensity modulation given by

$$T(x) = \frac{1}{2} + \frac{1}{2}\sin\left(\frac{2\pi x}{d}\right).$$ (6-38)

The sinusoidal grating is of period d and the factor $1/2$ makes sure that the transmitted intensity is positive and never greater than the maximum input intensity. If $f_i(x, y)$ and $f_t(x, y)$ stand for the input and transmitted intensities, then one can write

$$f_t(x, y) = f_i(x, y)T(x, y).$$ (6-39a)

If the input is uniform, then $f_i(x, y) = C$, and

$$f_t(x, y) = \frac{C}{2}\left[1 + \sin\left(\frac{2\pi x}{d}\right)\right].$$ (6-39b)

Assume some matte surface reflection (diffuse scatterer) and let ϕ be the angle between the incident ray and the normal to the surface. Since the incident rays are at angle α (with respect to the vertical) the shadow on the object shifts in the x-direction relative to the grating by $z(x, y)\tan \alpha$, where $z(x, y)$ is the depth sought. The reflected intensity f_r of the shadowed grating is then

$$f_r(x, y) = \frac{C}{2}\left[\cos \phi(x, y, z(x, y))\right]\cdot\left[\sin\left(\frac{2\pi}{d}\right)(x - z(x, y)\tan \alpha) + 1\right]. \quad (6\text{-}40\text{a})$$

Let us assume that the surface is smooth enough so that we can rewrite the reflected intensity as

$$f_r(x, y) = K\left[\sin\left(\frac{2\pi}{d}\right)(x - z(x, y)\tan \alpha) + 1\right]. \quad (6\text{-}40\text{b})$$

The reflected intensity then propagates back through the grating and yields the observed intensity f_o as

$$f_o(x, y) = f_r T(x, y)$$
$$= \frac{K}{2}\left[\sin\left(\frac{2\pi}{d}\right)(x - z(x, y)\tan \alpha) + 1\right]$$
$$\times \left[\sin\left(\frac{2\pi}{d}\right)(x + z(x, y)\tan \beta) + 1\right]. \quad (6\text{-}41\text{a})$$

where β is the angle the observer looks at the grating. The last equation can be rewritten as

$$f_o(x, y) = \frac{K}{2}\left\{1 + \sin\left(\frac{2\pi}{d}\right)(x - z(x, y)\tan \alpha)\right.$$
$$+ \sin\left(\frac{2\pi}{d}\right)(x + z(x, y)\tan \beta)$$
$$- \frac{1}{2}\cos\left(\frac{2\pi}{d}\right)(2x + z(x, y)(\tan \beta - \tan \alpha))$$
$$\left. + \frac{1}{2}\cos\left(\frac{2\pi z(x, y)}{d}\right)(\tan \beta + \tan \alpha)\right\}. \quad (6\text{-}41\text{b})$$

The last term in Eq. (6-41b) depends only on the depth $z(x, y)$, and thus each fringe of light or dark is a contour of constant depth. If two points are separated by N fringes, the difference in height Δ_z is given as

$$\Delta_z = \frac{Nd}{\tan \alpha + \tan \beta}. \quad (6\text{-}42)$$

Note that the period d has to be chosen such that the first three terms in Eq. (6-41b) are above the resolution limit of the observer's CSF, i.e., $(2\pi x/d)$ changes faster than $(2\pi z(x, y)/d)$.

The analysis yields topographic information only. Range information can be obtained using a base plane calibrated at known distance and implementing Eq. (6-42). The approach described so far cannot cope with sharp edges, and convex or concave surfaces may appear alike, because the sign of depth is missing. If the object were to move away from the observer, however, the fringes of convex surfaces will converge toward the center, and those of concave surfaces will move out. Processing Moiré patterns is expensive because a whole frame must be processed at one time. The approach, while not suited for real-time robotic manipulation, is still suitable for inspection and quality control with its maximum resolution of about 25 microns.

6.5.3. Range from Depth of Field

Vergence is another example of proprioception. Both convergence and divergence are disjunctive eye movements initiated when images move toward or away from a viewer. The goal is to keep things in focus and to correct for possible blur. Autofocus schemes measure depth pixelwise, by adjusting the optics, such that blur is reduced and best focus results. One can derive depth across the whole field of view, however, as a function of change in the gradient of focus. The following method (Pentland, 1987) is characteristic of structured light and is another example of active perception. The results are comparable to those obtained from stereo, (over a span of several meters), but relieves one of solving the correspondence problem.

Assume a thin lens and the corresponding equation

$$\frac{1}{u} + \frac{1}{v} = \frac{1}{f}, \tag{6-43}$$

where u is the distance between a point in the scene and the lens, v, the distance between the lens and the plane of a perfectly focused image, and f is the focal length. If a point at distance $u_o < u$ is projected through the lens of radius r, it focuses at a distance $v_o > v$, so that a blur circle of radius σ is formed. One can then write

$$\tan \theta = \frac{r}{v} = \frac{\sigma}{v_o - v}. \tag{6-44}$$

If we substitute the variable distance D for u_o and use the last two equations one obtains

$$D = \frac{f \cdot r \cdot v_o}{r \cdot v_o - f(r + \sigma)}$$

$$= \frac{f \cdot v_o}{v_o - f - \sigma F}, \qquad (6\text{-}45)$$

where F is the F-number of the lens. Thus, to recover depth, one needs to find σ. One strategy suggested by Pentland is to vary the aperture of the lens systems or, equivalently, the depth of field. Take one view and image it through two apertures. The corresponding images $f_1(r, \theta)$ and $f_2(r, \theta)$ are centered at some point (x_o, y_o) and yield corresponding Fourier transforms $F_1(s, \theta)$ and $F_2(s, \theta)$. No matching between the two images is required; they are already in registration, but adequate image resolution and high frequency scene content is still needed, as we will see next. Both images $f_i(r, \theta)$ result from blurring, i.e., convolving the original image f_o with (point-spread-function) Gaussians of width σ_i, i.e.,

$$f_i(r, \theta) = f_o(r, \theta) \otimes G(r, \sigma_i). \qquad (6\text{-}46a)$$

Recall from Chapter 2 that the Fourier transform of a Gaussian is a Gaussian and, using the scale property, one can then write

$$F_i(s, \theta) = F_o(s, \theta)G\left(s, \frac{1}{\sqrt{2\pi}\,\sigma_i}\right). \qquad (6\text{-}46b)$$

Then, if we define $F(s)$ as

$$F(s) = \int_{-\pi}^{\pi} F(s, \theta)d\theta, \qquad (6\text{-}47)$$

we obtain

$$\frac{F_1(s)}{F_2(s)} = \exp[s^2 2\pi^2(\sigma_2^2 - \sigma_1^2)]. \qquad (6\text{-}48a)$$

and

$$s^2 2\pi^2(\sigma_2^2 - \sigma_1^2) = \ln F_1(s) - \ln F_2(s). \qquad (6\text{-}48b)$$

The last equation can be rewritten as $As^2 = B$. If $\sigma_1 = 0$ (pinhole camera), then $A = 2\pi\sigma_2^2$ and D can be obtained from Eq. (6-45) for

$\sigma = \sigma_i$. In general, if we have access to several views, the problem becomes overconstrained and can be solved from knowing that $A_{ij} = 2\pi^2(\sigma_i^2 - \sigma_j^2)$. The overconstraint aspect allows a further check of the answer.

6.6.   Conclusions

In this chapter we advanced the concept of active perception and the important role exploration and mobility play in vision. Active perception is an essential ingredient for learning and adaptation. It helps the visual system manage computational complexity and exhibit robust behavior. Whether for need or curiosity, active perception brings to mind a rich and complex world and stimulates thinking. People and the environment are one ecosystem. They interact and influence each other, and in the process, they transform and remodel themselves.

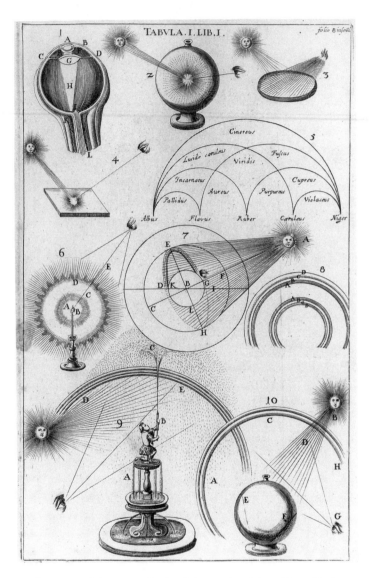

Anatomy of the eye and role of the sun in vision. (From Trabar, Z., Nervus opticus. . . . , 1675.) Reprinted courtesy of the National Library of Medicine.

7

Object Recognition

What being walks sometimes on two feet, sometimes on three, and some-times on four, and is weakest when it has the most?

The Sphinx's Riddle

Recognition is a basic biological function. It is crucial for biological systems to recognize specific patterns and to respond appropriately. Antibodies, for example, attack foreign intruders, our ears capture sound and speech, and animals locate edible plants. Failure to recognize can be fatal.

Recognition is the ultimate goal of any visual system. The whole body of knowledge presented so far and the system-like approach suggested in the preceding chapter (see Fig. 6.1.) are used to build an appropriate theory of invariant object recognition. *Recognize* means to classify and recover the original input. Specifically, we are interested in the design of robust and fault-tolerant systems that demonstrate graceful degradation—that is, relative insensitivity to missing or erroneous data and/or internal memory damage. Furthermore, these recognition systems must display geometrical invariance to affine transformations and perspective distortions.

7.1.   Human Recognition

Much has been learned about the internal workings of the human recognition system from psychophysical studies and studies of functional deficits due to brain damage. The possible loss of mental function following brain damage led Farah (1985) to perform a componential analysis on the recognition and imaging system, corresponding to known mental deficits reported in the literature. Farah's model involves two types of memory storage. The long-term visual memory (LTM) is the repository of stored patterns, while much of the processing is mediated through a visual buffer, usually referred to as short-term memory (STM). Patterns of activation are formed in the visual buffer by either the visual imagery process from LTM (Kosslyn, 1980) or by a perceptual encoding process. The presence of activation in STM may be detected directly, or an inspection process may read out structured patterns of activation for further processing, such as matching with LTM in the case of recognition tasks. The organization of LTM for recognition purposes is the topic of this chapter, while the (LTM, STM) pair and its role in learning visual patterns and setting memory traces is deferred to Chapter 8.

Observed visual recognition deficits can be assigned to the internal LTM structure or to the memory indexing scheme used for matching. Disregarding specific details of memory implementation, the solutions must consider the issue of *addressing* and *searching* through the memory.

Studies of brain-damaged patients suffering from visual memory functional disorders led Warrington (1982) to suggest a modular and hierarchical memory organization (Fig. 7.1.). Her experiments show that recognition impairment can still occur even when contour (or figure-ground) discrimination is preserved, thus suggesting "that the organization of contour information into coherent forms is not the critical factor." Such a finding allows for the possibility that figure-ground separation and/or image segmentation are not necessarily a prerequisite for object recognition. In Fig. 7.1., the right hemisphere system (RHS) handles perceptual organization and facilitates rapid visual identification, while the left hemisphere system (LHS) manages semantic categorization. Perceptual categorization is the matching and identifying of a physical object, while semantic categorization matches functional similarity. Performance on visual tasks, such as

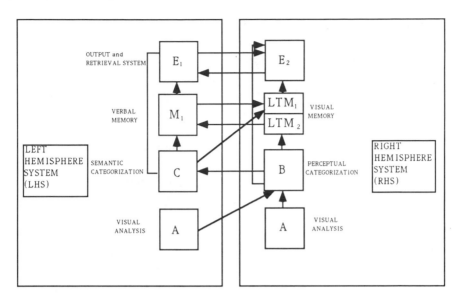

Figure 7.1.
Warrington's stage model of object recognition. Reproduced with permission from
Warrington, E.E., 1982, Neuropsychological studies of object recognition, *Phil Trans. R.
Soc. London B*, **258**.

the identification of objects seen from unusual views and under
changing illumination and/or overlapping, is impaired for human
subjects whose cerebral lesions are restricted to RHS. This suggests
that recognition, as discussed earlier, correlates well with perceptual
categorization. Furthermore, perceptual categorization is independent
of language and verbal hypothesis and can be achieved without
knowledge of "meaning."

Pinker (1985) considers four major classes of possible solutions for
memory representations: template matching, characteristic features,
structural methods, and Fourier analysis. We will briefly analyze each
method for its capabilities and shortcomings.

The template method, as the name implies, assumes that a 2D stored
iconic image is sufficient to recognize input patterns. This method fails
to accommodate invariance and also lacks fault-tolerant characteristics. The approach assumes isolated patterns and cannot cope with
overlapping objects. Furthermore, figure-ground separation is needed
to isolate the object being recognized. As we will see in Section 7.2.3.,

the template method can be redefined in the PDP sense and thus could accomplish fault-tolerant recognition tasks.

Another interesting evolutionary consideration related to template matching comes from the clonal-selection theory. Earlier theory, that of the template, could neither provide a host organism with enough information to respond to a wide spectrum of antigens nor explain why antibodies become better at binding their target antigens over time. The presently accepted theory is that the body is endowed with pre-existing antibodies that recognize antigens. The binding ability improves because the antigen replicates those cells carrying genetic mutations that promote the match between antibody and antigen. (See also genetic algorithms in Section 8.4.8.) The immune system can then be conceived as a kind of Darwinian microcosm. Antibody-producing cells, like any organism in an ecosystem, are subject to mutation and selection; the fittest survives—fitness in this case is the match between a cell's antibody and the antigen. Such a theory requires some kind of (chaotic) randomization (see also Chaos in Section 8.4.9.), where each cell can make antibodies with slightly different properties for which there is no precedent (Ada and Nossal, 1987). In the case of recognition the antigens are the invariants characterizing the environment, while the evolving networks that bind (resonate) to them are the memory patterns corresponding to antibodies. There is no place for fixed templates and/or inferential processes in such a theory, and fault tolerance is easily accommodated. Learning is paramount for such networks, and the invariant resonance brings to mind the Lie groups discussed in Section 3.3.2. and the DAM discussed in Section 4.2.1.

The structural method calls for logic, graphs, or syntax to encode pre-existing relationships between image components. Logic using resolution, search for graphs isomorphism, or syntactical parsing are then employed to reason and infer recognition. Shepard and Cooper's (1986) psychophysical experiments on mental rotation showed that 3D matching involves image transformations and that the time needed to match two 3D structures is proportional to the orientation differential between the structures. The necessity of performing image transformations suggests that geometrical information—size and orientation—is coupled in images, rather than factored out. With the geometrical information factored out, as would be the case with structural methods, graph isomorphism timings should stay constant, contradicting experimental findings.

The Fourier method is based on frequency analysis, and the magnitude and phase provide information regarding the identity ("what") and location ("where") of an object. The magnitude and phase can be associated in a way similar to that of the brightness and lightness profiles (see Section 5.1.1.) through the mechanism of DAM and an experimental system exploiting such an approach is described in Section 7.2.3. The inherent drawbacks in using Fourier analysis were discussed in Section 2.3. Cojoint s/sf representations, as suggested earlier, could be used to encode a canonical LTM along the lines suggested by our FOVEA system discussed in Section 3.1.2.

We continue our discussion along a philosophical/psychological dimension—the psychophysics of perception. One school of thought, called *atomism* or *elementarism*, started with Aristotle and continued much later with the structuralists. The atomistic view assumed a basic vocabulary of elementary sensations from which our perceptions are made. Following the analogy of a basic set of phonemes, as suggested by speech recognition, Biederman (1987) suggested a similar vocabulary made up of 36 solid geometric components called geons for a theory of recognition by components (RBC). Such a theory calls for the visual system to detect edge properties, such as curvature, collinearity, symmetry, parallelism, and cotermination. To make such a system fault tolerant, regularization (Pragnanz) constraints are introduced at the component level of the full object. Hoffman and Richards (1985) also suggest a vocabulary-based scheme—made up of primitive shape descriptors called "codons" for describing 2D plane surves. The codons correspond to boundary segments determined by patterns of inflection and extrema of curvature.

The concept of a basic vocabulary of perception has been disputed in the past on several grounds. Most important is the knowledge that a particular context transforms and determines sensations and percepts as discussed in the philosophy of Wittgenstein. Rejecting a basic vocabulary is also consistent with the theory of psychophysical dualism (Locke), where discrepancies between the physics of perception and of experience suggest that perception is made up of sensation (inputs) and "reflections." Reflections encode implicit constraints as the subject learns to structure the external world. Furthermore, unless one can group parts of a scene to eliminate many of the possibilities, the preattentive computation of relationships, as proposed by Biederman, seems unlikely (Ballard, 1988). Visual perception cannot be determined by a one-to-one relationship between sensation and perception. (Refer to Kanade's scheme shown in Fig. 5.5.) Rather, it is a holistic process, where not all the information resides in the optic array; philogenetic constraints aimed toward regularization and invariance are part of a constraining neural mechanism.

The three major perceptual theories—inference, Gestalt, and Gibson—attempt to explain perception in general and recognition in particular (Rock, 1984). *Inference theory*, associated with the empiricist view, argues that knowledge is acquired solely by sensory experience and association of ideas. The mind at birth is a blank slate or "tabula rasa," upon which experience records sensations. Both Berkeley and Helmholtz later argued that we learn to interpret percepts through a process of association. Helmholtz described the process as one of unconscious inference, such that "sensations of the senses are tokens for our consciousness, it being left to our intelligence to learn how to comprehend their meaning." Research by Hurlbert and Poggio (1988) on learning lightness pattern (see Section 5.1.1.), which exhibits lightness constancy is compatible with inference theory.

The Gestalt view originated with Descartes and Kant. For Descartes and Kant, the mind was far from being a tabula rasa. Kant argues that "the mind imposes its own internal conception of space and time upon the sensory information it receives." The Gestaltist of the twentieth century believed in holistic perceptual organization preordained by "given laws that govern unit formation and the emergence of a figure on a background." Such laws could correspond to fidelity constraints as those introduced through the penalty functional P (see Section 6.1.4.) and to non-accidental environmental constraints. Constancy of perception follows from such considerations. For Gestaltists, perceptions are the result of spontaneous interactions as a result of sensory inputs. Note the similarity to PDP models of computation—a justification for Gestalt's saying that "the whole is qualitatively different from the sum of its parts."

Gestalt can be reconciled with Helmholtz and inference theory. Recent research suggests that inference networks and associations can be mapped into PDP models and the aspect of past experience (nature or nurture) is embedded in both theories. Nativism spares us from starting anew each time, taking benefit from the regularities of nature. Meanwhile, nurturing, or adaptiveness, allows the organism to develop better survival skills over time.

Row assocus are built.

The third theory, that characterized by the stimulus (sensed data) view, claims that sensory input is enough to explain our perceptions. The theory seeks to associate percepts with physical stimuli. Gibson is largely credited with resurrecting this approach by suggesting specific forms of such associations. The texture gradient discussed earlier indicates the slant of the ground and how it recedes into depth. Neisser (1989) argues convincingly that such affordances reflect only the layout of the environment and self-perception, and not recognition. The critical question left is how the associations once defined are implemented. The claim that no information processing is required by the HVS to resonate to such sensory inputs is clearly out of sync with present cognitive theories of perception and knowledge.

7.2. Two-dimensional Object Recognition

Having already discussed some of the main perceptual theories we now consider 2D object recognition. The discussion will be carried out along conceptual lines and includes examples of specific applications/ implementations.

7.2.1. Two-dimensional Modeling

Two-dimensional recognition schemes for machine vision and robotic applications are usually model driven and encode the memory models using the LTM schemes (template, features, structural, and Fourier) discussed earlier in this chapter. Chin and Dyer (1986) set up a number of performance criteria such as speed, accuracy, and flexibility and show that "most industrial parts-recognition systems are model-based systems in which recognition involves matching the input image with a set of predefined models of parts. The goal of such systems is to recognize each instance of an object and to specify its position and orientation relative to the viewer." The environment in which such recognition systems operate is quite constrained by assumptions such as the following:

(i) Components "may be exactly specified, with known tolerances on particular dimensions and features" as those appearing in the recognition model;

(ii)   The environment (illumination, etc.) and the image formation process (high contrast to separate figure from background...) are well controlled;

(iii)   There are few viewpoint configurations for a given part with occlusions usually forbidden (a bin of parts is usually shaken and isolated parts appear on a conveyor belt under the eyes of a camera...).

Models are usually based on geometric properties (of an object's surface or boundary) describing shape characteristics. The 2D image representations are based on global features (perimeter, area, moments of inertia), local features (corners, line, or curve segments), and/or relational features (distance and adjacency). Matching is implemented via statistical pattern recognition, syntactical pattern recognition, and relational graph matching. Generic machine vision systems using such an approach are discussed in Section 7.2.2.

Each of these methods, with such coined names as the global feature, structural feature, and the relational graph approach, has shortcomings. The global feature method cannot handle noise and/or occlusion; the structural method lacks invariance for the extracted features and it is not clear that a suitable alphabet of features and/or an appropriate recognition grammar could be always derived. The relational graph method lacks invariance for the local and relational features and graph matching is a slow process. The structural and relational graph methods (Chin and Dyer, 1986) approximate the structural method (Pinker, 1985) discussed earlier and suffer from geometrical information being factored out of the 2D image representation. While most 2D recognition methods consider shape as derived from objects' silhouettes, others attempt to recognize objects based on surface rather than boundary information.

7.2.2.   Machine Vision Systems

We will discuss in this section the architecture of machine vision systems that recognize 2D (industrial) parts as needed for assembly and/or inspection applications. Crowley (1984) provides a good review of these systems and classifies them as first generation—binary systems taking after the early SRI vision module and implementing the global feature method and second generation—edge description techniques.

7.2.2.1. Binary Vision Systems

Binary vision systems detect and classify connected binary regions, called blobs, and provide information about their position and orientation. The steps performed to yield a 2D global features representation, as required by statistical classifiers, are:

(*i*) *Thresholding and Histogramming*

The assumption of high contrast resulting from a bright object on a dark background (backlight the object...) leads to a bimodal histogram (the graph showing how many pixels have a given gray-level intensity). A threshold is chosen such that it lies at the bottom of the valley between the two peaks (modes) of the histogram. Although seemingly simple, this processing step requires additional tinkering before an edequate threshold is found.

(*ii*) *Run Length Encoding*

For reasons of data compression and efficiency, the system keeps track of only those rows with pixels above the threshold and records the column pairs where the intensity goes above and below the threshold. A convex blob will need only one such pair for each row that it crosses.

(*iii*) *Connectivity Analysis*

The run length pairs are grouped together by a process of connectivity analysis resulting in a data structure made up of "blob descriptors." Since the process scans one row at a time, it must consider cases of insertion (if a run occurs for which no overlapping blob exists then create a new blob descriptor), continuation (if a run overlaps with another one on a previous row then add it to that run's corresponding blob descriptor), deletion (close the blob descriptor if there are no run lengths to further grow the blob), and merging (two blob descriptors are combined if a run is found to overlap with both of them).

(*iv*) *Feature Measurement*

Most of the features listed next are calculated during the connectivity analysis and will be passed to the statistical classifier. Such features include number of holes, area (in calibrated units—see lack of invariance to scale changes), center of gravity, second moments (major and minor axes of the best fit of an ellipse to the blob), orientation (the angle of the major axis), perimeter, minimum and maximum radii from the center of the blob to its boundary, and the directions of those radii relative to the x-axis.

Before concluding with the classification step, we would like to show how some of the features could be derived. Assume that both connectivity analysis as described and chain code work together. The chain code links "l_j" are $(0, 1, 2, \ldots, 7)$, their length is $t(\sqrt{2})^s$, (t is the (calibrated) known grid size unit, and $s = \mathrm{mod}2(l_j)$), and of angle $l_j \times 45°$ (referenced to the X axis of a right-handed Cartesian coordinate system). The links l_j can be expressed as $l_j = (a_{jx}, a_{jy})$, where $a_{jx}, a_{jy} \in [-1, 0, 1]$. (As an example $l_5 = (-1, -1)$.) Then assuming that $x_i = \Sigma\, a_{jx} + x_0$ and $y_i = \Sigma\, a_{jy} + y_0$, where (x_0, y_0) are the coordinates of the pixel starting the chain code, one can write

$$\text{Perimeter} = t(n_e + n_o\sqrt{2})$$

$$\text{Area} = \Sigma a_{ix}(y_{i-1} + \tfrac{1}{2}a_{iy})$$

$$\text{Width} = \max(x_j) - \min(x_j)$$

$$\text{Height} = \max(y_j) - \min(y_j). \qquad (7\text{-}1)$$

where n_e and n_o are the number of even and odd chain links, respectively. An alternative area definition, in terms of boundary elements, can be derived using Green's theorem. Assume Ω is a closed planar region the boundary of which is a piecewise smooth Jordan curve. If

$$f(x, y) = P(x, y)i + Q(x, y)j \qquad (7\text{-}2a)$$

is continuously differentiable on an open set that encloses Ω, then

$$\oint f(r)dr = \iint \left[\frac{\partial Q}{\partial x} - \frac{\partial P}{\partial y}\right] dxdy. \qquad (7\text{-}2b)$$

As a corollary, the Riemann formula equates the line and area integrals

$$\int_C [P(x, y)dx + Q(x, y)dy] \equiv \iint_\Omega \left[\frac{\partial Q}{\partial x} - \frac{\partial P}{\partial y}\right] dxdy. \qquad (7\text{-}2c)$$

Assume now that a region R over S contains a hole H, and it is surrounded by boundary $C = \{(x_k, y_k)\}_{k=1}^N$, where $(x_1, y_1) = (x_N, y_N)$, and that the boundary corresponding to H is C_1 given as $C_1 = \{(x_k, y_k)\}_{k=1}^M$. Then

$$\text{Area} = \text{Area } (R) - \text{Area}(H)$$

$$\iint_S \left[\frac{\partial Q}{\partial x} - \frac{\partial P}{\partial y}\right] dxdy = \oint_C (Pdx + Qdy) - \oint_{C_1} (Pdx + Qdy). \qquad (7\text{-}3a)$$

But the definition "Area $= \iint_S dxdy$" leads one to conclude that

$$Q = \tfrac{1}{2}x, \quad dx = x_{k+1} - x_k, \quad P = -\tfrac{1}{2}y, \quad dy = y_{k+1} - y_k$$

and as a result

$$\text{Area} = \frac{1}{2} \sum_{k=1}^{N} (x_k y_{k+1} - y_k x_{k+1}) - \frac{1}{2} \sum_{k=1}^{M-1} (x_k y_{k+1} - y_k x_{k+1}). \qquad (7\text{-}3b)$$

(v) *Object Classification/Discrimination*

A statistical tree classifier is trained on the global features derived in the previous step. It considers both the mean (μ) and variance (σ^2) for each of the (part) classes to be recognized and incorporates them in a multidimensional Gaussian probability function, $N(\mu, \sigma^2)$.

7.2.2.2. Edge Vision Systems

Edge vision systems operate under the assumption that edge elements correspond to objects' boundaries. In addition to the theoretical objections raised against edge detection, highlights and shadows due to uneven illumination, surface texture, and imaging artifacts (such as blooming) can further contribute to spurious edges. Furthermore, real physical edges could remain undetected due to uniform gray levels at the boundary between two surfaces.

Edges are detected by differential operators, which act as a filter. Smoothing, both before and after applying these mask operators, helps enhance the SNR and possibly avoids false edges.

Once edges are detected, they need to be grouped together into meaningful contours or boundaries. Shape characteristics and structural methods operate on those silhouettes (which link edges and can bridge occasional gaps). The contours can be approximated by either line or curve segments, and structural methods recognize parts and objects by using either syntactical pattern recognition or graph matching (isomorphism). The curve or line segments include a list of attributes, and each attribute is weighted according to its relevance in characterizing a given object. Confidence in recognizing a specific class instance accrues as a function of such weights. (See the system developed by Vamos (1977) for detecting industrial parts.)

Both binary and edge vision systems fail to retain local information and cannot cope with occlusion/overlap situations. The PDP approach to be presented in the next section overcomes such shortcomings with the distributed nature of computation and provides a general framework for developing invariant and fault-tolerant recognition systems.

7.2.3. Two-dimensional Invariant Recognition Using DAMs

The challenge of the object-recognition problem stems from the fact that the interpretation of an image is confounded by several dimensions of variability. Such dimensions include uncertain perspective, changing orientation and scale, sensor noise, object occlusion/overlap, and nonuniform illumination. Machine vision systems must not only detect the identity of objects despite such changes, but they must also be able to characterize these changes, because the variability of the image formation process carries much of the valuable information about the world surrounding us. Humans can recognize objects when viewed from novel orientations, moderate levels of noise, and occlusion, while new instances of a category can be readily classified. Available machine vision systems, as described in Section 7.2.2., lack such characteristics. They usually assume limited and known noise, even illumination, isolated objects, and the classes of objects are learned for a limited set of stable attitudes in space (i.e., viewpoints). Both industrial automation and target recognition (and tracking) are hampered by such restrictions. The goal of this section is to show how Wechsler and Zimmerman (1988, 1989, 1990a) approach the problem of recognizing 2D objects subject to changes in the image formation process.

The recognition process matches a preprocessed image representation from the short-term memory (STM) against long-term memory (LTM). We suggest the LTM organization in terms of distributed associative memory (DAM). The DAM is related to generalized match filters (GMF) (Caulfield and Weinberg, 1982) and synthetic discriminant functions (SDF) (Hester and Casasent, 1981). Like GMFs and SDFs, DAMs attempt to capture a distributed representation that averages over the statistical variations characteristic to the same class. DAMs allow for the implicit representation of structural relationships and contextual information—helpful constraints for choosing among different image interpretations. Finally, because information is distributed in the memory, the overall function of the memory becomes resistant to noise, faults in memory, and degraded stimulus key vectors.

Next, we will describe the invariant object-recognition system in detail. We begin by examining the preprocessing system that produces the vectors to be associated by the distributed associative memory. A block diagram of the system is shown in Fig. 7.2.

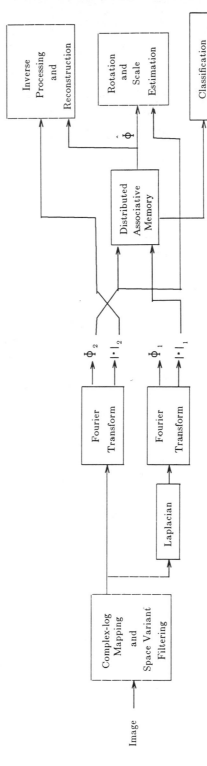

Figure 7.2.

Wechsler and Zimmerman's block diagram for invariant two-dimensional object recognition.

315

The goal of the preprocessing system is to produce a vector that is invariant to metric changes in the image. The image is conformally mapped so that rotation and scale changes become translation in the transform domain. Along with the conformal mapping, the image is also filtered by a space-variant filter to reduce the effects of aliasing. The conformally mapped image is then processed through a Laplacian in order to solve problems associated with the conformal mapping as discussed later. The Fourier transform of both the conformally mapped image and the Laplacian processed image produce the four output vectors. The magnitude output vector, $|\bullet|_1$, is invariant to linear transformations of the object in the input image. The phase output vector, ϕ_2, contains information concerning the spatial properties of the object in the input image.

A problem associated with the complex-log mapping is sensitivity to center misalignment of the sampled image. Small shifts from the center cause distortions in the conformally mapped image. The system assumes that the object is centered in the image frame. Slight misalignments are considered noise, while large misalignments are considered to be translations and are accounted for by changing the gaze in such a way as to bring the object into the center of the frame. The decision about what to bring into the center of the frame is an active function and should be determined by the visual task.

The second box in the block diagram is the Fourier transform. The Fourier transform of a 2D image $f(x, y)$ can be described by two 2D functions corresponding to the magnitude $|F(u, v)|$ and phase $\phi_F(u, v)$. The magnitude component of the Fourier transform, which is invariant to translation, carries much of the contrast information of the image. The phase component of the Fourier transform carries information about where things are placed in an image. Translation of $f(x, y)$ corresponds to the addition of a linear phase component. The conformal mapping transforms rotation and scale into translation, the magnitude of the Fourier transform is invariant to such translations, and as a result, the magnitude $|\bullet|_1$ will not change significantly with rotation and scaling of the object in the image.

The Laplacian that we use is a difference of Gaussians (DOG) approximation to the $\nabla^2 G$ function as given by Marr (1982). i.e.,

$$\nabla^2 G = \frac{-1}{\pi\sigma^4} [1 - r^2/2\sigma^2]\exp\left\{\frac{-r^2}{2\sigma^2}\right\}.$$

The result of convolving the Laplacian with an image can be viewed as a two-step process. The image is blurred by a Gaussian kernel of a specified width σ. Then the isotropic second derivative of the blurred image is computed. The width of the Gaussian kernel is chosen such that the conformally mapped image is visible—approximately two pixels in our experiments. The Laplacian sharpens the edges of the object in the image and sets any region that did not change much to zero.

The Laplacian eliminates the stretching problem encountered by the conformal mapping due to changes in object size. When an object is expanded the conformal mapped image will translate. The pixels vacated by the translation will be filled in with more pixels sampled from the center of the scaled object. These new pixels will not be significantly different than the displaced pixels, so the result looks like a stretching in the conformal mapped image. The Laplacian of the conformal mapped image will set the new pixels to zero because they do not significantly change from their surrounding pixels. The Laplacian also eliminates high-frequency spreading due to the finite structure of the discrete Fourier transform and enhances the differences between memorized objects by accentuating edges and deemphasizing areas of little change. As a result, cross talk (Section 4.2.1.) is diminished.

The DAM is content addressable, stimulus magnitude vectors are associated with response phase vectors, and the result of this association is spread over the entire memory space. Distributing in* this manner means that information concerning a small portion of the association can be found in a large area of the memory. New associations are placed over the older ones and are allowed to interact. The size of the memory matrix stays the same regardless of the number of associations that have been memorized. The previous discussion illuminates several properties of distributed associative memories, which are different from the more traditional ones about memory. Because the associations are allowed to interact with each other, an implicit representation of structural relationships and contextual information can develop; consequently, a very rich level of interactions can be captured. Since there are few restrictions on which vectors can be associated, extensive indexing and cross referencing in the memory occurs. With the information distributed, the overall function of the system is resistant to faults in the memory and degraded stimulus vectors.

Now, we will discuss computer simulations for the 2D invariant recognition system shown in Fig. 7.2. Images of objects are first preprocessed, and as a result, four vectors $|\bullet|_1$, ϕ_1, $|\bullet|_2$, and ϕ_2 are obtained. We construct the memory by associating the stimulus vector $|\bullet|_1$, with the response vector ϕ_2 for each object in the database. (If we were to use the Mellin transform, the important phase information would be lost. Translation invariance, as discussed later on, can be handled using a moving camera.) To perform recall from the memory, the unknown image is preprocessed by the same subsystem to produce the vectors $|\tilde{\bullet}|_1$, and $\tilde{\phi}_2$. The resulting stimulus vector $|\tilde{\bullet}|_1$ is projected onto the memory matrix to produce a response vector $\hat{\phi}_2$, which is an estimate of the memorized phase ϕ_2. The estimated phase vector $\hat{\phi}_2$ and the magnitude $|\tilde{\bullet}|_1$ are used to reconstruct the memorized object. The difference between the estimated phase $\hat{\phi}_2$ and the phase $\tilde{\phi}_2$ is used to estimate the amount of rotation and scale change experienced by the recognized object.

The database of images consists of twelve objects: four keys, four mechanical parts, and four leaves, as is shown in Fig. 7.3. Each object was photographed using a digitizing video camera against a black background. We emphasize that all of the images used in creating and testing the recognition system were taken at different times using various camera rotations and distances. The images are digitized to 256×256, eight-bit quantized pixels, and each object covers an area of about 40×40 pixels. The small object size, relative to the background, is necessary due to the nonlinear sampling of the conformal mapping. The orientation of each memorized object was arbitrarily chosen such that their major axis was vertical. The 2D images that are the output from the invariant representation preprocessing subsystem are scanned horizontally to form the vectors needed for memorization.

The first example of the operation of the recognition system is shown in Fig. 7.4. Figure 7.4a. is the image of one of the keys as it was memorized. Fig. 7.4b. is the unknown object presented to the system. The unknown object in this case is the same key that has been now scaled. Figure 7.4c. is the recalled, reconstructed image. The rounded edges of the recalled image are artifacts of the complex-log mapping. Notice that the reconstructed recall is the unrotated memorized key with some noise caused by errors in the recalled phase. Fig. 7.4d. is a histogram, which graphically displays the classification vector that corresponds to $S^+ |\tilde{\bullet}|$, where S^+ is the pseudoinverse of the stimulus

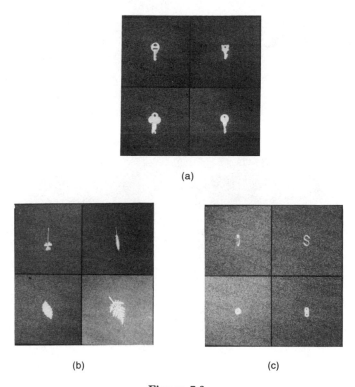

(a)

(b)   (c)

Figure 7.3.
Database for the two-dimensional invariant object recognition system. (a) keys; (b) leaves; (c) mechanical parts.

matrix. The histogram shows the interplay between the processed memorized images and the unknown image. The "six" on the bargraph indicates which of the 12 classes the unknown object belongs to. The histogram gives the minimum squared error (MSE) estimate of the image relative to the memorized objects. Another measure, the signal-to-noise ratio (SNR), is given at the bottom of the recalled image. SNR compares the variance of the ideal recall after processing, with the variance of the difference between the ideal and the actual recall. This is a measure of the amount of noise in the recall. The SNR does not carry much information about the quality of the recall image, because the noise measured by the SNR is due to many other factors such as misalignment of the center, changing reflections, and dependence between memorized objects—each affecting quality in a variety of ways. Rotation and scale estimates are made using the vector **D**

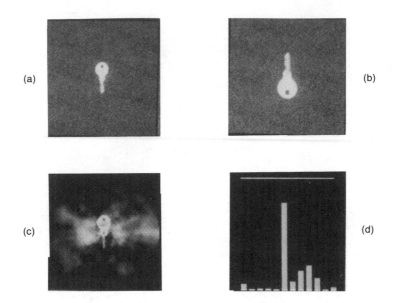

Figure 7.4.
Recall using rotated and scaled key. (a) Original. (b) Unknown. (c) Estimated rotation
90°. SNR = −3.37 dB. (d) Memory: 6.

corresponding to the difference between the unknown vector $\tilde{\phi}_2$ and
the recalled vector $\hat{\phi}_2$. In an ideal situation, **D** will be a plane whose
gradient indicates the exact amount of rotation and scale the recalled
object has experienced. In our system the recalled vector $\hat{\phi}_2$ is corrupted with noise, which means that rotation and scale have to be
estimated. The estimate is made by letting the first order difference **D**
at each point in the plane vote for a specified 45° range of rotation or
scale. However, the system could provide a perfect copy from its
archive upon request.

Figure 7.5. shows the result of randomly setting the elements of the
memory matrix to zero. Fig. 7.5a. shows the ideal recall. Fig. 7.5b. is the
recall after 30% of the memory matrix has been set to zero. Fig. 7.5c. is
the recall for 50%, and Fig. 7.5d. is the recall for 75%. Even when 90%
of the memory matrix has been set to zero, a faint outline of the pin
could still be seen in the recall. This result is important in two ways.
First, it shows that the distributed associative memory is robust in the
presence of noise. Second, it shows that the full memory is not

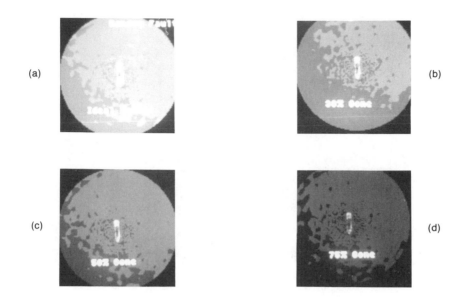

Figure 7.5.
Recall for memory matrix randomly set to zero. (a) Ideal recall. (b) 30% removed. (c) 50% removed. (d) 75% removed.

necessary, and as a consequence, a scheme for data compression of the memory matrix could be employed.

We will consider next the generic bin-picking problem as the search for the identity of objects O_i within a mix given by $\Sigma \, \alpha_i$, where α_i can be one of the following: (a) $T_i(O_i)$ for some geometric change T_i in object scale and orientation; (b) N for noise in both input and the storage memory; (c) A for ambient illumination; and (d) some object.

Solving the bin-picking problem is equivalent to unscrambling the mix of degraded objects and picking up the identity of the unknown objects one at a time. Figure 7.6. is an example of both occlusion and some geometric transformation of one of the objects in the database. The unknown object in this case is $O_4 = $ "S", which is rotated from the memorized S as given by the transformation T_4. A portion of the bottom was occluded by α_2. The problem is equivalent to recovering O_4 from the mix given by $T_4(O_4) + \alpha_2$. The resulting reconstruction is very noisy but has filled in the missing part of the bottom of the object. The

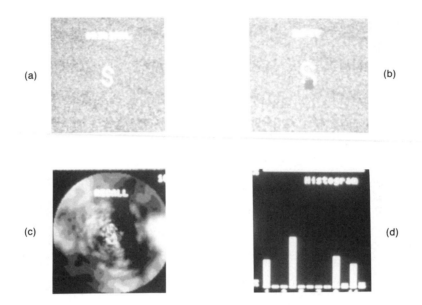

Figure 7.6.
Recall using rotated S part with occlusion. (a) Original. (b) Unknown. (c) Recall.
(d) Histogram.

noisy recall is reflected in both the SNR and the memory interplay
shown by the recall histogram.

Next, we will consider the overlapping case. The generic case is that
of $\Sigma\, T_i(O_i) + N$, i.e., the objects O_i are subject to some geometric
transformation, given by T_i, they possibly overlap, and the resulting
image might be further corrupted by some noise N due to the image
formation process. Figure 7.7. shows three overlapping parts, where a
pin sits on top, an S sits in the middle, and the key sits at the bottom.
The recall histogram again succeeds in identifying the correct elements as the pin (item one), the S part (item four), and the key (item
seven). The relative magnitude of the recall histogram elements is not
necessarily an indication regarding the order of the elements in the
pile. Additional factors such as relative size and centering should be
considered as well.

We describe next a series of experiments on a database made up of 15
classes of colored polyhedral objects (Wechsler and Zimmerman,
1989b). The objects were videotaped directly from above while sitting
on a flat surface. Each object class has at least one other class that

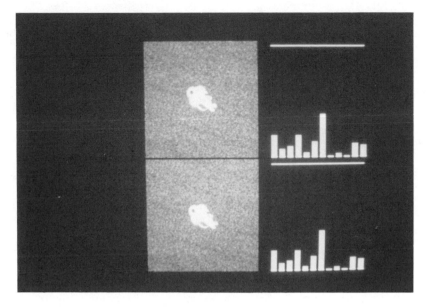

Figure 7.7.
Overlapped pin, S, and key.

represents the same shape in the image. Several of the classes also consist of objects of the same color. The stimulus vectors are now of the form $\mathbf{s}_i^T = [\mathbf{s}_{i_R}^T | \mathbf{s}_{i_G}^T | \mathbf{s}_{i_B}^T]$, where R, G, and B correspond to the red, green, and blue channels of the video recorder. The images used to test the system were taken at a different time than the ones used to construct the memory and are subject to specific transformations. Table 7.1. specifies the database (DB) and the transformations the polyhedral objects were subject to before the memory recall operation is performed. Each object, subject to the linear transformation specified in Table 7.1., was correctly recognized.

The next experiments were aimed at exploring the capability of our system when exposed to overlap situations, such as those encountered during bin picking for the polyhedral world. The database for this memory consists of six objects taken from the DB shown in Table 7.1. Three of the objects have a learned side view state, along with the standard top-down view used in the previous example. This memory is tested using three overlapping objects (the central object is supported by the other two). The camera moves, similar to a conveyor belt, samples the image in five distinct locations, and the recall histograms vary as the camera is moved. Each of these objects were memorized in

Table 7.1. Description of 15 classes of colored polyhedral objects and their corresponding transformation.

20b—20-sided medium blue; rotation approx. 30° right
20g—20-sided gray; rotation 15° left
12r—12-sided red; smaller with rotation; very dim
12y—12-sided yellow; smaller
10b—10-sided light brown; smaller rotated 20°
10o—10-sided orange; rotation 40°
8b—8-sided light blue; rotation 30°
8bl—8-sided black; smaller; extremely dim
8o—8-sided orange; larger
6b—6-sided dark blue; larger
6g—6-sided green; rotation 45°
6y—6-sided yellow; larger
4b—4-sided dark blue; rotation 45°
4g—4-sided light blue; smaller rotation 10°
4p—4-sided lavender; rotation 30°

two views—one from the top looking down and a second view from the side. The results show that, for points close to the central locations of the objects being viewed, the response is correct. For viewpoints between two objects, the response was dominated by one of the objects present in the input.

Polzleitner and Wechsler (1990), using known relationships between DAM and regression analysis, have shown how the selectivity of the recall histogram can be improved in an iterative way by discarding from further consideration recall vectors deemed to be insignificant. Such selectivity allows the system to focus on the significant associations and to reduce cross-talk effects, and also provides for a reject option. The reject option is important so the system instead of making a forced choice can adapt and possibly learn a new class of objects. Furthermore, the same approach facilitates both preattentive and foveal perception.

The experiments described in this section show the feasibility of building an invariant 2D object recognition system that enjoys the benefit of fault-tolerant behavior. Any (machine) vision system built to last should be able to sustain internal damage during its life span. The recognition system presented was designed with only 2D metric distortions in mind. It has shown some ability to deal with clustered 3D polyhedra. Further extensions toward 3D object recognition are discussed in Section 7.5, where we propose to use an approach based on characteristic views (Chakravarty and Freeman, 1982) or aspects

(Koenderink and Van Doorn, 1979). This suggests that the infinite 2D projections of a 3D object can be grouped into a finite number of topological equivalence classes.

7.2.4. *Shape and Size Constancy*

Humans have the ability to perceive a given object that they have seen before as always being of the same size and shape regardless of its orientation and its increasing or decreasing retinal image. This size and shape constancy is well studied in the psychophysical literature. Our interest in the topic is driven by how it might help to advance the knowledge of object recognition. According to Gibson (1950), perception of a stimulus object involves two components—shape and orientation. The shape is not experienced in isolation, but rather in a given orientation (slant, tilt). (One could further elaborate the concept of shape and, as already discussed in Section 7.1., think about it as incorporating geometrical information (size/rotation) as well.) The perceived (ecological) orientation combined with the apparent shape yields a constant shape. Other visual theories suggest that both accommodation (achieving a sharp retinal image—see Pentland's work described in Section 6.5.3.) and convergence (merging two disparate images into a single one) are the reasons behind size constancy. There is much evidence that biological vision can account for distance when computing object size from retinal size. Both Section 7.2.3. on 2D invariant recognition and the FOVEA system described in Section 3.1.2. suggest an alternative approach. Based on ecological information, the input is reprojected for the perceived slant and tilt. Then, the resulting image is subject to 2D invariant recognition as described in the preceding section. Active perception and ecological cues suggest the most likely perceived (slant, tilt) pair and thus can help with the reprojection process. Distance is perceived as well but would be the by-product of the recognition system, which not only identifies familiar objects but also locates them in space.

7.3. Three-dimensional Object Recognition

Three-dimensional modeling is an integral part of any system concerned with recognizing and locating objects. A lot of research has been limited to a polyhedral, nonoccluded world, within a clean and calibrated environment. Modeling assumptions are greatly simplified and most systems are (silhouette) edge based. They attempt, within the framework of the generate (hypothesize) and test paradigm, to recognize objects and to locate them in terms of rotation and translation. The real goal, however, that of fault-tolerant recognition, is still beyond the reach of the computer vision community. Fault-tolerant behavior will allow a vision system to deal with complex environments, where varying illumination, geometric distortions, noise, and occlusion are quite common.

Some general and useful suggestions regarding 3D modeling were made by Rosenfeld (1986). Specifically, there is a need for (i) concise and canonical models, (ii) explicit knowledge and control strategy (in order to yield generality), and (iii) improved, fast, and flexible indexing capability. Furthermore, as suggested by Kender (in Rosenfeld, 1986), representations should be iconic rather than symbolic. There is a need for better spatial reasoning tools and coupling representation and metric information together. Although the task of recognizing objects in 3D has been approached using either a depth/orientation map (such as the 2.5D sketch) by Marr and Nishihara (Marr, 1982), there are other approaches that attempt to accomplish the task from one monocular 2D view. As Ullman (1985) notes, "people seem to be good at the 3D from 2D recognition task, making it unclear whether 3D sensory input is actually necessary for recognition." Next we will review both suggested strategies and conclude by advancing an active approach, where 2D views are incrementally acquired and analyzed under the guidance of the recognition process.

Most 3D machine vision is model based, and we will discuss it along three dimensions—representation, alignment (correspondence and/or transformation), and matching. Alignment and matching are both examples of search problems and seek the *transformation and object model* pair that could bring the perceived scene object and the memory model into close correspondence.

7.3.1. *Three-dimensional Surface Representation*

Intrinsic data, like range (depth) and orientation, is often used to derive characteristics such as (point) features, (boundary) contours, and/or (planar or quadratic) surfaces. Point features are usually defined as curvature extrema and/or inflection points, characteristics to which the human visual system is sensitive. Contours are linked sequences of (point) features and/or edge elements and are usually used to build silhouettes. Planes or quadrics represent surfaces and are appropriate for modeling 3D objects in terms of 2D views. Planes are represented as $\mathbf{x} \cdot \mathbf{n} + d = 0$, where $\cdot$ is the inner product, $\mathbf{n}$ is the normal, and d the distance to the origin. Quadrics, first introduced in Section 3.3.1., can be compactly represented as $\mathbf{x}^T A \mathbf{x} + \mathbf{x} \mathbf{v} + d = 0$, where the principal directions of the quadric are the eigenvectors v_1, v_2, v_3 of the symmetric matrix A, and the center of the quadric (when it exists) is given by $c = -(1/2)A^{-1}\mathbf{v}$.

How one fits the planes or quadrics to the intrinsic data is important. Faugeras and Hebert (1986) view surface fitting as a minimization problem (see also Section 4.1.4.), where one attempts to reduce the residual error between the actual points and the surface fitted to them. For planes, the error e is defined as the distance between the m points x_i and the best fitting plane in the MSE sense. Specifically,

$$e = \min \sum_{i=1}^{m} (x_i \cdot \mathbf{n} + d)^2 = \min P(\mathbf{n}, d). \qquad (7\text{-}4)$$

The function P is homogeneous in $\mathbf{n}$ and d, and to avoid the trivial solution $(0, 0)$, one has to introduce the natural constraint $\|\mathbf{n}\| = 1$. To minimize e, one requires $\partial P/\partial d = 0$ and obtains $d = -\sum_{i=1}^{m} (x_i \cdot \mathbf{n})/m$. The best fitting plane P is

$$P = \sum_{i=1}^{m} \left(x_i \cdot \mathbf{n} - \sum_{i=1}^{m} \frac{x_i \cdot \mathbf{n}}{m} \right)^2. \qquad (7\text{-}5)$$

The direction of the best fitting plane is $\mathbf{n}_{\min}$, it corresponds to the smallest eigenvalue $\lambda_{\min}$ of the covariance matrix of the points x_i, and the error e is given as $e = \lambda_{\min}$.

The case of fitting a higher-order surface such as the quadrics is slightly different. Again, let us define an error e (to be minimized) as

$$e = \min \sum_{i=1}^{m} (\mathbf{x}_i^T A \mathbf{x}_i + \mathbf{x}_i \cdot \mathbf{v} + d)^2 = Q(A, \mathbf{v}, d). \qquad (7\text{-}6a)$$

The quadric can be further represented by the set $\{a_i\}_{i=1}^{10}$ with the following correspondence:

$$A = \begin{bmatrix} a_1 & \dfrac{a_4}{\sqrt{2}} & \dfrac{a_5}{\sqrt{2}} \\[2mm] \dfrac{a_4}{\sqrt{2}} & a_2 & \dfrac{a_6}{\sqrt{2}} \\[2mm] \dfrac{a_5}{\sqrt{2}} & \dfrac{a_6}{\sqrt{2}} & a_3 \end{bmatrix} \tag{7-6b}$$

$$\mathbf{v} = [a_7\ a_8\ a_9];\ d = a_{10}.$$

$Q(A, \mathbf{v}, d)$ is homogenous in its parameters a_i. There are several alternative constraints that one could use

$$\text{(i)} \sum_{i=1}^{m} a_i^2 = 1;$$

$$\text{(ii)}\ a_{10} = 1; \tag{7-6c}$$

$$\text{(iii) Tr}(AA^T) = \sum_{i=1}^{6} a_i^2 = 1.$$

Only the third alternative is invariant with respect to rigid transformations.

Define $\mathbf{R} = (a_1, \ldots, a_{10})^T$, $\mathbf{R}_1 = (a_1, \ldots, a_6)^T$, and $\mathbf{R}_2 = (a_7, \ldots, a_{10})^T$, and the constraint $\|\mathbf{R}_1\| = 1$. The functional Q to be minimized is quadratic and can thus be rewritten as $Q = \Sigma_{i=1}^m R^T M_i R$, where the symmetric matrix M_i is given as $M_i = \begin{bmatrix} B_i & C_i \\ C_i^T & D_i \end{bmatrix}$. Define the matrices B, C, and D as the sum of the $(6 \times 6)\ B_i$, $(6 \times 4)\ C_i$, and $(4 \times 4)\ D_i$ matrices, respectively. The original minimization problem can then be restated as $\min(R^T M R) = \min Q(R)$ with $\|\mathbf{R}_1\| = 1$. The minimum is found using Langrange multipliers, or equivalently

$$\begin{bmatrix} B & C \\ C^T & D \end{bmatrix}\begin{bmatrix} R_1 \\ R_2 \end{bmatrix} = \begin{bmatrix} \lambda R_1 \\ 0 \end{bmatrix}.$$

The minimum for R_2 is found as $\mathbf{R}_2 = D^{-1}C^T(R_1)_{\min}$, where $(R_1)_{\min}$ is the unit eigenvector corresponding to the smallest eigenvalue $\lambda_{\min}$ of the symmetric matrix $B - CD^{-1}C^T$, and the resulting error e is given as $e = \lambda_{\min}$.

7.3.2. *Gaussian Curvature Map*

Our search for adequate 3D surface representations takes us into differential geometry. In this section we consider curvature characteristics of 3D surfaces and their relevance for segmentation, i.e., image decomposition. The curvature is given in terms of the Gaussian (K) and mean (H) curvature, and it has been suggested (Terzopoulos, 1988) that an intrinsic map (K, H) representation might be the end result of middle-level vision, because of the invariance properties of K and H with respect to viewpoint.

Credit for introducing and popularizing 3D curvature characteristics largely goes to Besl and Jain (1986). The parametric representation of 3D surfaces S is given as

$$S = \{(x, y, z): x = h(u, v), y = g(u, v), z = f(u, v), (u, v) \subseteq R^2\}, \quad (7\text{-}7)$$

where all three parametric functions possess continuous second partial derivatives. To describe depth maps, as those derived from range data, one can use Monge patches for which $h = u$, $g = v$, and $f(u, v)$ is the depth function. It can be shown that a 3D surface is uniquely determined by certain local invariant quantities labeled as the first and second fundamental forms.

Assume $x = x(u, v)$ and $dx = x_u du + x_v dv$. Then the first fundamental form Φ_1 is defined as

$$\Phi_1 = dxdx = (x_u x_u)du^2 + 2(x_u x_v)dudv + (x_v x_v)dv^2$$

$$= E\,du^2 + 2F dudv + G dv^2$$

$$= [du\ dv] \begin{bmatrix} g_{11} & g_{12} \\ g_{21} & g_{22} \end{bmatrix} \begin{bmatrix} du \\ dv \end{bmatrix} = du^T[g]du, \quad (7\text{-}8)$$

where $[g] = \begin{bmatrix} E & F \\ F & G \end{bmatrix}$ is referred to as to the metric of the surface. $[g]$ is invariant under a parametric transformation—it depends only upon the surface itself and does not depend on how the surface is embedded in 3D. One can readily see that $\Phi_1 \geq 0$ and that $\Phi_1 \equiv 0$ iff $du = dv = 0$; thus, Φ_1 is positive definite, and the quantities E, G, $EG - F^2$ are all greater than zero, where $EG - F^2 = |x_u \times x_v|^2$. If β is the angle between x_u and x_v, then $\cos \beta = (x_u \cdot x_v)/|x_u||x_v| = F/\sqrt{EG}$; thus, the u- and v-parameter curves are perpendicular iff $F \equiv 0$.

The second fundamental form Φ_2 is defined as

$$\Phi_2 = -dx \cdot dn = [du\ dv]\begin{bmatrix} b_{11} & b_{12} \\ b_{21} & b_{22} \end{bmatrix}\begin{bmatrix} du \\ dv \end{bmatrix} = du^T[b]du, \qquad (7\text{-}9a)$$

where the unit normal is

$$n = \frac{x_u \times x_v}{|x_u \times x_v|}, \quad dn = n_u du + n_v dv, \quad (dn \perp n), \quad (x_u \perp n), \quad (x_v \perp n), \quad \text{and}$$

$$b_{11} = L = -x_u \cdot n_u = x_{uu} \cdot n$$

$$b_{12} = b_{21} = M = -\tfrac{1}{2}(x_u n_v + x_v n_u) = x_{uv} \cdot n$$

$$b_{22} = N = -x_v \cdot n_v = x_{vv} \cdot n.$$

Then one can rewrite the second fundamental form as

$$\Phi_2 = Ldu^2 + 2Mdudv + Ndv^2, \qquad (7\text{-}9b)$$

where the form Φ_2, now dependent upon how the surface is embedded in 3D, determines the shape of the surface. The normal curvature k_n to a curve C at a point $P \in S$ is defined as $k_n = \Phi_2/\Phi_1$.

As an example, assume that a sphere of radius a is given as $x = (a \cos \theta \sin \phi)\mathbf{i} + (a \sin \theta \sin \phi)\mathbf{j} + (a \cos \phi)\mathbf{k}$ for $\{0 < \theta < 2\pi, 0 < \phi < \pi\}$. Then $E = x_\theta \cdot x_\theta = a^2 \sin^2 \phi$; $F = x_\theta \cdot x_\phi = 0$; $G = x_\phi \cdot x_\phi = a^2$; $L = x_{\theta\theta} \cdot n = a \sin^2 \theta$; $M = x_{\theta\phi} \cdot n = 0$; $N = x_{\phi\phi} \cdot n = a$; and the curvature $k_n = $ constant $= 1/a$.

A number k is called a principal curvature iff k is a solution of the equation

$$(EG - F^2)k^2 - (EN + GL - 2FM)k + (LN - M^2) = 0. \quad (7\text{-}10a)$$

At an umbilical (singular) point $k = $ constant, and every direction is a principal direction; thus $k = L/E = M/F = N/G$. One can divide the preceding equation by $(EG - F^2)$ and obtain $k^2 - 2Hk + K = 0$, where

$$H = \frac{1}{2}(k_1 + k_2) = \frac{EN + GL - 2FM}{2(EG - F^2)}$$

$$K = k_1 k_2 = \frac{LN - M^2}{EG - F^2} \qquad\qquad (7\text{-}10b)$$

and (K, H) are called the (Gaussian, mean) curvatures, respectively. K

is invariant of the representation, and according to the sign of K, a point on a surface S is elliptic, iff $K > 0$; hyperbolic, iff $K < 0$; parabolic or planar, iff $K = 0$. Based on the $[g]$ and $[b]$ matrices used to define the first and second fundamental forms, one can define a "shape operator" $[\beta]$ as $[\beta] = [g]^{-1}[b]$. Then $K = \det[\beta]$ and $H = \text{Tr}[\beta]$, where Tr is the trace operator. For a sphere of radius "a", $(K, H) = (1/a^2, \pm 1/a)$, where the sign of H depends on the orientation of the sphere. We are now ready to introduce the label maps (K, H), and its corresponding 3D geometry (Besl and Jain, 1988) as shown in Fig. 7.8. The labels can be used for segmentation in a region-growing-like algorithm.

There are a number of invariant characteristics (Besl and Jain, 1986) that make the (K, H) curvature (intrinsic) map attractive for characterizing 3D surfaces. K and H are invariant to arbitrary transformations of the (u, v) surface parameters as long as the Jacobian of the (u, v) transformation is non-zero. Furthermore, (K, H) are invariant to rotation and translation. Finally, K is an isometric mapping of a surface, where arcs under an isometric mapping have the same length. The last property can be used to check if two 3D surfaces are congruent and match.

It is appropriate at this point to address the question of discrete surface curvature computation (Besl and Jain, 1988) for digital processing. Assume again Monge patches and that $x(u, v) = [u\ v\ f(u, v)]^T$. Then one can write

$$x_u = [1\ 0\ f_u]^T; \qquad x_v = [0\ 1\ f_v]^T; \quad x_{uu} = [0\ 0\ f_{uu}]^T$$
$$x_{vv} = [0, 0, f_{vv}]^T; \quad x_{uv} = [0\ 0\ f_{uv}]^T. \tag{7-11a}$$

The normal $\mathbf{n}$ can be approximated as

$$\mathbf{n} = \frac{x_u \times x_v}{\|x_u \times x_v\|}$$

$$= \frac{\begin{vmatrix} e_1 & 1 & 0 \\ e_2 & 0 & 1 \\ e_3 & f_u & f_v \end{vmatrix}}{\left\| \det \begin{vmatrix} e_1 & 1 & 0 \\ e_2 & 0 & 1 \\ e_3 & f_u & f_v \end{vmatrix} \right\|} = \frac{1}{\sqrt{1 + f_u^2 + f_v^2}} [-f_u\ -f_v\ 1]^T, \tag{7-11b}$$

	$K > 0$	$K = 0$	$K < 0$
$H < 0$	Peak L=1	Ridge L=2	Saddle Ridge L=3
$H = 0$	(none) L=4	Flat L=5	Minimal Surface L=6
$H > 0$	Pit L=7	Valley L=8	Saddle Valley L=9

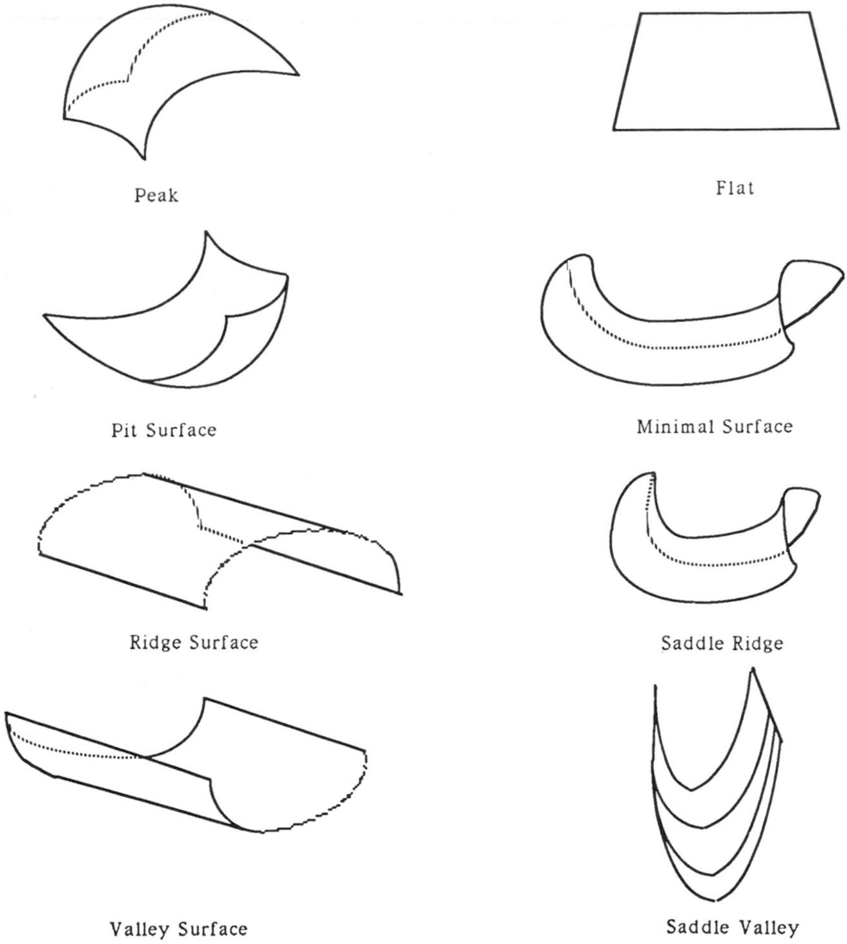

Peak

Flat

Pit Surface

Minimal Surface

Ridge Surface

Saddle Ridge

(b)   Valley Surface

Saddle Valley

Figure 7.8.

Gaussian and mean — (K,H) curvature map. (a) Surface labels; (b) geometry corresponding to (K,H) surface labels. Reprinted with permission from Besl, P. J., and Jain, R. C. (1986), Invariant surface characteristics for three-dimensional object recognition in range images, *CVGIP* **33**, 33-80, copyright Academic Press.

where e_i are the unit normals, **i**, **j**, **k**. The $[g]$ and $[b]$ operators are defined as

$$g_{11}\,(\,=E)=1+f_u^2\,(\,=x_u\cdot x_u)$$
$$g_{12}\,(\,=F)=f_u f_v\,(\,=x_u\cdot x_v)$$
$$g_{22}\,(\,=G)=1+f_v^2\,(\,=x_v\cdot x_v) \tag{7-11c}$$

$$b_{11}\,(\,=L)=f_{uu}/\sqrt{1+f_u^2+f_v^2}\,(\,=x_{uu}\cdot n)$$
$$b_{12}\,(\,=M)=f_{uv}/\sqrt{1+f_u^2+f_v^2}\,(\,=x_{uv}\cdot n) \tag{7-11d}$$
$$b_{22}\,(\,=N)=f_{vv}/\sqrt{1+f_u^2+f_v^2}\,(\,=x_{vv}\cdot n).$$

The Gaussian (K) and mean (H) curvature are then defined as

$$K=\frac{LN-M^2}{EG-F^2}=\frac{f_{uu}f_{vv}-f_{uv}^2}{(1+f_u^2+f_v^2)^2}=\frac{\det\,(\nabla\nabla^T f)}{\|\nabla f\|^4}. \tag{7-11e}$$

where ∇ is the gradient operator and $\nabla\nabla^T$ is the Hessian matrix operator $\begin{bmatrix} f_{uu} & f_{uv} \\ f_{uv} & f_{vv} \end{bmatrix}$.

$$H=\frac{f_{uu}+f_{vv}+f_{uu}f_v^2+f_{vv}f_u^2-2f_u f_v f_{uv}}{2(1+f_u^2+f_v^2)^{3/2}}, \tag{7-11f}$$

where the mean curvature H can be rewritten as

$$H=\mathrm{Tr}[\beta]=\mathrm{Tr}([g]^{-1}[b])$$
$$=\tilde{\nabla}\!\left(\frac{\nabla f}{\sqrt{1+\|\nabla f\|^2}}\right),$$

and $\tilde{\nabla}$ is the divergence operator

Next, we estimate the partial derivatives of 3D sampled surfaces, following the methods of (a) Besl and Jain (1986) and (b) Yang and Kak (1986), respectively.

(a)   First and second partial derivatives at a given location (u, v) are estimated by overlapping $N \times N$ windows centered at pixels (u, v) and fitting local quadratic surfaces. The points in the $N \times N$ window are associated with a position $(u, v) \in U \times U$, where N is odd and $U = \{-(N-1)/2, \ldots, -1, 0, 1, \ldots, (N-1)/2\}$. The quadratic surface fit is obtained by using the discrete orthogonal polynomials $\phi_0(u) = 1$; $\phi_1(u) = u$; $\phi_2(u) = (u^2 - M(M + 1)/3)$, where $M = (N - 1)/2$. The following $b_i(u)$ functions are normalized forms corresponding to the ϕ_i polynomials and are given by

$$b_0(u) = \frac{1}{N}; \quad b_1(u) = \frac{3}{M(M + 1)(2M + 1)} u;$$

and

$$b_2(u) = \frac{1}{P(M)}\left(u^2 - \frac{M(M + 1)}{3}\right),$$

where $P(M)$ is given as $P(M) = 8/45M^5 + 4/9M^4 + 2/9M^3 - 1/9M^2 - 1/15M$. The $b_i(u)$ are precomputed for any window and the Coon's surface estimate $\hat{f}(u, v)$ is one of the form $\hat{f}(u, v) = \Sigma_{i,j=0}^2 a_{ij}\phi_i(u)\phi_j(v)$ that minimizes the error $e = \Sigma_{(u,v)\in U^2} (f(u, v) - \hat{f}(u, v))^2$. The unknown coefficients are found as $a_{ij} = \Sigma_{(u,v)\in U^2} f(u, v)b_i(u)b_j(v)$. The first and second partial derivatives are found as $f_u = a_{10}$; $f_v = a_{01}$; $f_{uv} = a_{11}$; $f_{uu} = 2a_{20}$; $f_{vv} = 2a_{02}$ and the corresponding error e is computed as

$$e = \sum_{(u,v)\in U^2} f^2(u, v) - \sum_{i,j} a_{ij}\left(\sum_u \phi_i^2(u)\right)\left(\sum_v \phi_j^2(u)\right).$$

(b) The second method uses B-spline approximations for 4×4 patches given as

$$x(u, v) = UM_bPM_b^T V^T,$$

where $U = (u^3\ u^2\ u\ 1)$; $V = (v^3\ v^2\ v^2\ 1)$;

$$M_b = \frac{1}{6}\begin{bmatrix} -1 & 3 & -3 & 1 \\ 3 & -6 & 3 & 0 \\ -3 & 0 & 3 & 0 \\ -1 & 4 & 1 & 0 \end{bmatrix},$$

and $P = [P_{ij}]$ are the coordinates values over the 4×4 matrix. The corresponding partial derivatives are found as

$$x_u = \frac{1}{288} \begin{bmatrix} -1 & -6 & -10 & -6 & -1 \\ -2 & -20 & -52 & -20 & -2 \\ 0 & 0 & 0 & 0 & 0 \\ 2 & 20 & 52 & 20 & 2 \\ 1 & 6 & 10 & 6 & 1 \end{bmatrix} = x_v^T$$

$$x_{uu} = \begin{bmatrix} 1 & 6 & 10 & 6 & 1 \\ 0 & 8 & 32 & 8 & 0 \\ -2 & -28 & -84 & -28 & -2 \\ 0 & 8 & 32 & 8 & 0 \\ 1 & 6 & 10 & 6 & 1 \end{bmatrix} = x_{vv}^T$$

$$x_{uv} = \frac{1}{96} \begin{bmatrix} 1 & 2 & 0 & -2 & -1 \\ 2 & 12 & 0 & -12 & -2 \\ 0 & 0 & 0 & 0 & 0 \\ -2 & -12 & 0 & 12 & 2 \\ -1 & -2 & 0 & 2 & 1 \end{bmatrix}.$$

We are now ready to derive and make use of the (intrinsic) curvature map in qualitative (the label map $\langle K, H \rangle$) and quantitative ways. Once the partial derivatives are estimated, one can calculate K, and H (and thus determine their corresponding sign) and the additional quantities $\sqrt{g}$ and Q. The quantity $\sqrt{g}$ is related to the first fundamental form Φ_1 as $\sqrt{g} = \sqrt{EG - F^2} = \sqrt{1 + f_u^2 + f_v^2}$ and is equivalent to an edge magnitude map. The edge magnitude map can detect discontinuities in depth, corresponding to occluding boundaries, and thus help with image segmentation. The second quantity Q is a measure of flatness and is given as $Q = f_{uu}^2 + 2f_{uv}^2 + f_{vv}^2$.

The 3D intrinsic curvature maps such as $\langle K, H \rangle$ and $\langle \sqrt{g}, Q \rangle$ are tightly related to previous attempts to characterize shape by its high curvature points. Equivalently, the attempt to fit flat (smooth) surfaces, as would be the case for decomposing (segmenting) a given object (structure), leads to minimization problems, where the smoothness criterion is introduced via Lagrange-type multipliers. What form could such flatness criterion take? Koenderink and van Doorn (1986) consider the issue within the general framework of dynamic shape. The discussion on flatness criteria is carried out in terms of critical points and variational principles. For 2D curves, critical points are those for which $f_x = 0$. Extreme points are those for which $f_{xx} > 0$ (minima) or $f_{xx} < 0$ (maxima). An inflection point is such that $f_{xx} = 0$, or visually, the curve lies on both sides of the tangent. In the case of 3D, the critical points are those for which $f_x = f_y = 0$ (or equivalently, $f_x^2 + f_y^2 = 0$). Furthermore, critical points for which K, H are nonzero, are called nondegenerate. The Hessian of the quadratic form used to test for the existence of extrema was already defined as

$$H = \begin{vmatrix} \dfrac{\partial^2 f}{\partial x^2} & \dfrac{\partial^2 f}{\partial x \partial y} \\[2ex] \dfrac{\partial^2 f}{\partial x \partial y} & \dfrac{\partial^2 f}{\partial y^2} \end{vmatrix}.$$

Specifically, one has to seek the eigenvalues of $\det|H - \lambda I|$. A maximum results if the eigenvalues are negative; otherwise, the extrema is a minimum type. One can define the quantity Δ as

$$\Delta = \frac{\partial^2 f}{\partial x^2} \frac{\partial^2 f}{\partial y^2} - \frac{\partial^2 f}{\partial x \partial y},$$

and

(i)   the maxima points are those for which

$$\Delta > 0, \frac{\partial^2 f}{\partial x^2} < 0 \left(\text{or } \frac{\partial^2 f}{\partial y^2} < 0 \right);$$

(ii)   the minima points are those for which

$$\Delta < 0, \frac{\partial^2 f}{\partial x^2} > 0 \left(\text{or } \frac{\partial^2 f}{\partial y^2} > 0 \right);$$

(iii)   the saddle points are those for which $\Delta = 0$.

Some of the variational principles used include

(a) $V_1(f) = \iint (f_{xx}^2 + f_{yy}^2) \, dxdy$, but note that because the term f_{xy} is omitted, for a curved function such as $f_1(x, y) = xy$, the corresponding flatness index $V_1(f) = 0$!

(b) $V_2(f) = Q$, where Q is defined by Besl and Jain (1986) as

$$Q = \iint (f_{xx}^2 + 2f_{xy}^2 + f_{yy}^2)dxdy;$$

(c) $$V_3(f) = \iint [(f_{xx}^2 + f_{yy}^2)^2 - 2(1 - \mu)(f_{xx}f_{yy} - f_{xy}^2)]dxdy.$$

The term $(f_{xx}f_{yy} - f_{xy}^2)$ corresponds to the equation of a parabolic curve and stays invariant under perspective distortion. Koenderink and van Doorn (1982) point out the need to find adequate ways to provide for curvature maps that exhibit invariance characteristics. They attach relevance to parabolic lines with the following observations:

> In general, the surface of a smooth object can be divided into elliptic and hyperbolic areas. Elliptic patches (ovoids) enclose space (bound space either convex or concave), while in the case of a hyperbolic patch (saddle), the surface cuts the tangent plane (i.e., inflection case), and the patch cannot enclose anything. The dividing curves, between elliptic and hyperbolic areas, are called parabolic lines. Such lines are a family of nested, closed curves, which outline the elliptic patches or "objectlike" (in the sense of space enclosing) parts: bulges of material (convex) or pockets of air (concave). The hypothesis is that vision grasps shape as a hierarchical structure of elliptic patches, and this hierarchy is identical with that of the family of nested parabolic loops. Such work is related to the one contemplated by Klein and undertaken later on by Thom in conjunction with the catastrophy theory.

The corollary of this discussion is that one should search for those geometries that embed in a natural way the invariant characteristics the perceptual system is sensitive to.

A restriction of the 3D intrinsic curvature map discussed so far is the extended Gaussian image (EGI) (Horn, 1984). The EGI has been suggested for object recognition and 3D attitude in space determination. The Gaussian sphere used to represent the EGI is the equivalent of a histogram made up of needles (i.e., orientations). The technique works for convex object recognition if there is no occlusion. The nonconvex case is handled by using a set of needle histograms for each convex views. The EGI may not work well for every case because there are enough situations where two different 3D objects will have the same EGI (Besl and Jain, 1986).

7.3.3. *Three-dimensional Rotation*

We need to be concerned with not only how to represent the (intrinsic) data but also how to represent the 3×3 rotation matrix. The rotation matrix R is given as $R = (\mathbf{r}_1, \mathbf{r}_2, \mathbf{r}_3) = (r_{ij})$. The transformation R is rigid, lengths and angles are preserved, so $\mathbf{r}_i \cdot \mathbf{r}_j = \delta_{ij}$. In addition, the sense is also preserved, i.e., no reflection takes place, so the inner product $(\mathbf{r}_1 \times \mathbf{r}_2, \mathbf{r}_3) = 1$. Equivalently, these conditions can be given such that R is an orthogonal matrix, and that its determinant $|R|$ is equal to 1.

One way to represent rotation, by way of Euler's theorem, is in terms of some rotation angle σ around some orientation axis $\mathbf{n} = (n_1\ n_2, n_3)$, where n_1, n_2, and n_3 are the directional cosines of the rotation axis. Since $\mathbf{n}$ is a unit vector, only two of its components are independent, and thus there are still three degrees of freedom, two for representing $\mathbf{n}$, and one for the actual angle of rotation σ. The (Euler) rotation matrix is

$$R = \begin{bmatrix} \cos\sigma + n_1^2\,(1-\cos\sigma) & n_1 n_2(1-\cos\sigma) - n_3\sin\sigma & n_1 n_3(1-\cos\sigma) + n_2\sin\sigma \\ n_2 n_1(1-\cos\sigma) + n_3\sin\sigma & \cos\sigma + n_2^2\,(1-\cos\sigma) & n_2 n_3(1-\cos\sigma) - n_1\sin\sigma \\ n_3 n_1(1-\cos\sigma) - n_2\sin\sigma & n_3 n_2(1-\cos\sigma) + n_1\sin\sigma & \cos\sigma + n_3^2\,(1-\cos\sigma) \end{bmatrix}.$$

$$(7\text{-}12a)$$

Conversely, given a rotation matrix R, one can determine the axis $\mathbf{n}$ and the rotation angle σ. Specifically,

$$\sigma = \cos^{-1}\frac{\mathrm{Tr}(R) - 1}{2}$$

$$n_1 = -\frac{r_{23} - r_{32}}{2\sin\sigma}; \quad n_2 = -\frac{r_{31} - r_{13}}{2\sin\sigma}; \quad n_3 = -\frac{r_{12} - r_{21}}{2\sin\sigma}. \quad (7\text{-}12b)$$

Rotation can also be represented using the Euler angles θ, ϕ, Ψ. The angles θ, $0 < \theta < \pi$, and ϕ, $0 < \phi < 2\pi$ are the spherical coordinates of $\mathbf{r}_3$. Then assume l to be the intersection of the x-y plane and the plane defined by $\mathbf{r}_1$ and $\mathbf{r}_2$; Ψ, $0 < \Psi < 2\pi$ is the angle of vector $\mathbf{r}_2$ from l clockwise about $\mathbf{r}_3$. Let $R_X(\alpha)$, $R_Y(\beta)$, $R_Z(\gamma)$ be the rotation matrices corresponding to rotations by α, β, γ, clockwise, around the X-, Y-, and

Z-axes, respectively. The corresponding matrices are given by

$$R_X(\alpha) = \begin{bmatrix} 1 & 0 & 0 \\ 0 & \cos\alpha & -\sin\alpha \\ 0 & \sin\alpha & \cos\alpha \end{bmatrix}$$

$$R_Y(\beta) = \begin{bmatrix} \cos\beta & 0 & \sin\beta \\ 0 & 1 & 0 \\ -\sin\beta & 0 & \cos\beta \end{bmatrix} \qquad (7\text{-}13a)$$

$$R_Z(\gamma) = \begin{bmatrix} \cos\gamma & -\sin\gamma & 0 \\ \sin\gamma & \cos\gamma & 0 \\ 0 & 0 & 1 \end{bmatrix}$$

Then one can write the rotation matrix R as $R = R_Z(\phi)\,R_Y(\theta)\,R_X(\Psi)$ or as

$$R = \begin{bmatrix} \cos\theta\cos\phi\cos\Psi - \sin\phi\sin\Psi & -\cos\theta\cos\phi\sin\Psi - \sin\phi\cos\Psi & \sin\theta\cos\phi \\ \cos\theta\sin\phi\cos\Psi + \cos\phi\sin\Psi & -\cos\theta\sin\phi\sin\Psi + \cos\phi\cos\Psi & \sin\theta\sin\phi \\ -\sin\theta\cos\Psi & \sin\theta\sin\Psi & \cos\theta \end{bmatrix}.$$

$$(7\text{-}13b)$$

Representation of rotation in terms of Euler angles or rotation by some angle around some axis suffers from lacking an adequate composition rule. Given rotation R_1 followed by rotation R_2, both expressed in one of the two previously mentioned methods, it is not easy to calculate the total rotation about an axis. Kanatani (1986) presents an elegant methodology in terms of the Cayley–Klein parameters and quaternions. The geometry is that of polar projections of the unit sphere onto the x–y plane. Assuming that the projections of (x, y, z) and its rotated version (x', y', z') are $z(x, y) = x + jy$ and $z'(x', y') = x' + jy'$, respectively, $(j^2 = -1)$, then there is a conformal mapping of z into z' given as $z' = (\gamma + \delta z)/(\alpha + \beta z)$, where $(\alpha, \beta, \gamma, \delta)$ are the Cayley–Klein parameters such that $\gamma = -\beta^*$, $\delta = \alpha^*$, $\alpha\delta - \beta\gamma = 1$, and the angles are preserved under the transformation. Rotations now form a group of transformations $G(2)$, where the composition (multiplication) rule for two rotations given as $(\alpha, \beta, \gamma, \delta)$ and $(\alpha', \beta', \gamma', \delta')$ is $z'' = (\gamma'' + \delta''z)/(\alpha'' + \beta''z)$ for $\alpha'' = \alpha'\alpha + \beta'\gamma$, $\beta'' = \alpha'\beta + \beta'\delta$, $\gamma'' = \gamma'\alpha + \delta'\gamma$, and $\delta'' = \gamma'\beta + \delta'\delta$. The identity conformal mapping is given by $(1, 0, 0, 1)$ and the inverse transformation is given by $(\delta, -\beta, \alpha, -\gamma)$. The same rotation can be expressed in terms of homogenous coordinates (z_0, z_1) *as* $Z' = R_2 Z$ or

$$\begin{bmatrix} z'_0 \\ z'_1 \end{bmatrix} = R_2 \begin{bmatrix} z_0 \\ z_1 \end{bmatrix} = \begin{bmatrix} \alpha & \beta \\ \gamma & \delta \end{bmatrix} \begin{bmatrix} z_0 \\ z_1 \end{bmatrix}, \tag{7-14a}$$

where $\det(R_2) = 1$. One can further show that any 3D rotation can be expressed in terms of $G(2)$-type rotations as $Z' = R_2 Z R_2^*$. The R_2 matrix can be further expressed in terms of real parameters (q_0, q_1, q_2, q_3) as

$$R_2 = \begin{bmatrix} q_0 - i q_3 & -q_2 - i q_1 \\ q_2 - i q_1 & q_0 + i q_3 \end{bmatrix} = q_0 I + q_1 S_1 + q_2 S_2 + q_3 S_3 \tag{7-14b}$$

where

$$S_1 = \begin{bmatrix} 0 & -i \\ -i & 0 \end{bmatrix}, \quad S_2 = \begin{bmatrix} 0 & -1 \\ 1 & 0 \end{bmatrix}, \quad S_3 = \begin{bmatrix} -i & 0 \\ 0 & i \end{bmatrix},$$

and $q_0^2 + q_1^2 + q_2^2 + q_3^2 = 1$. The rotation can be expressed simply as $q = q_0 + q_1 i + q_2 j + q_3 k$, or as $q = \langle \mathbf{a}, b \rangle$ where $\mathbf{a} = (q_1, q_2, q_3)$ and $b = q_0$. The multiplication rule is the same as for real numbers except that $i^2 = j^2 = k^2 = -1$, $ij = -ji = k$, $jk = -kj = i$, $ki = -ik = j$. Alternatively, using the cross product $(\times)$ one can define multiplication as

$$q \cdot q' = \langle \mathbf{a}, b \rangle \cdot \langle \mathbf{a}', b' \rangle = \langle \mathbf{a} \times \mathbf{a}' + b \mathbf{a}' + b' \mathbf{a}, \, bb' - \mathbf{a} \cdot \mathbf{a}' \rangle,$$

where the quantity $q = \langle \mathbf{a}, b \rangle$ is called a quaternion and defines a group where the identity is given as $1 = \langle \mathbf{0}, 1 \rangle$. Any vector $\mathbf{v} \in R^3$ is expressed as $\langle \mathbf{v}, 0 \rangle$. The conjugate q^* of q is defined as $q^* = \langle -\mathbf{a}, b \rangle = q_0 - (q_1 i + q_2 j + q_3 j)$, and it is clear that $q^* q = q q^* = 1$. The norm of the quaternion is defined as $\|q\|^2 = q^* q = \|a\|^2 + b^2$. One can now represent any rotation R in terms of an axis $\mathbf{v}$ and angle of rotation θ as $R\mathbf{u} = q \cdot \mathbf{u} \cdot q^*$, where the mapping between the rotation $(\mathbf{v}, \theta)$ and the quaternion $q = \langle \mathbf{a}, b \rangle$ is defined as $\mathbf{a} = \sin(\theta/2)\mathbf{v}$ and $b = \cos(\theta/2)$. In terms of the Euler angles θ, ϕ, and Ψ, the quaternion q is given as $q(\theta, \phi, \Psi) = q_k(\phi) q_j(\theta) q_k(\Psi)$, or alternatively, one can write

$$q(\theta, \phi, \Psi) = \cos \frac{\theta}{2} \left(\cos \frac{\Psi + \phi}{2} + \mathbf{k} \sin \frac{\Psi + \phi}{2} \right)$$

$$+ \sin \frac{\theta}{2} \left(\mathbf{i} \sin \frac{\Psi - \phi}{2} + \mathbf{j} \cos \frac{\Psi - \phi}{2} \right). \tag{7-15}$$

7.3.4. *Alignment*

Canonical memory representations do not necessarily assume the same 3D posture as that of the perceived object. One needs to align 3D objects before they can be matched in a way similar to mental rotation and to functional theories of mental imagery (see also Section 8.2.1.). The transformation involved, in terms of 3D rotation and translation, can be performed on either the perceived object or the memory model. Alignment usually requires that corresponding points be found—a characteristic not only of object recognition, but motion analysis and target tracking as well. The search for the transformation can proceed independent of final recognition or be coupled in a symbiotic fashion to the recognition process. Next we will discuss several generic 2D and 3D alignment methods used in machine vision modeling.

The most general transformation sought between two sets of points $\langle \{p_{1i}\}, \{p_{2i}\} \rangle_{i=1}^{N}$ is given by $p_{2i} = Rp_{1i} + T + W_i$, where R is a 3×3 rotation matrix, T is a translation vector, and W_i is the noise vector. Seeking the pair $\langle R, T \rangle$ leads to two general classes of methods: iterative and noniterative. Two general comments are appropriate at this point before going into specific methods. First, we show that the search for the $\langle R, T \rangle$ pair can be decoupled (Lin *et al.*, 1986). Specifically, relate the centroids of two sets of points $\{p_{2i}\}$ and $\{p_{1i}\}$ as

$$c_2 = Rc_1 + T \tag{7-16a}$$

or equivalently,

$$\frac{1}{N} \sum_{i=1}^{N} p_{2i} = R\left(\frac{1}{N} \sum_{i=1}^{N} p_{1i}\right) + T, \tag{7-16b}$$

where N is the number of points in each set. Then

$$T = \frac{1}{N} \sum_{i=1}^{N} p_{2i} - R\left(\frac{1}{N} \sum_{i=1}^{N} p_{1i}\right). \tag{7-16c}$$

Next, define normalized sets of points $\{q_{1i}\}$ and $\{q_{2i}\}$ by subtracting the coordinates of the centroid, and writing

$$q_{\alpha i} = p_{\alpha i} - \frac{1}{N} \sum_{i=1}^{N} p_{\alpha i} \quad \text{(for } \alpha = 1, 2). \tag{7-16d}$$

(Note that no correspondence is needed to find the centroids.) Then one can obtain R from solving $q_{2i} = Rq_{1i}$, and the translation is subsequently solved as $T = c_2 - Rc_1$. The alternative, that of looking for the six motion parameters at the same time, using Newton–Raphson algorithms, may or may not converge to the correct solution, depending on the initial guess (Lin *et al.*, 1986). Most alignment methods also depend heavily on a specific correspondence process. As we have noted in the sections on optical flow derivation and/or stereo computation, the whole process of correspondence is highly questionable. What should be put in correspondence, and what to do if the correspondence tokens are missing, are problems that haunt the alignment process as well. In Section 7.5., we suggest that alignment can proceed even when no formal correspondence is established, and that active PDP type of computation facilitates finding the solution.

Another solution suggested by Lin *et al.* (1986) is that of a frequency-domain algorithm for determining the motion of a rigid object from range data without using correspondence. The algorithm takes advantage of the fact that a function and its Fourier transform experience the same rotation (the 2D case was shown in Section 3.1.1.), and that the computation of the FT does not require point correspondence. The FT along the rotation axis is unchanged so one searches for the rotation axis first. Once such an axis is determined, the rotation angle can be found. With the previous two comments in mind, we proceed to a discussion of alignment methods and begin with the iterative ones.

7.3.4.1. Iterative Alignment

We have already shown that the search for the alignment $\langle R, T \rangle$ pair can be decoupled, and that one seeks first to minimize error e given as $e^2 = \Sigma_{i=1}^{N} \| q_{2i} - Rq_{1i} \|^2$. (The translation is easily found afterwards as $T = c_2 - Rc_1$, where c_1 and c_2 are the centroids of two sets of points, before and after the transformation $\langle R, T \rangle$.) Huang *et al.* (in Lin *et al.*, 1986) developed a general 3D motion alignment method by first solving the 2D case when the corresponding sets lie on the $x - y$ plane. Write $q_{1i}(x_{1i}, y_{1i}) = (r_{1i} \cos \alpha_i, r_{1i} \sin \alpha_i)$ and $q_{2i}(x_{2i}, y_{2i}) = (r_{2i} \cos \beta_i, r_{2i} \sin \beta_i)$. Then the error e^2 to be minimized can be rewritten as

$$e^2 = \sum_{i=1}^{N} \| r_{2i} \exp[j\beta_i] - r_{1i} \exp[j(\alpha_i + \sigma)] \|^2, \qquad (7\text{-}17a)$$

where σ is the angle of rotation. If one examines $\partial e^2/\partial\sigma$ and $\partial e^2/\partial\sigma^2$ for $0 < \sigma < 2\pi$ then one finds that e^2 reaches exactly one local minimum (which is the global minimum) and one local maximum (which is the global maximum). The solution angles σ at which e^2 attains the minimum and maximum satisfy

$$\tan\sigma = \frac{\sum_{i=1}^{N} r_{2i}r_{1i}\sin(\beta_i - \alpha_i)}{\sum_{i=1}^{N} r_{2i}r_{1i}\cos(\beta_i - \alpha_i)} = \frac{\sum_{i=1}^{N}(x_{1i}y_{2i} - x_{2i}y_{1i})}{\sum_{i=1}^{N}(x_{1i}x_{2i} + y_{1i}y_{2i})}. \quad (7\text{-}17\text{b})$$

The second derivative $\partial e^2/\partial\sigma^2$ determines the global minimum. For the general 3D case where two sets of points do not lie on a plane, the following iterative procedure can be used. The rotation matrix R can be expressed in terms of Euler angles, or two components of the direction cosines of the rotation axis plus the angle of rotation. Furthermore, one can decompose the rotation matrix R into three successive rotations about the X-, Y-, and Z-axes as $R = R_Z(\phi)R_Y(\theta)R_X(\Psi)$, and then the error e^2 to be minimized can be rewritten as $e^2(\phi, \theta, \Psi) = \sum_{i=1}^{N} \|q_{2i} - R_Z(\phi)R_Y(\theta)R_X(\Psi)q_{1i}\|$. The approach is then to minimize e^2 with respect to one variable at a time and to keep the remaining two variables fixed. Each such minimization is then of the planar 2D type, which was treated before. The iterative procedure starts by initially guessing a $\langle\Psi_0, \phi_0\rangle$ pair and finding θ_1 as the solution to minimizing $e^2(\phi_0, \theta_0, \Psi_0)$. The 2D transformation sought is between projections of $\{q_{2i}\}$ and $\{R_X(\Psi_0)R_Z(\phi_0)q_{1i}\}$ on the x-z plane. In a similar way, one finds Ψ_1 and ϕ_1, such that one cycle of iteration yields the triple $(\phi_1, \theta_1, \Psi_1)$. The procedure is repeated k times until $e^2(\phi_k, \theta_k, \Psi_k)$ is below some threshold. Notice that in both of the previous examples of 2D and 3D alignment, the points must be carefully chosen such that they are on the object and in the same plane.

7.3.1.2. Noniterative Alignment

There are two main noniterative alignment algorithms, those based on singular value decomposition (SVD) and quaternions, respectively. The SVD algorithm (used by Arun, Huang, and Blostein (1987)), starts by assuming the sets $\{p_{1i}\}$ $\{p_{2i}\}$ and calculates then the centroids c_1, c_2 and the transformed sets $\{q_{1i}\}$ and $\{q_{2i}\}$, respectively. Then the matrix H is defined as $H = \sum_{i=1}^{N} q_{1i}q_{2i}^T$. The SVD of H is found as $H = UXV^T$, and the matrix X is defined as $X = VU^T$. Finally, calculate $d = \det(X)$. If $d = +1$, then the sought after matrix R is given as $R = X$. If $d = -1$,

X is a reflection. Should one of the singular values (λ_3, say) of H be zero, then the appropriate rotation is found by defining $X' = V'U^T$, where V' is derived from V by changing the sign of the third column. If none of the singular values of H is zero, a RANSAC-like technique is suggested (Fischler and Bolles, 1981).

Another method of noniterative alignment has been suggested by Faugeras and Hebert (1987) and is based on representing rotation as a quaternion rather than an orthonormal matrix, as was the case for the SVD method. Quaternions, introduced earlier, were shown to constitute a multiplication group; the relationships were expressed using either Euler angles or an axis and angle of rotation. Let us first consider the case of finding the optimal rotation between N pairs of planes given as $(\mathbf{n}', d')$ and $(\mathbf{n}, d)$ respectively, where $\mathbf{n}' = R\mathbf{n}$ and $T \cdot \mathbf{n}' = d' - d$. The minimization problem to be solved in order to find the optimal rotation is that of

$$e^2 = \min \sum_{i=1}^{N} \| \mathbf{n}'_i - R\mathbf{n}_i \|^2. \tag{7-18a}$$

However, remember that any rotation of axis $\mathbf{n}$ and angle θ can be represented in terms of quaternions as $R\mathbf{n} = q \cdot \mathbf{n} \cdot q^*$. The minimization problem can now be restated as

$$e^2 = \min \sum_{i=1}^{N} |q \cdot \mathbf{n}_i \cdot q^* - \mathbf{n}'_i|^2 \tag{7-18b}$$

subject to the constraint that $\|q\| = 1$. Since the quaternion module is multiplicative, i.e., $|q_1 \cdot q_2|^2 = |q_1|^2 \cdot |q_2|^2$ for any q_1 and q_2 and since $\|q\| = 1$, one can write

$$e^2 = \min \sum_{i=1}^{N} |q \cdot \mathbf{n}_i - \mathbf{n}'_i \cdot q|. \tag{7-18c}$$

The expression to be minimized is a linear function of q, and there are symmetric matrices A_i, such that the error e to be minimized can be written as

$$e^2 = \min \sum_{i=1}^{N} |q \cdot \mathbf{n}_i - \mathbf{n}'_i \cdot q|^2 = \min \sum_{i=1}^{N} q^T A_i q. \tag{7-18d}$$

If $B = \Sigma_{i=1}^{N} A_i$ then the minimization problem to be eventually solved is that of

$$e^2 = \min q^T B q \qquad (7\text{-}18\text{e})$$

subject to $\|q\| = 1$. B is symmetric and thus the solution is the vector $q_{\min}$ corresponding to the smallest eigenvalue $\lambda_{\min}$. A similar solution can be derived for finding the optimal rotation between two sets of points. Now one has to minimize

$$e^2 = \min \sum_{i=1}^{N} \| R x_i + T - x'_i \|^2$$

$$= \min \sum_{i=1}^{N} |q \cdot x_i \cdot q^* + T - x'_i|$$

$$= \min \sum_{i=1}^{N} |(q \cdot x_i - x'_i \cdot q) + T \cdot q|^2$$

$$= \min \sum_{i=1}^{N} |q^T A q + T'|^2$$

$$= \min V^T B V, \qquad (7\text{-}18\text{f})$$

where $T' = T \cdot q$, $V = (q, T')_{8 \times 1}$, B is symmetric and given as

$$B = \begin{bmatrix} A & C \\ C^T & N \times I \end{bmatrix},$$

$A = \Sigma_{i=1}^{N} A_i^T A_i$, $C = \Sigma_{i=1}^{N} A_i$. The solution is $T' = C q_{\min}/N$ and $T = T' \cdot q^*$, where $q_{\min}$ corresponds to the smallest eigenvalue $\lambda_{\min}$ of the symmetric matrix $A - C^T C/N$.

Arun et al. (1987) compared the iterative approach of Lin et al. (1986) with the SVD and the quaternions approaches. Some observations were made, among them that the quadratic matrix B that was introduced might have repeated eigenvalues when each of the two point sets are collinear. For such cases, it is obvious that $\langle R, T \rangle$ is not unique. The computer time requirements for SVD and quaternions were found to be comparable, while the time for the iterative method was much longer. However, 10% accuracy for the noniterative methods has to be considered (vs. seven-digit accuracy for the iterative method) if one contemplates reducing CPU time by a factor of two.

7.3.5.   *Matching*

Many recognition schemes suggested so far attempt to interpret a given scene by labeling usually isolated, nonoccluded objects. This amounts to search for legal (rigid) transformations and memory models such that the transformed models or objects map into each other. The recognition process, model based, requires an efficient computational strategy to cope with the large size of the search space. The main strategy is to employ specific constraints, limiting the size of the search tree, and pruning unpromising or dead-end paths. Such constraints usually relate to the surface (patch) decomposition of the 3D object and the specific relationships between patches (in terms of distance and angle separating them), as well as the characteristic of the patch itself. Next, we will discuss some of the matching strategies used for 3D object recognition.

Relational matching (or graph isomorphism) is an approach based on a probabilistic relaxation strategy. While the main constraint of rigidity is global in nature, the iterative relaxation process is basically local, and as such, rigidity is difficult to enforce. Furthermore, the stable state attractor yielded by relaxation depends heavily on the initial estimate, which means a local rather than global maxima could result. Grimson and Lozano–Perez (1986) work within a polyhedral world where objects are modeled using up to six degrees of freedom. Local constraints on the distances between surfaces, angles between surface normals, and the angles between the sensed points relative to the surface normals are used to discard inconsistent interpretations. Once the interpretation tree has been pruned by local constraints, actual transformations from model coordinates to sensor coordinates are calculated, and the system verifies that such transformations are feasible for the sensed data. There are two serious limitations with this approach: (1) objects are assumed to be isolated, and (2) the interpretation tree is relative to one model only.

Faugeras and Hebert (1986) have developed an interesting approach where the rigidity constraint is iteratively enforced through the recursive application of a Kalman filter. The estimation of an object's attitude in space is made recursively. Partial matches are grown at the same time that the rotation is estimated. Once a complete cycle is traversed, confidence about the identity of a potential model and transformation is reached. The cycle can resume with the estimated transformation as the starting alignment if the confidence is too low and iterate until the model is confidently identified.

Huttenlocher and Ullman (1987, 1988) make the following interesting suggestion on improving search process efficiency. They suggest a sequential strategy beginning with alignment, followed by the actual match. Using a small number of features to determine the 3D attitude in space in terms of position and orientation, their alignment method avoids structuring the recognition process as an exponential search through the interpretation tree. The alignment process as a novelty deals not only with rotation and translation, but with scale distortions as well. There is still a lingering question regarding the effectiveness of finding corresponding token features and the ability of the whole process to be fault tolerant to occlusion (some tokens might be missing) and/or errors in estimating such features or using them.

Ballard and Sabah (1983) factored out the alignment process from the recognition process as well and concerned themselves with finding the object frame (model) to viewer-frame transformation (scale, orientation, and translation). Their approach used the Hough technique, and as such, is characteristic of connectionist models as discussed in Section 4.2.4. Sequential estimation of rotation and translation is achieved through the use of corresponding accumulator arrays. The method is conceptually insensitive to noise and occlusion.

Hinton (1981) and Hrechanyk and Ballard (1982) did conceptual work on developing a general class of connectionist algorithms, where positive feedback and inhibitory links between corresponding units search for the most likely (interpretation, mapping) pair by clustering supporting evidence. Note that such an approach is quite appealing due to its parallel search for model identification and the corresponding alignment. Yet another major class of strategies for matching, generically known as expert systems (ES), are discussed later within the context of AI in Section 8.1.4.

7.4. Perceptual Modeling

Now we will begin to review work related to perceptual 3D object recognition and will suggest an active model that includes object recognition and localization in space. The task is constrained by the requirement for fault-tolerant behavior (subject to noise, geometric distortions and/or occlusion) and computational efficiency. We will consider several perceptual models and, in the next section, introduce our own. We present this discussion in terms of image representations, algorithm (strategy), and computational implementation. Next, we will consider the major efforts undertaken by Marr and Nishihara (in Marr, 1982), Koenderink and van Doorn (1979), and Huttenlocher and Ullman (1987, 1988) regarding 3D object recognition models and their implications.

7.4.1. *Marr and Nishihara*

This model consists of three levels of representation corresponding to the primal sketch (2D), 2.5D sketch (depth, orientation, discontinuities in depth)—the viewpoint-dependent needle map, and the 3D object-centered representation. Three-dimensional modeling assumes that canonical representations are object centered and embedded within a coordinate system centered upon the object. The fault-tolerance issue is referred to as the stability problem and is handled by using a coordinate system hierarchy (similar to that suggested by Johansson (1976) for deriving shape from motion), and limiting the range of the representational parameters. The hierarchical organization suggests parallel hierarchical memory indexing. Such a representation requires a specific alphabet of basic forms each 3D object can be decomposed into. Carving the perceived image at inflection/curvature points is a possibility, and the CAD/CAM-type of filling in with generalized cones/ cylinders completes the description.

There are major concerns as to how the system of coordinates or the frame of references should be aligned, and where their origin should be. One frame of reference is for the object itself, which is aligned with the object's axes of elongation. The axis of elongation depends on bilateral symmetry, gravitational forces, and the corresponding (up-down) retinal upright position. The frame of reference corresponding to the visual field can be given in terms of retinal position and/or head location. Clearly, interaction between locations within the retina and head system are crucial for active perception, feedback, and safe locomotion.

There are a number of issues that can be raised about such a model. Disregarding the relevance, if any, of using the primal sketch, the canonical representations suggested previously still have a long way to go toward coping with the stability problem. Decomposition is difficult to achieve in general, much more so when one has to cope with occlusion or when the axes of references are forshortened. The model contains no computational details about how to map the 2.5D sketch into the 3D object-centered representation. Finally, the model is bottom-up with no top-down influence or guidance. One will definitely agree with Pinker's (1985) contention that "given the enormous selection advantage that would be conferred on an organism that could respond to what was really in the world as opposed to what it expected to be in the world whenever these two descriptions were in conflict, we should seriously consider the possibility that human pattern recognition has the most sophisticated bottom-up pattern analyses that the light array and the properties of our receptors allow." Ullman (1985) also points out that, "we do appear to be extremely accurate perceivers even when we have no basis for expecting one object or scene to occur rather than another, such as when watching a slide show composed of arbitrary objects and scenes." However, it is well known that unrecognizable objects suddenly pop out if the perceiver is told about their supposed location and identity, and this indicates a strong interplay between bottom-up and top-down processes. This last issue is addressed in Section 7.5. and again in Section 10.2., where we discuss directed perception.

7.4.2. Koenderink and van Doorn

The Koenderink and van Doorn (1979) model is a serious attempt to address the 3D object recognition problem and has had a major influence on the CV community and on our thinking as well. Clearly, as was mentioned earlier, both the dynamic aspect of visual perception (as it relates to motion analysis or active perception of the exploratory type), and the geometric variability due to perspective transformations, are highly relevant to visual perception. To quote Koenderink and van Doorn: "...the most successful trompe-l'oeil paintings are arranged like peep-hole shows. The active observer is not deceived. This is because the perspective transformations on his retina can only be interpreted as corresponding to a plane surface, however skillful the pigments are arranged. The deformation of a complicated figure may be a phenomenon easy of comprehension, though the figure itself has to be left unanalyzed and undefined...." Both the kinematics of rigid bodies and the shear component of the motion parallax field, i.e., the deformation of the retinal images through the changing perspective, provide the sensory input for the process of 3D modeling and recognition. The shear component is local, while the rigid kinematics, perhaps ecologically and evolutionary motivated, enforce global constraints.

We have already discussed in Section 7.3.2. the relevance of curvature (or singularities/critical points) for describing one of several possible invariant intrinsic maps. Again to quote from Koenderink and van Doorn, "the structure of the singularities for a single view of an object is called "aspect." The aspect has invariance properties: for almost all vantage points the observer may execute any small movement without affecting the aspect. Such points are called stable vantage points, and almost all vantage points are stable. If the observer traverses some orbit, the internal model of the object must provide the necessary structure to predict the events about to occur." The aspects and their mutual order may be given in the form of a flow chart. The shear field is determined up to an isomorphism and most important of all, the aspect is a Gestalt-like characteristic of the visual input.

Visual potential (VP) actively models the 3D representation and recognition process. Specifically (Koenderink and van Doorn, 1979)

> the set of stable vantage points that yield a single aspect occupies a contiguous volume of space. We may picture the space surrounding an object as parcellated into discrete cells, from any cell the object is seen in a single aspect. If the orbit of the observer traverses the border surface between two such cells a visual event occurs; the set of singularities changes in one of the possible manners. Two such aspects can be said to be connected by that single event. That means that the set of all aspects has the structure of a connected graph: every node denotes an aspect, every edge a visual event. One may also say that an edge denotes an equivalence class of space paths that yield that event. This graph is called the visual potential of the object; to any orbit of the observer corresponds an edge progression of the visual potential. The potential contains all possible trains of events. Thus, the VP represents in a concise way any visual experience an observer can obtain by looking at the object when traversing any orbit through space. There is both a qualitative aspect, as if the VP were a scaffold for the quantitative information needed by the organism to interact with the object.

The VP shown in Fig. 7.9. is characteristic of a tetrahedron and exhibits three types of aspects where either one, two, or three faces are visible.

A major issue of concern when considering VP as possible 3D modeling schemes is that of complexity. The complexity of the VP can

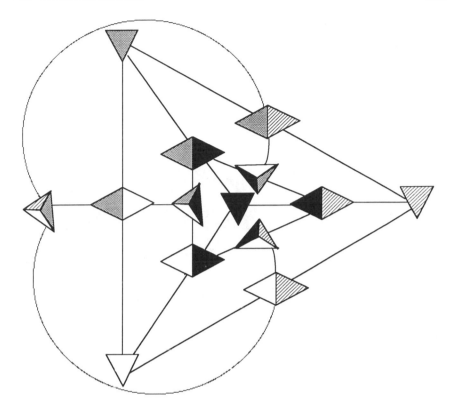

Figure 7.9.
Koenderink and van Doorn's visual potential of a tetrahedron.

be measured in terms of its diameter D, where D is the maximum number of nodes to be traversed in order to connect any arbitrary pair of visual aspects. For the tetrahedron shown in Fig. 7.9., the diameter D is equal to four. The VP corresponding to a sphere (or ovoid) is made up of one aspect only, and thus the diameter D is equal to zero. However, for most interesting shapes, D can get very large. Therefore, a major concern in using such visual potentials is how to reduce the complexity D of the VP graph and still enjoy fault-tolerant behavior, viewpoint independent recognition, and the Gestalt-like characteristics. It seems that PDP modeling, including parallel indexing of the memory, could handle the complexity issue by collapsing aspects into a reduced "isomorphic" visual potential, where incomplete information would not hamper the recognition process. Active perception would fuse information originating from several aspects and the constraints corresponding to the edges connecting them, in order to enhance the resulting 3D object identification.

7.4.3.   Huttenlocher and Ullman

We already mentioned the model suggested by Huttenlocher and Ullman in Section 7.3.5. There are four steps, which include alignment key extraction (detecting token features), alignment (indexing memory for transformation between input and 3D models), model filtering (making conjectures about an object's identity) and matching. Such a process might be ambiguous due to noisy or occluded input and errors introduced by the differential search for decomposition at (high curvature) critical points. There is a definite advantage in choosing alignment first and decreasing the size of memory models that make up the search space. Parallel indexing of the memory as well as incremental detection of the ⟨identity, transformation⟩ pair could, however, achieve the same result. The resulting approach is similar to that used by image understanding systems (Section 8.1.4.), which include prediction and verification steps.

7.5.   Parallel Distributed Computation

As we have suggested in Fig. 6.1. (Multilevel Visual Architecture), recognition is an iterative process occurring at the interface between middle and high level vision (see the MATCH box). Matching is mediated by active perception of both the exploratory (EP) and functional (FP) type. The recognition stage closes the interpretation loop, sets up both exploratory and functional perception (the top-down part of the recognition loop). Sensory inputs (affordances) provide new insights on the environment (the bottom-up part of the recognition loop).

Much of the research work on 3D recognition deals with surface reconstruction only (see Section 8.3.2. on shape from Multiple Cues) and forgets that ultimately functionality is paramount to the reconstruction process. How to actively incorporate the interpretation and recognition aspects into the reconstruction process is still an open problem. One has to work out the details of how the recognition process actually determines the EP (i.e., gaze control) and FP (i.e., resetting model's fitting parameters). This problem is crucial for computational vision and relates to many conceptual issues such as schematas (Neisser, 1976), mental imagery (Shepard and Cooper, 1986) and anticipation (Farah, 1985), and directed perception (Cutting, 1986). (Those concepts are discussed in Section 8.2. on Spatial Reasoning.) Wechsler and Zimmerman (1990b) have done initial work on the problem, which suggests that parallel distributed computation (PDC) facilitates the use of active perception toward 3D object recognition. Fig. 7.10. an instantiation of Fig. 6.1., concerns itself with how surface reconstruction and object recognition is mediated through both exploratory and functional perception. The recognition information was qualitative and it mostly decides on the need for additional (relaxation) iterations and/or the volume of space being analyzed. Additional iterations provide new affordances and ultimately recognition occurs.

The fundamental problem faced by all vision systems is the ambiguity created by the projection process. An object's projected shape in an image can change dramatically with small changes in the observer's

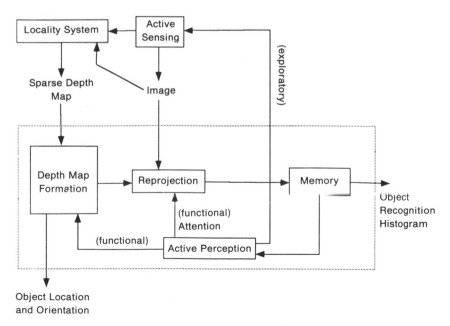

Figure 7.10.
Wechsler and Zimmerman's block diagram for 3D object recognition.

viewpoint. This is the basic difficulty in creating a machine vision system that can respond robustly in an unconstrained 3D environment. Wechsler and Zimmerman's (1990b) approach to this problem defines a structure enabling the vision system to actively engage its interpretation of the surroundings using a distributed system modeled as visual potentials.

The Wechsler and Zimmerman (WZ) computer vision system is essentially a feedback loop. Recognition should affect depth interpretation and depth interpretation affects recognition. The approach includes implicit dynamics to actively engage the environment and object-centered representations. The approach combines object-centered and viewer-oriented views through active perception. Input to the system is a 2D image and a sparse depth map (a viewer-centered map of the distances to visible surface points). A sparse depth map can be obtained from locality systems, which are modular units that use cues such as stereo, shading, or motion to make estimates of depth. The output from the system is a dense depth map and the corresponding classification of the visible surfaces. The WZ system, in effect, balances object recognition with an understanding of the visible space.

The 3D object recognition system consists of four main components, which are graphically displayed in Fig. 7.10. It begins with the formation of the depth map. The depth map formation system blends information from visual distance cues and recognition coefficients from the memory, appropriately "filtered" through the active perception system, to create a smooth dense depth map. The intensity image and the depth map are used by the reprojection system to produce a distorted image—a flattened, frontal version of all visible surfaces. This flattened image is then analyzed by the memory system. The memory system used is similar to the rotation and scale invariant memory described in Section 7.2.3. The volume of space being examined by the reprojection system, the scale of the features, and the stability of the depth map are under the guidance of the active perception system. The recognition system instantiates the middle- and high-level part of the multilevel visual architecture shown in Fig. 6.1. Each module of the recognition system is described in the following subsections.

7.5.1. *Depth Map Formation*

The depth map formation system uses a relaxation process to blend location and recognition information. Often this information is sparse or imprecise across the field of view, so it is necessary to employ relaxation in order to interpolate and to smooth. The depth map formation system consists of three planes—confidence, distance, and thin membrane, which are in registration with the input image. The thin membrane is the dense depth map composed from the anchor points, held in the distance plane, and initialized by the locality systems.

The confidence plane receives information from the memory (filtered through the active perception system) and the locality systems. The locality systems produce sparse distance information, irregularly spaced across the field of view. Each point in the confidence plane indicates the confidence in the perceived distance value. When confidence is high, the thin membrane is frozen and not allowed to change. When confidence is low, the thin membrane can change a great deal. The distance plane contains the anchor points used by the relaxation process. High confidence from the recognition system for an area of the image will solidify the depth interpretation in that area. The depth map is formed from the sparse depth information in the following way. First, the data is interpolated using a thin membrane. The energy function used to enforce the constraints among the grid points is similar to that used to implement the snakes described in Section 6.1.3. on functional perception. The energy function is composed of two terms. One enforces the constraint that the derived depth map be close to the actual range data; the other tries to find the flattest membrane that will fit the data. The stiffness of the membrane is slowly reduced, which results in a membrane interpolation that captures the surfaces indicated by the data. This is done using successive relaxation steps.

The depth map formation is shown below in Fig. 7.11. The initial input to the depth map formation system (Fig. 7.11b.) is a sparse depth map corresponding to the original cube (Fig 7.11a) from which 50% of the distance information has been randomly deleted. Note the output after 10 and 20 iterations, respectively (Fig. 7.11c. and d). The errors in the depth map occur at large (and sudden) jumps in distance that correspond to the object's boundaries.

7.5.2. Reprojection

When an object rotates about an axis perpendicular to the image plane, its projected shape will rotate but will not change shape. On the

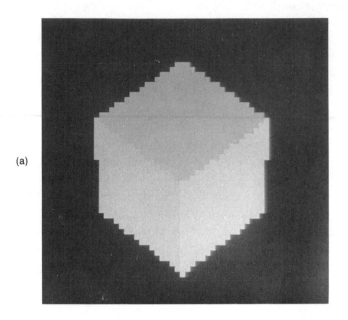

(a)

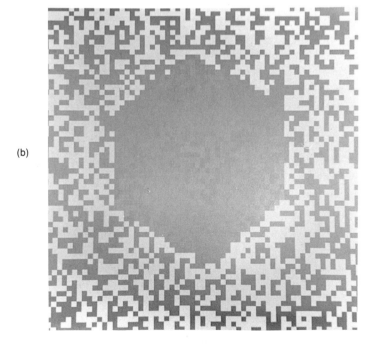

(b)

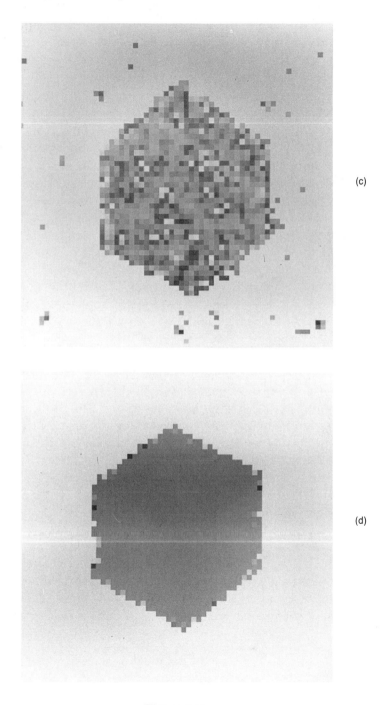

(c)

(d)

Figure 7.11.
Depth map formation.

other hand, if the same object rotates about an axis parallel to the image plane (rotation in depth), its projected shape will change dramatically due to the appearance of new surfaces, foreshortening, and perspective distortions. Foreshortening occurs when the surface being viewed slants away from the viewer. Perspective projection occurs in all imaging devices and results in parallel lines on the surface becoming converging lines in the image plane. The WZ computer vision system tries to minimize this problem by "reprojecting" the surfaces being viewed to a flat, frontal position. The reprojection system must satisfy several criteria. First, the reprojection function should compensate for the distortion of a single disconnected planar surface. This requires that surface markings, such as writing, be invariant to the slant of the surface. Second, the reprojection function should topologically map an object in an image to a characteristic shape. Third, the reprojection function should have the ability to zoom in on an area of the image—limiting the spatial extent of the processing. Following this criteria, the whole system should not only be able to recognize an object from its characteristic surface shapes but should also be able to recognize surface markings.

The reprojection function assumes each pixel is an area of intensity at a distance given by the dense depth map (thin membrane). The area of the pixel is first expanded horizontally and then vertically from the center axis outward. Each pixel is expanded according to the relative distance between itself and its neighbor along the direction of expansion. The reprojection will correctly compensate for the distortion of a single planar surface. It is nonlinear and memoryless. Changes occurring to either the image or the depth map are immediately incorporated into the reprojected image.

Let the input information to the reprojection system be from a single plane slanted in space about the x-axis. The slanted plane will be vertically foreshortened in the image. If the plane were vertical, the projected unit distance, h, of the object's surface covered by the center two adjacent pixels is given by $h = d/f$, where f is the focal length, and d is the distance from the lens to the surface of the plane. The distance along the slanted plane covered by the center pixels is

$$v = \frac{d}{f \cos \theta - \sin \theta},$$
(7-19)

where θ is the slant angle of the plane and its magnitude cannot be greater than arctan(f). The vertical distance v is larger than h for positive θ, smaller than h for negative θ, and equal to h for $\theta = 0°$. To rotate the plane to a frontal position in the projection plane, the image needs to be expanded or contracted by the ratio v/h. For a surface of arbitrary slant and tilt, the image is expanded in the vertical direction and then expanded in the horizontal direction using gradient space (p, q) information.

The reprojection function is a homolographic projection of the visible surfaces in the image. Homolographic projections result in mappings where visible object surfaces maintain their relative 3D size. When this projection is performed on a world globe, the result would be equivalent to Lambert's Equal Area Map first developed in 1722. This mapping is still commonly used for polar regions.

Figure 7.12. is an example of the reprojection function. Figure 7.12a. shows two sides of a cube viewed from a close vantage point. The faces (top and bottom) are digitized images (of Zimmerman and Wechsler) that have been mathematically pasted onto the surface of the cube. Figure 7.12b. is the result from the reprojection function. Notice that both the foreshortening and perspective distortion of the faces have been removed.

Curved surfaces could be treated as multifaceted planar surfaces. All visible surfaces can be approximated to within a finite error by planar surfaces if the size of the planar surface is small enough. Essentially, the reprojection function "sees" each pixel as a planar frontal surface in 3D space, which has been projected to form the image being analyzed. The fundamental assumption for our computer vision system is that a 3D object is represented as a collection of 2D reprojections of the object from widely varying viewpoints.

7.5.3. Memory

Recognition takes place when processed information is matched with memory. When the correlation is high, both recognition and location of the object will occur. The WZ system assumes the matching, and thus the storage of models is done in terms of stored 2D views. The structure of the mapping implied by the proposed model is viewer-centered with the relative depth information placed implicitly within the stored model. As more dynamics are added to the system, these views will be dynamically "tied" together to form the whole object,

(a)

(b)

Figure 7.12.
Reprojection of a cube.

and will amount to viewer-oriented rather than viewer-centered representations.

The input to the distributed associative memory is the flattened characteristic views of the object. A characteristic view of a polygon corresponds to those points of view that have a specified number of planar surfaces present in the projected image. For example, a cube will have three characteristic views—one side present, two sides present, and three sides present. The system is designed such that it will recognize the object and recognize the pattern printed on the object's surface.

The output from the memory is a classification vector, which measures how well the input view matches the stored views. It is used by the depth map formation system, appropriately "filtered" by the active perception system, to adjust (raise or lower) the confidence plane.

7.5.4. Active Perception

The ultimate goal of the Wechsler–Zimmerman (WZ) computer vision system is to interpret the scene with respect to what is there and where things are. This requires a definite interaction between location and recognition. The best way to examine the system is to separate it into automatic and active parts. The automatic parts are functional and include the construction of the depth map, reprojection of the image, and recognition by the memory. The active components control the search process by specifying the volume of space to be analyzed by the reprojection system, choosing new fixation points, and determining the balance between reliance on cues from the locality systems and the need to smooth the thin membrane. The active perception system accomplishes this by changing the parameters of other systems using (memory) recognition information.

The volume of space under inspection will change within the parameters of attention point, depth point, and depth gradient. The attention point is the 3D point along the reprojected image's line of sight. The depth point is the point along the line of sight where the expansion or contraction of the reprojection function stops. The depth gradient is the maximum gradient marking where the expansion or contraction of the reprojection function stops as well. Notice that the volume of space being processed at any moment is determined by these parameters and by the environment under inspection, and it corresponds to the attention element of (functional) active perception.

The response from the memory should be used to modify the shape of the depth map. For example, if the locality systems were sending information that the object under scrutiny is flat, but the memory system's best guess is that the object should be curved in a specific way, then after a certain amount of time the active perception would release the distance restrictions enforced by the locality systems. This could be done by lowering the confidence in those distance estimates given by the locality systems. The reduction in confidence increases the importance of the recognition system allowing the depth map to become appropriately curved. The ability to override the different cues to depth is critically important both because the cues may be invalid for a variety of reasons and because the task may require it. One such task would be to recognize the person's face from a photograph. Stereo and motion cues could indicate that the photograph is a flat surface with blotches of color on it, while the recognition system wants to recognize the 3D face that the photograph represents. The modification of the depth map corresponds to the functional aspect of dynamic vision as described in Section 6.1.3.

Finally, the active vision system will alter the interpreting process; it will choose new viewpoints to be examined. If the recognition system cannot converge, then the system should physically change its position in space, and reinterpret the surroundings. Choosing new viewpoints corresponds to the exploratory aspect of active perception. In the WZ recognition system, the idea of active vision is not only a process of motor control, but is a process for engaging the environment in search for interpretations of the surrounding environment.

Figure 7.13. demonstrates the total system operation. A database of seven viewpoints was used to create the memory for this example. The database consisted of one view of a cube, two views of a pyramid, one view of a plane, and three views of a tetrahedron. The object to be recognized is a rotated cube (Fig. 7.13a.—the original cube pattern is shown in Fig. 7.11a.) The depth map for the rotated cube had 50% of its elements randomly deleted, similar to that of Fig. 7.11b. The depth map was interpolated with a thin membrane using 30 iterations of the relaxation process. The resulting reprojected image is seen in Fig. 7.13b. The ragged edges on the reprojected image are due to the large distance jumps at these points. The recognition histogram, shown in Fig. 7.13c., indicates that the system identified correctly the novel input.

(a)

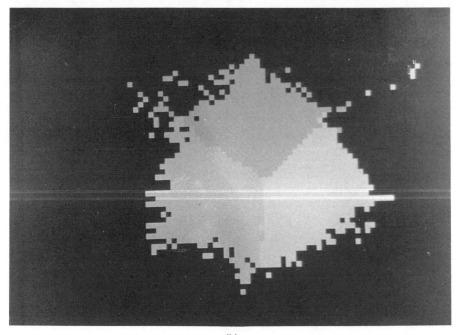

(b)

Figure 7.13.
Three-dimensional object recognition.

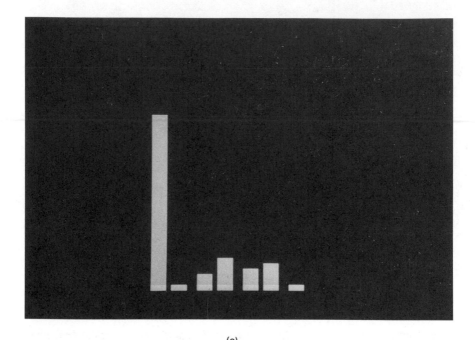

(c)

Figure 7.13. (*Continued*)

7.6.   Conclusions

Parallel distributed computation in general, and active perception in particular, seem to be crucial for object recognition. The question, yet to be answered, is how the functionality of the task and (partial) recognition information can guide the interpretation process and how the exploratory and functional aspects of active perception can be appropriately set.

Cavanagh (1989) suggests the existence of a variety of shapes whose internal representations appear to depend on the task being performed. As an example, a rigid, wire-frame object would appear to create a veridical 3D representation when a task of static 3D localization is used, a non-rigid 3D representation when the same shape is rotated in a structure-from-motion (SFM) task, and essentially a 2D representation (total loss of constancy) when the shape is used as the mouth of a schematic face and the judgement is one of facial expressions. Cavan-agh (1989) concludes that several internal representations are created simultaneously and these are accessed in a manner that depends on the task requirements.

Marr (1982) argued that it is impossible for early vision to segment a scene into objects. The question to be asked then is what comes first, segmentation or recognition? According to Cavanagh (1989), many image contours cannot be unambiguously identified until the object to which they belong has been identified, and the extra borders will seriously disrupt the segmentation of object parts into volumetric units—geons (Biederman, 1987) or generalized cones (Marr, 1982). There is in fact strong evidence that initial memory contact is based on 2D views (see also Sections 7.3 and 7.5), that the visual system operates on viewer-centered representations and not 3D object models when accessing memory, and that such 2D views mediate recognition (Cavanagh, 1989). Furthermore, within the context of the partitioning and frame problems, which involve successive abstractions, how can one decide what information can be thrown away as irrelevant and how can one partition the image into meaningful entities before it has been determined what objects are actually present in the scene (Fischler, 1989)? Segmentation might then become apparent once specific objects are recognized and the occupancy grids (see Section 8.3.4.) are filled in.

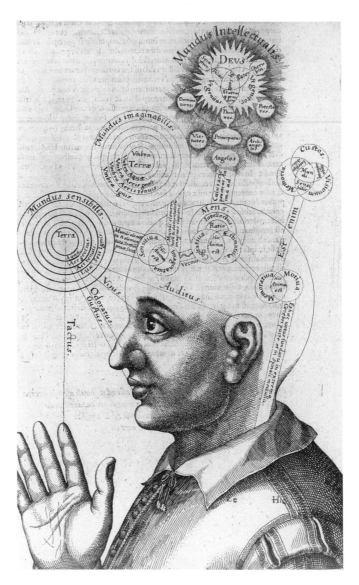

Brain: three worlds—of senses, imagination, and intellectuality
(From Fludd, R., ca. 1617.) Reprinted courtesy of the National
Library of Medicine.

8

Intelligent Systems

People only see what they are prepared to see.

Ralph Waldo Emerson

Philosophers have struggled to define what intelligence is since the dawn of civilization. There are, however, a number of qualities characteristic to intelligence that most of us agree upon, including learning from experience and adapting to our environment. Furthermore, one could add the ability to fuse disparate sources of information and cross-modal integration, and in the context of computational vision, the quality of spatial reasoning. The research needed to unravel the mysteries of the brain has to be synergetic, and many branches of science could make some contribution toward elucidating what lies behind the nature of human intelligence.

Psychology, a branch of philosophy, involves the study of the phenomena of mental life, including perception, language and thought, reasoning and emotions, and needs and beliefs. The cognitive approach to psychology involves the use of knowledge (representations) and resembles computer information processing. The epistemology of the nature and origin of knowledge includes intentionality beyond mere (mental state) representations. Intelligence has much to do with meaningful activity. As Wittgenstein wittily suggests, one should not ask for meaning, but for use instead. This leads straight to the concept of purposeful activity. The pragmatist view of Peirce considers the effects that, "we conceive the object of our concept to have" for in those effects lie "the whole of our conception of the object." Such a view is similar to the concepts of acting and break-down discussed in Chapter 6. Pragmatism leads to verification theories and the "hypothesize, predict, verify" structure of expert vision systems as discussed in Section 8.1.3.

Pragmatism and verification theories lead also to logical positivism, which is based on linguistical analysis (to determine the meaning of statements) and criteria for empirical verification. Logical positivism by itself fails, however, the test of reality. Statements can rarely be verifiable in isolation, and as Heraclitus remarked long ago, there is unity in change and permanence (of meaning) is an illusion. To be useful for perception, logical positivism needs to be augmented, at least by intentionality and task-driven behavior.

There are two major computational schools of thought competing on understanding what intelligence is and how to build intelligent machines. The AI (artificial intelligence) approach is based on symbolic processing and logic, while the PDP approach, as discussed in Chapter 4, is based on distributed computation. Logic operates in terms of symbols and propositions, but recent research on mental imagery seems to suggest that spatial reasoning involves images rather than mere symbols. Symbolic processing is also deeply rooted in atomistic philosophy and in the certainty of a given proposition. Much of modern science and philosophy speaks instead the language of uncertainty and relativism. The definition of concepts is, however, context dependent and can vary, while the observer himself is not passive but actively affects his observations.

8.1.   Artificial Intelligence

There are many definitions of AI. One defines AI as dealing with problem solving and decision making continuously faced by humans in meeting the world surrounding them. Allen Newell points out that it is not enough to have a lot of information; too much can be overwhelming. AI is also concerned with the effective use of knowledge (K). A subject S then understands knowledge K if S uses K whenever appropriate. It is clear that K has to be discriminating enough, and that "filtering" constraints are most valuable in discarding dead ends as soon as possible.

So far, a major limitation in building AI systems has been that most available knowledge is of the surface rather than the deep type. Surface knowledge is empirical, narrowly focused, involves a large number of heuristic rules of thumb, and cannot be easily extended. Deep knowledge, characteristic of basic theories and first principles in well-established sciences, such as physics and chemistry, is still lacking.

Simon (1982) considers AI as a normative science concerned with optimal engineering design. He notes the principle of bounded rationality, i.e., that people have to accept "good-enough", possibly suboptimal solutions, because computationally there is no other choice. Such a principle is incorporated into imperative logic, the paradigm of procedural rationality. Specifically, given a set of constraints and (fixed) parameters, *find* those values of the implementation variables that maximize the utility of the AI system design. The AI system design is then equivalent to an optimal search of the solution space. As an alternative to logic and as a reflection on AI as a normative science, Simon has recently suggested viewing thinking as heuristic search through a problem space that models the problem domain.

Newell (1988) suggests a cognitive model such that:

- All processing must be local; locality of information is a fundamental law of nature.
- The processors must be limited in their information capacity.
- The bandwidth of a communication between the processors must be limited.
- Processing must be hierarchical (for reasons of stability) and massively parallel.

Newell also describes a cognitive architecture that has four levels —neural, cognitive, rational, and social. As one goes from the neural to the social layer, the granularity of processing becomes larger, and the processing itself slows down.

There is still resistance to the very idea of creating a machine capable of duplicating human intelligence. Dreyfus (1979) following on earlier rebuttals, suggests that human beings have an intuitive intelligence, including common sense reasoning, which reasoning machines cannot match. Alan Turing has postulated that a machine could be considered intelligent when a human questioner cannot tell whether a response comes from the machine or a human being. This test has been recently undermined by Searl (1984) who argues against the idea of AI being able to duplicate human intelligence on grounds of functionality and intentionality. Specifically, a system might excel in symbol manipulation according to prespecified rules, yet, for all that, it might be unaware of the meaning of the symbols, what they stand for (intentionality), and thus the system could not be considered having a mind of its own and corresponding mental states.

The debate on the feasibility of creating artificial intelligence came again to center stage with the recent publication of *The Emperor's New Mind* (Penrose, 1989). Notwithstanding the tongue-in-cheek cheerful analogy to an earlier story by Andersen called *The Emperor's New Clothes*, Penrose presents a strong case that AI still has a long way to go before it can claim success and that AI is in dire need of a new methodology. To make his case, Penrose considers among many other examples, Gödel's theorem. The theorem states that for any formal and consistent system concerned with arithmetic, there are propositions that cannot be proved or disproved. Penrose concludes by saying that judgments involving mathematical truths are not necessarily algorithmic and that consciousness, intentionality, and insights are needed to comprehend the full implication of truths such as Gödel's theorem. Penrose addresses the mind-body problem charging that AI takes a radical dualistic position that treats only the (algorithmic) mind and disregards the (machine) body it has to run on. Finally, Penrose, a well-known mathematical physicist, suggests that new physics yet to be discovered—quantum gravity—are needed to fill gaps in quantum mechanics and to explain, among other things, the workings of the mind.

8.1.1.   *Representation and Reasoning*

We will briefly consider the issues of representation and reasoning as they relate to building knowledge-based systems (KBS). The interested reader can find an extensive literature on the topic, including the *AI Handbook* (Barr, Cohen, and Feigenbaum, 1981), *Artificial Intelligence* (Rich, 1989), *Artificial Intelligence* (Winston, 1984), *Building Expert Systems* (Hayes-Roth, Waterman, and Lenat, 1983). The methodology discussed follows Abramovich and Wechsler (1986).

A major problem within the AI community is the lack of a common language or formal methods (Dijkstra, 1976) for describing the knowledge-based systems being built. Without a uniform framework for describing AI systems, it is difficult to understand the conceptual architecture and functionality behind a system's numerous implementation details, sometimes called AI paradigms. A common framework could also help in the building stages as well, because it would prevent early jumps to the implementation level, and the resulting need later on for expensive backtracking.

What should the answer be when asking "What kind of a system is KBS-X?" The answer we hear all too often is "KBS-X is a rule-based system." This may be correct, although one does not gain too much knowledge about KBS-X from this information only. What should a good answer be then? Should we describe KBS-X in terms of its operating domain, the generic task it performs, the AI paradigms used to build the system, or its programming language? Ideally, it should be possible to describe it on any of these levels. It would have been very easy to answer this question if it would have been addressed when the KBS-X was originally built. The reason the problem comes up is that the jump to the implementation level is made too quickly. This can happen because of an oversight, but more often because of the misunderstanding of what AI tools are and what they are not.

We suggest a framework that provides a uniform way for describing AI problems and implementing KBS. Essentially, it provides several levels of problem description, each level being realized in terms of the level below it. Figure 8.1. shows graphically the five levels of the model. The first level is the task level, which has been proposed and described by Chandrasekaran (1985). The main concept is that there is a set of generic tasks in terms of which a problem can be formulated. The next level is the AI paradigm level, which contains a set of AI paradigms that can be integrated in various ways to realize the tasks of the first level. The third level contains different programming paradigms, similar to the ones found in the so-called expert system building tools of today. The fourth level is a language level and provides

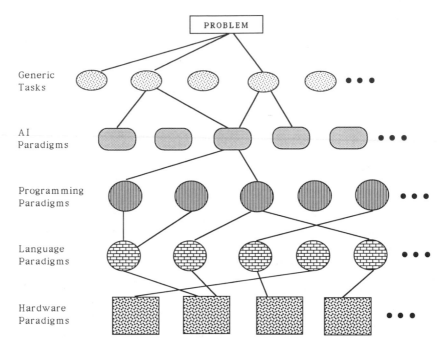

Figure 8.1.
Multiple description levels of knowledge-based systems.

programming language support to implement programming paradigms. The fifth level is a hardware implementation level—how the language level is implemented physically.

In terms of this model, the problem we are currently facing is that first and second levels are often by-passed, and that the design problem is mapped directly into some set of programming paradigms. The reason for this is that currently available AI programming tools (e.g., KEE, LOOPS, ART, etc.) provide a set of paradigms at the third level only. Some tools provide only one paradigm, like OPS-5—a rule-based language with forward chaining, which greatly reduces design flexibility. The design framework is related to Marr's characterization of information processing tasks, which includes three levels of description: computational theory (task and strategy), representation and algorithm, and hardware implementation. Both the previous framework and Marr's agree on the highest and lowest levels: the task level and computational theory level are essentially the same thing—they both describe what has to be computed and why—the hardware level is concerned with physical implementation in both cases. The difference is in the intermediate levels: in our framework, KBS representations and algorithms are modeled through three different levels of abstraction—AI paradigm level, programming paradigm level, and language level.

8.1.1.1. Generic Tasks (Level One)

Chandrasekaran (1985) suggests that a complex real-world task can be decomposed into a set of generic tasks. There is a representation language and control regime for each generic task, and the input/output information is specified. Some of the conceptual tasks mentioned are classification, knowledge-directed information passing, hypothesis matching, and object synthesis by plan selection and refinement. In our formulation, generic tasks do not need to have the knowledge representation and control regime fixed in advanced at the generic task level. For example, if the task is design, it is not necessary for the control regime to be top-down as suggested by Chandrasekaran; it could be bottom-up or a combination of both. The same is true for knowledge representation—it is suggested to be a hierarchy mirroring the object structure, but it could be a functional decomposition instead. Therefore, there is no need to prescribe specific control strategies and knowledge representation schemes at the generic task level.

8.1.1.2. AI Paradigms (Level Two)

The most important advantage of using a "natural" paradigm is the leverage that can be gained through *elision*. This means that each paradigm contains an implicit set of assumptions about the manner in which it is used; therefore, if the assumptions are confirmed, some things do not have to be explicitly stated—they are implicitly incorporated into the paradigm. If a given concept is not a member of the "clear concept set" as defined by the paradigm, it cannot be expressed without some tampering with the concept. Therefore, the choice of the right set of paradigms is extremely important. This principle applies to levels two and three.

Paradigms on this level represent different knowledge structures, problem-solving strategies (for search and control, because search is ubiquitous in AI and affects the performance of most KBS), and

support algorithms that are used to implement the generic tasks. The
set of AI and programming paradigms constitutes the AI toolbox—an
integrated set of AI tools.

The important point to remember is that only an abstract specification of a paradigm exists—its implementation is done at the lower
levels. Some of the paradigms at this level are:

(a) *Knowledge Representation Schemes*
     Frames
     Semantic Networks (SN)
     Production Systems (PS)
     Logic
(b) *Problem-Solving Strategies (Control)*
     Means-ends analysis
     Data-driven
     Goal-driven
     Blackboard
(c) *Support Paradigms*
     Truth Maintenance System (TMS)
     Resource Allocation

Next we illustrate how second level paradigms could be modeled, by
using some of the paradigms listed, and analyzing them in some detail.
In the following discussion we are not attempting to present a complete
theory behind the paradigms. Instead, we concentrate on defining the
"right" level of abstraction on which a given paradigm has to be
described.

Frames.   It has been some time since Minsky (1975) has introduced
the notion of frames. Since then, this concept has had many different
interpretations, some of which hardly resemble the original description. The two issues that are subject to much confusion are "What" do
frames represent, and "How" are frames implemented. In particular,
the inclusion of object-oriented programming in a typical "expert
system-building tool" has made things even worse. The terms "frame-
based" and "object-oriented" are often used as synonyms, and there is
some danger when concepts on a different level of abstraction are
confused. Object-oriented programming is only a very powerful programming paradigm useful to implement frame-based systems. According to the original notion, a frame is a structure that represents a
stereotypical situation. The structure has slots that can be filled in

with symbolic values or point to other frames. The slots can also have
default values. It is clear that the previous concept can be realized in
many different ways. For example, Hayes (1981) has shown an elegant
way of translating KRL, one of the first representational languages to
be designed based on the idea of frames (Bobrow and Winograd, 1977),
into predicate logic.

Semantic Networks (SN).   The term "semantic networks" has
been used to represent a variety of things ranging from a model of
human memory to a set of cells with pointers. SN represent concepts in
terms of graphs, whose nodes encode the main characteristic aspects,
while the links indicate the relationships between those aspects.
Visual potentials, as discussed in Section 7.4., are one example of a
semantic network.

We are primarily interested in establishing the right level of abstraction of various paradigms and, therefore, do not go into theoretical
issues dealing with the "semantics of semantic networks". Brachman
(1979) introduces what he calls a "missing level" in the semantic
network analysis. The new level, the epistemological level, is intended
to fill the gap between the conceptual and logical levels, by providing a
set of knowledge-structuring primitives on top of particular knowledge
primitives. The AI paradigm level, semantic networks have to be
modeled along both conceptual and epistemological levels, where the
former level would deal with establishing conceptual objects and
relations, while the latter one would define the constructs to structure
them. KLONE (Brachman, 1978) provides a good representational
language along these lines.

Production Systems (PS).   The production rule-based system paradigm is another classical example of how levels of abstraction could
be confused. By saying that the production system paradigm is "natural" for a particular problem, we mean more than the fact that the
knowledge can be encoded as a set of rules. What one implies is that
there are some attributes of the problem domain, which correspond to
some "good" properties the production system paradigm has to offer.
For example, if the knowledge is diffuse, and it is easily separated from
the manner in which it is used, a production system might be a good
solution. The same idea applies to control as well; if the control is
loose, the production system is appropriate. The choice of PS is then
decided at a higher level than level three, which provides for choice

between forward and backward chaining strategies. The production system paradigm is very different from the rule-based programming paradigm, which is just one of the ways to implement the PS paradigm. One could use a look-up table for the production system paradigm, or even a neural network implementation (Cruz–Young and Tam, 1985).

Data-Driven Search. Data-driven search is a general control strategy. As an AI paradigm, it has to be specified without any commitment as to how it is actually realized and what kind of representation it operates on. For example, data-driven search can be implemented at the programming level through forward chaining, a graph search, or some other form of search explicitly tied up to the knowledge representation scheme. At the AI paradigm level, data-driven strategy can be defined as a state-space search given some initial data. The data is used to generate more data until some (termination) condition is met. Data-driven search corresponds to bottom-up strategies used in computational vision, and characteristic of low-level vision.

Goal-Driven Search. Goal-driven search is the dual strategy to data-driven search, corresponds to top-down strategies, and is characteristic of high-level vision, where the goal is to verify some conjectured interpretation of the sensory information acquired so far.

Logic and Theorem Proving. Knowledge about the world can be expressed using logical representations and processed using appropriate inference mechanisms. The vocabulary of logic includes terms as objects, variables defined over such objects, and functions that take and return objects. Predicates defined over such terms taken as arguments yield either *true* or *false* with respect to some world model. Literals, possibly negated predicates, can be recursively connected using the universal set of Boolean operators (*and, or not*), and the quantifiers of the universal (for all $\forall$) or existential (there is $\exists$) type, to yield a rich knowledge representation language. Reasoning is straightforward using resolution, which is a generalization of modus ponens, and includes proof by refutation. Resolution states that *if* $E_1 \rightarrow E_2$ and E_1 *then* E_2, where E_1 and E_2 are logical expressions, and $\overline{E}_1$ is the negation of E_1. Note that the premise can be rewritten using disjunctives as $\overline{E}_1 \vee E_2$ and E_1, and that a false statement cannot be derived from a true one. Despite the beauty and the conciseness of logic, its use as an AI paradigm is predicated, among other things, on our acceptance of Boolean rather than "fuzzy" logic, where the truth of a statement can vary between *false* (0) and *true* (1). As more information comes in, the truth of a statement is likely to change, and it does so by degree. Resolution can be very inefficient, unless filtering constraints guide the unification process, which is responsible for choosing those statements participating in the modus ponens step, and the particular binding of variables. There is a strong conceptual link between logic as previously described (level two) and rule-oriented implementation (level three). As a consequence, similar considerations should be given when choosing rule-oriented programming paradigms.

Truth Maintenance System (TMS). TMS is a general purpose module, and it oversees the inference mechanisms. The three basic functions for TMS are to (1) record inference results; (2) provide explanations; and (3) assist in resolving contradictions and keep the knowledge consistent. Some examples of TMS are found in Doyle (1982) and DeKleer (1986). For integration into an overall AI system, the TMS must have its interface with the inference engine clearly defined. TMS provides the status of facts on request and responds to contradictions using (demon) interrupts. TMS in general need not know how knowledge is represented within the rest of the system, just as the system need not know about the insides of the TMS.

8.1.1.3. Programming Paradigms (Level Three)

The programming paradigms usually found in most of the so-called tools for building knowledge-based systems are:

- object-oriented
- procedure-oriented
- access-oriented (based on annotated values)
- rule-oriented
- logic-oriented (basically—Prolog).

There would definitely be some other paradigms. For example, a parallel processing paradigm, characteristic of neural networks, would allow, regardless of the underlying hardware architecture, the simulation of parallel processing algorithms. Such paradigms are just software engineering (SE) tools that have been applied primarily to AI applications and, in fact, there are claims about AI as being just a new

software environment for exploratory programming of complex tasks
(Sheil, 1983). It is just a matter of time before these tools will be widely
used in standard software engineering applications. LOOPS (Bobrow
and Stefik, 1983), which combines object-oriented, access-oriented,
procedure-oriented, and rule-oriented paradigms, is one example of
such an approach.

Object-Oriented. The central idea behind object-oriented programming is to provide a uniform way for combining structure and
behavior into a single unit. This is closely tied to the idea of data
abstraction, i.e., defining a data type in terms of the operations that are
performed on it, and software modularity. As a consequence, a programmer can build a whole system before making decisions about the
physical data representation. Should the specifications change later,
the only piece of code that has to be modified might be the object itself.
All the other modules remain unaffected because the internal structure
of the object is transparent to them and the interface with the object is
through some standard protocol. An additional source of power is the
fact that the message receiver is determined at run time; if various
entities have the same protocol (but different implementations of it), a
calling program need not make any distinction between those entities.
Another powerful idea behind object-oriented programming is inheritance, which allows a programmer to create entities that are almost
like some other entity. All the properties and protocols are inherited,
so that only the specialization needs to be defined. The "child" entity
can override any of the inherited information. Most object-oriented
languages allow multiple inheritance as well, thus allowing one to
create "mixins", or hybrids of several entities. Since inheritance is
also one of the attributes common to frames, object-oriented programming is a good way to implement frames. Last, but not least, an
additional benefit to be derived from object-oriented programming is
that it forces us to write better code.

Access-Oriented. In some ways access-oriented programming is
the opposite to object-oriented programming. In object-oriented programming, a side effect may occur when one object receives a message
from another, while in access-oriented programming a message may
get sent as a result of a side effect. Access-oriented programming is
based on the concept of value annotation. The main idea is that an
annotation can be associated with data. In particular, data entities can

have demons attached to them, which can be triggered on different types of access (read or write). The main advantage of this type of programming is that annotations are transparent and can be annotated themselves. If used appropriately, access-oriented programming can significantly improve the ease of writing, debugging, modifying, and reading software. Access-oriented programming is a natural way for implementing AI paradigms used to model such tasks as monitoring and simulation. It is also good for realizing various constraint-propagation algorithms.

8.1.1.4.   Programming Languages (Level Four)

The programming language level provides language primitives to realize the programming paradigm level. In most cases this would be some dialect of Lisp (e.g., Interlisp, Zetalisp). Sheil (1983) gives a good overview of tools at this level. Languages for parallel computation are also starting to appear, and OCCAM is one example.

8.1.1.5.   Hardware Implementation (Level Five)

This level deals with the physical implementation of the programming language level, and it can pose severe constraints on all the levels above it. With the emergence of the fundamentally new hardware architectures, we may be forced to rethink the representation schemes and the algorithms we use, and their implementation. A good example is related to the Connection Machine described by Hillis (1985). Its fundamentally new parallel architecture forced the introduction of a special-purpose language, Connection Machine Lisp, an extension of Common Lisp, which is designed to support a parallel programming paradigm. The repertoire for parallel and distributed computation, as is the case with neural networks, encompasses a wide range of options: digital (from signal-processing accelerators to distributed systems like the Hypercube), analogue chips, and optical. (See also Chapter 9.) Researchers are even contemplating molecular computing, which is characterized by implicit programming and depends on evolution by variation and selection (i.e., self organization). Recognition, as implemented by proteins such as enzymes and antibodies, is a function of binding specific molecules based on their shape. The concept of an adaptive match filter, like the DAM paradigm, fitting a "key" stimulus into a "lock" memory is thus far reaching in its implications.

The problem of "framework navigation"—identifying a set of paradigms at all levels, most appropriate for a particular problem—becomes very important. In other words, given a problem X, one has to find a "path" (or several candidate paths) through the graph of Fig. 8.1. Could this task be automated? The model introduced so far is not sufficient for performing the task of framework navigation, because it merely defines levels of abstractions and the type of concepts that occupy those levels and does not in itself have any knowledge of how to provide a mapping from a concept on one level to the level below it. What one needs is the capability to map from a node on the level N to a set of nodes on the level $(N + 1)$. To do this, one needs to allow nodes in this conceptual graph to have attributes and that their values not be computed until the graph is traversed (of course, there can be defaults). Attributes form their own dependency network in which their values can propagate to other attributes. Usually the propagation will be downward, from the higher level concepts to the lower ones. The goal of the navigator is to find one or more "satisfying" paths from the top to the bottom of the graph, such that all the constraints are satisfied. The graph search will be a heuristic search with more than one solution; as a heuristic search there is no guarantee that the best solution will be found. Both knowledge acquisition and learning can enhance the graph search.

8.1.2. Blackboards

The blackboard is one of the major AI paradigms. Conceptually it involves several agents known as knowledge sources (KS) cooperating, according to their intrinsic capabilities, toward solving some problem. It is like a jigsaw puzzle, where each of the participants in the game can contribute when he/she knows how to fill in.

The blackboard approach has its roots in speech understanding systems such as HEARSAY (Erman *et al.*, 1980). Speech is just another form of a signal, different from imagery, but nevertheless sharing in the need to be parsed, interpreted, and understood at possibly different levels of abstraction. The blackboard can be realized using rule-oriented or table look-up approaches characterized by the availability of condition-action pairs. The knowledge sources, i.e., the specialist agents, permanently monitor the blackboard for a condition that will satisfy the requirements for their activation. Once this condition is detected, the knowledge source is activated, and in turn may change the current state of hypotheses on the blackboard by generating a new hypothesis or modifying an existing one. This could provide the stimulus for the activation of another knowledge source, and so on. Selective attention is given to those knowledge sources that are evaluated as contributing most to the interpretation. It is the scheduler which establishes a priority for each action and then activates those KS having the highest priority. The scheduler itself could be replaced by a separate, scheduling blackboard implementing the meta-level of control and knowledge discrimination. The blackboard approach structures the solution space hierarchically, and thus provides for reasoning at different levels of abstractions. In the HEARSAY system, the levels were those of phrase, word-sequence, word, syllable, segment, and acoustic parameter. The knowledge and control are separated, and the same separation holds between the agents and their communication via the blackboard. This is why object-oriented programming (level three) can be also appropriate to implement the blackboard.

The blackboard can be thought of as focusing its attention on pending time-dependent activities, like choosing or verifying some hypothesis. According to their specialization, a KS agent is selected and becomes operational on the attended event. Once execution is completed, the cycle repeats itself until the problem is solved. The approach, suitable to large solution spaces made up of imprecise but independent sources of knowledge, is one way to implement multisensory data integration, a task known also as data fusion. The solution gradually emerges, and both the least commitment strategy and intelligent backtracking help in efficiently finding such a solution. No agent will commit itself unless there is enough evidence to warrant taking an incremental step in building up the solution. Failures lead to backtracking and their implications, clear to the system, prevent conceptually related faults from being repeated. The blackboard seeks the most credible overall interpretation and accrual of constraints reduces the uncertainty inherent in the input data. Generic tasks (level one) involving complex and/or ill-structured problems usually require an AI paradigm (level two) such as the blackboard. Finally, the blackboard is just one conceptual example of distributed problem-solving systems (DPSS). Such systems exhibit robustness and fault tolerance, and they might be realized (level five) using massively parallel architectures.

8.1.3.   Expert Systems

Expert systems (ES) are computer programs that use knowledge (K) and inference procedures to solve problems requiring significant human expertise. The knowledge necessary to perform at such a level, plus the inference procedures used, can be thought of as a model of the expertise of the best practitioners of the field. The field of knowledge engineering (KE) focuses on ways to bring expert knowledge to bear on problem solving. While preceding chapters suggest ways for both machine and human vision to operate, it is clear that we still have a long way to go before fully understanding the mechanisms of human vision and building comparable expert vision systems.

Building expert systems requires introspection of the process of human expertise. Such introspection, usually implemented via proto-col analysis, is known as knowledge acquisition. Humans usually perceive and interpret imagery with little effort, even that novel imaging techniques like NMR require specialized training. But, if asked how we do it, most of us will return a blank, puzzled stare. There has been a shift from generality (GPS—general problem solver) toward narrow and specific expert systems. The use of extensive but specia-lized knowledge presents distinctive advantages, but it clearly lacks flexibility. Successful expert systems are finely tuned for specific applications, and they cannot be easily retrained for new tasks. With regard to computational vision, the expertise sought is usually that of image interpretation and understanding. Visual expertise might be needed in remote or hostile environments or for automation. Most visual expert systems are application dependent and draw from surface rather than deep knowledge. There are major limitations affecting expert systems in general, and machine vision systems in particular, such as the limited ability to process spatial and temporal knowledge, lack of common sense reasoning, and difficulty handling inconsistent and/or incomplete data. The same expert systems also fail by not knowing their own limitations and collapsing rather than displaying graceful degradation in face of unforeseen imagery.

There are a number of basic strategies for implementing visual expert systems, which correspond to the AI paradigms (level two) discussed earlier. The strategies are of the bottom-up (data-driven), top-down (goal-driven), and blackboard type. The top-down strategy is akin to template matching (match filter) and is characteristic of the hypothesize-and-test approach. One usually combines the data- and goal-driven strategies into a loop such that the hypotheses are triggered by the actual imagery being perceived. Whatever strategy is chosen, it has to eventually bridge the gap between the knowledge used to model and represent the world and the actual input imagery. The blackboard approach is characteristic of distributed processing and was discussed in the preceding section.

Next we consider ES, which are roughly classified as the precompiled or functional type, even though the dichotomy is not as clear cut as it might sound.

8.1.3.1. Precompiled Expert Systems

Image understanding consists largely of matching scene knowledge to actual structures in the image being analyzed. The knowledge might assume the form of a semantic network with specific relational constraints between the nodes. Assuming that the nodes stand for specific entities, the ensuing search to match them against the segmented regions of a given image, subject to specific constraints embedded in the network, can be perceived as an instantiation of relaxation. The search can also be thought of as that corresponding to a Markov chain, and optical-character recognition (OCR) systems, discussed in Section 3.2.4., are characteristic of this approach. Markov chains provides a powerful approach for using contextual information and enjoy the benefit of the Viterbi algorithm for finding the MAP interpretation. Going from OCR to scene interpretation involves moving the search from 1D to 2D. The ARGOS system (Rubin, 1980) implements just such a 2D search. The knowledge concerning the outline of a city is precompiled into a network of entities such as rivers, bridges, roads, buildings, sky, and so on, and their corresponding relationships. Labeling the image amounts to searching through the images nodes (segmented regions) and finding a path consistent with the precompiled network.

Ikeuchi and Kanade (1988) suggest ways to "automatically compile object and sensor models into a visual recognition strategy for recognizing and locating an object in 3D space from visual data". Note that, as already shown in Fig. 5.5., one has to model both the objects and the image formation process. The compilation process needs to determine those object and sensor characteristics critical for recognition, and the specific control sequence needed to cope with the variability in object appearance. Again note that such precompiled knowledge is concep-tually similar to the visual potentials (VP) made up of characteristic views. The control sequence is engraved on the links connecting the views and stands for what we labeled as dynamic vision. The recogni-tion strategy assumes the form of an interpretation tree seeking both recognition and attitude in space (location and orientation). The actual implementation makes use of object-oriented programming. The precompiled approach could benefit from several enhancements. First, the modeling could be done in terms of distributed representations, in order to withstand occlusion. Next, as we already mentioned in Chapter 6, dynamic vision involves exploration and is thus tightly linked with learning and self adaptation. The optimal control sequence is ultimately the result of learning and depends on many factors, such as functionality of task, physical and geometrical reasoning, and frequency of object appearances.

8.1.3.2. Functional Expert Systems

Most expert systems in vision employ a mixed strategy including both bottom-up and top-down elements. The generic structure of such systems is given as

1. Feature extraction (data-driven/bottom-up)
2. Hypothesis generation (feature interpretation, high-level model associations)
3. Prediction (model-driven/top-down)
4. Verify (low-level associations)
    If "credibility" < threshold (interpretation) *then* go to (2)
5. Final interpretation.

We call the expert systems characterized by this structure "funct-ional" due to the specific "task and model functionality" implemented by the (hypothesis generation, prediction, and verification) cycle.

The blackboard approach can be used as a "situation board" for generating new hypotheses for interpreting the raw data. The HASP/ SIAP system described by Nii (1984) interprets sonar data, but similar ideas, using analogy reasoning, can be employed for image interpreta-tion as well. The goal of HASP is to associate the sonar information to specific ships/vessels. Specialized knowledge is available in the form of control rules to select and apply appropriate knowledge sources.

Filtering (pruning) out is essential for computational efficiency and is implemented using metaknowledge. Metaknowledge, i.e., knowledge on knowledge (ad infinitum ...) structures the knowledge, makes it discriminating, and usually uses production rules for its implementation.

Associating low- and middle-level data with specific hypotheses comes naturally for the associative memories described in Chapter 7. The DAM allows specific interpretations to be triggered (recalled) by known stimuli, even if they are incomplete or imperfect in their quality. The generic model can then be rewritten as

1. Short-term memory (STM) encoding looks up long-term memory (LTM)
2. Secondary checks (predict and verify)
3. *If* "credibility" < threshold (interpretation) *then* go to (1)
4. Final interpretation

Such a model preserves metric and spatial properties, and the LTM is content-addressable; it can be searched in parallel and is also fault tolerant. Furthermore, the LTM allows for modularity-specific interface between imagery and concurrent perceptions, i.e., it facilitates the process of multisensory data integration.

8.1.4. Image Interpretation and Understanding Systems

The main recognition strategy is that of < hypothesize, predict, and verify > and is iterative. Memory is indexed via specific image features and a subset of the most likely models is selected. Confidence in such models is probably not high enough to guarantee identification, but it nevertheless suggests (predicts) what features the systems should seek in order to increase its confidence. The appearance of sought-after features is predicted for specific locations according to, among other things, reprojection and geometric considerations. The expert system proceeds to verify that the predictions hold. The recognition cycle, traversed until enough confidence in some specific model is gained, allows for different conjectures (hypotheses) and predictions to be considered at different iterations. Some parallel realization of such an approach is highly desirable.

A major expert system operating along the lines suggested is ACRONYM (Brooks, 1981). It is a model-based image interpretation system whose primary domain of application has been the recognition of airplanes in aerial views of airports. Various features of the system have also been used for simulation of robotic systems, reasoning about how to grasp objects, and extraction of 3D information from images. The four key elements in ACRONYM are modeling, prediction, description, and interpretation. Modeling in ACRONYM refers to the user description of objects, the storage of object descriptions, and the hierarchy involved in defining classes of objects. The primitives for describing objects are generalized cones. A generalized cone is the volume swept out by the cross section as it translates along the spine, while being deformed according to a sweeping rule. To describe complex objects, a number of cones may be combined and the relationship between them described in such a way as to generate the objects of interest, as it is the case for CAD/CAM systems. The coordinate system used is object oriented rather than viewer oriented, i.e., each object has its own coordinate system. ACRONYM then uses a complex geometric reasoning system to transform the object coordinate system into the image coordinate system. The models entered by the user are stored in a data structure called the "object graph". The nodes of the object graph are objects and subobjects. Directed arcs in the object graph define relationships between nodes in the graph. Subpart arcs represent the coarse to fine hierarchy of complex objects, and affixment arcs describe the spatial relationships between subparts. Associated with the object graph is a "restriction graph", which contains information defining constraints on object parameters such as size and number of subparts.

Since the object graph deals with objects defined in a 3D world, and because images are instances of objects in two dimensions, a mapping procedure is required to provide a bridge between the corresponding representations. This is done in ACRONYM using the "predictor" and "planner" module. This module consists of approximately 280 production rules that are executed according to a backward chaining control strategy. Clearly, once a hypothesis is triggered, it has to be justified (validated) by evidence on the ground. The application of these rules produces a data structure called the "observability graph". The nodes of this graph are observable image features, and the arcs in the graph define relationships between features. Consider the case of an airport observability graph. The graph would contain nodes for runways, taxiways, and airplanes. The node for airplanes could be an observability graph for a generic class of airplanes.

The image features predicted by ACRONYM are two-dimensional projections of generalized cones called ribbons and ellipses. Relationships are also predicted between image features and include observable relationships such as colinearity, connectivity, and relative spine angle. ACRONYM uses an edge finder to find instances of ribbons and ellipses in the two dimensional image. These ribbons and ellipses are represented by a data structure called the "picture graph" in which the nodes correspond to identified ribbons, surfaces, and curves. Arcs between nodes represent spatial relationships between the nodes such as intersection, collinearity, and containment. The picture graph contains no 3D information.

Finally, the "matcher" tries to instantiate each node of the observability graph by finding sets of nodes in the picture graph that have similar properties. As instantiations are made, an "interpretation graph" is built. Relaxation techniques are used to resolve inconsistencies between the graphs. In experiments using this system, models for generic wide-body aircraft, such as L–1011 and Boeing 747, were described to the system. These aircrafts were situated in the San Francisco airport, and an aerial photograph was taken of the airport and input into the ACRONYM system. The system then labeled instances of the aircrafts in the airport photograph.

Kuan and Drazovich (1986) refined the ACRONYM system by providing a multilevel approach for both modeling and analysis. The interpretation process proceeds by comparing 3D image features with object models in a coarse to fine hierarchy. The radar target classification (RTC) system (McCune and Drazovich, 1983) is another example of using the multilevel analysis approach. There is a specific link between the nature of data being processed and the process itself. As representations become more compact and easier to understand by humans, the links connect ⟨raw radar signal to signal processing⟩, ⟨2D radar image to pattern recognition and image processing⟩, ⟨image features to knowledge-based scene interpretation⟩. The RTC system works as follows:

1. The system starts its operation by using a bottom-up approach where easy-to-find features are extracted. (Occluding and concave edges help to locate the outline of model components, while convex edges help in locating such components.)

2.   The features extracted index the memory by triggering the most likely associations in the form of hypothesized models. Contextual (logistic) information can be fused as well into the interpretation process in order to prune out unlikely alternatives.

3.   Prediction is top-down and suggests additional image features to be observed in order to enhance a specific hypothesis.

4.   The system verifies that the predicted features appear in the image.

The cycle, consisting of steps two through four, is iterated until a classification (interpretation) decision is reached. The indexing step two uses a heuristic matching process that employs interpretation rules as those found to help the decision-making process of expert human analysts. One example of such a rule could be "*If*: (1) the image is a side view of a ship, and (2) the highest superstructure of the ship in the model library, and the ship in the image are in similar locations; *Then*: there is evidence (with certainty 0.5) that the library and image ship are the same." It is interesting to note that the production rules could contain in their "then" section an active module searching (exploring) for additional information. Exploratory search for missing data is guided by the hypotheses themselves, and as such, it is characteristic of active perception as discussed in Section 6.1.2.

Both model-based recognition and (exploratory) active perception are useful ingredients for the robotics world as well. Kent and Albus (1984) at the National Bureau of Standards (NBS) showed how "the world model could function as the interface between the sensory and control systems, transforming control actions into attention-generating sensory predictions on the one hand, and transforming sensory data into feedback for the guidance of control actions on the other. At every level of the hierarchical organization, the interaction between the interpretative and modeling processes attempts to reconcile observations with expectation. The attention of the interpretative processes may be directed first to those items most likely to disconfirm the expectancy, i.e., to prune out unlikely alternatives" (i.e., validation is predicated upon the possibility of refutation.) The models act as match filters by integrating sensory inputs and filtering the relevant information from noise. There is permanent interaction between interpretative and modeling processes, which work together at reconciling observation with expectation. Selective attention, the specific instantiation of the prediction step, helps in narrowing down the reconcilation process.

8.2. Spatial Reasoning

Intelligent behavior is intrinsically linked to mobility. One needs to locate and to recognize objects. Assuming that the memory stores canonical object representations, not necessarily isomorphic to the current view entertained by the observer, recognition involves matching the identity of the object and defining the transformation, which will map the internal representation into the sensed view (or vice versa). The recognition process, as we argued earlier, is an iterative process, where (flexible) matchings and (alignment) transformations alternate. Crucial to recognition is the mapping between viewer-centered representations and object-centered representations. Such mappings are mediated by the (ecological) environment through internalized prewired constraints (and/or learning) and active perception. We briefly discussed the ethological aspects in Section 5.3., and we have also seen that it is advantageous to precompile much of the knowledge for efficiency reasons, while learning and calibration would be valuable when one has to deal with environmental changes difficult to anticipate.

In this section we are concerned with those internalized constraints, which mediate the mappings referred to previously. (Learning is deferred to Section 8.4.) The constraints to be discussed belong to spatial cognition and concern recognition and spatiotemporal layout. Additional constraints include those of the physical and functional types. Physical constraints involve the mechanics of stability and locomotion, while functional constraints involve the frequency and the potential use of some specific object. There seems to be a heterarchical organization of such constraints, but its details still have to be worked out.

8.2.1. Mental Imagery

Mental imagery is our vivid experience of imagining "seeing" as if we were actually engaged in sensing and perceiving. Defining the meaning of many concepts in the preceding sentence could keep philosophers busy, but for us, mental imagery is analogue to actual perceptual input, except that it originates within ourselves. Keep in mind that mental imagery is not of the photographic type and that it is not merely retrieved but actually generated. Finke (1985) surveys contemporary theories relating mental imagery to perception and concludes that imagery truly resembles perception in at least some fundamental respects. The theories surveyed next are those of the structural, functional, and interactive type.

8.2.1.1. Structural Theories of Imagery

Structural theories suggest that imagery is of the analogue type, and (retinotopically) preserves the spatial and pictorial properties of the physical objects surrounding us. Kosslyn (1980), one of the main proponents behind this theory, strongly suggests that imagery "seems to represent distance by embodying spatial extent, not describing it", and that the same imagery serves a functional role for tasks like navigation. Specifically, (anticipatory) expectations enhance our performance, because we can imagine ahead of time the places we are most likely to move through.

There is a large body of experimental evidence supporting structural imagery. Distance to objects affects the time it takes to scan images, images overflow if expanded too much, and subjectively, smaller images are more difficult to scrutinize.

8.2.1.2. Functional Theories of Imagery

Functional theories of imagery suggest that the transformation (alignment) of mental images is essential to object recognition (see also Sections 7.3.4. and 7.4.). Shepard and Cooper (1986) provide strong empirical support in favor of such a theory through the use of mental chronometry methods. Such methods use behavioral data and are grounded in the assumption that the time required to solve a spatial visual problem is related to the internal representations and processes used to solve that problem. Mental rotation experiments show that the time needed to compare two 3D structures is proportional to the angle separating them. As with any theory, there are alternative explanations to account for the alignment and comparison process as well— tacit knowledge being one of them.

Another alternative is provided by Rock *et al.* (1989). They describe experiments showing that subjects are unable to perform mental rotation unless they "make use of strategies that circumvent the process of visualization." They further suggest that "the linear increase in time required to succeed in mental rotation tasks as a function of the angular discrepancy between the figures compared is the result of increasing difficulty rather than of the time required for rotation." Reported experimental results on mental rotation could be then explained in terms of familiarity with the figure and practice (i.e., learning) and availability of landmarks for recognition, while at the same time the results question if an object-centered representation is achieved at all.

8.2.1.3. Interactive Theories of Imagery

According to Finke (1985), interactive theories suggest that "perception itself, could be improved through appropriate visualization." The analogy that immediately comes to mind is that of simulating behavior —anticipating and having specific expectations and thus being better prepared to handle the input when it is actually sensed and perceived. Such theories are challenged especially by alternative eye movement explanatory accounts. Still, conceptually, such theories are akin to and fit well with the aspect graphs discussed in the preceding chapter. Remember that the (entity) nodes are connected by links specific to the exploratory acts characteristic of dynamic vision. The links, either prewired or learned, tell the visual system what is the most beneficial exploration for enhancing the recognition act.

The previous theories of mental imagery are complementary. The theories suggest that imagery is analogue (vs. symbolic), isomorphic to the real-world objects, and that it serves an anticipatory function needed to enhance performance. Such an anticipatory function is also consistent with our present understanding of top-down priming, also known as task and model-driven processing, which facilitates perception by selectively priming the appropriate neural mechanisms. As an example, letter comparison experiments show that physical appearance (and not only meaning) can be anticipatory and facilitate the comparison. (Note that such priming can be done along a hierarchical dimension, with more abstract concepts at the top of the hierarchy. Then, mental imagery will be enlarged beyond the visual/spatial representation, and it will include the physical and functional characteristics alluded to earlier as well.)

Mental imagery, because of its supposed analogue nature, is strongly opposed by proponents of propositional (logic) representations. As we already stated, however, propositional representations fail to overcome the complexity issue. How many primitive types and relationships are we willing to accept in our vocabulary? Then, like shifting sands, the meaning of each of the primitives cannot be fixed in advance and is dependent on the context, which includes the observer as well. Finally, fault-tolerance and impreciseness are not easily accommodated by symbolic representations.

One aspect that makes mental imagery appealing is the efficiency of its usage. Mental imagery implicitly stores the relationships between the different entities appearing in the image. The structure of functional expert vision systems makes use of DAMs, which fit well with analogue mental imagery. If both mental imagery and perception are compatible in their isomorphic, retinotopic-like representations, concurrent processing is facilitated, and visual performance is enhanced. Another argument used to challenge the analogue nature of mental imagery is the role language and logic play in perception. While nobody questions the usefulness of either language or logic, it is quite clear that mental processes are hierarchical, that perceptual categorization requires neither language nor verbal hypothesis (Warrington, 1982) and that logic is a very slow way to implement perception. What one needs is fast reaction to visual stimulus and less "thinking" in interpreting it. Vision capability is not restricted to humans, and if one takes the time to look around, one will be amazed at the agility of a squirrel jumping from one tree to another. If logic were responsible for such performance, why would it be restricted to vision only? It seems that nature took good care of us and endowed us with the hardware needed to optimally "resonate" to external affordances. Most visual tasks do not require too much reasoning. We reason when we plan how to navigate from one place to another one or how to pass over to the fast lane.

8.2.2. *Visual Maps for Robotics*

Mobile robots like people need to build and maintain models of the environment. The models, known as visual maps, enable the robot to perform its main tasks—navigation and manipulation. There are a number of issues to be considered when generating such maps, such as spatial resolution, errors, uncertainty, and last but not least, the very nature of the map itself. Many visual maps employed use the projected 2D image as a global map within an absolute coordinate system. As Brooks (1985) forcefully argues, however, if the visual map is to be useful, it has to faithfully represent the complexity of the 3D world rather than an impoverished 2D projected image. A useful visual map would in fact be a collection of local maps and their relationships. As the observer moves, new information becomes available, and the view of the world can be appropriately updated.

Tsuji and Zheng (1987) suggest a number of criteria according to which local maps should be generated. Foremost among them is the requirement that local maps should be functionally suited to the robot-specific task. There is no need for a unified structure, but rather for a distributed collection of maps, varying in scope (size) and resolution, and appropriate to plan navigation and/or manipulation activities. Navigation requires only coarse representations of free (void of obstacles) space. Such maps do not have to include a detailed representation of objects. Landmarks layout, if available, helps in correcting errors caused by sensors or uneveness of motion, and its price is low when compared to other means such as inertial navigation. Manipulation, on the other hand, requires precise object models, but the extent of the map is quite limited.

Brooks (1985), Ballard (1988), and Asada *et al.* (1988), strongly suggest that the "global map" should be made up of local visual maps and the geometrical relations between them. Sensor maps, whose origins are fixed to the moving sensor, are generated at locations close to each other, and then integrated to yield local maps containing object-centered representations. Such an approach draws much of its flavor from the dynamic aspect of active perception and is motivated by the need to cope with complexity. The reliability of the map increases as the observer moves along, and the sequence of motion most beneficial has to be learned during actual trials. As Ballard points out, the task of the observer is to relate information from viewer-oriented (rather than centered) frames into object-centered frames, and the frames are selected to suit information-gathering goals. An active observer knows its own motion and viewing geometry, so the observer is able to infer how surfaces orient with respect to him or her. This results in an object-centered frame of reference whose origin is the fixation point of the camera (eye) movement system. Assuming that information gathering and processing may be the responsibility of different sensors, the methodology that emerges is opportunistic planning (as you go along), distributed representations (of visual space), and distributed processing (among sensors and modules). Not surprisingly, blackboards as discussed earlier, are appropriate paradigms to build, fill in, and maintain an adequate representation of the visual space surrounding a mobile observer.

Limited spatial resolution of the sensor leads to error and uncertainty. The error in depth estimation is usually proportional to distance from the object, and the density of points is usually inversely proportional to the distance. Scarcity of points in distant areas requires interpolation, and both kinds of errors introduce uncertainty. Uncertainty can be handled through the gradual accumulation of evidence. Andress and Kak (1988) make use of blackboards and Dempster-Shafer interval logic (see Section 8.3.3.) to update the system's belief in filling in the visual map.

A number of recent robotics projects point out the most likely direction for future research. Two major themes dominate—distributed computation over time in terms of representation and processing, and the ecological link connecting the robot and its environment.

Tsuji and Zheng (1987) start by assuming a flat floor (ground support surface), and that the local map is made up of the projection of an infinite plane containing the floor. The interpretation needed for navigation purposes is carried out in a perspective world rather than a 2D map void of any depth information, and, as a consequence, the spatial resolution of the map matches that of the sensors. There is no need to interpolate and/or transform within any of those local maps, and the use of a field of vision rather than the top view of the world (corresponding to the 2D map) is more natural. Any map is labeled with free, occupied, and uncertain invisible space, and as more information comes in, uncertain space should be labeled as either free or occupied, and invisible space becomes visible. Occupied space likely to cause collisions can occur not only on ground but can be caused as well by hanging objects. Finding the height of objects then serves a dual purpose and can be performed as follows. Assuming that two cameras separated by a base b, of focal length f, and having their optical axes parallel and horizontal, image a point $P(X, Y, Z)$ at (x, y), $y \neq 0$, the disparity d between the corresponding projections is given as

$$d = \frac{fb}{Z} = \frac{by}{h - h_o}, \tag{8-1}$$

where h and h_o are the heights of P and the lens center from the floor. Clearly, floor points are those whose disparity is given as $d = -by/h_o$, while the other points are of height h given as $h = (by/d) + h_o$. As the robot moves, the perspective world changes, and this amounts to moving up from the bottom row of the image. Planning an actual path can proceed as in Section 6.4.2., available environmental maps (filled possibly with landmarks) can be dynamically used as in Section 7.4., within the context of aspects graphs, and uncertainty reasoning continuously updates the free space using geometric reasoning to connect between successive maps. Asada *et al.* (1988) further suggest that sensor maps viewed at locations close to each other can be integrated into a local map using a Cartesian system of coordinates attached to some object in the common views. If the (reference point) object disappears from the view, a new local map is generated and built-up as before. The relationship between local maps, recorded in the global map as the equivalent of "rubbery and stretchy relational maps," originally suggested by Brooks (1985), is obtained by matching geometrical properties of the obstacles.

Visual maps can be constructed and used at different conceptual levels in order to gain efficiency. Kuipers and Levitt (1988) suggest four hierarchical levels, the sensorimotor, procedural, topological, and metrical type, respectively. The sensori-motor level implements the input-output relations between the robot and its environment. Procedural knowledge includes sensorimotor schemes for tasks such as route following and/or site finding. Topological information includes fixed entities and landmarks linked by topological relationships, while the metric levels quantify the relationships.

Navlab (Thorpe *et al.*, 1988) is an environment created to accommodate mobile robots operating in outdoor environments. The software environment implements the conceptual equivalent of a blackboard. The central database, called the local map (LM) is managed by the local map builder (LMB). Processing agents, modular in design, access the LM for storage and retrieval via an LMB interface, responsible for all the communication and synchronization. The LM, LMB, and LMB interface constitute the CODGER (communications database with geometric reasoning). The organization is characteristic of heterarchical systems, because knowledge is processed by different modules and at different levels of abstraction. The authors note that CODGER is a refinement of a blackboard, which they label "whiteboard." The modules are continuously running programs, and their synchronization is implicitly achieved by the data retrieval facilities, which can suspend or awake processes according to the need.

8.2.3. Spatial Cognition

Spatial reasoning encompasses everything that allows a mobile and intelligent observer to find its way safely in the surrounding environment and to manipulate it to its advantage. Spatial cognition is that aspect of spatial reasoning involved in organizing the perceptual spatial structures and continuously updating the visual maps of the environment. As we saw in the preceding section on visual maps, geometric transformations between the detected structures in successive local maps facilitate the updating of visual maps into a global map. Furthermore, as we saw in Chapter 7, canonical object representations make available spatial information as a by-product of the recognition process. This is a direct result of metric information being encoded implicitly within the representation itself. Finally, the actual motion of the observer and the flux of relationships between visual and proprioceptive information helps find the transformations needed to determine the global map. Again, it is active perception, in the form of motion and/or eye movements that makes problems (possibly) ill-posed and combinatorially expensive, computationally feasible.

Neisser (1976) explains spatial cognition as the result of a Perception-Control-Action (PCA) cycle that mediates between cognition, perception, and reality. The PCA cycle (see Fig. I.1) consists of cognitive structures, called schemata, that are anticipatory and "ready" the observer to know what to seek and how. The ensuing exploration amounts to directed perception (see Section 10.2.), which senses the present environment for available information. The sensed information (affordances) can modify the schemata, and the cycle is repeated. The object schemata can be further *embedded* in cognitive maps, which describe the world surrounding the observer, and an upgraded and all-encompassing PCA cycle can then be defined. It will consist of cognitive maps, locomotion and action, and world affordances, which "wait" to be picked up. The emphasis on *embedded* rather than *successive* stages or levels of processing is the work of Neisser. The object schemata and the cognitive maps coexist, just as real objects do in the real world. The interactions with the environment occur simultaneously and suggest concurrent processing.

The same PCA cycle also can be thought of conceptually as consisting of direct perception of the veridical optical array (as in Gibson's work) information processing needed to accept such information, and testing and validation of visual conjectures and anticipations, as is the case with AI expert systems. Anticipations do not need to be highly specific, so the need for sensory perception is not pre-emptied, and the possibility for matching is enhanced. Sensory perception is primed and guided according to the already existing schemata. Thus it becomes clear that the observer plans the observation process. The PCA cycle primes sensory perception according to well-defined plans. Planning is discussed at length in the next section, but note that planning and acting (sensing) are interleaved rather than separated. The advantage of interleaving planning and acting is all too clear if one remembers that we have to perform in a dynamic and ever changing world.

Another advantage provided by anticipations seems to be computational. If processing is interleaved, sensory perception and subsequent computation might be skipped, and a corresponding decrease in computation will result. Maybe affordances could be processed only at "decision points" deemed crucial for both spatial, temporal, and functional continuity. Anticipations, as embedded in the schemata, record the non—accidental characteristics of the world and its events in terms of probabilistic structures. As such, anticipations could possibly help with the frame problem in robotics. Specifically, how can a robot keep track of the side effects of contemplated actions? One possible solution will assume that, probabilistically, things remain unchanged, unless something has changed according to available schemata.

Schemata are similar to the frames discussed in Section 8.1.1. While frames, their slots notwithstanding, are usually static repositories to fill-in information, the schemata are plans to acquire new information. What to do when unexpected stimuli occur or spatial and temporal continuity is broken is an open problem for both the perceptual cycle and the frame models. The perceptual cycle is not restricted to vision. It has been borrowed by computational linguistics, labeled as conceptual dependencies, schema, and scripts, and used to comprehend sentence or event structure (Schank and Abelson, 1977). The planning aspect of such a cycle can also help with natural language understanding (Wilensky, 1983). The PCA cycle can use more than just visual information. It can also use auditory affordances and language as referential information. Concurrent activities take place and are integrated according to specific perceptual cycles. Such integration is much more than data fusion as practised today (see Section 8.3.) and is the result of both genetically available (at least ethologically) cognitive maps and further cognitive development. Schemata offer the possibility of developing only along certain lines, but the nature of such development is determined by the interaction between the organism and its environment. As is the case with chaos (discussed in Section 8.4.7.), nature has to be flexible enough to deal with unexpected contingencies and, toward that end, thought of mobility, exploration, and adaptation. As such, the schemata start as genotypes and evolve as phenotypes. (According to *Webster's*, "the genotype stands for a class or group of individuals sharing a specified genetic makeup, while the phenotype represents the visible properties of an organism that are produced by the interactions of the genotype and the environment".)

The PCA cycle explains selectivity and attention as inherent to the process of information pick up. Neisser quotes Mackworth and Bruner, and Eleanor Gibson, respectively, in saying that, "visual search must develop from a state in which the gaze is controlled by the nature of the stimulus and its intrinsic features to one in which it is an instrument of thought, and that attention changes from being captured to being exploratory". He continues by saying that "the encounter between cognition and reality is unpredictable, but in the long run such encounters must move us closer to the truth". This is so because the PCA cycle is self-correcting and can accommodate novel information. The exploration characteristic of the PCA cycle is ultimately responsible for those situations where cognition and reality mismatch and require us to think and to reason.

Spatial cognition depends largely on recognition and tracking, and to accomplish these goals, the visual system needs to permanently update the transformations mapping the memory contents into the continuously changing view of the world. Wechsler and Zimmerman (1989, 1990) (see Sections 7.2.3. and 7.5.) show how active and distributed processing can help achieve such goals. Ballard (1988) considers active vision as well in the context of a spatial memory holding object-centered frames (OCF) and their relationships. The transformations sought map the OCF into views dynamically established as fixation-point frames (FPF). The mappings are found using connectionist models as those introduced in Section 4.2.4. Such mappings emerge when enough evidence supporting a particular transformation has accrued as a result of matching the OCF and FPF components. Hinton and Parsons (1988) bring supporting evidence in favor of scene-based representations (SBR) based on mental rotation-like experiments. The paradigm they suggest would map retina-based representations (RBR), which are viewer centered, into canonical object-based representations (OBR). Recognition results, and the appropriate OBR is then mapped into the SBR, which explicitly encodes the relationship of the object to the scene. The OBR is held fixed and the map between the OBR and SBR is continuously altered until matching occurs. Such an approach is quite similar to the one espoused in Fig. 6.1. and further developed in Section 7.5.

8.2.4. *Planning*

Planning is a major AI area of research, and it is concerned with the allocation of resources for finding and implementing an optimal sequence of activities. There are many situations where planning is essential for dealing with complex visual tasks, usually related to recognition, navigation, and/or manipulation. So far, we have seen how one can plan for an optimal, collision-free path, and we also considered how one could plan activities in an unstructured environment. The latter task led us to suggest the blackboard as one of the main AI paradigms, and we briefly discussed how such a paradigm could be used for signal interpretation and understanding.

We have considered recognition as one of the major visual tasks and discussed ways to solve it. It seems that present research is largely converging on a solution that attempts to map canonical object-centered representations into viewer-centered representations (or vice-versa). We introduced the concepts of active perception and visual potentials and showed that the dynamics of vision can facilitate the recognition act. The dynamics, embedded onto the links connecting the characteristic views making up the visual potential, can mimic a planned and active recognition sequence. The sequence, if optimal, can reduce the complexity of both the (visual potential) representation and the (dynamic) recognition algorithm. The dynamics of motion present the viewer with viewer-oriented aspects, and these are matched with the canonical views stored into the memory.

The planning discussed amounts to what we define in Section 10.2. as directed perception and results from both adaptation (learning) and (possibly) prewired constraints. The constraints, nurtured on us by nature, include physical aspects related to stability and locomotion, functional elements related to the potential use of the objects to be recognized, and the frequency of occurrence.

The internalized constraints that mediate our perception can be thought of as metaknowledge. It seems only appropriate to quote from Shepard and Cooper (1986) that

> The view of the nature and origin of the human mind that seems to be emerging resembles that of the empiricist philosophers in this one respect: It looks to what comes into the mind from outside (i.e., from the world) as the ultimate source of knowledge. However, the view to which we are led differs from the traditional empiricist view in two fundamental respects: First, it sees that a major part of this knowledge has been internalized through biological evolution and so, for each individual, is prewired rather than acquired. And second, it sees that this innate portion of each individual's wisdom about the world—unlike the knowledge about concrete, specific objects and events, which presumably has to be learned through individual experience—reflects very abstract and general constraints governing, for example, the conservation, projection, and transformation of objects in 3D Euclidean space.

Constraints help in planning because they reduce the size of the search space by filtering out those solutions that violate basic assumptions about nature's regularities. Pure empiricist views suffer when one considers the many illusions we can experience. Furthermore, following Kant's earlier insight that "mind interprets reality in terms of specific structures," Sperry (1956) has shown that retina and optic nerve reconnections, after a salamander's eye has been dissected from the head and replaced, can occur only along predetermined paths (see also Section 5.3. on ethology), and that learning plays no role in such rewiring.

Schemata are plans for actions. According to Gibson, they pick up information, are altered by information, and use information for further planning and acting. Sensory information filters in and becomes part of the plan for future action, while the schemata themselves, as pre-existing structures, facilitate learning, or in Piagetian terms, facilitate accommodation. We can start comprehending the world and its events only to the extent that we can anticipate them. The schemata structures are the precompiled plans for anticipations, while the discrepancies between anticipations and affordances provide the basis for future accommodation and action.

Much of the work on planning has been done in the context of robotics research. SHAKEY is an early 1970s example of planning within the context of autonomous robotics (Nilsson, 1984). The control generates a complete plan and then executes it. To achieve a goal, SHAKEY uses STRIPS (Fikes and Nilsson, 1971), which implements a very general problem solver (GPS) technique, that of means—end analysis, or, to a limited extent, primitive precompiled plans. If the plan fails to achieve the goal, the sequence of backtracking, (re)planning and execution is iterated until the goal is achieved or resources are exhausted and the robot gives up.

The drawbacks of SHAKEY are two-fold. First, GPS search is exponential, and second, SHAKEY cannot plan and sense at the same time. If some catastrophic event were to occur, SHAKEY will still plan how to get some bananas hanging from the ceiling. Such drawbacks led to another class of models called reactive planning (Georgeff and Lansky, 1987), which is concerned with timely and robust response. Such behavior is achieved by using mostly precompiled routines, appropriately coupled and triggered by sensory inputs. The relevant questions to ask about reactive planning are how is concurrent control implemented and how is the precompiled knowledge structured. True concurrence avoids central control, continuously monitors the affordances, shares resources, and is fast in response. Structuring of knowledge can be hierarchical (Albus, 1981), heterarchical or embedded (Neisser, 1976) and/or stimulus–response pairs, as is the case with reflexes implemented as production rule-based systems. Hierarchical planning assumes that actions are hierarchically embedded and are driven by anticipations originating at higher levels of decisions, while heterarchical control assumes that there is mutual support between the different levels.

Lumia and Albus (1988) consider planning as well in the context of space robotics. Following the work of Kent and Albus (1984) (see end of Section 8.1.4.) they suggest a system architecture that will support the evolution of robot control from teleoperation to autonomy. The control system architecture, conceptually related to the PCA Cycle (see Fig. I.1 in the Introduction), is "a three legged hierarchy of computing modules, serviced by a communication system and a global memory. The task decomposition modules perform real-time planning and task monitoring functions; they decompose task goals both spatially and temporally." (Lumia and Albus, 1988.) The world modeling modules can refer to the global memory and can thus remember, estimate, predict, and evaluate sensory information coming in. The world model, according to Lumia and Albus, corresponds to a blackboard, and could also fulfill an anticipation role, as discussed by us earlier, by predicting the results of hypothesized actions.

Turau (1988) discusses the model of an autonomous mobile robot, which could be used in a assembly manufacturing cell (AMC). According to Kak *et al.* (1986), such an assembly cell, "should consist of one or more robots, sensory elements, a motion control system, and supervisory intelligence. The cell should be knowledge driven and be aware about the objects and plans is suposed to deal with. Sensory feedback gives the integrated and knowledge-based assembly cell enhanced flexibility and lessens the need for high-priced part feeders and precision fixturing." Design flexibility and autonomy preclude complete preprogramming and can come only when a mobile robot, endowed with the capability for dealing with unanticipated events, can handle such events through appropriate accommodation and recovery routines.

The system architecture suggested by Turau consists of a supervision module (SM), which coordinates and controls the activation of the entire system and the interaction between the modules. The modules involved are those of planning (P), error analysis (EA), error recovery (ER), and database (CAD) about objects and the layout of the AMC. The centralized aspect of the database ensures data integrity and dynamic updating. The SM activates the planning module and as a result an executable plan, consisting of a sequence of elementary operations, is produced.

Object oriented programming (OOP) concepts, consisting of messages and objects, facilitate modularity and, "decouples the abstract task from the actual realization of the solution" (Turau, 1988). Specific examples of elementary operational objects include approach, depart, transfer, insert, detach, grasp, measure, dock, and move. The operation of such modules is conditioned upon the availability of preconditions, monitoring rules, and fast sensory feedback loops to assess the state of execution (see the similarity to the old general AI problem solver (GPS), known also as means-end analysis).

In case an object fails to fulfill a message request and cannot carry out its task, the EA module is activated to "reason" about what might have led to the specific failure, and has to initiate appropriate ER modules. Error analysis is based on the availability of geometrical, functional, and causal modeling. Geometrical modeling knows about objects, their relations, and the AMC, while functional modeling is aware about the robot and its expected behavior. The causal model contains properties of the robot that are causally related in that the value of one property is determined by the values of other properties and are represented as rules.

Hypotheses regarding reasons for failure have to be corroborated, according to causal modeling, using newly acquired sensory information. The error analysis process cycles, until it understands the reasons for failure or until it has exhausted the list of possibilities and it has to give up and abort the task. Once an error has been detected, it is up to the ER module to proceed and use special recovery routines or ask the planning module to generate a new plan.

Brooks' (1986, 1990) research on building robots is characteristic of distributed visual strategies and considers the need for robots to deal with cluttered, unconstrained and dynamically changing environments. He adopts an evolutionary method of building complex autonomous agents, where the components are simple, distributed computational elements. The perceptual components, tied into behavior generating networks, extract only those aspects of the world which are relevant to the particular task for which they are tuned to respond. The strategy used by Brooks, labelled as "subsumption architecture" is hierarchical in nature, and consists of task layers such as explore, monitor, and avoid and/or recognize objects.

The layers in the subsumption architecture implement a fixed-topology network and consist of simple finite state automata (FSA). There is no central locus of control, and the FSAs are data-driven by the inputs and messages they receive. The layers are combined through the mechanisms of suppression and inhibition. Suppression and inhibition, for predefined time constants, occur on the input and output lines, respectively, of the FSAs.

The "subsumption architecture" challenges the traditional view that intelligent systems do task decomposition by function. Instead, Brooks (1988) suggests decomposition by activity, that representations are just implicit, and that each activity, or behavior, individually connects sensing directly to action. The same view also further enhances the concept that, computationally, vision is truly distributed in representations, processing, and strategies. Multiple and distributed strategies lead naturally to fault-tolerance, fast processing, and the development of reactive (human or machine) visual systems.

Ginsberg (1989) challenges universal planning in uncertain domains on the grounds that even if the compile-time costs of the analysis are ignored, the size of the corresponding look-up table will grow exponentially with the complexity of the domain. As a consequence, reactive systems and subsumption architectures of the (sense, act) type are unlikely and have to be replaced by (sense, think, act) as our original perceptual cycle in Fig. I.1. would suggest. Reactive plans can nevertheless be defended if implemented as caches that trade time for space complexity (Scoppers, 1989) or if implemented using distributed and hierarchical concepts.

Whitehead and Ballard (1988) also suggest the paradigm of distributed behavioral strategies. Their approach, which is highly concurrent, uses data streaming as the primary mechanism for achieving responsive behavior. Their perceptual cycle implements a (priority) hierarchical drive control (Albus, 1981; Kent and Albus, 1984), consisting of (learned) precompiled routines, which provide for a simple but powerful distributed decision making mechanism. Whitehead and Ballard's model is conceptually similar to that of Minsky (1986), which describes "a society of tightly coupled units whose cooperative interaction acts to control the robot. Each unit encodes a single piece of control information, which by itself is useless but when combined with others becomes powerfully robust."

The cooperating units, called AUs (activity units), can be either tonic AUs or phasic AUs. Tonic AUs are always active, while the phasic AUs are active only when triggered by other AUs. Tonic AUs implement sensing and (demon) monitoring and are continuously activated by streaming data. Phasic AUs are precompiled routines used to implement specific behavior. As affordances change, so does the robot's behavior. The data flow nature of the sensory inputs provides for reactive behavior, while contextual memory provides for spatial and temporal continuity. (i.e., an object continues to exist even if it disappears temporarily from the field of view, unless there is causal evidence to suggest that it has been destroyed.)

The data flow (streaming data) model is characteristic of opportunistic planning. Because behavior is sensory based, the robot can seize on opportunity when it arises. Opportunistic behavior is also useful because "competition for resources and control is the basis of all intelligent decision making. There can never be a strict ordering of complex behaviors. Whether to perform one behavior or another depends on the situation; it should not be made during construction and frozen for all time. What is needed are flexible mechanisms for making those decisions," (Whitehead and Ballard, 1988).

Whitehead and Ballard (1989) have recasted planning as a choice between acting and thinking. Our own PCA cycle suggests that acting and thinking routines should be interleaved, and that the choice of what routine to run and when to run it is predicated upon the availability of anticipations. Anticipations, similar to look-ahead strategies used by the master chess players, can estimate the cost benefits involved at any time in either acting or thinking/planning, without actually executing any of them. Anticipations can not only reduce the costs involved in actual sensory processing but the learning costs involved in adaptation as well.

As an example, Whitehead and Ballard developed an adaptable reactive (rule-based) system (ARS) based on ideas suggested by Holland *et al.* (1986). An ARS is defined by the set $< S, C, E, D >$, where S consists of atomic propositions describing the current state of the world and C is the set of actions the robot can execute. For each state in S or W (the set of world states) there is an associated utility which measures the benefit of being in that state. The estimates are updated using the bucket brigade algorithm (Holland *et al.*, 1986). D is the set of production rules that maps sentences constructed out of atomic propositions in S and connected by (AND, OR, NOT) into actions from C, and weights them by calculating priorities so it can guide the control cycle.

The ARS model consists of an ACTION and ANTICIPATION algorithm. For each trial, the ACTION algorithm will

1. Determine the active rules, i.e., those whose conditions are satisfied;

2. Determine the bidders (most specific rules) and probabilistically pick an action to execute based on current priorities and priorities variances;

    An active rule is a bidder if there is no other active rule whose condition is more specific. The priority of a rule is adjusted ("reinforced") to approximate the expected utility estimate of the state that results from executing that rule. As the robot adapts, general rules are replaced by more specific and robust ones, whose reinforcements tend to be less variable. The utility of a state is estimated as the expected reward over all future sequences (of a given length) of actions starting in that state (bucket brigade algorithm). Choosing the action to execute is done by assigning a probability that equals the normalized sum of the prioritized bids received for that action;

3. If the variance associated with the chosen action is less than some preselected threshold, i.e., if the action is consistent in leading to rewarding futures, then execute the action, otherwise,

(a) anticipate the effects of executing the action;

(b) use the results of the anticipation to adjust the priorities and variances of active rules;

(c) go to Step 2.

For each anticipation, the ANTICIPATION algorithm will

1. Predict the state resulting from executing an action based on the current perceived state and knowledge about the effects of actions; or

2. If the certainty about anticipation is less than some preselected threshold, then quit the anticipation and return a degraded utility value; or

3. If the variance of the utility of the predicted state is low enough, then quit the anticipation and return the utility of the predicted state and its variance as new estimates of the action's priority and variance, respectively; or

4. Use the current predicted state as the basis for another anticipation.

   Low variance leads to high certainty, so states whose utility is characterized by low variance are stable enough and no additional anticipation is required. As one would have expected, initial experiments showed that anticipation indeed increases the learning rate and improves performance. The question that still awaits an answer is how predictions can also be learned and if the costs associated with such learning outweigh the gains resulting from anticipations. ARS systems are intuitively appealing and they deserve further investigation.

8.3. Data Fusion

Biological vision is modular in design, and by analogy, a similar approach for computational vision has been suggested throughout this

book. The modularity aspect, handy for both the design itself and the
"divide and conquer" approach when dealing with computational
complexity is also useful to achieve fault-tolerant behavior. Integrating different modules suggests distributed processing, where the distributed aspect extends over a combined spatiotemporal dimension. The
blackboard approach mentioned earlier is a possible candidate, which
could be potentially implemented using either symbolic AI or PDP type
of processing. Data fusion is concerned with integrating information
originating from different modules, coordinating their activities, allocating resources as needed, and incrementally enhancing the reliability of interpretation through the process of evidential reasoning.
Evidential reasoning handles an imperfect and imprecise world and is
mostly involved with the calculus of uncertainty.

8.3.1. *Multisensory Data Integration*

Next, we consider examples of how both biological and machine
vision approach the data fusion problem from a conceptual viewpoint.
Calculus of uncertainty and its use for robotics is deferred to Sections
8.3.3. and 8.3.4.

Psychophysical and physiological experiments suggest that biological vision is modular in design. Psychophysical tests of human infants
suggest a modularity of sensitivity to different cues. According to
Yonas *et al.* (1987), the development of an infant follows a definite
pattern of sensitivity. Retinal size and motion parallax are early cues
used by infants to discriminate depth. These are followed by binocular
parallax and pictorial depth. Staggered development of the ability to
use different depth cues indicate that different processing is occurring
in parallel on the same visual stimulus. Recent work in neurosciences,
reviewed by Van Essen and Maunsell (1983), shows that there are a
large number of well-defined subdivisions in the visual cortex. The
evidence suggests the existence of at least two major functional
streams, one related to the analysis of motion, and the other related to
the analysis of form and color. These functional streams are independent in many respects, and the processing of information is continuous
and parallel. The question that immediately comes to mind is, if these
processes are independent and modular, how are they integrated to
maintain a continuous, complete, and coherent perception of the
world?

Biological perception of space is not limited to cues within the single modality of vision but can also be driven by entirely different modalities, as we have seen in Section 5.3. on ethological images. An example of the influence of extravisual sources in spatial orientation tasks can be found in the barn owl. In barn owls the auditory system plays a crucial role in prey capture. Hunting at night when visual cues are of little value, owls rely heavily on their sense of hearing to localize prey and guide their attack. The owl has cells in its optic tectum (the main visual center) that have topographic representations of visual space that are bimodally sensitive to auditory and visual stimulation (Knudsen, 1982). The topographic representations of visual space depend on point-to-point projections from the retina. The map of auditory space is an emergent property of higher-order processing. The auditory system must derive its map from the relative patterns of auditory input arriving at each ear (Knudsen and Konishi, 1978). Despite different ways of deriving the spatial information, the visual and auditory maps were found to be remarkably similar and closely aligned. Other studies of bimodally sensitive cells have found the interactions between the modalities can be nonlinear and complex (Hartline *et al.* 1978).

Sensory inputs, properly mediated by internalized constraints, have to be integrated into a coherent view of the world. This is not always the case, however, as reports of sickness occuring in cockpit simulators have recently shown. Visual and auditory cues can create conflicting motion sensations, such as those experienced by pilots simulating gyrations. Further conflict can result between affordances and mental expectations based on experience, even though slight disparities can influence learned behavior later on.

Multisensory data integration plays an increasing role in computational vision, and for tasks such as automatic target recognition (ATR), autonomous land vehicles (ALV) and automation and robotics. Recent advances in sensory technology makes multisensory data integration an active research area worth pursuing (Kak and Chen, 1987). The methodologies suggested include statistical pattern recognition (SPR), symbolic AI enhanced by evidential reasoning, and PDP. Different modalities and different cues within the same modality have to be combined. Evidential reasoning has to account for the errors and the uncertainty intrinsic to the sensor itself, and/or the algorithm used to derive information, and has to cope with the sensors' inherent redundancy and complementarity of information.

Henderson *et al.* (1987) report some of the main concepts of multisensory data integration related to manufacturing automation. The conclusions, however, appropriately instantiated, apply to many other domains as well. The major areas considered include those of models, sensors and signals, and distributed systems.

The models specify the function, the operation mode, and the response performance of the sensors. Modeling includes the geometry and the physics of the environment, and how they affect the image formation process. The sensors provide the system with raw signals that have to be interpreted according to how the world is modeled. The sensors have to be characterized in terms of performance, including parameters such as dynamic range, sensitivity, calibration, and accuracy. An important aspect of the modeling process involves how to represent the information conveyed by the sensor and the dependency of the representation on the task being modeled. Multilevel representations, such as those used in the blackboard approach, are highly desirable, and they usually include intrinsic data (range and reflectance, etc.) and symbolic features (edges, corners, etc.).

Highly relevant to any model is how the control strategy responsible for the integration of the sensory systems is implemented. The control can be (i) centralized, leading to sequential procedure calls; (ii) weakly centralized, leading to a central control unit that initiates parallel processes, and (iii) asynchronous, leading to autonomous modules (Henderson *et al.*, 1987). The asynchronous model reminds us of the whiteboards referred to earlier and is characteristic of active perception. There are major issues to be resolved, including information exchange between sensory systems and integration of conflicting or complementary information. This amounts to how the (white or black) boards are actually implemented, suggesting that object-oriented programming might be an appropriate solution. The asynchronous control strategy and the availability of different sensory systems also suggest that a distributed processing system (DPS) approach might offer distinct advantages for multisensory data integration.

Next, we suggest using PDP in the form of a distributed associative memory (DAM) to integrate information from multiple sources. A source, amounting to a 2D function extracted from an image that carries some distinct information about the 3D scene, can take on two forms: direct or derived. A direct source is one where the information is directly in registration with the iconic image. Examples would be infrared images, range images, and spectral band images. A derived source is one that requires some type of preprocessing before the spatial information is made explicit—characteristic examples are stereopsis and optical flow derivation. Integration means combining information from several distinct sources to produce a response that is possibly quite different from the input information. An example of integration would be the ability to extract reliable viewer-centered depth information from the binocular and motion information contained in a set of images. In this case the sources, stereopsis and optical flow, are distinct, but both carry geometric information about the scene.

Our approach (Wechsler and Zimmerman, 1989a) assumes that the output of the sources vary consistently for a given input. Let $S = [\mathbf{s}_1|\mathbf{s}_2|\dots|\mathbf{s}_n]$ and $R = [\mathbf{r}_1|\mathbf{r}_2|\dots|\mathbf{r}_n]$ be the stimulus and response matrices, respectively. Each stimulus vector is made up of elements coming from the different sources $\mathbf{s}_i^T = [\mathbf{s}_{i_1}^T|\mathbf{s}_{i_2}^T|\dots|\mathbf{s}_{i_j}^T|\dots|\mathbf{s}_{i_l}^T]$ where $s_{i_j}^T$ consists of elements from the jth source. The DAM is constructed in the same fashion as shown earlier, i.e., $M = RS^+$.

The integration of sources in this manner has several positive properties. First, it is quite general, and as such it is easily extensible to multiple sources. If new sources need to be added they simply augment the previous stimulus vector structure. Second, prioritizing the influence that a single source can have on the output can be done by adjusting the size and/or the quantization of elements that source donates to the stimulus vectors. Statistical and/or factor analysis could "prioritize" sources according to both their relevance and reliability. Learning, in the form of adaptive systems (see Section 8.4.), could further enhance the prioritization aspect.

Looking for a general method for combining multiple distinct sources into a unified interpretation is a problem that has been examined from many angles (see the *Int. Journal of Robotics Research* special issue ⟨Vol. 7, No. 6⟩ from December 1988 on Sensor Data Fusion.) Statistical approaches such as Garvey and Lowrance's (1983) threat assessment for battle management, which uses the Dempster and Shafer model for source integration, need a lot of a priori information about the statistical variations between the different source measurements and are computationally very expensive. Hybrid approaches, such as Waxman and Duncan's (1986) stereomotion, use extensive knowledge about how the sources used to develop their algorithm vary and, as a result, are not general or easily extensible to the addition of other sources. Both of these methods need higher level (AI) processing to handle conflicts in source interpretation.

8.3.2. Shape from Multiple Cues

There are many cues that help to determine the shape of a specific object. With the exception, perhaps, of textural information, the methods developed so far are brittle and quite restrictive in their assumptions, mainly because most problems of this type are ill-posed. A case has even been made that the locality of analysis might fail to grasp holistic patterns about to emerge and thus prevent the appropriate interpretation. Furthermore, the modularity of processing might suggest that the derivation of visible-surface representations could be the result of several cues (constraints). Aloimonos (1988) criticizes the general class of solutions for "shape from X," because the assumptions either are not general enough or are not present in the world, the constraints do not guarantee uniqueness, and even if an algorithm is proved to have a unique solution, the resulting solution might be unstable.

Data fusion is a generic class of methods whose goal is to merge information and to enhance the reliability of the solution (or alternatively to decrease the uncertainty/ambiguity). Sandini *et al.* (1990) research is characteristic of "shape from multiple cues", where the cues could be shading, motion, stereo, texture, contour, and so on. Such research could help in providing the much sought after redundancy needed for robotics applications. The application developed by Sandini *et al.* (1990) provides each modality with a (depth, uncertainty) pair of images. During the integration process, the individual depth images are used to produce and maintain a "voxel" representation of the environment hedged by the uncertainty map. The two modalities to be integrated are stereo and motion, and the parameters sought are depth and the associated confidence. The confidence (or equivalently the measure of uncertainty) takes the form of stochastic Gaussian variables of known variance. The equations estimate depth from stereo (triangulation) and motion (parallax) and compute a depth variance. The volumetric integration further divides the space (seen, occluded, unknown), and specific upgrading rules modify the confidence in the depth measurements. Exploratory strategies, characteristic of active perception, help choose where to look next and increase confidence in labeling the volumetric space.

The goal of "shape from multiple clues" method is surface reconstruction. We have already seen several examples where motion and active perception help in surface reconstruction. Matthies *et al.* (1987), Shmuel and Werman (1989) (see Section 6.3.4. on depth from motion), and Wechsler and Zimmerman (1990) (see Section 7.5. on directed perception and distributed computation for 3D object recognition) have shown how gaze control, data acquisition, and functional perceptual processing can enhance visual processing.

Abbott and Ahuja (1988) suggest the integrated use of focus, camera vergence, and stereo disparity for surface reconstruction. (The intersection of the optical axes of two cameras determines the point of vergence. Note that the disparity near the vergence point is almost nil. See also Fig. 5.3. and the discussion on kinetic depth in Section 6.3.2. and diagrammed in Fig 6.6). Their motivation is that different clues have to be activated at different times, according to their functionality (e.g., the specific part of the visual field and the range are determined by camera vergence and focus parameters), and that the (partially) reconstructed surface map guides the movement of the camera. Remember that each clue yields a depth estimate of its own (see as an example Section 6.5.3. on how one can derive range from depth of field) and that the clues are functional in different ways (e.g., focus accuracy decreases with depth).

Abbott and Ahuja argue that "at any given time during imaging, sharp images can be acquired only for narrow parts of the visual field, capturing a limited depth range. However, real scenes are wide and deep; therefore the camera vergence and focus must be controlled to scan the scene to acquire complete data. The global surface map of the scene must be synthesized from local, partial maps obtained using local data". (See the analogy in Section 8.2.2. to the problems encountered in deriving global visual maps for robotics.) "To accomplish this, the acquired images can be analyzed for stereo correspondence, and a surface map obtained from the associated disparities. This partial surface map may then be used to direct movement of the camera to new, unmapped portions of the scene. Thus a cooperation between camera motion and image analysis, or equivalently between image acquisition and 3D surface extraction, is necessary".

The active perception aspect, which guides camera motion and data acquisition, involves iterations of target selection and surface estimation in the selected target area. The first step, that of *target selection*, depends among other things on proximity to foveal fixation, smooth extension of surface already reconstructed, avoidance of areas within the visual field which can not be imaged from both cameras, and computational optimization with respect to the number of fixation points being considered. As we have discussed in Section 7.5., there is an intrinsic link between active perception and recognition. Therefore, to the criteria considered by Abbott and Ahuja for target selection, we would add recognition parameters, which could indicate how scene understanding and interpretation proceeds. Furthermore, adaptation resulting from learning could encode optimal camera motion along the edges connecting characteristic views within the already discussed aspect graphs (Koenderink and van Doorn, 1979). The point we want to emphasize is that task functionality and performance (in terms of recognition) are quite relevant in determining optimal camera motion. The second step, that of *surface estimation and reconstruction*, is implemented using optimization techniques similar to those described in Section 6.1.3. on functional perception. The functional being minimized consists of (i) a fidelity term, which attempts to bring in close agreement the estimates provided by the left and right camera focus and vergence versus the present estimated depth, and to enhance the vergence quality in terms of calculated disparity and blur; and (ii) qualitative terms which measure both the smoothness and the sharpness of the reconstructed image.

Malik (1989) argues that shading, contour, and texture constraints are closely coupled, and that partitioning their analysis into separate shape from X modules, which interact only at the output stage, is a bad idea. Such an approach leads to algorithms where, in order to isolate the effect of one factor, one has to assume that the other cues are known and constant. As an example, shape from shading algorithms need boundary conditions that are available only after the contours in the line drawing have been labeled. Most likely, what is needed is the availability and integration of algorithms that adapt to the image conditions, instead of completely distinct algorithms (Alomoinos and Shulman, 1989).

8.3.3. Uncertainty Analysis

Much of the information we have about the world is imprecise, incomplete, and not totally reliable. Knowledge about the human brain and its functions is still sketchy, but there has already been much effort to impose on it a logical system and to justify by analogy a similar system for artificial intelligence. Traditional logic will have us believe that symbols and some type of logic, maybe first order, suffice to make up for our perceptual (and intellectual) capabilities. Imprecision is almost ignored—the most those willing to face the reality of imprecision will admit to is the existence of some "fuzziness" in the symbols themselves and their processing. Symbolic perception cannot be justified by the human use of logic as a formal system. Logic is used by most of us for well-defined problems, where there is no impreciseness.

Science history teaches us that while qualitative science has some merits, there is no escape from quantitative and ultimately statistical sciences. There is a measurement process and even the measurement is not totally predictable and can be influenced by the observer. The influence can vary from physical constraints (Heisenberg) to individual ones (Wittgenstein).

The PDC approach, distributed in terms of both representation and processing, seems to offer a natural way for dealing with a nonreliable world, due either to impreciseness or incompleteness. While we have dedicated much of this book to how PDC does this, the rest of this section describes how AI does it. Next, we will consider the Bayesian probability, Shortliffe–Buchanan belief system, and the Demster–Shafer interval logic. Issues associated with these models include their definition, plausibility (i.e., do the results make sense), and some comparisons in order to bring some perspective on these models.

8.3.3.1. Probability

The probability function P is defined over some domain Ω and takes values over the interval [0,1]. For any event $e \in \Omega$, the corresponding probability is $P(e)$, and it is nonnegative. The probabilities corresponding to false (F) and true (T) are given as $P(F) = 0$ and $P(T) = P(\Omega) = 1$. If the intersection of two events e_1 and e_2 is empty (i.e., they have nothing in common and $e_1 \cap e_2 = \varnothing$) then $P(e_1 \cup e_2) = P(e_1) + P(e_2)$. For any event e_1 and its negation $\bar{e}_1$ the relationship $P(e_1) + P(\bar{e}_1) = 1$ holds. From set algebra it follows that if $e_1 \rightarrow e_2$ (i.e., $e_1 \subseteq e_2$) then $P(e_1) \leq P(e_2)$. This last statement, however, does not necessarily agree with our common sense reasoning.

We define $P(H|E)$ as the probability of some hypothesis H, assuming that the truth of the evidence E has been already established. The Bayes theorem says that

$$P(H_i|E) = \frac{P(E|H_i)P(H_i)}{\Sigma\,P(E|H_k)P(H_k)},\tag{8-2}$$

where H_i is one of the hypotheses H_k, $k = 1, \ldots, n$. The Bayes model does not stand up to the common sense requirements in several ways. Independence requirements are hard to establish, but they are essential to the Bayesian model. All possible outcomes must be disjoined, and there is no way to accommodate the equivalent of "none of the above" and/or to leave some possibility for further allocation as more information comes in. Computationally, the model is expensive to implement, even more so when one has to account for updating all the probabilities, if and when new knowledge becomes available.

8.3.3.2. Belief Models

Belief models and certainty factors were introduced by Shortliffe and Buchanan (1984). For any event E, its belief is defined as $B(E)$, and like the standard probability, it satifies the requirements that $B(F) = 0$, $B(T) = 1$, and if $E_1 \rightarrow E_2$ then $B(E_1) \leq B(E_2)$. Any belief function obeys the relationship $B(E) + B(\bar{E}) \leq 1$. One can then define $MB[H|E]$ and $MD[H|E]$, the measures of increased belief and disbelief in the hypothesis H based on evidence E, as

$$MB[H|E] = \begin{cases} 1 & \text{if } P(H) = 1 \\ \dfrac{\max[P(H|E),P(H)] - P(H)}{1 - P(H)} & \text{otherwise} \end{cases}$$

$$\tag{8-3}$$

$$MD[H|E] = \begin{cases} 1 & \text{if } P(H) = 0 \\ \dfrac{P(H) - \min[P(H|E),P(H)]}{P(H)} & \text{otherwise} \end{cases}$$

Both MB and MD are belief functions yielding, as an example, $MB[H|E] + MB[\bar{H}|E] \leq 1$. Also, note as an example, that if there is a decrease in the probability of the hypothesis H, then $MB[H|E] \equiv 0$ and $MD[H|E] > 0$. The certainty factor of some hypothesis H based on evidence E is defined as $CF[H|E]$, and it is given as

$$CF[H|E] = MB[H|E] - MD[H|E].\tag{8-4a}$$

To avoid situations where only one event of negative evidence could overwhelm the combined evidence accumulated so far, it has been suggested to use an alternative definition than the one given in Eq. (8-4a). Assume that $CF[H|E_1, E_2] = CF(E_1, E_2)$, $CF[H|E_1] = CF(E_1)$, and $CF[H|E_2] = CF(E_2)$. Then

$$CF(E_1, E_2) = \begin{cases} CF(E_1) + CF(E_2)[1 - CF(E_1)] & \text{if} \quad CF(E_1), CF(E_2) > 0 \\[2mm] \dfrac{CF(E_1) + CF(E_2)}{1 - \min[|CF(E_1)|, |CF(E_2)|]} & \text{if} \quad CF(E_1)CF(E_2) < 0 \\[2mm] -CF[-CF(E_1), -CF(E_2)] & \text{if} \quad CF(E_1), CF(E_2) < 0 \end{cases}$$

$$(8\text{-}4b)$$

The certainty factor is usually heuristically defined, and for small $P(H)$, the certainty factor $CF[H|E]$ is almost equivalent to $P[H|E]$. The range of the certainty factor is $[-1, 1]$, and for a certain hypothesis H (or $\bar{H}$), the certainty factor is given as $CF = 1$ (or -1). Lack of evidence is recorded as MB, MD, and CF being all equal to zero. Note that based on Eq. (8-4a), $CF[H|E] + CF[\bar{H}|E] = 0$, when evidence E confirms hypothesis H. How do we combine evidence using such a belief system? As evidence is incrementally acquired, one can define for any two evidences E_1 and E_2 the corresponding MB, MD, and CF as

$$MB[H|E_1, E_2] = \begin{cases} 0 & \text{if } MD[H|E_1, E_2] = 1 \\ MB[H|E_1] + MB[H|E_2](1 - MB[H|E_1]) & \\ & \text{otherwise} \end{cases} \quad (8\text{-}5a)$$

$$MD[H|E_1, E_2] = \begin{cases} 0 & \text{if } MB[H|E_1, E_2] = 1 \\ MD[H|E_1] + MD[H|E_2](1 - MD[H|E_1]) & \\ & \text{otherwise} \end{cases} \quad (8\text{-}5b)$$

and,

$$CF[H|E_1, E_2] = CF[H|E_1] + CF[H|E_2](1 - CF[H|E_1]). \quad (8\text{-}5c)$$

Note that the definitions are characteristic to monotonic reasoning, which is not necessarily too realistic. Unless MD becomes true with the certainty of one, MB never decreases. However, there are many situations where two events, each on its own, confirms some hypothesis H, but when taken together disconfirm the same hypothesis.

Combining hypotheses H_1 and H_2 is done according to the following definitions

$$MB[H_1 \cap H_2 | E] = \min(MB[H_1 | E], MB[H_2 | E])$$
$$MB[H_1 \cup H_2] | E] = \max(MB[H_1 | E], MB[H_2 | E]).$$

$$(8\text{-}6)$$

The belief system as described seems appropriate when statistical data such as inverse probabilities is lacking, and when conditional independence is plausible.

8.3.3.3. Interval Logic

Interval logic, known also as the Demster–Shafer model of uncertainty, was introduced by Shafer (1976) to answer some of the shortcomings of traditional probabilistic models. For the domain of definition Ω, there is a mass probability MP defined over the power set of Ω (known as 2^Ω) and which takes values in the interval $[0, 1]$. Clearly, the events in the power set are not necessarily disjoint, and the observer is thus not forced to choose between distinct alternatives. Furthermore, the observer is allowed to keep some of the mass probability unallocated and available for future distributions, as more information comes in and the reliability increases. The unallocated mass is assigned to an event θ defined as "unknown." As it is the case for probability, $MP(\emptyset) = 0$ and if $E \in 2^\Omega$ then $\Sigma\, MP(E) = 1$.

Interval logic derives its name from the way confidence in some evidence A is defined. Rather than sometimes arbitrarily deciding on some probability $P(A)$, the Demster–Shafer model allows for an interval of confidence defined as $[SP(A), PL(A)]$, where $SP(A)$ and $PL(A)$ stand for the support and plausibility of A. Clearly, the residual given as $[PL(A) - SP(A)]$ indicates the initial ignorance or reluctance to make a commitment. The support of A draws all evidence leading one to believe in A with certainty, while plausibility, as the name implies, includes all evidence that does not preclude A. Formally, one defines the support and plausibility as

$$SP(A) = \sum MP(E),$$

$$(8\text{-}7a)$$

where $E \to A$, i.e., $E \subseteq A$, and

$$PL(A) = 1 - SP(\bar{A})$$
$$= 1 - \sum MP(E),$$

$$(8\text{-}7b)$$

where $E \to \bar{A}$. Clearly, the true probability of A, if one indeed exists, lies somewhere between $SP(A)$ and $PL(A)$, and as processing goes on, the uncertainty of the interval given in terms of its length shrinks.

How is information combined within the framework of interval logic? Given two mass probabilities $MP_1(A)$ and $MP_2(A)$, build a Cartesian table whose logical entries correspond to $MP_1(A_i) \cap MP_2(A_j)$. Assume that errors are restricted to measurements only, and that evidential independence holds for the conjunction of propositional events A_i and A_j. Start by defining a measure of compatability X between conflicting events A_i and A_j such that $X = \Sigma \, \alpha_{ij} < 1$, where $\alpha_{ij} = MP_1(A_i)MP_2(A_j)$. Then one proceeds to find the combined new mass probability $MP(A)$ as $MP(A) = MP_1(A) \oplus MP_2(A)$ given as

$$MP(A) = (1 - X)^{-1} \sum_{\substack{F \subseteq MP_1 \cap MP_2 \\ F \subseteq A}} MP(F), \qquad (8\text{-}8)$$

where F are compatible logical entries in the intersection table $MP(A)$, and $\Sigma \, MP(A) = 1$. Note that the intersection of two mass probabilities is restricted in its scope when compared to the original mass probabilities being combined, and that the method lends itself easily to spatio-temporal processing by its ability to evolve new mass probabilities according to some prespecified generation rules.

Zadeh (1986) shows that interval logic can be viewed as an instance of inference from second-order relations, that is, relations in which the entries are first-order relations. The second-order relations correspond to relational databases made up of relations whose elements are atomic. Furthermore, he notes that the normalization process, where the null values are not counted and their (incompatible) mass is redistributed, can lead to counterintuitive responses, where conclusions can be drawn about specific events even that conflicting evidence about their existence is suppressed within the incompatibility measure X.

One can make a number of comparisons between interval logic and either Bayesian probability or belief systems. Bayesian probability allows for singleton events only and handles lack of information by assigning $1/n$ as initial probability to each of them. Interval logic, on the other side, starts by allowing for the unknown mass and defining $MP(\theta) = 1$. Shortliffe and Buchanan (1984) show that interval logic and belief systems yield similar results for two events, both confirming or disconfirming some hypothesis, but not for conflicting events. (If one tries to convince oneself, remember that the intersection between any event E and the unknown/unassigned event θ becomes more specific, i.e., is E.) By examining the certainty factor (CF) as given by Eq. (8-4b), one sees that two (conflicting) evidences of opposite sign result in a CF whose sign is that of the CF corresponding to the strongest evidence. In contrast, interval logic will reduce the support for both hypotheses, H and $\bar{H}$, and such behavior seems more rational. If applied to the case of one evidence confirming H to degree d and the other evidence disconfirming H to the same extent, the resulting CF is zero, i.e., no evidence at all, while interval logic yields a belief of $d/(1 + d)$ in both H and $\bar{H}$.

8.3.4. Certainty Grids for Robotics

There is inherent uncertainty associated with any physical space in which the mobile robot moves. The uncertainty is due to both the internal mechanics of the robot and to its perceptual capabilities. Moravec (1987) suggests that certainty grids have to be overlapped over the physical space, and that sensor fusion is needed to enhance the knowledge of cells making up the grid. One should note that the physical space is made up of the immediate workspace appropriate for manipulation and of those surroundings corresponding to navigation.

The effective management of uncertainty (Stephanou and Sage, 1987) is increasingly recognized as a major issue in the design of intelligent systems. For robotics applications, two forms of uncertainty can be distinguished. Configurational uncertainty is the deviation of the position, orientation, or geometry of a workpiece in the real environment from that in the model. Indeterminacy is a second form of uncertainty, which means that even when an exactly known operation is performed on a workpiece in precisely known configuration, the outcome cannot be predicted uniquely. In complex manipulation (e.g., with multifingered hands) or navigation tasks, both forms of uncertainty need to be dealt with in a computationally efficient manner. This duality requires the effective integration of perception and motor algorithms within the context of a unified world model. The interval logic as suggested by the Demster–Shafer model, and discussed in the preceding section, seems appropriate for such tasks because of its ability to combine different sources of information (Andress and Kak, 1988).

Stephanou and Lu (1988) introduce the concept of generalized entropy as a measure of the uncertainty in a knowledge source and show that pooling of evidence using the Demster–Shafer rule of combination decreases the total amount of generalized entropy in the knowledge source. The decrease in entropy corresponds to the focusing of knowledge, and it is used as a measure of consensus effectiveness. An important ability of any mobile robot is to coarsen and refine sensory information, so any task is processed at the most relevant level of precision. During an early exploratory phase, reasoning is initiated at a coarse level. This is followed by an active sensing phase, during which attention is focused on the most task-relevant systems, but whose output may be below the desired specificity level. Additional sensory evidence is acquired at a higher precision level when necessary, and the active sensing scheme is aimed at minimizing the overall entropy.

Certainty grids, in terms of their occupancy, have been strongly advocated by Elfes (1989) for mobile robot perception and navigation. Geometric reasoning uses geometric modeling to interface between low-level sensing and processing and (symbolic) high-level reasoning. As Elfes argues, such geometric reasoning by default leads to a wide gap between the high-level and low-level information processing layers because it requires early instantiation of models, it lacks the capability to handle sensor uncertainty and errors (e.g., positional drift), and it relies heavily on the adequacy of the precompiled world models. The occupancy grid framework, on the other hand, employs probabilistic sensor interpretation models and Markov random field (MRF) representation schemes and places emphasis on sensing rather than a priori geometrical models. Such a framework relies on additional sensing to disambiguate interpretation and leads naturally to active sensing strategies.

Incremental (and multisensory) data acquisition and integration is accomplished using the iterative definition of Bayes' theorem for MRF of order 0, i.e., for the case where the occupancy cells can be estimated as independent random variables and probabilistic sensor modeling is available (Elfes, 1989). MAP (maximum a posteriori) estimation labels the cells as occupied, empty, or unknown. Multisensory data integration, for the case when separate occupancy grids are maintained for different sensors (e.g., sonar and stereo), can be accomplished using opinion polls. Each sensor has functional characteristics of its own,

which make it suitable for specific tasks. (See the analogy to clues'
specificity in surface recovery as discussed in Section 8.3.2. on shape
from multiple clues.) Stereo, as an example, will fail to detect surfaces
(e.g., walls) because of a lack of high contrast image features that can
be correlated, but it has high angular resolution. Sonar is functionally
complementary to stereo, in that it detects surfaces well but lacks good
angular resolution. Integration of stereo and sonar leads then to
improved behavior and upgraded occupancy grids. The methodology is
general enough that the separate grids can include precompiled maps,
if and when known. Ecological and landmark information (and less
dead–reckoning) could further enhance this very attractice approach.

The certainty grids are similar to the space labeling employed by
Sandini *et al.* (1990) when fusing depth and motion information (Section 8.3.2.). As another example, the Stanford Cart produced (stereo)
depth measurements from several images taken as a camera moved at a
right angle to its direction of view. Image pairings led to different
baselines some short, some long, where long baselines enjoy the benefit
of less uncertainty (at the possible expense of failing to find the
correlation features in both images). Probability distributions corresponding to such pairings can then be combined into a certainty grid
with respect to depth information. The same certainty grid approach is
suitable for labeling space as free or occupied for navigation and
manipulation tasks, with the grids being represented hierarchically
and according to increasing coarseness. Since the accuracy of the
sensor drops with range, the grids can be quite economical in their
computational requirements. As suggested in Section 8.2.2. (on visual
maps for robotics), there is much need to keep the uncertainty grids in
correspondence with the actual visual maps and to employ maps that
go beyond 2D projected images and faithfully record the full complexity of the 3D world.

8.4. Learning

Visual learning is a specific form of biological adaptation for a complex
organism within its surrounding environment. To quote Simon (1983),
"learning denotes changes in the system that are adaptive in the sense
that they enable the system to do the same tasks drawn from the same
population more efficiently the next time." Learning methodology and

theory draws from many areas such as neurosciences, psychology and psychophysics, artificial intelligence (AI), statistical pattern recognition (SPR), neural networks (NN), and automatic control theory. Many problems in learning do indeed belong to automatic control (AC) if one remembers that "the very problem of AC as a whole can be considered to be the problem of assigning the input situation to one or another class, and to generate the optimal response as a function of that class" (Aizerman *et al.*, 1964).

An important distinction should be made between those situations where the optimal response is known and learning amounts to the system being able to produce it, versus the situations when the optimal response has yet to be identified. Supervised learning with a teacher corresponds to the first case, while self-organization, i.e., unsupervised learning, corresponds to the latter case. Supervised learning is characteristic of error-correction strategies, where the system seeks equilibrium through negative feedback. Self-organization, conceptualized as discovery, is one of the main characteristics of evolution, and it is characteristic of extreme-search strategies. Note that error-correction strategies are a subset of extreme-search strategies, and that extreme-search solutions could be thus used for error-correction type of problems, but not vice versa.

Most present machine and visual learning is of the supervised type and involves error-correction strategies. Sutton and Barto (1981) consider learning a result of an adaptive system (AS) interacting with its environment (E). The AS acts on the E while stimuli (S) from the E are made available to the AS. Open loop (OL) learning is characteristic of those situations where the AS has no influence over the E. Most of the problems encountered in SPR, error-correction strategies, and classical conditioning are of the OL type. Gradient descent methods implementing least mean square (LMS) (Widrow and Winter, 1988), correspond to error-correction, and they have been shown by Sutton and Barto (1981) to be similar to the Rescorla and Wagner (1972) theory of classical conditioning. Organisms only learn when events violate their expectations, and the difference measured is that between the expected outcomes and the actual outcomes of behavior. Learning is of the closed loop (CL) type if the AS has a modulatory influence over the E, and the (stimuli) signals S the E generates. Learning of the CL type is characteristic of both biological and physical systems such as the thermostat.

Cybernetics, a novel scientific discipline introduced by Wiener (1948), has been concerned from the very beginning with the possibility that the complexity of human minds can be mechanically duplicated using among other things, feedback loop control and information theory concepts.

Learning is usually thought of and modeled as the modification of synaptic strengths (w_{ij}) connecting pairs of neurons i and j. Hebb (1949) suggested correlation learning as a possible hypothesis for synaptic plasticity. It amounts to concurrent (pre- and postsynaptic) neural activity increasing the synaptic efficacy. Neurochemistry findings such as the long term potentiation (LTP) and NMDA receptors back up the previous view. Concurrent activity, however, might lead to spurious memories and because the synaptic strengths can grow exponentially, the memory can saturate very quickly. Temporal decay (or leaking), i.e., *forgetting*, and hard clipping (sigmoid) functions are introduced in such models to alleviate the problems. Synaptic plasticity is also supported by recent neurofindings within the hippocampus (Lynch, 1986) and also by developmental studies on the emergence of orientation columns (Linsker, 1988). Synaptic plasticity is not restricted to concurrent activity only. Concurrent differential change rather than activity could be held responsible for changing the synaptic weights, and the drive-reinforcement theory (DRT) (Klopf, 1987) is one good example of such a differential learning mechanism. The implication then is of an organism that responds to novelty. The novelty can be also quantified using cross-covariance matrices, characteristic of stochastic processes, or optimal control like the Kalman filter (KF).

Theories characteristic of self-organization include neural Darwinism (Edelman, 1987), and topological maps (Kohonen, 1987). Edelman suggests that selective processes similar to the evolutionary ones are responsible for the development of the nervous system, and that synaptic plasticity results from the closed loop interaction of the organism with the environment. The system has to cope with the consequences of its own behavior, and there is an analogy to the causality involved in the Law of Effect. Note that the Law of Effect allows for the organism to seek novel inputs (and behavior), and that the consequences of an act, and not its mere repetition, determine whether the new behavior will be learned or not (Holland, 1975). Such a dichotomy is characteristic of the one involving the homeostat (keeping behavior in equilibrium) versus heterostat (maximizing new inputs and/or new behavior), and leads naturally to the use of entropy as a novelty measure. Topological maps are feature maps similar to the retinotopic (visual) or tonotopic (auditory) organizations found within the cortex and result from the availability of extensive lateral inhibition followed by the corresponding enhancement of the most responsive (processing elements) neuron(s).

Training strategies are complementary to learning for achieving successful adaptive behavior. The concept of closed loop adaptation and the reciprocal influence between the organism and the environment strongly suggest that efficient supervised and unsupervised training depend upon continuously evaluating the performance of the learning organism. Adaptive behavior operates on its environment, and such operant behavior is complementary to the traditional respondent or classical behavior. Evaluation involves assessing the types of errors and deciding on specific (remedial) training presentations. A full theory of how such an evaluation can proceed and what is the best training strategy is still lacking. There is, however, some indication of possible ways to assess the learning behavior and to suggest ways of improving it. Early AI research (Winston, 1975) showed the importance of using both positive and near-miss ("negative") examples for learning structural descriptions appropriate to the blocks world. Recent research in neural networks suggests that training for connectionist models benefits from the use of misspelled words for their ultimate recognition and recall. Together, these suggestions point to the need for a representative repertoire of training cases, covering the true reality of the world, which is noisy, imprecise, and incomplete.

Little attempt has been made in the past to study the effects of training parameters on neural networks. From the psychological literature, however, we know that parameters such as size of the training set, number and order of presentations, and similarity of items within a set are all important determinants of learning performance (Bourne and Restle, 1959; Hintzman, 1974; Slamecka, 1987). In addition, the conclusion one draws about the efficiency of a particular learning strategy is based, in part, upon the metric used to evaluate performance and to assess both the degree and the type of error (or novelty) involved in the input. Anderson et al. (1977) discuss the performance of associative memories and report on probability learning and the evidence for known psychological data on overshooting, recency data, and probability occurring. Other known phenomena that could be investigated include stimulus overshadow and blocking, and the chronometric fact that the reaction time should be longest for stimuli near the decision boundary.

Major issues to be considered when using specific learning models include their theoretical capability, fault tolerance, memory capacity, scaling, plasticity characteristics, learning and training strategy, and computational complexity. Theoretical capability amounts to identifying (stimulus-response) mappings that can be learned by the network. The single-layer (Perceptron) network (Rosenblatt, 1961) fails to detect image connectiveness, and it is further restricted because it can only partition the prototype (cluster) space with hyperplanes (Minsky and Papert, 1987). As a result, multi-layer networks, which are able in principle to learn any mapping, were introduced. Capacity is related to the amount of information that the system can learn before saturation and cross-talk occur. Fault tolerance is concerned with the system's ability to withstand imprecise and incomplete data and also to allow for internal memory faults. As the size of the input increases, the learning becomes faster. However, learning performance slows down when the number of examplars on which the network is trained is too large, and this suggests that overlearning is not necessarily beneficial. Plasticity characteristics are instrumental in deciding the temporal degree to which adaptive behavior can be modified over time. Stability prevents a system from ever adapting again, while full plasticity leads to permanently changing behavior. Plasticity characteristics, integral to the learning strategies being employed, determine *what, how,* and *when* to modify the learned behavior.

Learning strategies are tightly coupled to training strategies for both the supervised and unsupervised mode of adaptation. Learning has to be evaluated, and the training strategy needs to be optimal with respect to the observed behavioral errors. A loop connects learning and training. A major research thrust is to develop and assess the actual implementation of such a loop for tasks related to categorical perception and image recall.

In this section, we consider different learning models, their capabilities (and limitations), and their potential applications to vision. As one could expect, learning involves some type of storage, and indeed, we start by discussing memory organization.

8.4.1. Memory

Since the dawn of civilization, philosophers have been intrigued by memory and its workings. Lashley (1950) is among the first to bring strong evidence in favor of a distributed memory. The relation of errors in maze learning, to the extent of cerebral damage in rats, seemed to preclude localization, the topological maps referred to earlier not withstanding. Mountcastle (1978) also describes a cerebral function that is implemented through distributed processing. As we discussed in Chapter 7 and in Section 8.2. on mental imagery, the memory is not monolithic, but rather a host of functional capabilities (Kolb and Whishaw, 1985; Gazzaniga, 1988). There seem to be different types of memories and different subsystems that could account for the observed learning behavior.

A dichotomy of memory appears along several dimensions. It can be either procedural or declarative, explicit versus implicit, and semantic versus episodic (within some personal spatiotemporal context of past events). The dichotomy could be an indication of the difference between skill and fact learning, and it might be traced along an evolutionary path. Procedural learning accounts for the modification of behavior that takes place when a new skill is mastered. Declarative memory seems to be responsible for cognitive learning by integrating new information with past experiences, creating associations between different sensory inputs and perhaps even fusing them, and filtering incoming stimuli according to their novelty and the motivation and/or emotional state of the organism.

Mishkin and Appenzeller (1987) report experimental results suggesting that the hippocampus might be responsible for consolidation, the memory process that creates permanent LTM traces. The findings seem to indicate that cortical sensory inputs activate two parallel circuits, one based in the amygdala, and the other in the hippocampus. Both anatomical structures are located within the limbic system and send signals to the basal forebrain, which through its many synapses to the cortex could close the loop. The basal forebrain provides feedback by releasing neurotransmitters back to the sensory areas of the cortex.

Complementary processes of consolidation can be generically labeled as "forgetting." They are responsible for memory reorganization and integration of recent experience with past knowledge. Too much information can overburden the memory and can lead to spurious memories such as those characterizing cross-talk phenomena. Winson (1985) suggests that reorganization takes place during REM sleep, and is typical of off-line processing. Recent neural network research (Hopfield *et al.*, 1983) also suggests that unlearning (i.e., learning with a reversed sign) has a stabilizing effect in collective memories of the correlation type. Unlearning or forgetting can be easily incorporated into neural models through decay ("leaking") terms, and it is essential to nonmonotonic reasoning.

Kanerva (1988) develops a computational theory of sparsely distributed memories, which he locates within the cerebellum. Sensory inputs are represented as long binary sequences such that the distribution of Hamming distances between them allows new associations to be established and recalled holistically from memory. Specifically, Kanerva considers 1,000-bit patterns spanning a memory space $N = \{0, 1\}^n$ whose size is thus 2^{1000}. The distribution of the space N is binomial, and one can thus derive that 0.001 % of the space N is within 451 bits of any binary pattern, and that all but 0.001 % of the patterns are within 549 bits. These parameters and the fact that the mean distance between any two binary patterns is 500 bits leads to the establishment of new associations. As it is the case with chaos (see Section 8.4.6.), this very possibility of establishing new connections, which are sometimes startling and random but within prespecified constraints, is essential for adaptation and learning, as well as discovery. Kanerva's model includes several interesting concepts. He suggests that reading or writing are not from just one memory location, and he shows the usefulness of distributed storage (in terms of multiple storage and retrieval) to generalization by building archetypes from the pool of retrieved data. Memory retrieval is then the result of an iterative process that, depending on the initial retrieval, might not converge, as is the case many times with human memory. Another useful concept, usually forgotten by those involved in learning research, is that there is no distinction between training and testing. We continue to adapt while we are tested (by our environment) and we can also take advantage of the perceptual cycle (see Section 8.2.4.) by testing how closely the (anticipated) words read from memory agree with the unfolding situation.

Kosslyn (in Gazzaniga, 1988) emphasizes the role of mental imagery and the modular and distributed functional architecture of the memory. He suggests that the skills and abilities that have been viewed as unitary and undifferentiated, have in fact a rich and complex underlying structure, and that the main memory functions are those of *retain*, *generate*, and *transform*. One of the systems hypothesized by Kosslyn, called *categorical look-up*, is assumed to actively search a memory of the associative type in order to disambiguate between inputs and dynamically seek those characteristics deemed to be crucial for recognition.

Kosslyn expands on early work of Van Essen and Maunsell (1983) and indicates the existence of two cortical visual systems. The ventral pathway, responsible for shape processing, originates in the occipital lobe and ends in the inferior temporal lobe. The dorsal pathway, which extends from the occipital lobe to the parietal lobe, is probably responsible for locating positional information. There are two bottom-up processing paths corresponding to shape and positional information, and a third path, which is both active and top-down. The active path, known as categorical look-up, is assumed to search a memory of the associative type in order to disambiguate between inputs and then dynamically seek those aspects judged to be crucial for recognition. The categorical look-up subsystem is supported by the coordinate look-up and the category to coordinate conversion subsystems. Attention is actively shifted according to the three subsystems mentioned. Many concepts introduced are already being used in experimental recognition systems, such as those discussed in Chapter 7.

Another very interesting brain characteristic is the critical period, a sensitive interval period during which animals possess special learning capabilities. Beyond this period the ability to learn within a specific domain ceases, and whatever has been learned constitutes the only basis for future behavior. A common example from psycholinguistics is categorical perception, which during the critical period creates appropriate phoneme structures for later discrimination. Japanese native-speakers who have not been exposed to other languages during the critical period have difficulty recognizing the different phonemes "r" and "l" as distinct categories. Thus, the critical period is essential to the development and self-organization of the brain neuronal connections and should be considered in any learning strategy.

Hirst and Gazzaniga (in Gazzaniga, 1988) express the hope that PDP models of memory architecture could bridge the gap between neuroscience and psychological research, and thus "unite mind and brain by modeling psychological function in physiological terms". Philosophical objections aside, we are still a long way from fulfilling such a prophecy. However, the concept of a modular and distributed memory architecture made up of many interacting subsystems seems well founded and entrenched, to the extent that it is already used to model many computational vision systems.

8.4.2. *Structural Learning*

We are all aware of the ubiquitousness of learning in enhancing computational vision capabilities, but relatively little effort has gone into actually using learning in vision systems. Active perception and mobility within a dynamic world are predicated upon the availability of learning abilities. Before the renewed interest in neural networks research, much of the learning was restricted to structural and statistical methods. (Statistical methods are conceptually similar to neural network learning.) Structural learning, the topic of this section, is predicated on the availability of symbols, and it includes symbolic (logic) and syntactical methods. Syntactical methods attempt to do categorical perception with respect to either objects or textures, and they do that by using formal language theory.

As discussed in Chapter 1, categorical perception corresponds to identifying corresponding grammars, possibly of the error-correcting type, and then parsing the inputs in order to check for their acceptance by a specific class grammar. The research on syntactical learning stalled in the early 1980's largely because of the difficulty in identifying an appropriate vocabulary of primitives and of learning a grammar suitable for more than just trivial examples. Structural learning tried to overcome these drawbacks by suggesting AI representations and reasoning mechanisms using symbolic primitives embedded in semantic networks. The generality provided by such computational mechanisms (of representation and processing) did not make much headway either and still begs the question from where do the symbols come. We briefly review some major structural methods and conclude with a discussion on the appropriateness of symbolic methods for learning.

Winston (1975) considered the problem of learning structural descriptions from examples. According to Piaget, human cognition is a slow and gradual (staged) process that allows an organism to adapt to a complex environment. A good training sequence of (positive) examples and near misses (i.e., characteristic negative examples) of the concept to be learned should allow for the gradual adaptation of a visual machine. Concept learning relies heavily on good training sequences and belongs to supervised learning. The underlying representation used by Winston is that of semantic networks because of their inheritance capabilities. The property of inheritance, implemented through the use of links of the IS-A and A-KIND-OF type, is very important, since one thus avoids redundant representations and can achieve generalization. The procedure starts with preliminary scene analysis leading to region labeling and describing the scene using semantic networks (SN). The network is made up of nodes (entities) corresponding to scene regions, their conceptual assignment is via IS-A or A-KIND-OF links and there are connecting relationships like SUP-PORTED-BY, IN-FRONT-OF, LEFT-OF, etc. Once the SN for the concept to be learned has been developed, it undergoes a cycle of modifications until it accurately describes the concept under consideration. Modifications have to be done for two reasons. First, one needs to generalize the concept description in order to include other instances of the concept. The second reason restricts the description in order to exclude near misses. A near miss is an object that is not an instance of the concept, but it is very similar to such instances, differing in only a small number of significant characteristics. The cycle of modifications starts by performing scene comparisons. The SN of the concept has to be compared with incoming instances of both the positive and negative (near miss) type. The process that decides which nodes have the same function in the net is called matching and results in the establishment of links between pairs of corresponding nodes in the two networks being compared. The corresponding nodes are merged and yield a new net called a skeleton network, supplemented by comparison notes (C-NOTES), which clarify the similarities and/or differences between the two nets. Such notes are analyzed using the inheritance characteristics of the SN and might be replaced if a generalization is feasible. (As an example, if one node IS-A brick and another one IS-A wedge, merging will occur and will point to that object that is the generalization of both brick and wedge.) The cycle ends by deriving a new SN model and is repeated with a new scene description until the teacher deems learning performance as appropriate.

The availability of new sensors and the interest in active perception and interpretation of real-world data led to recent research in 2D (Connell and Brady, 1987) and 3D (Walker *et al.*, 1988) image representations and learning. The image representations are again of the semantic network (SN) type, and they combine both symbolic and numerical range values information. Connell and Brady reason and learn about their 2D world using analogy (Winston, 1980). There is much similarity between their AI approach and the PDP methodology concerning the major issues to be handled, such as those of redundancy, plasticity, and visual hierarchies. The concepts of distance and novelty are crucial for perceptual categorization. Generalization in AI is usually the result of induction using different "drop" heuristics (Michalski, 1983). Connell and Brady show that a new approach that they call *ablation* could enhance the generalization process. Specifically, if two objects belong to the same class, the difference between their representations must be irrelevant and is thus deleted. To prevent over generalization and still allow for novelty to be acknowledged, there is a provision for learning disjunctions when the differences are thought "significant" according to some metric. As new training data comes in, however, disjunctions can be merged into one class, if the incoming data can smooth over the differences observed earlier and conceptually fill in the gap between the disjunctions. The same approach can lead to the formation of hierarchical structures, and it also suggests that training should proceed in small steps. Another interesting observation is that form (representations) usually follows function, and that as a consequence (task) functionality could be of much help in learning.

Inductive learning (Michalski, 1983) amounts to knowledge acquisition using generalization such that the resulting descriptions are appropriate to positive examples, but not to near misses. While deductions and syllogism are always true, induction can fail. Note that induction goes beyond the given evidence and somehow attempts to make guesses, while deduction merely reformulates the evidence. If all of the positive examples cannot be covered by the same description, disjunctions of partial prototypes are introduced. Generalization using logic can easily lead to combinatorial explosion. Metaknowledge in the form of seed (clause) selection based on statistical evidence can indicate what is most typical of those positive examples, and sequentially consider only those logical characteristics most likely to yield the generalization sought after. Annotated predicate calculus of the inductive type, using first-order-logic (FOL) or frame representations (implemented via object orient programming), where descriptions activate routines to derive information (if and when triggered by demons) is employed to generalize and generate models. The seed selection mentioned leads to the two-tiered concept representations suggested by Michalski (1989).

In the traditional representation ("one-tiered"), any concept is defined by specifying basic features that cover all instances of the concept. It is often assumed that these instances can be described by a single conjunct. In a more general approach, a description consists of several conjuncts linked by disjunction. Each such conjunct contributes to the accuracy of the description, depending on how many examples it covers or explains. If these conjuncts are put in the order of decreasing coverage of examples, consecutive terms contribute a decreasing amount of the total value of the function. A concept description that includes rare or exceptional events will typically have a number of conjuncts that cover only a small number of events. A complete description may, therefore, be overly complex, difficult to comprehend, and costly. To deal with this and related problems, Michalski proposes a two-tiered concept representation. The complete concept description is split into two-parts: The basic concept representation (BCR) and the inferential concept interpretation (ICI). The BCR defines the concept simply and explicitly by characterizing the typical or ideal concept cases, either in terms of attributes observed in the examples or in terms that are constructively learned during concept formation. The prototypical instances of the concept can, therefore, be classified by simple matching with the BCR. Anomalies, exceptions, and context-dependent cases are handled by the ICI, which involves a reasoning process. The ICI deals with exceptions by inferring that they are instances of the concept (concept *extending*), or that they ought to be excluded from the description in the BCR (concept *shrinking*). The ICI uses production rules that may be deductively chained. A simple form of ICI is to define a certain similarity (or distance) measure to classify examples that are similar to those covered by the BCR which, can be implemented using flexible matching.

The idea of two-tiered representation can be illustrated with the concept of *chair* (Michalski, 1989). The *Random House* dictionary gives the following definitions:

1. A seat, esp. for one person, usually having four legs for support and a rest for the back and often having rests for the arms. 2. a seat of office or authority. 3. a position of authority, as of a judge, professor, etc. 4. the person occupying a seat of office, esp. the chairman of a meeting. 5. see electric chair. (...)

The description indicates several meanings, but does not tell when each meaning is applicable. It makes no distinction between the typical meaning and context-dependent meaning. It is rather hard to comprehend, and it is incomplete. A two-tiered representation of the chair concept could have the following form:

BCR:
A piece of furniture typically used for sitting by one person. Usually consists of a seat, four legs, and a backrest. (A picture of a typical chair or a description of the relationship among the parts may be included.)

ICI:
The number of legs may vary from one to four
the shape, the size, the color, and the material of all components can
  vary as long as the function defined in the BCR is preserved
chair without the backrest → stool rather than chair
chair with arm-rests → chair specializes to armchair
context = museum exhibit → chair is not used for seating any more
context = toys → Dimensions can be much smaller, but other physical
    properties are preserved. Does not serve for sitting by normal
    persons but by correspondingly small dolls.
context = execution → specializes to electric-chair

This simple example illustrates several important features of the two-tiered representation. Typical examples match the BCR; therefore, it is easy to identify them. The ICI involves metaknowledge, e.g., showing which properties in the BCR are crucial and which are not; and context-dependent knowledge, showing how the properties change in different contexts. In general, contexts can be hierarchically organized, and the ICI inference rules may chain.

The quality of the two-tiered representation is higher than the quality of the dictionary definition, if used in an AI system. First, the accuracy is improved, since the two-tiered description is more complete and consistency has not changed. Second, comprehensibility is somewhat greater, because the prototypical properties of the chair are separated from its possible modifications and specializations. The same two-tiered representation suggests that structural learning needs eventually to cope with the impreciseness and context-dependent meaning of concepts.

The two-tiered representation seems closely related to the philosophy of essentialism and the views expounded by Hilary Putnam. Essentialism would define normal membership (i.e., the BCR) for a given conceptual class in terms of a constant meaning, where the constancy of the meaning is related to some essence. How far the ICI may then deviate from the norm and still belong to the concept is relegated to an arbitrary decision. The crucial question, "But how is this essence supposed to be determined" cannot be answered, and Ayer (1984) judges "there to be more loss than profit in any attendant talk of essence or necessity or possible worlds."

Symbolic learning has developed two tracks—empirical (or similarity-based) and explanation-based learning (EBL). Empirical learning, which tacitly assumes that learning can proceed without a priori knowledge of the domain under investigation, belongs to the inference type of perceptual theories discussed in Chapter 7. EBL, an analytic learning method, "attempts to formulate a generalization after observing only a single example, but in contrast to empirical learning techniques, requires that a learning system be provided with a great deal of domain knowledge at the outset" (Ellman, 1989). EBL is thus rationalistic and should enjoy definite advantages over pure empiricist methods.

Searl (1984) deals at length with the appropriateness of symbolic manipulation and along the way makes a strong critique of AI's claims at duplicating mental functions. The presentation is centered around the "Chinese room argument." Try to imagine an English speaker who is locked in a room with a rule book such as (production rules-based systems) for manipulating Chinese symbols for machine translation. In principle, the monolingual English speaker could pass the Turing test for understanding Chinese, i.e., the translation will be undistinguishable from that of a native Chinese speaker, because he will produce correct Chinese symbols in response to Chinese questions.

However, it is clear to all that the English speaker does not know what any of the symbols mean, and the analogy to a functionally similar computer program is straightforward. The computer program can indeed manipulate symbols, but it does so without any reference to their meaning or interpretation, except that given from the outside by the programmer. Compare such behavior with that of a human where there are mental contents made up of thoughts, motivations, and emotions (see also Mishkin and Appenzeller (1987) for human memory organization). The computer program does possess abstract sets of rules for symbol manipulation, but those rules lack any connection with a physical medium, biological or otherwise. Syntax is not semantics and is much less than what we have defined earlier as episodic memory. Searl concludes by suggesting that symbolic reasoning amounts to dualist philosophy by denying the biological characteristics of mental phenomena. If symbolic reasoning is deficient, can one make any alternative suggestion, and what role if any could episodic mental imagery play?

There are even more prosaic reasons to doubt symbolic reasoning as the foundation of visual perception. What is the right level of abstraction of the symbols being used, and where do they come from? Structural reasoning assumes that symbols are already given, that there is no impreciseness, and that the domain of application is well defined and usually that of a toy (block) world. The positive and near-miss examples considered earlier belong to an idealized world where there is very little noise, geometric distortion, or occlusion, and where the designer is already operating at the high-level end of processing.

Symbolic AI seems to be grounded in logical positivism where the meaning of a proposition is identical with its empirical verification. Analytic philosophy, as espoused by Wittgenstein, has been suggested to overcome the limitations embedded in logical positivism, including the verifiability principle carried out through logical analysis. Analytical philosophy assumes that the meaning of words comes from their use in a "language game." There is no way to find a single meaning, and it is futile to find a Platonic essence. (But see intrinsic image characteristics and their "pure" essence as Spinoza would have defined them.) To reach the full meaning, the words or symbols would have to be referred to through many indexed cross references, and would have to allow for varying degrees of relationships. Distributed computation and multi-system integration seems to be, at least for now, the only feasible alternative, even if the present neural networks implementations might have very little to do with it.

Von Neumann (1958) also made a strong case that the language of the brain is not the language of mathematics. Both mathematics and logic can be thought of as languages used only as forms of expression. When we use them, however "we may be discussing a secondary language built on the primary language truly used by the central nervous system. Thus, the outward forms of our mathematics are not absolutely relevant from the point of view of evaluating what the mathematical or logical language truly used by the central nervous system is. However, whatever the system is, it cannot fail to differ considerably from what we consciously and explicitly consider as mathematics". There is a mathematical language of the brain, but we have yet to find it.

8.4.3. Perceptrons

Categorical perception is a major learning issue related to the development of appropriate visual taxonomies needed during image retrieval. Adaptive pattern recognition (APR) was suggested in the early 1960s for the design of pattern classifiers (PC). The existence of suitable PC is predicated on good clustering characteristics of the object classes that have to be discriminated. In reality, however, clusters usually overlap, due to noise or insufficient measurements, and they might not even be conceptually separable. We seek in this section PCs made up of an input layer and an output layer and label them as single layer networks (SLN) because there is just one layer beyond the input layer. We are interested to see how such SLN can be designed, how they learn to do categorical perception, and what their limitations are.

The Perceptron (Rosenblatt, 1961) is an example of a SLN where the input is $\mathbf{x} = (x_i)_{i=1}^n$, and the output is one "neuron" whose response is given by

$$y = \sum_{i=0}^n w_i x_i = \mathbf{w}^T \mathbf{x}, \tag{8-9}$$

where w_i are the synaptic strengths connecting input unit i to the output unit, w_o is acting like a threshold, and $x_o = 1$. If $y \geq 0$, the pattern sensed is assigned to class c_1; if $y < 0$, it belongs to c_0. Categorical perception for the multiclass case ($m > 2$) can result if the output layer is made up of m units, responses $(y_j)_{j=1}^m$ are calculated, and the pattern is assigned to category c_j if $y_j > y_l \; \forall j \neq l$. The Perceptron implements a linear decision function that for $m = 2$ is $\mathbf{w}^T \mathbf{x} \geq 0$. The only nonlinearities allowed by SLN are those of the input, which still leave the Perceptron linear in an augmented space.

The learning algorithm for the Perceptron is iterative and consists of supervised training on both binary inputs and outputs. If one assumes that $m = 2$, and given known training sets belonging to class c_0 and c_1, c_1, choose an initial weight vector $\mathbf{w}(0)$. Learning amounts to deriving the synaptic weight vector $\mathbf{w}$ such that it can perform the two class discrimination task. The iterative learning algorithm adjusts the weights at each cycle (t) as given next

(a) If $\mathbf{x}(t) \in c_1$, $\mathbf{w}^T(t)\mathbf{x}(t) < 0$, and ε is the magnitude of correction,

$$\mathbf{w}(t + 1) = \mathbf{w}(t) + \varepsilon \mathbf{x}(t). \tag{8-10a}$$

(b) If $\mathbf{x}(t) \in c_0$, $\mathbf{w}^T(t)\mathbf{x}(t) \geq 0$,

$$\mathbf{w}(t + 1) = \mathbf{w}(t) - \varepsilon \mathbf{x}(t). \tag{8-10b}$$

(c) If no classification error occurred, there is no change in the weight vector, i.e.,

$$\mathbf{w}(t + 1) = \mathbf{w}(t). \tag{8-10c}$$

The algorithm is of the reward and punishment type, and it can be shown that it eventually converges, i.e., it finds a weight vector $\mathbf{w}$ that categorizes all pattern inputs correctly.

We recall from Chapter 4 that gradient descent techniques can be used to find the minimum of a given function. The negative of the gradient vector points in the direction of maximum decrease of the function, as its argument increases. Assume the function $J(\mathbf{w}, \mathbf{x}) = (|\mathbf{w}^T\mathbf{x}| - \mathbf{w}^T\mathbf{x})$ whose minimum is $J(\mathbf{w}, \mathbf{x}) = 0$ when $\mathbf{w}^T\mathbf{x} \geq 0$. Finding the minimum of $J(\mathbf{w}, \mathbf{x})$ amounts then to learning the synaptic weight vector $\mathbf{w}$, assuming that the patterns $\mathbf{x}$ are known and are given as the training sequence. The gradient descent technique finds the minimum of $J(\mathbf{w}, \mathbf{x})$ by incrementing $\mathbf{w}$ in the direction of the negative gradient, and it is given as

$$\mathbf{w}(t + 1) = \mathbf{w}(t) - \varepsilon \left\{ \frac{\partial J(\mathbf{w}, \mathbf{x})}{\partial \mathbf{w}} \right\}_{\mathbf{w} = \mathbf{w}(t)}. \tag{8-11}$$

If the patterns are linearly separable, one can derive the reward and punishment algorithm (Eq. 8-10) corresponding to the Perceptron by choosing $J(\mathbf{w}, \mathbf{x})$ as

$$J(\mathbf{w}, \mathbf{x}) = \tfrac{1}{2}(|\mathbf{w}^T\mathbf{x}| - \mathbf{w}^T\mathbf{x}) \qquad (8\text{-}12)$$

and substitute $\partial J/\partial \mathbf{w} = \tfrac{1}{2}[\mathbf{x}\,\text{sign}(\mathbf{w}^T\mathbf{x}) - \mathbf{x}]$ in Eq. (8-11).

The Perceptron is conceptually similar to ADALINE (adaptive threshold element) devised by Widrow in the early 1960s. Assuming binary elements, the Perceptron Eq. (8-10) can be rewritten as

$$\mathbf{w}(t+1) = \mathbf{w}(t) + \varepsilon[c(t) - c]\mathbf{x}(t), \qquad (8\text{-}13)$$

where $c(t)$ is the true class identity for $\mathbf{x}(t)$, and c is the class response provided by the Perceptron. One can readily observe that Eq. (8-13) yields Eq. (8-10a,b) for

(a) $\mathbf{x}(t) = c(t) = 1$ but $c = 0$ (i.e., $c = c_0$)
(b) $\mathbf{x}(t) = c(t) = 0$ but $c = 1$ (i.e., $c = c_1$)

ADALINE's adaptive behavior, known also as the Widrow–Hoff algorithm, is based on least-mean square (LMS) error-minimization, which is an iterative approximation of MSE. Assume that one seeks to get the class identity c as close as possible to its true value, i.e.,

$$c = \mathbf{w}^T\mathbf{x} + e \qquad (8\text{-}14a)$$

and that to minimize the error e amounts to minimizing the functional $J(\mathbf{w}, \mathbf{x})$ given as

$$J(\mathbf{w}, \mathbf{x}) = E[(c - \mathbf{w}^T\mathbf{x})^2], \qquad (8\text{-}14b)$$

where E is the expectation value. The weight vector $\mathbf{w}$, which results from iterative adaptive behavior, is given as

$$\mathbf{w}(t+1) = \mathbf{w}(t) - G(t)\nabla_m J|_{\mathbf{w} = \mathbf{w}(t)}, \qquad (8\text{-}14c)$$

where $\mathbf{w}(0)$ is some initial estimate, $G(t)$ is a gain matrix positive definite like $G(t) = \varepsilon(t)I$ for $\varepsilon(t) = 1/t$. The LMS solution is then

$$\mathbf{w}(t+1) = \mathbf{w}(t) + \varepsilon(t)[c(t) - \mathbf{w}^T(t)\mathbf{x}(t)]\mathbf{x}(t), \qquad (8\text{-}15)$$

Note the similarity between the Perceptron (binary) solution (Eq. 8-13) and the Adaline (continuous) solution (Eq. 8-15), and that both solutions are particular instantiations of gradient descent methods.

The single-layer networks research came to a halt in the late 1960s.

Note that a binary Perceptron made up of n binary inputs and one binary output could potentially implement 2^{2^n} logical functions corresponding to the 2^n input patterns. However, only a subset of those functions are realizable and correspond to linearly separable inputs. Minsky and Papert (1987) showed back in 1969 that "simple" functions such as exclusive-Or (XOR) or parity are not realizable, and that features such as connectiveness are also beyond the power of the "simple-minded" Perceptrons. A curtain fell over much of neural network research and stayed that way until the early 1980s. The nonlinear MLN and back propagation, the topic of next section, were the "adaptive response" of neural networks research to the failed Perceptron and the linear SLN.

8.4.4. Back Propagation

Multi-layer networks (MLN) are feed-forward networks, made up of an input layer, $(N-1)$ "hidden" units layers, and an output layer. MLN are thus made up of N layers, while perceptron-like devices are made up of one layer only, according to the definition. MLNs are equivalent to the connectionist models discussed in Section 4.2.4., and they were suggested as a replacement for the single-layer networks (SLN) due to their theoretical capability to implement any input/output mapping and thus to overcome the limitations characteristics of the Perceptron. As important to the acceptance of MLN was the rediscovery and popularization of a suitable learning algorithm, known as back propagation or delta-rule by Rumelhart et al. (1986). The algorithm was able to handle the credit (reward or punishment) assignment problem, even though it lacks proof of convergence.

The back propagation (BP) algorithm can be generically described as follows. Assume a MLN architecture made up of processing elements (PE) where the synaptic strengths between $PE(i)$ and $PE(j)$ are given as w_{ij}, and where the output of $PE(i)$ is a function of its total input.

Given some MLN with some initial setup of the synaptic strengths, BP seeks in a supervised mode those weights $\{w_{ij}\}$ that map input patterns $\mathbf{x}(k)$ into outputs $\hat{y}(k)$ for $k = 1, \ldots, m$. Learning such a mapping corresponds thus to "conditioning" the MLN to m stimulus-response pairs. There are two passes to the BP algorithm. First the pattern $\mathbf{x}(k) = (x_1(k), \ldots, x_n(k))$ is presented to the input layer and passes through the network yielding at the output layer $y(k)$, usually different from the desired output $\hat{y}(k)$. The error $[y(k) - \hat{y}(k)]$ is used in the second pass, starting at the output layer and back-propagating it all the way to the input layer through the intermediate hidden layers. The credit assignment works by modifying the weights w_{ij} as a result of minimizing the observed errors. A suitable penalty cost function at the output layer could be $E_k(W)$ given as $E_k(W) = \| y(k) - \hat{y}(k) \|^2$, where W is the matrix of synaptic weights. BP consists of many supervised cycles of presenting the input patterns $\mathbf{x}(k)$, observing their output $y(k)$ and adjusting the synaptic weights. The adaptation ceases when the observed errors go below some preestablished threshold θ. As an example, one could require that either $\Sigma_k \, E_k(W)$ or $\max_k \, E_k(W)$ is less than θ, where k spans the set of associations to be learned.

The specific BP algorithm suggested by Rumelhart, Hinton, and Williams (1986) works then as follows. Assume that the input and output at a $PE(j)$ are x_j and y_j, and that they are given as

$$x_j = \sum_i w_{ij} y_i \qquad\qquad (8\text{-}16a)$$

$$y_j = f(x_j) = (1 + e^{-(x_j + \theta_j)}) \qquad\qquad (8\text{-}16b)$$

where the (sigmoid function) thresholds θj are learned as are the synaptic weights. The cost or energy function to be minimized is

$$E = \tfrac{1}{2} \sum_k \sum_j (y_j(k) - \hat{y}_j(k))^2$$

$$= \tfrac{1}{2} \sum_j (y_j - \hat{y}_j)^2, \qquad\qquad (8\text{-}17)$$

where k denotes the class pattern to be learned, and y_j and $\hat{y}_j$ are the actual and desired outputs. The minimization is done individually on each class so the class index k can be dropped. Equations (8-16) and (8-17) constitute the feed-forward pass, and the weights w_{ij} are modified in the second back propagation pass. At the output layer N one can observe that

$$\frac{\partial E}{\partial y_j^{(N)}} = y_j^{(N)} - \hat{y}_j^{(N)} \qquad\qquad (8\text{-}18a)$$

and as a consequence

$$\frac{\partial E}{\partial x_j^{(N)}} = \frac{\partial E}{\partial y_j^{(N)}} \cdot \frac{\partial y_j^{(N)}}{\partial x_j^{(N)}}$$

$$= [y_j^{(N)} - \hat{y}_j^{(N)}](1 - y_j^{(N)}) y_j^{(N)}$$

$$= \delta_j^{(N)}. \tag{8-18b}$$

The change in the cost function E as a result of changing the weights $w_{ij}^{(N-1)}$ between layers $(N-1)$, and N is then obtained as

$$\frac{\partial E}{\partial w_{ij}^{(N-1)}} = \frac{\partial E}{\partial x_j^{(N)}} \cdot \frac{\partial x_j^{(N)}}{\partial w_{ij}^{(N-1)}}$$

$$= \frac{\partial E}{\partial x_j^{(N)}} \cdot y_i^{(N-1)}$$

$$= \delta_j^{(N)} \cdot y_i^{(N-1)}. \tag{8-19}$$

Using the chain rule again

$$\frac{\partial E}{\partial y^{(N-1)}} = \sum_j \frac{\partial E}{\partial x_j^{(N)}} \cdot \frac{\partial x_j^{(N)}}{\partial y_i^{(N-1)}}$$

$$= \sum_j \frac{\partial E}{\partial x_j^{(N)}} w_{ij}^{(N-1)}$$

$$= \sum_j \delta_j^{(N)} w_{ij}^{(N-1)}. \tag{8-20}$$

The change in the energy function E as a result of changing the weights $w_{ij}^{(N-2)}$ can then be obtained from

$$\frac{\partial E}{\partial w_{ij}^{(N-2)}} = \frac{\partial E}{\partial x_j^{(N-1)}} \frac{\partial x_j^{(N-1)}}{\partial w_{ij}^{(N-2)}}$$

$$= \left[\frac{\partial E}{\partial y_j^{(N-1)}} (1 - y_j^{(N-1)}) y_j^{(N-1)} \right] y_i^{(N-2)}$$

$$= \left[(1 - y_j^{(N-1)}) y_j^{(N-1)} \sum_j \delta_j^{(N)} w_{ij}^{(N-1)} \right] y_i^{(N-2)}$$

$$= \delta_j^{(N-1)} y_i^{(N-2)}. \tag{8-21}$$

This process can be repeated for any hidden layer provided the outputs y_i^α ($\alpha = 1, \ldots, N - 1$) are known. The weights are then adjusted at each cycle t as

$$\Delta w_{ij}(t) = -\varepsilon(t) \frac{\partial E}{\partial w_{ij}} (t), \tag{8-22a}$$

which corresponds to the LMS algorithm, or alternatively (for speedup reasons) one can use

$$\Delta w_{ij}(t) = -\varepsilon(t) \frac{\partial E}{\partial w_{ij}} (t) + \gamma(t)\Delta w_{ij}(t - 1), \tag{8-22b}$$

where γ is an exponentially decay factor ($0 < \gamma < 1$), which takes into account earlier gradients as well, and thus can push learning faster if it is consistent. The thresholds θ_j are adjusted according to

$$\Delta\theta_j(t) = -\varepsilon(t)\delta_j + \gamma(t)\Delta\theta_j(t - 1). \tag{8-22c}$$

The MLN can implement any continuous mapping (Kolmogorov, 1957). The mappings can be thought of as yielding decision boundaries in some multidimensional space. While the SLNs yield only hyperplanes as the separating boundaries (and thus failed to solve the XOR problem) three-layer networks can be trained to yield any arbitrary decision boundary and in principle could solve any mapping problem. The level of performance of MLN results from the nonlinearities available within the output nodes. Sigmoidal nonlinearities (Eq. 8-16b) yield decision spaces bounded by smooth curves versus line segments as it is the case with hard limiting nonlinearities of the on/off type (Lippmann, 1987). A basic question on the appropriate MLN architecture remains in terms of the number of hidden layers and units within such layers. Heuristics, based on a priori knowledge regarding the features most likely to influence the mappings, are used in the determination of the hidden layers. The BP requires many cycles before it converges, and the convergence is not known to hold. The BP, such as the LMS algorithms it reduces to, can settle in a local rather than global minima, which is another problem with which to cope. It is interesting to point out that MLN can be trained using an extended Kalman filter (KF) (Singhal and Wu, 1989) and that BP is a degenerate form of the same extended KF (Ruck et al., 1990).

Karp (1988) reviews connectionist models and their computational complexity with regard to specific computational problems. He shows that it is *NP-hard* to determine whether a SLM can learn to correctly classify input patterns from a given set of examples. It is also *NP-hard* to determine whether a general linear threshold network will reach a stable state for a given input, and symmetric networks may take exponential time to reach a stable state.

8.4.5. *Correlation Learning*

Hebb (1949) suggested correlation learning as a possible hypothesis for the synaptic plasticity characteristic of classical conditioning. The learning matrix (Steinbuch, 1961), is an early attempt to implement such a hypothesis by associating stimuli to specific responses. The learning matrix uses the conjunction learning rule which is the analogue of the Hebbian rule of increasing the synaptic efficacy as a result of concurrent neural activity. The learning is characteristic of correlation matrix approximations (of DAMs), known also as associative mappings and is the topic of this section.

Anderson *et al.* (1977) consider the issue of correlation learning within the context of categorical perception. Assume first a simple association model between activity patterns s_k and r_k, such that the m stimulus patterns are orthogonal, i.e., their inner product $(s_i, s_j) = s_i^T s_j = \delta_{ij}$. The memory buildup is given as

$$M = \sum_{k-1}^{m} r_k s_k^T, \tag{8-23a}$$

and recall corresponding to an "unknown" activity pattern s_j is

$$M s_j = \sum_k r_k s_k^T s_j = r_j. \tag{8-23b}$$

If however, the stimulus activity patterns are not orthonormal, i.e., $s_i^T s_j \neq \delta_{ij}$, then the recall includes in addition to the response r_j crosstalk noise. A major contribution made by Anderson *et al.* (1977) is the demonstrated relationship between correlation learning and factor analysis. As a consequence, correlation learning is suitable to model-known psychological phenomena such as probability learning and thus can predict overshooting and recency data quite accurately. To see the statistical equivalence assume that each activity pattern is presented

k_i times, for $i = 1, \ldots, m$. It follows that the memory M is given as

$$M = \sum_{i=1}^{m} k_i s_i s_i^T, \tag{8-24a}$$

if the autoassociative model is considered, i.e., if $s_i = r_i$. The recall is then obtained as

$$M s_j = k_j s_j, \tag{8-24b}$$

i.e., s_j is an eigenvector of M, and the eigenvalues correspond to the frequency of presentation. Further assume that the activity patterns are normalized so that the average input $\Sigma\, k_i s_i / \Sigma\, k_i = 0$. The activity patterns s_i can be thought of as instances of a random variable S such that COV(S), the covariance of S, is given as

$$\text{COV}(S) = \sum p_i s_i s_i^T - [\sum p_i s_i]^T [\sum p_i s_i]$$
$$= \sum p_i s_i s_i^T, \tag{8-25}$$

where we used the fact that the average input is zero, and that p_i is the frequency of occurrence for each activity pattern and is given as $p_i = k_i / \Sigma\, k_i$. If we look at Eq. (8-24), we observe that the memory M is proportional to the covariance matrix COV(S) as given by Eq. (8-25), is positive semidefinite, and its eigenvalues are greater or equal to zero. It can be further shown that any input is a linear combination of the eigenvectors of the memory M, and that the eigenvectors, if ranked according to their corresponding eigenvalues, orderly account for the differences between inputs as their rank will indicate. The last question addressed by Anderson *et al.* (1977) was related to the significance of feedback on a system as described. It turned out that the eigenvalues act as gain control, and that the high-ranked eigenvectors grow faster than the lower ranked and could be conceptually thought of as the equivalent of "features." Assume that the activity at time t is given as $a(t)$ and that as a result of feedback the activity at time $(t + 1)$ is

$$a(t + 1) = a(t) + M a(t)$$
$$= (I + M) a(t), \tag{8-26a}$$

where the memory M is given in terms of an orthonormal (eigenvector) base $\{e_i\}$ as

$$M = \sum \lambda_i e_i e_i^T. \tag{8-26b}$$

λ_i are the corresponding eigenvalues, and the activity $a(t)$ can be expressed as

$$a(t) = \sum a_i e_i. \tag{8-26c}$$

It follows then that

$$a(t + 1) = (I + M)a(t)$$
$$= [\sum (1 + \lambda_i)e_i e_i^T] \sum a_j e_j$$
$$= \sum a_i(1 + \lambda_i)e_i, \tag{8-27a}$$

and as a consequence, $\|a(t + 1)\|^2 \geq \|a(t)\|^2$. Activity then grows without limit and thus needs to be confined to some maximum firing rate, say A. The containment effect leads to the model being referred to as the "brain-state-in-a-box" (BSB) model and can be easily implemented as

$$a(t + 1) = \max[-A, \min\{A, (I + M)a(t)\}]. \tag{8-27b}$$

For the saturation model, the final state is always one corner $A(\pm 1, \pm 1, \ldots, \pm 1)$ of the 2^N possible ones.

The CAN model we have used for optimization problems in Section 4.1.4. is also applicable to correlation learning because of the way it sets the synaptic weights. Assume m classes and that the examplars x^k are binary (± 1) and consist of n units. The connection memory matrix M is made up of synaptic weights w_{ij}, which are defined as

$$w_{ij} = \begin{cases} \sum_{k=1}^{m} x_i^k x_j^k & i \neq j \\ \\ 0 & \text{otherwise.} \end{cases} \tag{8-28a}$$

Hopfield (1982), like Hebb (1949) and Anderson et al. (1977) before him assumes that the examplars are almost orthogonal. The energy function E at time t is $E(t)$ and is defined as

$$E(t) = -\frac{1}{2} \sum_{i,j}^{n} w_{ij} x_i(t) x_j(t) \tag{8-28b}$$

with the property that $\Delta E = [E(t + 1) - E(t)]$ is a monotonically decreasing function. The CAN model, which is described by an Ising spin physical system and is defined by Eq. (8-28), will eventually settle in a minimum energy state. The steady minima energy states correspond to the stored states, and they create local attractors emulating a content-addressable memory (CAM). To see that the stored states indeed correspond to the minima, imagine that any input is a n-dimensional point in space and that there is a vector pointing to it. The length of the vector is always greater than that corresponding to the corners of the n-cube. Another way to see that the stored states correspond to minimal energy states is to consider the recall stage. If $y_i(t)$ is the output of node i at time t and x_i is the ith element of the input pattern then the outputs are first initialized as

$$y_i(0) = x_i \qquad 1 \le i \le n, \tag{8-29a}$$

and the system then iterates until it converges using the update equation

$$y_j(t + 1) = f_H\left[\sum_{i=1}^{n} w_{ij} y_i(t)\right], \tag{8-29b}$$

where f_H is a (sigmoid) hard limiting nonlinearity function. This process continues until the outputs remain unchanged with further iterations. The node outputs then represents the pattern that best matches the unknown examplar. If the patterns stored are almost orthogonal, the recall sifts out only the original associated pattern; otherwise, it sifts out crosstalk noise as well. (To see this, substitute the synaptic weights as given by Eq. (8-28a) into the recall Eq. (8-29b) and observe that minima energy corresponds to the perfect recall.)

   The number of patterns N that can be stored and accurately recalled has been shown by Hopfield (1982) to be bound by $0.14n$ if an error rate of 5 % is accepted. For perfect recall $N \le n/(4 \log n)$ (Abu–Mostafa and St. Jacques, 1985), where n is the number of nodes in the net. Spurious attractors corresponding to the synthesis of new metastable states may result if the bound is exceeded and thus lead to erroneous recalls. As the number of stored states exceeds the bound, the CAN model enters a chaotic stage where the number of synthesized metastable states increases at an exponential rate and dominates the network's behavior.

   An alternative to the CAN, which enjoys the benefit of being the optimum minimum error classifier when bit errors are random and independent, is the Hamming net. The Hamming net is made up of two subnets. The first subnet [Eq. (8-30)] calculates the matching score between the input and the m examplar patterns as n less than the Hamming distance (i.e., the number of mismatches between input and examplar bits). The upper net [Eq. (8-30b)], called MAXNET, is made up of m nodes, which inhibit each other with weights ε, and in a similar way to the Kohonen self-organization topological maps [see Eqs. (8-37)] yields the maximum among the matching scores. Formally, the weights in the lower and upper net are assigned as

$$w^1_{ij} = \frac{1}{2} x^j_i; \qquad \theta_j = \frac{n}{2} (1 \leq i \leq n; 1 \leq j \leq m) \qquad (8\text{-}30a)$$

$$w_{ij} = \begin{cases} 1 & i = j & (1 \leq i, j \leq m) \\ -\varepsilon & i \neq j, & \varepsilon < \dfrac{1}{m}. \end{cases} \qquad (8\text{-}30b)$$

The lower net is initialized with the unknown pattern and the outputs $y_j(0)$ are given as

$$y_j(0) = f_H\left(\theta_j + \sum_1^n w^1_{ij} x_i \right). \qquad (8\text{-}31a)$$

The upper net then iterates according to

$$y_j(t+1) = f_H\left(y_j(t) - \varepsilon \sum_{i \neq j} y_i(t) \right) \qquad (8\text{-}31b)$$

until convergence is achieved (always the case for $\varepsilon < 1/m$), and the maximum node corresponding to the "winner-take-all" class is found. Beyond being the optimum classifier, the Hamming net also enjoys additional advantages when compared with the CAN. The memory capacity N is such that $N \gg n$ (compare that against Hopfield's $N < 0.14n$) and from the construction of the net it is obvious that much less connections are needed. The CAN requires n^2 connections, while the Hamming net requires only $(n + m - 1)m$ connections, where n is the size of the input and m is the number of classes. Hence, the hardware efficiency is much improved for the Hamming net.

The distributed associative memory (DAM) (Kohonen, 1987), one of the best-known approaches to implement classical conditioning, was introduced in Section 4.2.1. and used extensively for 2D and 3D object recognition as detailed in Sections 7.2.3. and 7.5., respectively. The mapping to be learned corresponds to ⟨stimulus → response⟩ pairs given as ⟨$\mathbf{s}(k) - \mathbf{r}(k)$⟩, where m is the number of associations. If the dimensionality of $\mathbf{s}$ and $\mathbf{r}$ is n_1 and n_2, respectively, then stimulus and response matrices $S(n_1 \times m)$ and $R(n_2 \times m)$ can be defined for $k = 1, \ldots, m$. Learning the mapping amounts to finding a memory matrix M such that $R = MS$. The solution involves minimizing the norm $\| MS - R \|^2$ and yields

$$M = RS^+, \tag{8-32a}$$

where S^+ is the (generalized inverse) pseudoinverse matrix of S, while the recall is given as

$$\mathbf{r} = M\mathbf{s}, \tag{8-32b}$$

where $\mathbf{s}$ is an unknown stimulus pattern. The case when $R \neq S$ is called heteroassociative, while the associative case corresponds to $R = S$. Note that if the stimulus vectors are independent, the memory solution M is given as

$$M = R(S^T S)^{-1} S^T, \tag{8-32c}$$

which reduces to

$$M = RS^T \tag{8-32d}$$

for orthonormal stimulus vectors.

The memory models defined by Eqs. (8-32a) and (8-32d) correspond to the generalized inverse memory (GIM) and outer product learning (OPL), respectively. OPL, also referred to as correlation learning [see Eq. (8-23a)], enjoys better performance than GIM for a large sample of the perceptual learning tasks. Assume that n is the input space dimensionality and m is the number of stored associations. When m approaches n, the stimuli vectors $\mathbf{s}_i$ are likely to become dependent. Hence, even small amounts of input noise can cause the GIM approach to misclassify. To see that this is indeed the case, imagine the geometry of the 3D space corresponding to $n = 3$. Error-free classification is performed, as it should be, in 3D. But with small input errors, classification is likely to be performed in a lower dimensionality subspace, i.e., 2D, and this leads to misclassifications. In contrast, OPL classifiers seek the maximum inner product between the input and each of the prestored stimuli vectors. The OPL, which performs angular classification, is thus likely to outperform the traditional DAM implementations based on GIM and linear decomposition, and to "saturate" less.

Correlation learning is general, and DAM learning in particular, is closely related to statistical analysis. Kohonen (1987) notes that the memory solution, given above in Eq. (8-32c), corresponds to linear regression. Furthermore, linear regression is related on its own to the best linear unbiased estimator. If a linear statistical dependence between stimulus and response vectors **s** and **r** can be established, recall amounts to estimation, and the LMS solution is given as

$$M = \sum \text{COV}(r, s)\text{COV}(s, s), \tag{8-33}$$

where $\text{COV}(r,s)$ and $\text{COV}(s,s)$ are the covariance matrices corresponding to the $\langle \mathbf{r}, \mathbf{s}^T \rangle$ and $\langle \mathbf{s}, \mathbf{s}^T \rangle$ vector combinations.

The recall given by Eq. (8-32b) can be thought of as being made up of two components $\hat{r}$ and $\tilde{r}$ such that

$$\mathbf{r} = \hat{r} + \tilde{r}, \tag{8-34}$$

where $\hat{r} = RS^+ \mathbf{s}$ is the projection of **s** on the subspace L spanned by the linear associations already learned and stored by M, while $\tilde{r}$, the novelty of the attempted recall is the projection of **s** on the subspace $L^\perp$ orthogonal to L. The novelty $\tilde{r}$ is crucial to any learning system in deciding if the stimulus is characteristic of a new class or not, and if the system as a consequence should build up a new category or not. More than just simple metrics are involved in this decision, and evaluation of past learning performance and utility criteria probably are most influential.

The learning memory M can be obtained using iterative solutions as well. Such incremental solutions benefit from their ability to add/delete associations without recalculating the generalized inverse (S^+) from scratch, an expensive proposition. (Note that back propagation is not incremental, and that the weights in the network have to be recalculated each time an addition/deletion is requested.) The memory solutions shown previously are the exact ones, if they exist, and they can be assessed according to properties like numerical stability (Mohideen *et al.*, 1989). The iterative solution characteristic of LMS instantiations is given as

$$M(t + 1) = M(t) - \varepsilon(t)[M(t)\mathbf{s}(t + 1) - \mathbf{r}(t + 1)]\mathbf{s}(t + 1), \tag{8-35}$$

where $\varepsilon(t) \to 0$ for $t \to \infty$.

Drive-reinforcement theory (DRT) (Klopf, 1987), belongs to the class of real-time learning mechanisms, the best known being that associated with Hebb (1949). Reinforcement learning is characteristic of those cases which can be described by two acts: $x \Rightarrow a$ and $(x,a) \Rightarrow U$, where x is some input, a is possibly some motor control act, and U is the utility of performing the mapping (x,a). Clearly, U, the utility (motivation) is a parameter to be designed and then tested for its efficacy during training. DRT is a differential learning mechanism and may be considered an extension of related earlier differential mechanisms such as that of Rescorla and Wagner (1972) and Sutton and Barto (1981). As a real-time mechanism, it is capable of allowing a system to learn in an unsupervised mode. Network activity is given by (remember Eq. (4-1))

$$y_i(t) = \sum_{k=1}^{n} w_{ik}(t)x_k(t) - \theta_i(t),$$

where $y_i(t)$ represents the post-synaptic activity level at time t of the ith processing element (PE) (neuron), $w_{ik}(t)$ is the value of the weight associated with the synapse connecting the ith and kth PE at time t, $x_k(t)$ is the activity level associated at time t with the kth PE, and $\theta_i(t)$ is the threshold value for the ith PE at time t and for simplicity considered to be fixed in time and to have the same value for all PEs in the system. The synaptic weights $w_{ik}(t)$, can be either positive (for excitatory), or negative (for inhibitory) synapses. Because negative activities do not make sense biologically, $y_i(t)$ and $x_k(t)$ are constrained to lie in the open region bounded by zero and some arbitrary maximum value. The assumption of a finite maximum value corresponds to the biological observation that there is a maximum firing frequency of a few hundred Hertz associated with typical biological neurons. The learning mechanism corresponding to DRT is given by the following difference equation:

$$\Delta w_{ik}(t) = \Delta y_i(t) \sum_{j=1}^{\tau} c_j |w_{ik}(t-j)| \Delta x_k(t-j), \qquad (8\text{-}36)$$

where $\Delta w_{ik}(t) = w_{ik}(t+1) - w_{ik}(t)$, $\Delta y_i(t) = y_i(t) - y_i(t-1)$,   and $\Delta x_k(t-j) = x_k(t-j) - x_k(t-j-1)$. The equation states that, in DRT, the change in the efficacy of a synapse on the ith PE during the interval from time t to time $(t+1)$ is the result of the product of the change in the postsynaptic activity level of the ith PE and a weighted sum. Each term of the sum is proportional to the product of the efficacy of that synapse and the change in the activity of the PE feeding that synapse during a preceding time interval indexed by the integer j. The proportionality constant, or learning constant, c_j depends on j. In order to understand the significance of c_j, it will be necessary to further examine the nature of the "index" j and the parameter τ. In a biological context, the index j of the last equation represents the interstimulus interval (ISI) of a conditioning experiment. In the context of training ANS, j represents the sampling interval, that is, the interval between the temporally associated input vectors presented to the system. The parameter τ, then, represents the number of previous sampling intervals that are important in determining the change in synaptic weight w_{ik}. In a biological context, j is found to be about 500 ms and τ about five. In the same context, c_j is a constant that reflects the effectiveness of each time interval in determining the change in efficacy of the relevant synapse. In an ANS, however, j, τ, and c_j all become system parameters that can be varied to optimize the training and operation of a DRT network for visual learning. The ability of DRT to predict the shape of the learning curve found for a number of animal learning experiments is improved if a final constraint is placed on changes in synaptic efficacies. In particular, it is found that only those changes in presynaptic activities which are (a) positive, and (b) relatively rapid, are effective. Again, in the context of training and operation of an ANS utilizing DRT, these rules may or may not lead to improvements.

8.4.6. Self-Organization

Self-organization is characteristic of unsupervised learning and represents an important stage of adaptive behavior.

Kohonen's (1987) feature map self-organizes and produces an orderly neural arrangement that reflects the characteristics of the sensory input, like the frequency underlying the tonotopic maps found in the auditory pathway. Assume n inputs and m outputs connected by weights w_{ij}, initially set to some small random values. Upon presentation of new input $x_i(t)$ the distance d_j to each output node is calculated as

$$d_j = \sum_{i=1}^{n} (x_i(t) - w_{ij}(t))^2, \tag{8-37a}$$

and the node j^*, which yields the minimum distance, is found as $d_j^* = \min_j d_j$. Define $N_j(t)$ as the neighborhood of node j including the node as well. The update stage is given as

$$w_{ij}(t + 1) = w_{ij}(t) + \varepsilon(t)(x_i(t) - w_{ij}(t)) \tag{8-37b}$$

for $j \in N_{j*}(t)$, $1 \le i \le n$, and the gain factor $\varepsilon(t)$ decreasing with time. An example of gain factor could be a Gaussian whose width decreases with time. The feature map is conceptually similar to the k-mean clustering algorithm, where k corresponds to the neighborhood $N_j(t)$. The map preserves topological order, represents hierarchical relations characteristic of high-dimensional spaces, in the 2D cluster space provided by the m output nodes, and is useful for data compression.

A basic issue underlying self-organization is that of determining what principles, if any, lead to the development of neural circuitry. Linsker (1988) considers multi-layer networks (MLN) with feed-forward connections, linear summation response and Hebbian learning, and derives optimization principles leading to the development of cortex cells. The development is predicated on the preservation of information when signals are transformed at each processing stage. The model could account for the development of a hypercolumnar architecture made up of cells whose sensitivity is with respect to contrast and orientation. The specific Hebbian model allows for positive feedback leading to consensus and can be described as

$$\dot{w}_{ij} = \sum_j Q_{ij} + \left[k_1 + \frac{k_2}{n} \sum w_{ij} \right], \tag{8-38a}$$

where w_{ij} is the synaptic weight from cells x_j to cells y_i upstream along the cortical pathway, and k_1 and k_2 are combinations of the constants a_{1-5} describing the activity of cell y_i at snapshot π as

$$y_i^\pi = a_l + \sum_j x_j^\pi w_{ij} \tag{8-38b}$$

$$(\Delta w_{ij})^\pi = a_2 x_j^\pi y_i^\pi + a_3 x_j^\pi + a_4 y_i^\pi + a_5 \tag{8-38c}$$

and Q_{jl}, the covariance of input cells j and l, is given as

$$Q_{jl} = \langle x_j^\pi - \bar{x}) \times (x_l^\pi - \bar{x}) \rangle, \tag{8-38d}$$

where $\bar{x}$ is the ensemble average of the input at a synapse. The weights w_{ij} lie within the interval $[w^-, w^+]$ to prevent saturation. The model described, if iterated, leads to the development of oriented center-surround cells and also allows for plasticity if the ensemble statistical properties Q_{jl} were to change. The Hebb-type rule exhibits specific optimization properties, such as maximizing the output activity variance $\langle (y_i^\pi - \bar{y})^2 \rangle$ if one rewrites the CAN energy function as

$$E_1 = -\frac{1}{2} \sum_{j,l} Q_{jl} w_{ij} w_{il}, \tag{8-39a}$$

i.e, substitutes a covariance matrix Q_{jl} for the symmetric matrix T of synaptic connections. Define an energy function $E = E_1 + E_2$, where

$$E_2 = -k_1 \sum w_{ij} - \frac{k_2}{2n} \left(\sum w_{ij} \right)^2 \tag{8-39b}$$

such that $-\partial E / \partial w_{ij} = \dot{w}_{ij}$ and note that E_1 can be rewritten as

$$E_1 = -\tfrac{1}{2} \langle (y_i^\pi - \bar{y})^2 \rangle. \tag{8-39c}$$

We observe again the relationship between network models and the covariance matrix characteristic of statistical and factor analysis. Such a relationship can be further extended if one remembers that stochastic analysis, like the optimal Kalman filter, is predicated upon gain factors defined in terms of covariance matrices as well. Variance maximization, subject to the constraint $\Sigma_j w_{ij}^2 = 1$, can be achieved as well if one employs the rule

$$\dot{w}_{ij} \propto y_i^\pi (x_j^\pi - y_i^\pi w_{ij}), \tag{8-40a}$$

where $\propto$ stands for proportionality, $\langle x_j^\pi \rangle = 0$, and

$$y_i^\mu = \sum x_j^\pi w_{ij}, \tag{8-40b}$$

where the expression given by $\Sigma\, x_j^\pi w_{ij}$ is the projection of the input onto some $\mathbf{w}$ such that the variance of y_i^π is maximized. Such behavior, characteristic of principal component analysis (PCA) (Oja, 1982), can identify the most relevant discriminant "features". Finally, Linsker shows that the same Hebbian rules provide for maximum information preservation. Assuming that the cell activity y_i^π can be modeled as

$$y_i^\pi = \sum_j x_j w_{ij} + \gamma^\pi, \tag{8-41a}$$

where y_i is a Gaussian of variance V, γ is Gaussian noise of variance B, and uncorrelated to the input (i.e., $\langle \gamma x_j \rangle = 0$). The information rate R, defined as the amount of information that knowing y_i conveys about x_j, is given as

$$R = \frac{1}{2} \ln \frac{V}{B},$$ (8-41b)

where the ratio (V/B) corresponds to the SNR (signal-to-noise) ratio. R is maximized when the activity cells y_i display maximum variance V. If noise is introduced on each connection w_{ij} then Eq. (8-41a) is rewritten as

$$y_i^\pi = \sum_j (x_j^\pi + \gamma_j^\pi) \, w_{ij},$$ (8-41c)

the information rate R reads

$$R = \frac{1}{2} \ln\left(\frac{V}{B} \sum w_{ij}^2\right),$$ (8-41d)

and R is maximized when the synaptic weights are chosen so as to perform PCA on the signals x_j.

The adaptive resonance theory (ART) (Grossberg, 1988) is another model of unsupervised learning and is conceptually similar to leader clustering algorithms. The net is made up of a lower and upper net and performs both classification and recall. The lower net corresponds to short-term memory (STM) and is made up of nodes x_i $(i = 1, \ldots, n)$. There are output nodes y_j in the upper layer corresponding to m classes $j = 1, \ldots, m$. Like in the Hamming model, bottom-up weights b_{ij} make up the MAXNET network and choose the best match j^*. The best match recalls from long-term memory (LTM) the corresponding examplar pattern which yields top-down weights t_{ij*}. The pattern is matched against the input, and if the fit is above some vigilance factor, the system enters the learning phase. Learning consists of updating the weights t_{ij*} based on the old values, and the actual input, i.e., $t_{ij*}(t + 1) = t_{ij*}(t)x_i$, assuming binary inputs. If the fit is below the vigilance factor, the examplar j^* is suppressed, the next best match is found, and the net operation proceeds as before. Provision is made to include new examplars when no match is found. The model can be sensitive to noise, and its operation depends on the vigilance factor, i.e., the perennial question of deciding what should be construed as novelty and what should not. There are many additional features to enhance the model just described, and a continuous ART model has also been suggested.

The discrete and continuous ART models provide the capability for both classification and examplar recall through the use of several subsystems. As discussed earlier, the memory function is not unitary, but made up of many subsystems as well. There are additional PDP models made up of several subsystems. Counter propagation network (CPN), characteristic of these models, has its first stage made up of Kohonen's feature map operating in the unsupervised mode to achieve classification. (Remember that the weights corresponding to the feature map are trained to resemble in their use a statistical look-up table whose entries are the identity of the input patterns.) Following the self-organization stage, supervised associative learning such as in back propagation or the Grossberg's outstar net, is employed to learn appropriate LTM examplars. The outstar net assumes that the connections are given by w_{ij} and that the input layer is made up of m classes, where j^* was the winner take all of the unsupervised stage. Then, for any unit i at the output, the weights are adjusted as

$$\Delta w_{ij}(t) = -\varepsilon_1(t)w_{ij}(t-1) + \varepsilon_2(t)y_i, \qquad (8\text{-}42)$$

where y_i is the ith unit of the expected output examplar, and $\varepsilon_1(t)$ and $\varepsilon_2(t)$ decrease with time. Such multistage network architectures have the potential to generalize and build hierarchical organization due to the clustering capability of the unsupervised stage.

8.4.7. Chaos

The learning we have described so far is brittle and far from satisfactory. Yet, learning is one of the basic human characteristics, and it is only natural to ask if there are alternative ways to look at the learning problem and to solve it. The *Bible* tells us that there was chaos before God began to create things, and the same chaos is then referred to by the Greeks as some infinite and empty space that predated everything yet to come. In this section we briefly consider chaos, while evolution is discussed in the next section. The treatment is light because very little is known about how the mathematics of chaos and evolution work, but some insights might nevertheless be gained, and lead to advancements in learning and adaptation sometime in the future.

Much of our behavior is predicated upon our ability to derive information from our environment and to measure it. We have already seen that there is a limit to our ability to measure when we considered the joint distributions of space/spatial-frequency like the Wigner distribution. We related such uncertainty to quantum mechanics and the well-known uncertainty principle. We have also mentioned that the observer affects the measurement by the very process of measurement and we have also discussed how interpretation, not only the measurements of the world surrounding us, is affected as well by the schema or cognitive structures we possess. The uncertainty referred to leads to information being made available within some statistical framework. But how about predicting or forecasting phenomena whose behavior is modeled by dynamical systems, starting from some initial conditions? If everything was predictable from the very beginning, this would be a large blow to the very notion of free will. As we are about to see, this is, fortunately, not the case, and it seems that nature took care to allow for change in the form of chaos.

Dynamical systems are modeled in terms of states and the dynamics of state transition. Consider as an example the pendulum. It is a dissipative system, due to friction, which will eventually come to rest. The states are given by $\theta(t)$, the angle at time t between the pendulum and the vertical, and the dynamics are given as

$$\ddot{\theta} + \alpha\dot{\theta} + \sin\theta = A\cos(\omega t), \qquad (8\text{-}43)$$

where α is a damping constant, and the driving torque is of amplitude A and frequency ω. As long as the pendulum is not driven over the top, it will eventually come to rest in what is known as an attractor. (The attractor for a frictionless pendulum is a loop.) If the pendulum is driven over the top, however, depending on the initial conditions, it can switch either left or right and continue its motion.

Two observations are in order at this time. First of all, note the sensitive dependence on initial conditions, which thus makes it impossible to make long-term predictions, because in practice we have only finite accuracy in setting the conditions on which everything is dependent. This observation became known as the butterfly effect because a butterfly flapping its wings in Beijing could potentially affect the weather in Washington, D.C. Any effect, however small it might be, has thus the potential to reach macroscopic proportions. Errors grow too fast and make any prediction impossible. It is not noise, uncertainty (as in quantum mechanics), or any degree of freedom which leads to such unpredictable behavior, but the property of nonlinear systems of separating initially closed time trajectories (the dynamics of state transitions) exponentially fast. This brings us to the second observation, that nonlinearity is necessary but not sufficient for making dynamical systems unpredictable. The unpredictability referred to previously is due to random elements, and the corresponding behavior is known as chaos. Chaotic behavior prevents us from associating cause to effect as we are accustomed. Still, chaos is deterministic, and it can result from rules where chance does not play any role.

Chaotic behavior can be discrete, as in the case of population growth, or continuous, in the case of the pendulum as discussed before. The long-term population growth behavior can lead to unchecked growth or extinction. The attractors characterizing dynamical systems define a basin of attraction, in a similar way to the CAN model. There are two types of attractors: predictable and chaotic. The predictable attractors are few and take the form of fixed points (nearby states remain close as they evolve over time), limit cycles, and/or torus. Random elements lead to the development of chaotic attractors, which seem to be fractals. Informally, chaotic systems mix the state space as if it were dough, i.e., stretching and folding it onto itself. As a result, no causal relationship between past and future can be established. Note that the beating of the heart, if normal, should be chaotic, according to recent medical research!

Skarda and Freeman (1987) studied the olfactory system of the rabbit and suggest that chaotic behavior underlies the neural background activity when recognition is of concern. Chaotic states define basins of attractions for each of the recognition states the olfactory system has become conditioned to. The system can move from one state to another one via bifurcations, and most interestingly, there is a chaotic well that allows the system to escape from known attractors and thus respond appropriately to novelty. The mathematics of chaos are only starting to be understood, but they seem able to provide the organism with a deterministic way in which to incorporate change as needed by evolution. The recent discovery of chaotic attractors challenges reductionist views because complex behavior, such as chaos, can result from simple interactions only, and because behavior cannot necessarily be inferred when causality apparently breaks down. Chaos can, however, provide the organism with significant advantages. The butterfly effect, similar to the genetic algorithms to be discussed in the next section, provides access to novelty. Furthermore, chaos could organize the randomness and impose structure where none would be available otherwise. The neural Darwinism discussed in the next section does the same when it subjects variability to evolutionary control.

8.4.8. *Evolution*

Evolution works by change, and it is a characteristic of closed-loop systems as defined earlier. The intelligent system implements some adaptive strategy that is continuously tested by its environment. As suggested by psychobiology, the organism and the environment mold each other in the ensuing process, and evolution results. Evolution is not necessarily a smooth and continuous process. Punctuated equilibrium sees evolution as a discontinuous process, where revolutionary change might be the norm rather than the exception. Revolutionary development is thus not confined only to science but to the very organism responsible for such change as well. Adaptive behavior of evolutionary type is then much more than just learning known facts. It is about discovering things, it involves seeking extrema rather than merely solving error-correction problems, and it is ultimately responsible for "progress" to occur. If very little is known about learning, much less is understood about how dramatic changes responsible for discovery can be induced within the organism.

Genetic algorithms (Holland, 1975; DeJong, 1988) have been suggested for those cases where an acceptable a priori solution is not known. They are characteristic of the widely known AI paradigm of generate-and-test and are thus similar to simulated annealing. The solution or control space is defined as genetic material the organism can choose from and then measures the performance for selected points within the same space. Throughout the whole process a reproductive plan maintains and updates the genetic pool. The (individual) solutions at a given moment in time are characterized by their performance index. The next generation or control space is generated such that the number of offspring an individual solution is allowed to have is proportional to its performance index. Genetic operators, available to the adaptive system, define what the offspring would look like, and specific instances might include crossover, mutation, and inversion. Think of solutions as made up of segments and segments as made up of genes. Crossover constructs new solutions from existing high performance control points by concatenating some of their segments. Mutation, which is less likely in nature, will modify one or more of the genes making up the solution. Reproductive plans characteristic of genetic algorithms have been applied mostly to optimization problems and are further discussed in Section 9.4. as an application area for shared memory systems such as the butterfly.

Edelman and Reeke (1982) and Edelman (1987) consider learning akin to evolution and appropriately label it as neural Darwinism. They developed a simulation system, along such evolutionary concepts, which is made up of two subsystems, DARWIN and WALLACE. DARWIN, made up of a repertoire of features or their aggregates, can selectively choose those characteristics deemed most successful in recognition tasks. DARWIN achieves group selection by employing overlapping and redundant feature representation on the one side, and by updating the synaptic strengths using appropriate amplification functions on the other side. Limited generalization and geometric invariance can be achieved when WALLACE performs contour tracing across the input array, and thus implements global feature correlation. The two subsystems are fully interconnected and employ parallel networks for their implementation. Neural Darwinism is selective rather than instructive and corresponds to closed-loop systems whose behavior is determined by their performance within the environment. In the case of neural Darwinism, performance is measured by the capability of the system to select those groups of features and their relationships that can eventually accomplish the recognition task.

The intellectual history of evolution makes fascinating reading. Gould (1989) recounts the events related to deciphering the Burgess shale, a limestone quarry (from British Columbia) rich in life's remnants after the Cambrian explosion. Concluding that there is no ladder of evolution, he says "replay the tape a million times and I doubt that anything like *Homo sapiens* would ever evolve again." The punctuated equilibrium eliminates species by "lottery," maybe by chaotic processes, even while evolution, driven by contingency and opportunism, tends toward increasing system complexity.

Evolution is very appealing, both as a theory of brain development and for explaining adaptation. The question that still needs to be answered, however, is the amount of computation and time needed to achieve significant adaptation and development. Early AI attempts have also considered evolution only to abandon it quickly. Perhaps, after all, punctuated equilibrium and chaos are nature's way of accelerating development.

8.5.   Conclusions

We have seen that the latest stages of visual processing are involved
in reasoning and learning. Taken together, those stages correspond to
what we usually label as intelligence, even though it is not at all clear
how "intelligence" is actually implemented. While it is clear that
parallel and distributed computation (PDC) is also employed at this
stage, we still do not know what cognitive structures are actually
involved and their relationships. Many interesting suggestions have
been made, but extensive and convincing results have yet to come in.
Intelligent systems developed so far are quite brittle when facing
novelty not envisioned by the designers of the system, and most of them
require extensive training within narrow domains.

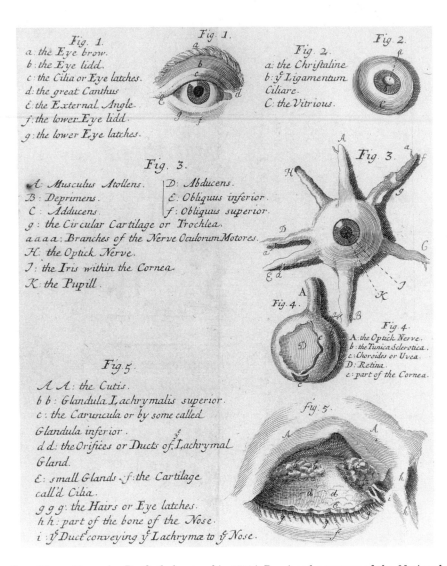

Eye. (From Kennedy, P., *Opthalmographia* 1713.) Reprinted courtesy of the National Library of Medicine.

9

Parallel Hardware Architectures

Look with all your eyes, look.

From *Michael Stroganoff* by Jules Verne

Where there is one there is no power, where there are two the power increases and the enemy does not multiply.

Romanian Proverb

We have emphasized the importance of PDP-type computation in computational vision. After examining the organizations appropriate for visual tasks in terms of representations and algorithms, we conclude by discussing current parallel hardware architectures. Here, we introduce a taxonomy of parallel hardware in terms of the computation model employed, the topology of the network (logical layout of the processing elements (PE) nodes and the communication paths connecting them), and the PE themselves. Additional issues to consider are scaling, granularity, flexibility (or reconfiguration), and balancing loads between the (memory, CPU) and (memory, input/output) pairs. Scaling is understood as the hardware scheme's ability to handle different input sizes. Granularity means considering the tradeoffs between fine versus coarse architectures, i.e., uniformity versus power and specialization. Reconfiguration is that characteristic enabling a

463

given hardware organization to be flexible enough to accommodate different computational tasks. The crucial problem affecting a parallel scheme's effectiveness, however, centers on the communication paths interconnecting the processing elements, which we discuss later. The taxonomy of parallel hardware considers, therefore, vector processing, pipelines and systolic arrays, single instruction multiple data (SIMD), multiple instruction multiple data (MIMD), data flow and functional design, and artificial neural systems (ANS). Good reviews are available in Hwang and Briggs (1984), Reeves (1984), and Uhr (1987).

9.1. Network Topology and Data Communications

Network topology and data communication paths can be discussed according to complexity (and cost), efficiency (versus overhead), and fault tolerance. A logical, rather than physical, viewpoint follows.

The crossbar network, in which every PE connects to every other, is the simplest but most flexible network. There is no competition for communication resources, but its complexity of $O(n^2)$ for n nodes is too great to make it practical except for small input sizes. Another scheme, the ring network, implements data communication on a circular bus. Because only $1/n$ of the bus bandwidth is available to each PE, such a scheme is appropriate only when communication requirements are relatively small as compared against those of processing. The near-neighbor (NN) mesh, as its name implies, is appropriate for low-level visual tasks, where updating pixel values is dependent on the immediate and finite neighborhood. The complexity of such a scheme is clearly of the $O(n)$ type.

The next two schemes, the shuffle exchange and perfect shuffle, are characteristic of applications that implement sorting algorithms and have a complexity of the order $O(n)$. Specifically, assume a graph $G(V, E)$, where the set of vertices V is made up of processing elements PE(i), for $i = 0, 1, \ldots, n - 1$, and $n = 2^m$ for some m. The edges E connect between even numbered processors and the next higher odd PE. Therefore, for i even there is an edge $E(i, i + 1)$ and for i odd the corresponding edge is $E(i, i - 1)$. When $m = 3$, the shuffle exchange connections are $0 \leftrightarrow 1$, $2 \leftrightarrow 3$, $4 \leftrightarrow 5$, and $6 \leftrightarrow 7$. The perfect shuffle is characterized by edges E of the type $E(i, 2i\langle \mod n - 1 \rangle)$ but note that the last edge is $E(n - 1, n - 1)$ rather than $E(n - 1, 0)$. The perfect shuffle PS can be defined as $PS\langle a_{n-1}, a_{n-2} \ldots a_1 a_0 \rangle = \langle a_{n-2} a_{n-3} \ldots a_0 a_{n-1} \rangle$, where $\langle a_{n-1} a_{n-2} \ldots a_1 a_0 \rangle$ is the binary representation of the source PE. When $m = 3$, the PS connections are $0 \leftrightarrow 0$, $1 \rightarrow 2$, $2 \rightarrow 4$, $3 \rightarrow 6$, $4 \rightarrow 1$, $5 \rightarrow 3$, $6 \rightarrow 5$, and $7 \leftrightarrow 7$.

The binary k-cube or the hypercube is again characteristic of sorting applications, but is also relevant to AI applications involving marker propagation and inheritance. The backbone of the Connection Machine, it will be discussed under SIMD architectures. The hypercube connects $n = 2^k$ PEs, where $n = 0, 1, \ldots, 2^k - 1$, such that the wire connections link PEs whose binary code representations differ in exactly one position. Clearly, any two nodes cannot be further apart than k steps. The scheme allows communication among distant nodes, and there are $k = \log(n)$ connections per processor. Both the cube-connected cycle (CCC) and the perfect shuffle (PS) can be thought of as emulators of the hypercube for tasks, such as divide and conquer. The CCC can be conceptually described as 2^k processors arranged into 2^{k-r} rings of size 2^r at the corners of a $(k - r)$ cube. As r increases, the number of connections decreases, and the CCC tends to become a ring. Conversely, as r decreases, the number of connections increases and the CCC tends to become a (hyper)cube. Another alternative for emulating a hypercube is to use multistage networks, known as the Ω-network. There are k stages, each of them implementing a perfect shuffle, and the network as a whole is a shuffle exchange.

The next major interconnection scheme is the butterfly, which consists of $(k + 1)$ horizontal (or vertical) blocks of $n = 2^k$ nodes. The jth processor PE on block i is labeled as v_{ij}, where $0 \leq i \leq k, 0 \leq j \leq n$. The edges E connect such that $E(v_{ij}, v_{i-1,l})$ for $i > 0$ if $l = j$ or if the binary representation of l differs from that of j in only the ith place from the left. The connecting scheme's complexity is $O(n \log(n))$, and we will show its use in computing the discrete Fourier transform (DFT). The butterfly scheme can be emulated through the cube-connected cycle (CCC) and the hypercube. Let's assume that (as an example) $k = 3$ and that block O is identified with block k so block l is linked with block k. For our example, the PEs are now v_{ij}, $i = 1, 2, 3$, and $j = 0, 1, 2, \ldots, 7$. The first step of the conversion (emulation) includes the edges $E(v_{1j}, v_{2j})$, $E(v_{2j}, v_{3j})$ and $E(v_{3j}, v_{1j})$. The second step connects v_{ij} to v_{il} iff j and l differ in the ith bit from the left in their corresponding binary representation. Therefore, while the butterfly connects v_{ij} and $v_{i-1,l}$ in just one step, the CCC requires two steps. First v_{ij} links with $v_{i-1,j}$ and

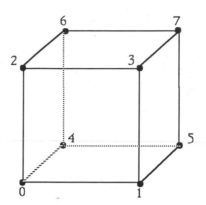

Figure 9.1.
Hypercube.

second, $v_{i-1,j}$ links with $v_{i-1,r}$. Clearly, if one were to collapse each CCC cycle (block), one obtains the hypercube (i.e., the three-binary cube with nodes $0 \leq j < 8$. The corresponding hypercube is shown in Fig. 9.1., and one can easily retrace the emulation procedure by expanding the j nodes and linking them as we did in steps one and two of the preceding procedure. Finally, the $(k + 1)$-cube can be easily obtained from connecting two k-cubes.

The computational complexity of the butterfly, shuffle exchange, and hypercube are equivalent, and their diameter scales logarithmically with n. Furthermore, if complexity is defined in terms of wiring (Ullman, 1984), then those organizations require almost $O(n^2)$ area, and the wire area dominates the processor areas. Such organizations enjoy a definite advantage over the mesh (crossbar) and/or the tree with respect to communication speed.

The butterfly interconnection scheme yields the FFT in $O(n \log(n))$ as compared with $O(n^2)$ needed for the DFT. Remember that given a sequence $X(m)$, its corresponding DFT is given as $Y(n)$, where $0 \leq m, n < 2^{\alpha}$, for some α. Specifically,

$$Y(n) = \sum_{m=0}^{N-1} X(m) W_N^{nm}, \tag{9-1}$$

where $N = 2^{\alpha}$ and $W = \exp(-j2\pi/N) = \sqrt[N]{-1}$. Note that W^{nm} are orthogonal and that there is computational redundancy in the powers of W, i.e., $W^{nm} = W^{(*)nm}$, $W^{(N/2)n} = (-1)^n$ and $W^{2km} = W^{km}$. The N-point

FFT can be recursively calculated as two $(N/2)$ FFT, because $Y(n)$ can be rewritten as

$$Y(n) = \sum_{m=0}^{(N/2)-1} \left[X(m) + (-1)^n X\left(m + \frac{N}{2}\right) \right] W_N^{nm}, \qquad (9\text{-}2)$$

and the even and odd coefficients are obtained as

$$Y(2k) = \sum_{m=0}^{(N/2)-1} \left[X(m) + X\left(m + \frac{N}{2}\right) \right] W_{N/2}^{km} \qquad (9\text{-}3\text{a})$$

$$Y(2k+1) = \sum_{m=0}^{(N/2)-1} \left[X(m) - X\left(m + \frac{N}{2}\right) \right] W_N^m W_{N/2}^{km}. \qquad (9\text{-}3\text{b})$$

There are $\alpha = \log(N)$ iterations corresponding to the butterfly blocks $i = 1, 2, \ldots, \log(N)$. The last iteration computes a two-point DFT. The multipliers for the odd terms are given as $-W_{2^i}^{\beta}, 0 \le \beta \le 2^{i-1} - 1$, where i is the (block) iteration number, and they can be prestored at the corresponding PEs. Note that each iteration shuffles the even and odd terms, and that the FFT coefficients are obtained in an ordering corresponding to a bit-reversal process. For $N = 8$, the corresponding input and output sequences are $[X(0), X(1), \ldots, X(7)]$ and $[Y(0), Y(4), Y(2), Y(6), Y(1), Y(5), Y(3), Y(7)]$. The butterfly scheme needed to obtain the FFT for $N = 8$ is shown in Fig. 9.2.

The last interconnection scheme we will mention is the Banyan network given as $B(f, s, l)$, where there are l levels, and each node (switch) has a spread (the degree in) s and a fanout (the degree out) f. The Banyan network is a partially ordered graph, where there is exactly one path from every base (fanout = 0) to every apex (spread = 0) node. Using node duplication, as an example, one can easily derive

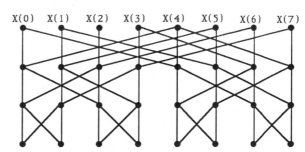

Figure 9.2.
Butterfly for N = 8.

the $B(2, 2, 2)$ from a three-level tree. The Banyan network requires $0(n$ $\log(n))$ links for complete interconnections. Both the k-cube and tree networks can be obtained from Banyan by combining communication links. The corresponding homomorphic network has a lower hardware cost but could lead to contention.

9.2. Pipelining

The first architecture that we consider is pipelining, which provides overlapped parallelism in space and time and is akin to a factory assembly line. The task has to be properly divided into subtasks. The stages of the pipeline then execute the original task in a concurrent fashion. In designing and analyzing pipeline architectures, one can consider the setup of the pipe and the delay associated with flushing out the first results. The stage elements of the pipe are most likely to complete their tasks at different times. Apropriate job sequencing, data buffering, and/or collison prevention compensate for the time differentials. The expected CPU speedup cannot be achieved unless the usual gap between CPU itself and memory is bridged, usually through interleaved memory organizations.

We consider three major classes of pipeline architecture. The first one, vector supercomputer processing, involves arithmetic processing and is included mainly for completeness. We are most interested, however, in multifunctional pipelines and systolic architectures, as well as their image processing and computational vision applications.

9.2.1. Vector Supercomputers

Vector-processing architecture can be found among both attached scientific processors and supercomputers.

The attached processors (AP) enhance the floating-point (FP) and vector-processing capabilities of the host computer. Multiplicity in the processor organization and concurrency of operations is achieved through software development of specific μ coded packages, usually for matrix operations. Furthermore, the interface between the AP and the host might require data reformatting and programmed direct memory access (DMA).

Supercomputers or vector-processing architectures are unhampered by the overhead usual for loop-control mechanisms. Four basic vector operations are handled: (i) vector to vector (such as $A(I) = A(I)**2$), (ii) vector to scalar (such as $s = \Sigma A(I)$), (iii) (vector × vector) to vector (such as $A(I) = B(I) \pm C(I)$), and (iv) (vector × scalar) to vector (such as $A(I) = \text{constant} * A(I)$). Furthermore, basic compare operations that yield Boolean vectors are then used as masks for compress and/or merge tasks. Basic design issues include fast memory access using vector registers (and base registers, offsets, and vector lengths), the setup and flushing time already mentioned, and how the theoretical peak performance degrades to the average speed when the system is faced with a mixed load of tasks including the I/O bottleneck. The CRAY generation is characteristic of vector processing, and its performance reflects setting multiple pipelines. A front-end (host) computer sets up computations and retrieves the results. The host is connected via multiple I/O channels to a CPU made up of a computation unit (CU) and an interleaved memory. The CU includes functional unit pipelines of the vector, floating point, scalar, and address type. Internal registers, such that the pipelines can operate on preloaded operands, are in the register-to-register transfer mode. Such registers include vector, scalar, and address blocks of registers. The vector registers are linked directly to the memory, while the scalar and address registers interface the memory via the scalar-save and address-save registers, respectively. Further pipelining is achieved through instruction buffers.

9.2.2. Multifunctional Pipelines

Multifunctional pipeline processing is characteristic of low-level image processing, such as operations that have to be performed in real time. The corresponding hardware organization can also be conceptualized as SIMD and/or MIMD—SIMD when local (neighborhood) operators are applied across the whole input image, and MIMD when regions of interest once detected require specific but different operations.

Mathematical morphology (Serra, 1982) has been developed to analyze texture and to perform visual inspection based on set theory and corresponding image transformations, such as dilation and erosion. Textural parameters are then determined for the distribution of the underlying random sets under specific transformations. The Cytocomputer and Genesis–2000 architectures developed by Sternberg (1985) constitute a pipeline implementation for mathematics corresponding to morphological analysis. The Genesis pipeline stages can be of the arithmetic logic unit (ALU) type, geometric logic unit (GLU) type (delete blobs that differ from a prespecified shape and/or intensity), count and locate unit (CLU) (count the number of occurrences for preprogrammed sets and locate them in space), and a CPU.

The GLU is the stage most concerned with morphological analysis. Dilation ($\oplus$) and erosion ($\ominus$) of element X are defined with respect to a structuring element A as

$$X \oplus A = \bigcup_{a \in A} X_a \tag{9-4a}$$

$$X \ominus A = \bigcap_{a \in A} X_a, \tag{9-4b}$$

where X_a represents the translation of set X to the points $a \in A$. Multifunctional pipeline architecture take advantage of the structuring element chaining property, which states that if $A = A_1 \oplus A_2$ then

$$X \oplus A = [(X \oplus A_1) \oplus A_2] \tag{9-5a}$$

$$X \ominus A = [(X \ominus A_1) \ominus A_2] \tag{9-5b}$$

For example, if the structuring element A is a hexagon, then A can be implemented by a three-stage pipeline, each stage dilating the original image X by a line segment of prespecified length corresponding to the three primary orientations of the hexagonal grid. Alternative definitions of the structuring element and the corresponding pipeline yield substantial savings in the iterations required to implement dilation or erosion. A high level language uses the basic primitives of dilation and/or erosion, where basic operations like *implode* and *clone* can be defined. The *implode* operation yields a binary 1 only at those pixel locations where the structuring element A can be centered, such that its underlying image mask is all 1.

Another representative example of multifunctional pipeline architecture is the pipeline image processing engine (PIPE) (Kent, 1985). Developed at the National Bureau of Standards (NBS) for robotic applications, PIPE involves recognition and spatial layout of objects in space as needed for guidance and servoing. Pipeline architecture is well fitted to the task, because the robot processes a continuous stream of images using several operand stages. Specific functional requirements are to handle different events extending and interacting over time, to build up hypotheses, and to use them to filter and process forthcoming input, for multiresolution analysis, and/or real-time response. PIPE is just one functional element in the NBS robot system and its main concern is "low-level" image processing. The pipeline operation is set up using video and/or range images and possibly processed by a conformal mapper (CONMAP) image transform (for geometric invariance). The iconic output of the pipeline is mapped into localized symbolic information by ISMAP (iconic to symbolic mapper), where the extracted features eventually interface to the host memory, the repository of knowledge representation (world models and newly acquired information), and the task decomposition and control module. The pipeline contains eight modular processing stages (MPS). Each stage can perform two separate and independent SIMD operations. Two separate buffers for each MPS allows one of them to mask and/or define regions of interest for specific algorithms, and the resulting operating mode is then that of MIMD. The system connectivity is quite complex and provides for forward paths (standard pipelining), recursive paths (for feedback and relaxation), backward paths (for knowledge-based and temporal processing), and global distribution (via "wild card" buses). The pipeline is indeed multifunctional and furthermore, it is reconfigurable according to intermediate results, hypotheses, and tasks. Such an engine provides adequate sensory information to drive a robot system guided by the active perception.

9.2.3. Systolic Processing

Systolic architectures expand on pipelines by allowing the inputs and the (partial) results to flow from stage to stage. Kung (1982) provides a good introduction as to why and how to use systolic architecture. Data flows from the host memory through several PE stages before returning to the memory. Advantages provided by such an architecture include the likelihood of balancing the I/O and the computation, and second, the modularity and reconfigurability of a (possibly) multidimensional (patterned) architecture. A systolic architecture could implement a basic operation such as convolution. Systolic pipelines can be thought of as (semi) systolic if there are three I/O ports per PE and global data communication or (pure) systolic if there are four I/O ports per PE without global data communication. With the semisystolic design, where PEs are of the accumulator-multiplier type, convolution is implemented in terms of a set of given weights $\{w_i\}$ and inputs $\{x_i\}$, $i = 1, \ldots, n$, as $y_i = \Sigma_{j=1}^{k} w_j x_{i+(j-1)}$, where $\{y_i\}$ are the outputs. One should assume that the inputs are broadcast over the bus and made available to all the PE at the same time. The weights are resident at each PE. Then a typical PE computes $y_{out} \leftarrow y_{in} + w \cdot x_{in}$, and a typical cycle computes and transfers (flows) partial results. If $k = 3$, the partial computations, including the setup and flushout, look like (even that alternative designs are feasible) those shown in Fig. 9.3.

For the pure systolic design the results stay, while the inputs and weights move in opposite directions. The typical PE computes $y \leftarrow y + w_{in} \cdot x_{in}$. Synchronization ensures that the $\{x_i\}$ meet $\{w_i\}$ and requires the input and weight data streams to be separated by two cycle times. The systolic output (dashed) path makes up for the lack of global communication and flushes out the results.

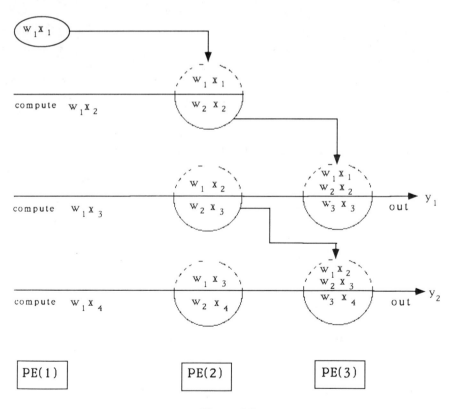

Figure 9.3.
(Semi) systolic design for convolution of order three.

Figure 9.4.(a, b) shows a typical computation where $y_i =$ $\sum_{j=1}^{3} w_j x_{j+(i-1)}$, i.e., $y_1 = w_1 x_1 + w_2 x_2 + w_3 x_3$, $y_2 = w_1 x_2 + w_2 x_3 + w_3 x_4$, and $y_3 = w_1 x_3 + w_2 x_4 + w_3 x_5$. Note that the output results are flushed out in the right order.

The Wigner distribution, the cornerstone of joint spatial/spatial-frequency representations discussed in Section 2.3., has also been considered for fast systolic implementation by Durani (1983). The particular instantiation of the discrete Wigner distribution (DWD) implemented is

$$W_{f,g}\left(n, \frac{m\pi}{M}\right) = 2 \sum_{k=-N+1}^{N-1} \exp\left[-j\left(\frac{2\pi km}{M}\right)\right] \cdot f(n+k)g^*(n-k), \quad (9\text{-}6a)$$

where $\{f(n)\}$ and $\{g(n)\}$ are two complex data sequences of length $(2N-1)$, $m = 0, \pm 1, \ldots, \pm(N-1)$, and $M = (2N-1)$. The equation can be rewritten as

$$W_{f,g}\left(n, \frac{m\pi}{M}\right) = 2 \sum_{k=-N+1}^{N-1} \omega^{km} \gamma(n, k), \quad (9\text{-}6b)$$

where $\omega = \exp[-j(2\pi/M)]$ and $\gamma(n, k) = f(n+k)g^*(n-k)$. The three basic cells are the delay, multiplier, and cordic types. The cordic cell is characterized by $\gamma_{out} \leftarrow \gamma_{in}$, $\omega_{out}^m \leftarrow \omega_{in}^m$, and $W(n, m, k) \leftarrow W(n, m, k-1) + (\omega_{in}^m)^k \gamma_{in}$. The complete architecture for calculating the DWD includes a characterizing array yielding $\gamma(n, k)$, and a cordic array where each horizontal pipeline calculates the equivalent of a DFT for $\{\gamma(n, k)\}$, at the specific frequency ordinate m. The characterizing data $\{\gamma(n, k)\}$ sets up the horizontal arrays through vertical pipelining, leaving a one cycle lag between the DFT of successive frequencies. Each cell in the horizontal array is indexed by the running parameter k, and the data wavefronts $\gamma(n, k)$ reach the horizontal pipeline delayed accordingly. As an example, if one were to use a cordic array for $N = 2$, then $k = -1, 0, 1$, and $m = -1, 0, 1$. The first term is $W(n, m) = W(n, -1)$ for some n and the frequency $m = -1$, and it is given by

$$W(n, -1) = (\omega^{-1})^{-1} \gamma(n, -1) + (\omega^{-1})^0 \gamma(n, 0) + (\omega^{-1})^1 \gamma(n, 1). \quad (9\text{-}7)$$

The generic systolic architecture discussed so far has been implemented as a systolic array computer called WARP (Annaratone *et al.*, 1987). The prototype WARP array includes 10 programmable PE cells

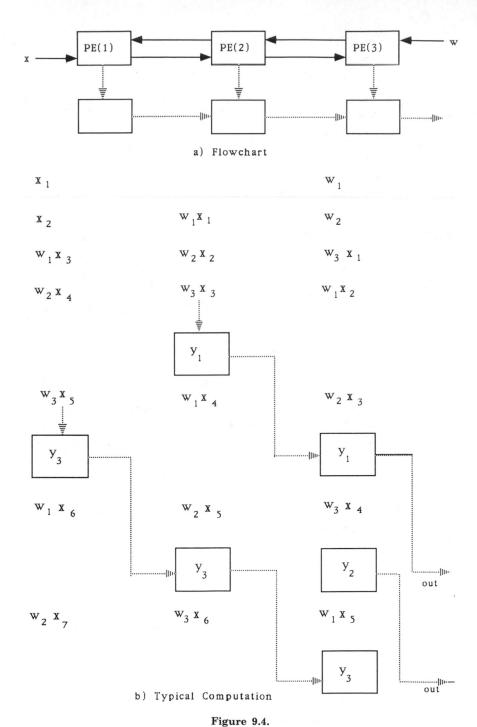

a) Flowchart

b) Typical Computation

Figure 9.4.
(Pure) systolic design for convolution of order three. (a) Flowchart. (b) Typical computation.

474

to compute at up to 10 MFLOPS and operates as a fast processor attached to a host machine. There is extensive parallelism at the (pure systolic) array level and the (data cross bar) cell level. The choice for topology was that of a linear array, due to its simplicity and low requirements on the I/O (two end cells) bandwidth. Programmable temporal multiplexing allows the WARP, however, to simulate a 2D systolic array as well. The WARP machine is a flexible machine that operates on problems requiring either fine or large-scale parallelism. The high bandwidth between adjacent cells provides fine-grain parallelism, while powerful PE cells allow large scale parallelism as well. Another dimension along which WARP can make its choice is that of local versus global processing. As we have seen so far, systolic architecture is usually most fit for local operations. The availability of large memories, however, enables the WARP to perform global operations as well. A high degree of parallelism is achieved by problem partitioning of input, output and by the very useful approach of multifunctional (heterogenous versus homogenous) pipelining. The heat equation that we discussed in Section 2.2., in the context of pyramidal structures, can be then iteratively solved using relaxation (over the 2D grid). Specifically, if a cell performs the kth relaxation step in row i, then the preceding and next cells perform the $(k-1)$ and $(k+1)$ relaxation steps on rows $(i+2)$ and $(i-2)$, respectively. WARP has been mostly used for robot navigation tasks and signal processing applications.

9.3. Single Instruction Multiple Data (SIMD)

Next, we will discuss parallel architectures characteristic of low-level visual tasks and local processing. Such PDP architectures assume that a single instruction operates on multiple data in large input arrays, and that there are no interactions during a given cycle between the participating local neighborhood data arrays.

9.3.1. Cellular Logic Array

Cellular logic array (CLA) constitutes the basic SIMD architecture. It is a bit serial internal PE architecture using a nearest neighbor (NN) interconnection mesh. The thrust of the architecture is fast arithmetic computation. A characteristic CLA example is the massively parallel processing (MPP) developed by NASA for processing remotely sensed data, such as those acquired by LANDSAT satellites. The MPP organization includes a NN select unit (to allow a bit plane to be shifted in any of the four connected directions), a Boolean processor whose operands are a P register (accumulator-like) and the data bus, and a reconfigurable arithmetic logic unit (ALU) with a variable length shift register (for circulating partial results as those derived during multiplication) and a full adder. Another SIMD machine is the ICL Distributed Array Processor (DAP), which consists of a square matrix of 64×64 PEs, each of them with 16 K of RAM and interconnected using the NN mesh. Such an architecture is well fit to handle the Markov random fields (MRF) used by Geman and Geman (1984) for image restoration (See Section 4.2.3.). Murray *et al.* (1986) discuss a parallel version to implement the MRF model and point out that a simple broadcast of the (identical) processing steps to all pixels might cause them to oscillate, as would be the case for an alternating pattern of black and white pixels. Geman and Geman (1984) suggested for this very reason the MIMD machine operating in an asynchronous fashion. Such a scheme will allow a pixel to change its configuration if and only if its neighbors broadcast the fact that they do not need to undergo any change. A SIMD scheme, which is much simpler to implement, can be still used, however, if PEs neighboring in the MRF sense do not change simultaneously. Using a "chequer-boarding" mask and its logical complement, the DAP can then change every other row and column synchronously, requiring only two SIMD cycles. The speedup is thus only half that of the total array, but the more costly MIMD scheme is unneccessary. Line processes can be handled as well through separate vertical and horizontal line locations. The corresponding SIMD restoration algorithm, at each temperature T, as suggested by Murray *et al.* (1986) for a DAP linked to its host, an ICL 2980, is shown in Fig. 9.5.

NEWIMAGE is a temporary array initially filled with the restored image—step (b) suggests random gray-level changes, and step (d) sets a logical mask according to the local energy change where EXP is an array and RAND is an array of random numbers uniformly distributed over $(0, 1)$. The other statements can be easily interpreted in a similar fashion.

A useful extension to the basic CLA type is the parallel hierarchical system of the pyramidal type, which yields a multifunctional SIMD architecture. Uhr (1987) comments on such "very large retina-like base

(a) NEWIMAGE = RESTORED

(b) NEWIMAGE(CHEQUER) = RANDOM

(c) Compute the DELTAE array

(d) ACCEPTED = DELTA $\leq$ 0 or
     EXP($-$ DELTAE/T) $\geq$ RAND

(e) RESTORED(ACCEPTED) = NEWIMAGE

(f) NEWLINES = VLINES

(g) NEWVLINES(CHEQUER) = .NOT.NEWVLINES

(h) Compute DELTAE

(i) Evaluated ACCEPTED as in (d)

(j) VLINES(ACCEPTED) = NEWVLINES

(k) Perform steps (g) $-$ (j) for horizontal lines

(l) CHEQUER = .NOT.CHEQUER
     Flip the chequerboard and repeat (a)–(k).

Figure 9.5.
SIMD restoration using Markov random fields.

of array-linked computers (PEs) that are in turn linked to a converging sequence of higher layers of successively smaller numbers of PEs." Specifically, he makes a strong point on the relevance of pyramid-like architectures by stating that the CLA type of SIMD organizations are slow at global computations due to the cost of circumventing the NN interconnection scheme. However,

> pyramids superimpose a good global interconnection network over the basic NN topology. They thus combine the virtues of arrays with those of trees, at the same time eliminating most of the drawbacks of each. They appear to be appropriate structures for a variety of tasks where information is frequently transformed, converged, and diverged, for example, multigrid approaches to numerical problems, and perceptual recognition.

The basic CLA arrays of different sizes are then integrated with "logarithmically converging tree links that pull distant parts of these arrays close together." (Compare this type of "togetherness" with that offered by interconnection networks, such as the Connection Machine discussed in the next subsection.) Two major examples of pyramid-like structures are the PAPIA (Cantoni and Levialdi, 1987) and the GAM pyramid (Schaefer *et al.*, 1987) systems.

9.3.2. *Interconnection Networks*

We consider next the Connection Machine (CM) (Hillis, 1985; Tucker and Robertson, 1988) as the major example of interconnection networks. The rationale for the CM comes partly from semantic networks (SN), knowledge representation languages such as NETL, and marker-propagation algorithms (Fahlman, 1979). Most of the knowledge available in such a network is implicitly stored, and one needs to use inference to make it explicit. We recall that a SN is a graph whose vertices stand for concepts and/or features, while the edges represent relationships linking the concepts. The process of inference across a SN is equivalent to pattern matching and masked retrieval operations. Fast performance can be achieved if truly parallel algorithms are emulated on some underlying machine. The hardware requirements for such performance are a large number of PE (one PE per concept), and programmable connections, such that the topology of the machine is reconfigurable to match that of the problem to be solved. The software virtual communication scheme implemented on the CM is well suited to both low- and high-level image processing applications. Both nearest neighbor (NN) mesh (for depth from stereo derivation) and tree communications links (for object recognition) can be made available. The CM can thus be thought as embodying a general communication scheme whose topology can be dynamically optimized for a specific application. Furthermore, the large number of bit-serial PE is characteristic of a fine grained architecture, while the flexibility and efficiency needed for well-structured problems, where most of the inherent parallelism is with respect to the data, led the CM to be a SIMD-like machine.

The organization of the CM includes up to four front end hosts (VAX or Symbolics) running UNIX, LISP, and/or a parallel dialect of FORTRAN. A 4×4 cross-point switch called the NEXUS interfaces between the host(s) and up to four sequencers (micro controllers). Each sequencer can control 16 K PE. High bandwidth I/O connects the processors to mass storage. The CM enjoys several characteristics of a virtual machine. First, the sequencer can map a virtual machine parallel instruction set (PARIS) into a RISC set on the PEs. Furthermore, virtual processors can be created by splitting the local memory allocated to PEs into regions and having the PEs sliced among them. The PEs are characterized by ALUs for bitwise processing, 4 to 16 K bytes of local memory, and eight masks (one-bit flag register) as appropriate for marker-propagation type of applications. The communication network allows for broadcast (to all PEs), a cardinal NEWS 2D grid (NN explicit mesh), which can be generalized to n-dimensions, and a hypercube interconnection scheme where a binary 12-cube connects 4 K PEs such that each 16 PE cluster lies at the vertex of the cube.

The power and versality of the CM is derived largely from its general and flexible communication scheme. The NEWS grid allows all PEs to communicate in the same direction during one instruction cycle. Addresses are implicit, there is no danger of collision, and the scheme is fast for regular local operations. The hypercube scheme is used by a router for linking across the CM. It is implemented as switched message packets, which include PE addresses and data. There is always the need to make routing and buffering decisions, and thus the router communication scheme is much slower than NEWS. High performance has been achieved by sustaining floating-point rates in excess of five gigaflops (10^9 operations per second).

Now we will describe some specific applications of CM. First, we consider computing the depth (intrinsic) map from stero. A broadcast to all pixels can load both the left and right image on corresponding PEs, such that one PE is allocated to each pixel. Then, in parallel across all the PEs, the DOG edge detection step is performed. Disparity is calculated by sliding the right image over the left image using the NEWS scheme and ANDing the edge patterns detected by the DOG step. A degree of correlation is calculated by counting in small neighborhoods over the Boolean maps resulting from ANDing the edge patterns. Finally, each PE selects the shift of maximum correlation, and a 2D best-diffusion model fills in for nonedge PE location in order to provide a smooth rather than sparse depth map.

Object recognition is probably the major computer vision application implemented on the CM because of its conceptual similarity to search and retrieval over large databases. The description of an object in terms of features is distributed over PEs, with each feature localized at a specific PE. The general strategy taken to solve the recognition problem is hypothesize-and-test. Detected features can hypothesize and trigger a specific object model. The models, in turn, suggest additional features to be tested, and the verification step follows in order to enhance the identity of the input and/or to remove the model from further consideration. Note that the parallel retrieval scheme of object models based on characteristic features is akin to the Hough transform (template matching/match filter) and connectionist PDP models as discussed in Section 4.2.4.

We finally take the example of image segmentation between fore-ground and background. Many algorithms that solve such problems operate on pyramidal structures. The choice of a pyramid comes from its ability to combine both local and global processing through its mixed interconnection schemes, those of NN mesh and tree, respective-ly. The CM has a reconfigurable topology and thus could match that of a pyramid. Kjell and Wang (1987) emulated pyramidal segmentation algorithms on the CM. To configure a CM as a pyramid, one can use one of the following two approaches. First, assuming a $n \times n$ array of PEs with local memory, node $N(i, j, k)$ (pixel location $\langle i, j \rangle$ at level k) is allocated to the PE located at $\langle i*2**k, j*2**k \rangle$. An alternative scheme would run the CM in the virtual mode by creating virtual PEs. For this same array, work with $n \times 3n/2$ PEs such that level 0 of the pyramid requires 2/3 of the original PEs, level 1 requires 1/6 PEs, and so on. The second scheme suffers from neighboring pixels being further apart, and indeed, the simulation results showed that the performance of the first scheme is superior. Extensive time is spent on iteration when one implements pyramidal algorithms linking parents and offspring. One could speedup the process by taking advantage of the extensive data parallelism offered by the CM and substitute concurrency to iteration through the use of masked operations as appropriate for the pattern sought.

9.4. Multiple Instructions Multiple Data (MIMD)

The multiple instruction multiple data (MIMD) architecture is most appropriate to high-level visual tasks that require different algorithms to be executed on disjoint data sets. MIMD is concerned with interac-tive processes that share resources and is thus characterized by asynchronous parallelism. Loosely coupled and tightly coupled PEs are the two main subdivisions within MIMD. Multicomputers is the first subclass, and its operating mode is message passing, while the second subclass is labeled multiprocessors, which derives its functionality from PE sharing memory. As we move from SIMD to MIMD, note that SIMD and MIMD are equivalent in that they can simulate each other. An SIMD machine could interpret the PE data as different instructions, while the MIMD could execute only one instruction rather than many across the PE array. There are no strict boundaries between architectures as we have seen so far, and the basic question is that of efficiency and cost for solving a specific problem. We proceed by looking into the multicomputers and multiprocessors classes, respectively.

9.4.1. Message Passing Multicomputers

Message-passing multicomputers are lattices of PE nodes connected by a message-passing network. The basic computational paradigm is that of concurrency of processes, where processes are instances of programs. The PE include private memory, and there is a global name space (PE#, process#) for variables across the multicomputer. The $(N \times N)$ network is the binary n-cube or mesh and facilitates locality of communication between the N nodes. Multiprogramming operating systems available at the PE and coordination through message-passing facilitate concurrency. Thus, the multicomputers constitute a physical and logical distributed system.

Athas and Seitz (1988) provide a good review and taxonomy for such systems. According to the grain size—medium (Mbyte of memory per PE) or fine (Kbyte of memory per PE)—different architectures can be defined. Most present systems are of the medium grained type, such as the COSMIC CUBE, INTEL iPSC/1, and the TRANSPUTER, which are characteristic to first generation medium-grained systems. Typical systems include up to 64 nodes, with 32 Mbytes, and some of their application include matrix operation, (partial) differential equations (PDEs), and finite element analysis. The projections for future systems include faster PE and communication networks, making their performance almost as fast as that of the supercomputers.

The TRANSPUTER architecture is made up of programmable VLSI components. Each element includes on one chip a PE, memory and appropriate communication links. The OCCAM programming language for the TRANSPUTER is implemented through point-to-point channels and also allows links between processes on individual elements. The advantages accrued from a point-to-point communication network are no contention on the dedicated link, smooth upward scaling, and unsaturated communication bandwidth. Downing and Bennett (1988) used the TRANSPUTER for high-level feature extraction (using chain code techniques) directed toward object recognition. The network simulated is that of a ternary tree structure, with a *demand* process executing on all nodes. The process consists of a package of image processing routines, whose results can be exchanged through two data packet routing and distribution processes, *up* and *down*, operating in parallel with the *demand* process. The chain code outlining the boundaries of interest, is generated in parallel by demand processes operating on variable sized rectangular subimages. The results are passed to the root *supply* process, which then invokes a joining algorithm operating on adjacent subimages. Another application for the TRANSPUTER was suggested by Richards (1988) for multilayers networks using hidden layers. The neural learning scheme called back propagation (or delta-rule) was discussed in Section 8.4.4. and renders itself nicely to parallel implementation.

Two forms of parallelism could be exploited. First, one can partition the training set, since the gradient array results from summing over the individual patterns in the training set. Second, there is always the possibility of partitioning the matrix algebra. A toroidal configuration that takes a 2D lattice and connects opposite edges facilitates partitioning. The training set could be split among the rows, and the weight matrix could be distributed within a row. Assuming that the torii are given in terms of arrays $(m \times n)$, where m is the number of rows and n is the number of columns, one can easily see that as m increases (at the expense of n) the communication overhead increases as well due to the need for exchanging information on the gradient arrays.

We briefly consider the possible equivalences and differences between the multicomputers and other architectures. The message-passing and shared-memory for MIMD are equivalent. Message-passing could be implemented in a global shared memory, while the shared memory type could substitute message-passing protocols for assignments involving shared variables. The MIMD architecture could easily implement the SIMD message passing organizations, such as the connection machine (CM) discussed in Section 9.3.2. but the equivalence is basically one way because the CM lacks concurrency of processes. Most work on message-passing has been done in the context of medium-grain multicomputers. Message-passing and shared memory architecture make multicomputers and multiprocessors almost equivalent. As the number of nodes increases, however, only the multicomputers are still feasible. One critical question, however, that has yet to be fully addressed and solved is that of creating an adequate software environment to implement concurrency. The programming languages (PL) for fine-grained concurrency could be either static, or even better, dynamic. The static PL assumes that the number of processes and the connectivity of the network is known before the program executes. Communication channels synchronize through the *send* and *receive* operations. The OCCAM programming language on the TRANS-PUTER is characteristic to the static approach. The dynamic approach allows processes to be created and destroyed on demand. We are then talking about true concurrency involving message-driven programs and object-oriented programming. A main characteristic of such an approach as suggested by Athas and Seitz (1988) is the reactive property, where objects are normally at rest until a message arrives. Just a short conceptual path lies between dynamic PL for fine-grained multicomputers and the data flow architectures to be discussed in the next section.

9.4.2. Shared Memory Multiprocessors

Shared memory multiprocessor systems operate over a single global address space and are thus characteristic to physically distributed systems. There is a $(N \times M)$ switch network between N processors and M memories, an interrupt signal interconnection network between the PE, and a network corresponding to the interface between PEs and I/O channels. Shared memory multiprocessors are asynchronous systems and synchronization can be achieved through shared variables and/or mutual exclusion implemented through semaphores. One still has to cope with specific problems related to memory competition.

The BUTTERFLY parallel processor developed by BBN is an example of a shared-memory architecture. One PE and (one to four Mbyte) memory are located on a single board called a processor node. The memory of the processor nodes forms the shared memory of the system. Interprocessor communication is achieved through the butterfly switch, whose topology was discussed in Section 9.1. as it applies to computing the FFT. The system scales up to 256 processor nodes and a fully configured machine displays a peak performance of 128 MIPS, up to 1024 Mbytes of memory, and interprocessor communication of 32 Mbit/sec for an overall bandwidth of 8 gigabits/sec. The system clearly displays fault tolerance characteristics as well due to its modularity.

Potential applications for the BUTTERFLY include road following, as it applies to the autonomous land vehicle (ALV), object recognition, and data fusion of intrinsic images. Processing nodes share the road, the 3D object descriptions, and/or specific intrinsic images. Another interesting application suggested by Suh and Van Gucht (1987) is related to genetic algorithms. Such algorithms, originally suggested for learning by Holland (1975), generically solve global search optimization problems. The search is conducted over a collection of structures rather than a single search space in order to avoid getting trapped in a local minima. (The same concept applies to simulated annealing as described in Section 4.1.5.) The standard genetic algorithm iterates by creating new search generations. Suggested structures or solutions at a given iteration step are evaluated and then a *survival of the fittest* step selects those solutions most likely to succeed based on their present performance. After the selection step, recombination is implemented through operators, such as cross-over (breeding), mutation, inversion, or local improvement. The genetic algorithms resemble natural evolution and can be easily parallelized using either a centralized or distributed genetic scheme. The schemes differ in the execution of the selection step. The centralized scheme uses a master processor for selection and synchronization, while the distributed scheme makes the selection step local. The centralized algorithm is unreliable, if either the master or any of the slaves asked to provide the evolution (requested by the master) fail, and synchronization delays are also likely to occur. The distributed genetic algorithm performs as well as any standard genetic algorithm, is reliable, and is more natural due to its very asynchronous behavior.

Connectionist models were discussed in Section 4.2.4. and shown to be appropriate to recognition problems dealing with partial, uncertain, and/or conflicting knowledge. Feldman *et al.* (1988) implemented such models on the BUTTERFLY for matching between real images of toy objects and stored topological models. Such PDP modeling is characteristic to winner-take-all solutions where simultaneous comparisons are needed (and are hopefully implemented in parallel), and there is much mutual activation and inhibition between the processing units. The use of a parallel architecture like the butterfly reinforces the possibility of merging *neural networks*, *learning*, and traditional *representation* and *inference* methods implemented using either production rule-based systems and/or logic.

9.5. Data-Driven Computation

We conclude our discussion on parallel hardware architectures with a look toward the future. One of the goals of this book is to advance parallel and distributed processing with respect to image representations and processing. The basic neural model of computation presented in Section 4.1. is built around a PE labeled as a neuron, and the corresponding network connecting PEs and operating in a parallel and asynchronous mode. Such a model departs from the control-driven scheme characteristic of Von Neumann computation and defines a new data-driven computational scheme, also known as data flow computing. We consider in this last section the main characteristics of data-flow computing, and some of its possible applications, and finally take a look at neurocomputing.

9.5.1. *Data-Flow Computing*

Two basic flows of computation are those driven by control or data, respectively. The traditional control flow (CF) has a centralized control scheme embedded in a program counter (PC), which makes the execution flow according to a program of instructions. The data-flow (DF) scheme proceeds by instructions activated by the availability of data operands, and it is thus not too different from the fine-grained multicomputers we considered in Section 9.4.1. The program sequencing for CF includes the basic sequential flow and GO-like statements and/or parallel concurrency instructions such as FORK and JOIN. The sequencing for DF is solely determined by the data dependency between instructions. Shared-memory cells for data transfer between instructions operate in the CF mode while the DF allows for data tokens to be transferred between connected (in some data-flow graphs like the PETRI NETS) instructions nodes as the computations carried out at specific nodes proceed in time. The last scheme, that employed by the DF, enjoys then the advantage of no side (or locality) effects. The CF employs a centralized, synchronous control while the DF is asynchronous, and lends itself easily to parallel implementations. The DF is conceptually built around operation/instruction packets and data tokens, both of them including specific destinations for the results and successor instructions. Such an architecture is similar to distributed systems of the multicomputers type.

There are two basic organizations for implementing the data-flow architectures (Veen, 1986). The static DF allows one token for each arc, and timing or synchronization is achieved using control tokens as acknowledgements (ACSs). Alternatively, dynamic DF allows for more than one token per arc, tags the tokens, and achieves synchronization through matching. Both organizations use a pipeline ring structure where the units are of the memory (M) type (holding instructions), processors (P) type (for parallel execution), a routing network (RN) (of the packet-switching type) and an I/O unit (which is used as well for matching when the dynamic DF model is employed).

The main design issues related to data-flow computation are concerned with the decomposition (or compilation) of a given program into some data-flow graphs exhibiting true parallelism and/or concurrency and an appropriate programming language (PL). The major advantages in using DF is its inherently high concurrency, modularity, easily upward scaling, and fit for VHSIC implementations, and its functional programming style and the locality of side effects. The disadvantages are that the data-driven mode at the instruction level can lead to excessive pipeline overhead, and that there is the likelihood of wasting computational resources (processors), if there is not enough parallelism and data storage. Both the memory access and the packet switching networking run the risk of becoming a bottleneck for the whole system. Finally, arrays or other complex data structures are not easily handled by DF architectures.

Most data-flow applications have been to design array processors for signal processors (SP). The wavefront array processors (WAPs) are of a mixed design, including both systolic and data-flow concepts, and are discussed at length by Kung *et al.* (1987). Most of the SP applications involve convolution and/or algebraic operations, which are regular, local, and usually recursive in their characteristics. WAPs are similar to systolic arrays except that there is no global time reference. Data transfers in systolic arrays are controlled by a global "beating heart," and as a result synchronization for large arrays is quite difficult, and slow downs could result due to the slowest amongst the PEs. While systolic architectures consider correct timing, WAPs are concerned with correct instructions sequencing, which is achieved by tracing and pipelining the computational wavefronts as defined by the data-flow graphs (DFG). The interconnection network (IN) linking the PEs achieves synchronization through handshake-like protocols. Each wavefront sweeping across the array corresponds to one recursion being traced, while successive recursions are pipelined one after the other. One could easily observe the WAP concept by considering matrix multiplication, where the product of two matrices A and B could be expressed as the sum ("recursion") of outer products between the columns $\{a_i\}$ of A and the rows $\{b_j\}$ of B. A wavefront starts at $PE(1, 1)$, propagates at $PE(1, 2)$ and $PE(2, 1)$, then to $PE(1, 3)$, $PE(2, 2)$, and $PE(3, 1)$, and so on. Computation starts at $PE(1, 1)$, which first computes $c_{11} = a_{11} \cdot b_{11}$, while the generic computation at cell (i, j) at wavefront (recursion) $(k + 1)$ is given as $PE(i, j) = c_{i,j}^{(k+1)} = c_{i,j}^{(k)} + a_i^{(k+1)} \cdot b_j^{(k+1)}$. Several distinct advantages can be drawn from using WAPs. First, when compared against SIMD in the context of data-dependent operations, the WAPs allow the host to select the appropriate data and to pipe it across the array. There is no global clock, and thus one avoids hard choices regarding clock implementation and scheduling of operations for the PEs. The PEs can operate at their full speed; fault tolerance at run time results because of the asynchronous nature of the design, and the whole architecture lends itself easily to wafer integration due to the shorter links connecting the PEs. The Inmos TRANSPUTER has been used as the PE building block for some of the WAPs due to its channel communication scheme.

Koren *et al.* (1988) look at and contrast the array processors (systolic or wavefront) against pure data-driven (VLSI) architectures for arbitrary rather than regular structured algorithms. In both systolic and WAPs architectures, the computation advances according to a predetermined, fixed sequence that is data independent. Such an architecture is appropriate for regular algorithms that exhibit cycles and/or dependency within their data-flow graphs. The pure data-driven data-flow architectures are appropriate for arbitrary algorithms with no internal regularity. Such an architecture is then akin to fine-grained multicomputer parallelism.

Our discussion on data-flow concepts points to the great similarity between DF and neural networks (NN), WAPs, and MIMD of the multicomputer type.

9.5.2. *Neurocomputing*

We have discussed PDP modeling and neural networks in Chapter 4. It is highly rewarding to start our discussion by considering Kohonen (1987) on neural networks and neurocomputing. Neurocomputing is the functional term characterizing the analog computing mode, coined by Hecht–Nielson (1986) as artificial neural systems (ANS). Again, one first has to determine the main goals for such systems. The most meaningful behavioral tasks according to Kohonen are sensory functions and automatic motor control, in compliance with the true statistical nature of the environment and its events. For such tasks involving estimation and prediction, statistical invariance (Section 3.2.) and neural learning (Section 8.4.) are possible solutions. We quote Kohonen on ANS as being "massively parallel interconnected networks of simple (usually adaptive) elements and their hierarchical organizations which are intended to interact with the objects of the real world in the same way as the biological nervous systems do."

What should be the characteristics of such ANS? As we have argued all along, the ANS should support both local and distributed computations. The neurons, as we stated earlier, are PE, which are much more than just simple logic gates, and the degree of parallelism exhibited by ANS is much higher than that ever contemplated by data-flow computation. Neurocomputing should be conceptually thought of as analog computation, even that it may be emulated, as it is actually done, using electronic or optical means. Analog computation simulates dynamical systems, defined through synaptic connections, rather than execute a specific program. This view is a major departure from the traditional Von Neumann concept of computation. Another major difference between digital and analog computation as assigned to ANS, comes from Kohonen (1987). "It is the discreteness, specifity, and stability of the state sequences (in a finite automata), which facilitate the recursive definition of various functions to an arbitrary degree of accuracy in digital computers. Such a specificity and stability is never met in neural circuits. The feedback functions and the relaxation processes in the neural networks serve different purposes, the computing process, by its nature, is closer to the stochastic approximation, which aims at iterative solution of approximation problems. The recursions in neural circuits, if they occur at all, must be short."

We conclude this section by briefly considering actual neurocomputing and ANS implementations. Suggested architectures can take the analog, electronic, electrooptical (optical links and electronic PE), and optical (both PEs and links are optical) forms. ANS are best described as analog devices, and some of the first implementations, such as that of Hopfield and Tank (1986) (for the TSP) at Bell Laboratories were indeed analog. Such devices map incoming signals into voltages, weights are represented by connecting resistors, and the changing results are then currents, which could be added together to yield the output. Going from an analog device to a chip is not easy, because it is virtually impossible to build accurate resistors on silicon wafers (Hecht–Nielsen, 1986). Jackel et al. (1988) built a "hybrid" chip for handwritten digit recognition. The synaptic weights are stored digitally on the chip in a bank of static RAM cells. The weights store templates corresponding to the digits to be recognized. For the low accuracy dot products required in template matching, analog electronic circuits enjoy a clear advantage over digital circuits: the analog circuits can be more compact and faster, because the summation in the dot product is computed "for free" just by adding the component currents in a summing wire (Jackel et al., 1988).

By far, ANSs largely assume an electronic form and are characterized by an architecture where neural network behavior is emulated by fast coprocessor boards plugged into a conventional host machine. TRW Mark III is an example of such a coprocessor attached to a VAX host computer. It can implement networks of up to 64 K PEs and approximately 10^6 synaptic weights, which can be trained and updated at the rate of about half a million per second. The TRW Mark V is made up of slave processors (from 1 to 16), each of them consisting of a Motorola 68020 32-bit floating point processor (multiplier/ALU) and a master processor built around the Motorola as well. The performance is up to 6×10^6 floating point connections per second (12.5 MHz) and 64 Mbytes of RAM for storing states and/or weights.

The SAIC SIGMA–1 is another example of an attached processor able to provide $5 - 8 \times 10^6$ interconnects/sec. Cruz–Young et al. (1985) have built a medium-scale VLSI-based ANS coprocessor called NEP (network emulation processor) attached to an IBM–PC. The resulting combination, coined PAN (parallel associative network), has been used for storing AI rules (as local minima of energy functions) and then running parallel rule-based searches. An example of an electrooptical ANS is the attentive associative memory (Athale et al., 1986). The DAM being implemented is of the correlation type, i.e., $M = SS^T$, where S is the stimulus matrix, and the ANS provides for nonlinearities. A channel-dependent nonlinearity can suppress spurious correlations and emphasize the similarity of the input to one of the stored vectors. A second nonlinearity reflects the a priori knowledge about the nature of the data, and thus skews the response toward a preferred output. The first nonlinearity is based upon the ratio of autocorrelation to correlation, known to be of the order of $\sqrt{n}$, where n is the size of the vector, while the second nonlinearity could be a hard clip (winner-takes-all in order to prevent cross-talk). The system is then made up of two optical vector-matrix (OVM) multipliers connected by LED through electrical nonlinearities devices. The first OVM implements the standard retrieval operation $s = Ms'$, and the first nonlinearity is a sigmoid function operating on the recall and biasing it toward expected values of correlation rather than autocorrelation. The second OVM reprojects the resulting vector onto the data space, and then the second nonlinearity, in an attentive manner, makes a final choice regarding the identity of the data vector. Experimental ANS of the optical type have been implemented by Psaltis and Farhat (1985) and Anderson *et al.* (1986), among others.

9.6.   Conclusions

We have shown that there are parallel hardware architectures corresponding to different levels of visual processing. Such architectures provide the appropriate implementational counterpart for the parallel representations and algorithms presented in the preceding chapters. The concept of parallelism thus permeates the whole stream of visual processing.

Woman stooping to pick up a handkerchief. (From Muybridge, E., The Human Figure in Motion, 1887.) Reprinted courtesy of the National Library of Medicine.

10

Epilogue

Science is not the steady, cumulative acquisition of knowledge that is portrayed in the textbooks. Rather, it is a series of peaceful interludes punctuated by intellectually violent revolutions ... in each of which one conceptual world view is replaced by another....

From *The Structure of Scientific Revolution* by Thomas Kuhn, University of Chicago Press, Chicago, 1962.

We have presented an overall and integrated view of computational vision. The reader should be well aware that our view is by no means conclusive, however, and that the study of vision is still in its infancy. We suggest a need for a revolution in our conceptualization of the whole computational vision problem, if we want to move forward.

10.1. Plastic Arts

Perception is continuously mediated by our operational structures. In modern art, many times there is very little to perceive but much to (re)structure. Both modern art and science gave up the concept of an ultimate reality, and thus they came to perceive and understand the world not in concrete terms but as abstract relations. Visual rendition of both our environment and our own mental imagery is complementary to visual interpretation. This is because the artist always keeps in mind the inseparability of seeing and interpreting, and because the mission of art is not merely to copy life but to express it.

Perception is mediated by our own knowledge of the world, and one can always be reminded that where there is no perception, there is no knowledge; and that where there is no knowledge, there is no perception. The knowledge we refer to is much more than mere photographic information, and it includes episodic memories, as discussed earlier. Knowledge can be structured in different ways, however, and it is the role of the artist to reflect and sometimes even to advance new ways about how we should look around us. If you were to take a leisurely stroll in any of the world's great art museums, you will not only enjoy and relish the exhibits, but maybe also ask yourself how all of the different artistic styles evolved across time, and what lies behind the many artistic movements. Suzi Gablik (1977) considers this very question in her provocative book, *Progress in Arts*, while Robert Hughes (1982) captures the advent of modern art in a book entitled *The Shock of the New*.

The question Suzi Gablik addresses in her writing is related to the many changes in artistic styles we have witnessed so far. She suggests that such changes are related to cognitive growth and can be understood using developmental psychology. Accordingly, the history of art parallels the dynamics and the structure of our own mental processes. It traces an evolutionary path starting with figurative and iconic imagery and leading to the nonrepresentational and abstract imagery of today.

As ontogeny recapitulates philogeny, so too the history of the arts follows the evolution of mental development. Even more specifically, the history of the arts follows genetic epistemology, a term used by Gablik to denote the genesis and development of systems of knowledge. The episteme is an epistemological space specific to a particular period, or what one would call a *spirit of age*. Development projects itself in different forms used to represent "reality." To quote Gablik, "evolution of human cognition has led to changes in the way we experience and represent the world. Art not only relates to the development of knowledge but presupposes it." The artist by the mere act of representation goes beyond perception. The artist does not merely copy his or her subject but uses his knowledge to reconstruct it. Quoting from

Piaget and Inhelder's work on childrens' conception of space, Gablik suggests that "the image is a pictorial anticipation of an action not yet performed, and a reaching forward from what is presently perceived to what may be, but is not yet perceived." Mental imagery also presupposes previous knowledge and uses it for anticipation.

Indeed, by examining plastic arts, one gains additional insights for future computational vision research. After all, early Chinese and Renaissance artisans were aware of and used the Craik–O'Brien illusion in order to create artificial visual boundaries or to enhance the actual boundaries as in chiaroscuro. At the turn of the century, Cezanne exhorted his fellow artists to "treat nature in terms of the cylinder, the sphere, the cone, all in perspective." How similar that exhortation is to much of CAD/CAM, the generalized cylinders, and image interpretation and understanding systems such as ACRONYM. Later on, George Braque tried to undo the process of recovering the 3D shape, and his painted billiard tables attempt to convey an impression of flatness. It is illuminating to quote Braque: "I said good bye to the vanishing point. And to avoid any projection towards infinity I interposed a set of planes, set one on top of another, at a short distance from the spectator. It was to make him realize that objects did not retreat backwards into space but stood up close in front of one another. Cezanne had done away with distance and that after him infinity no longer exists. Instead of starting with the foreground I always began with the center of the picture. Before long I even turned perspective inside out and turned the pyramid of forms upside down so that it came forward to meet the observer."

Art involves both ability and adaptation as they relate to competence and the possibility of integrating different visual experiences. It further involves the capability to describe, manipulate, and infer about the world surrounding us. Such capabilities presuppose that in the ensuing process the artist transforms both the self and his or her world. Throughout time, development in arts meant the choice of new invariances to depict reality and to restructure properties at higher levels of abstraction. Artistic development also parallels the self-organization of cognitive systems leading to complex interactions between the organism and its environment. Early art starts with figurative depiction, where the artist fixes his or her attention on single features, neglecting the other ones and their relationships. To faithfully depict reality, however, the artist needs to consider transformations that include compensatory and conservation processes and reach the stage of operational thinking. As Piaget notes, to know is to assimilate reality into systems of transformations. Furthermore, one needs to transform reality in order to understand how a certain state is brought about. Mental structures involved in perception and cognition are dynamic rather than static. And it is only through their continuous functioning and permanent interaction with the environment that they develop, stabilizing and making coherent sensory experiences, and compensating internally for external change.

Gablik suggests that development in the arts amounts to decentrating and enhancing the objectivity of experiencing reality. Decentration allows for the assimilation of reality into our structures, which she calls schemata, and for accommodation, where operational thinking can transform reality. According to her, art history has experienced three major stages of development, which correspond to the enactive, iconic, and symbolic modes, respectively. One finds many parallels between developments that occurred in art history and those developments we have witnessed in computational vision.

The enactive stage is characteristic of ancient and medieval art. It corresponds to a preoperational and static level of development, whereby space is organized in egocentric terms. Spatial characteristics of the enactive stage do not allow for depth representation, and size and/or distance are not preserved. The iconic stage is characteristic of the Renaissance, and it corresponds to the concrete-operational level of development. Space becomes organized and separates the observer from his world. It allows for coordinate systems and provides for projective and Euclidean relationships. Perspective takes a major thrust forward, but it does not allow for chance or randomness in the viewing process. It took another 500 years before Cezanne and Cubism introduced multiple and shifting viewpoints. Active and directed perception (to be discussed in the next section), allow for multiple viewpoints as well as their interactions. Finally, the symbolic stage is characteristic of (post)modern art history and corresponds to the formal-operational level of conceptual development. The artist eventually wins freedom from always having to refer to some specific empirical reality. Modern art works primarily through transformations and attempts to do many things with limited raw material. Geometric thought and reasoning become independent of objects and grow increasingly deductive in character. The viewer is an active participant in the process of deciphering the hidden message, and the final interpretation, if at all possible, is mediated by our own mental structures and emotions.

Technological innovation, among other things, led to the birth of the futurist movement in Italy after the turn of the century. Artists such as Boccioni would depict movement and would allow the sculptured human body to spring forth into space. The Futurists' interest in motion led them to borrow from multiple-exposure photography and the sequential images created earlier by Muybridge. Architects such as Frank Lloyd-Wright would later incorporate movement into their work and express it as fluid inner space. "Falling Water," one of Lloyd-Wright's masterpieces, leaves the viewer uncertain as to where the structure starts and where nature begins. The concept of a fixed boundary then becomes meaningless. For Ya'acov Agam, a leading exponent of kinetic art, there is no fixed imagery, but only perceptual motion. Agam insists that "his non-static art and his geometric imagery is in no way abstract, but is the most realistic of all, because in reality nothing is fully visible from one vantage point." Things are perceived in stages and evolve continuously. Agam deplores that "much of the art fought time to preserve momentary existence, and to freeze and perpetuate the present for eternity."

The development of art history fully reflects the importance of motion as a major component of visual expression. The descriptive value of lines for defining contours of objects lost its relevance. For Goya, "there are no lines in nature, only lighted forms and forms which are in shadow, planes which project and planes which recede." To quote again from Gablik, "each stage in the history of art manifests a denser and more authoritative manipulation of bodily movements and gestures; the images cluster, ramify, and precipitate toward becoming more corporal, fluid, and tangible. Modern artists achieve a synthesis of experience, movement, and perception over time which is unknown in earlier art." The modern artist thinks wholly about the total volume of a figure and synthesizes the profile and frontal views. Figures drawn by Picasso and Bacon converge and deflect themselves around surrounding space. The same figures, rather than being scattered and disconnected as was customary previously, coalesce with each other into one image. The aspects graphs, dynamic vision, and the visual 3D object recognition loop suggested in Chapters 6 and 7, are the computational counterparts to those used by the artist for whom multiple appearances not only succeed, but confirm, continue, and complement each other.

Early art was mimetic in its rendering. There was only one true account of the world, and art's success was measured by how close it could represent it. The meaning of the artistic work was rarely in doubt. However, modern art goes much beyond testable reality and defies the logic of scientific discovery. It attempts to find something new, that which is yet unknown. Note that this dichotomy (also considered in Chapter 8) is again between error-correction strategies and maxima search. Some of the modern art movements, for example Deconstructionism, go even further. They defy rules and cast many doubts on basic assumptions, such as those involving a unifying wholeness or fixed meaning. Deconstructionism tears apart the fabric of known artistic creativity and interpretation and attempts to free the artist from conventions. A sense of unity is replaced with a sense of fluidity and countless possibilities. One then perceives chaos, even though order lurks behind.

Art and science developed a partnership during the Renaissance that was beneficial to both of them. Today, the arts seem to have left the visual sciences behind. One can only conjecture as to what could happen if the arts and science were to rejoin. As Suzi Gablik reminds us, "art is not a recommendation about the way the world is, it is a recommendation that the world ought to be looked at in a given way." There are many ways and corresponding art styles to see the world —each of them true and different. The monolithic view of the world becomes a thing of the past, and choice of expression becomes essential to appropriately render and understand the world.

10.2.   How Direct Is Perception?

We will now leave the plastic arts and great art museums to stroll briefly into philosophers' ivory towers. The question we address now is related to the very act of perception and its meaning. Much of the discussion is centered around the directness of the perceptual act, which is relevant to how one approaches computational vision theory development. Ullman (1980), Cutting (1986) and Neisser (1989) provide engaging discussions on the topic.

Two basic philosophical lines about how perception works are the direct and indirect types. Few philosophers or scientists, however, have ever subscribed to either one in a pure and unqualified form. Direct visual perception (DVP) simply states that the relation between stimuli and percepts is immediate, or according to Gibson, that stimulation is a function of the environment, and that perception is a function of stimulation. Indirect visual perception (IVP), is mostly a cognitive approach involving some sort of inference and problem solving. The debate between DVP and IVP goes back at least as far as to Descartes, Locke, and Berkeley. The conceptual qualifiers, over which much of the debate was fought, include, according to Cutting, judgment and inference, speed of response, learning, mediation, suggestibility, awareness, physical distance, decomposition, and information efficacy. DVP is thought of as fast, innate, and it does not require any thought-like processes. It seeks invariants and resonates to them. Regarding the inference aspect, Cutting correctly points out that "if all premises of the inference are either sources of information in the optic array or derived from the way that the visual system is built, then this type of inference is not different from Gibson's direct perception." Note also that almost everyone agrees that the role of perception is not to create information, but to extract it, to interpret it, and to make it explicit and useful.

Visual perception is often undetermined by the stimulation available in the optic array. It is then clear that additional constraints are needed to make up for any lack of visual information. The additional constraints can be phylogenetic and result from the visual system's adaptation to nature's regularities during the course of evolution. The stimulus used to perceive is, therefore, more than the actual retinal stimulus as visual interpretation becomes regularized and thus feasible. If the constraints are phylogenetically determined, it would make sense, for efficiency reasons, to have them precompiled rather than interpreted and learned anew.

"That not all that is seen is learned" has been shown by studies suggesting that infants can perceive objects in depth with no previous coordinated visual and tactile experience. For Paul Klee, the well-known Swiss painter, "the eye follows the paths that have been laid down for it." Consequently, there is a place for learning in such a scheme, such as when we acquire new abilities through differentiation. The possibility for both *nature* and *nurture* then follows quite naturally from what we know from ethology about innate mechanisms and critical periods. Furthermore, both the availability and the need for internalized constraints shows that DVP is not so unmediated after all. There is an inseparability between seeing and interpreting, as the observer affects the observed, and also because we see it as we interpret it. Perception is mediated by our knowledge of the world, and general purpose and intelligent visual systems will emerge only to the degree that the richness of the surrounding world can be assimilated and further accommodation becomes feasible. The mind resonates appropriately only when it is fully aware of the world it has to resonate to.

There seems to be apprehension, to say the least, in introducing a concept such as resonance. However, to the extent that a concept is described in terms of meaning, representation and algorithm, and implementation, there is no radical departure from the computational approach started by Marr. Resonance in terms of PDP is clean and straightforward, and no magic is involved either. Modern physics and immunology are accustomed to resonance. Physicists who search for exotic particles call their work "resonance hunting," while immunologists seek antibodies that will resonate to foreign intruders and ultimately destroy them.

The directness of perception is a matter of degree, and it can be further discussed according to mappings between information and object properties. DVP and IVP assume one-to-one and many-to-many mappings and are conceptually characteristic of deductive and inductive processing, respectively. DVP is relatively fast and works through the invariants and affordances the environment provides. IVP, on the other hand, is relatively slow and works through many cues and constraints needed to untangle information and eliminate unlikely conjectures about the true nature of the information being mapped. Many expert vision systems today, as discussed in Chapter 8, are merely instantiations of the IVP approach and exhibit as a consequence both brittleness and narrowness in their scope. There is another approach to perception, appropriately called *directed* visual perception (DRVP) by Cutting (1986), which implements a many (information)-to-one(object properties) mapping. Thus, there is the possibility of being selective about which information source one can use, and one can still hold to the advantages provided by direct perception. Different invariants, based on their specific information efficacy, thus can be selected and used for different scenarios. Another way to look at the three perceptual approaches discussed, according to Cutting (1986), is to realize that in DVP, information is rich and specifies both process (what has to be picked up) and interpretation (of objects and events). IVP, alternatively, assumes that the stimulus is impoverished, and consequently underspecifies process and interpretation. Only DRVP underspecifies process but overspecifies interpretation. The perceiver has then the choice of what information to pick up, and both adaptation and task functionality can further enhance the ultimate visual performance. Exploration, which is characteristic of active perception, is clearly an essential ingredient for DRVP, because it allows the observer to "intelligently" attend to only those affordances most likely to be successful in sifting through the information available in the optical array. Note that many of those affordances are ecologically motivated and correspond to the niche we have built for ourselves along evolution.

Another dimension one can consider in assessing the three perceptual approaches is the directionality of processing. DVP and IVP are primarily bottom-up and top-down, respectively. DRVP is, instead, a loop, both bottom-up data-driven and (possibly functionally) top-down (task and model-driven) constrained. The integrated model suggested in Fig. 6.1. indeed presupposes a DRVP approach. Our parallel and distributed computation (PDC) modeling and 3D recognition results presented in Section 7.5. follow the DRVP approach, where the choice of process is engraved onto the connections linking the characteristic views into a coherent visual potential (VP). Furthermore, a PDC approach provides for both fast and robust parallel indexing. The look-up is definitely not a mere detail left for implementation, and if implemented using PDC, it speeds up visual processing.

In summary, much of the debate on the directness of perception revolves around the role cognitive processes play in perception. Some of the misunderstandings might be rooted in the use of the term *computation*. Quoting from Cutting, "the Latin root *computare* means 'with thinking,' and although I do not wish to isolate perception from cognition, it seems clear to me that most perception is not done with thinking; it is done on its own terms. And because our surroundings are rich, our perception must be directed." Vision is indeed computational, and it provides a "mobile and dynamic" observer with the capability to decide what and how much one needs to be exposed to, and from that experience what and how much information to "intelligently" pick up and process, so it can correctly interpret the surrounding world.

10.3. Conclusions

There must exist a book which is the cipher and perfect compendium of all the rest: some librarian has perused it, and is analogous to a God. I pray the unknown God that some man—even if only one man, and though it have been thousands of years ago!—may have examined and read it. If honor and wisdom and happiness are not for me, let them be for others.

From *Library of Babel* by Jorge Luis Borges

This book extends the domain of discussion on what vision is and how it might be implemented. It involves both mathematics and engineering and attempts to organize and integrate them as a science of visual computation. The need for experimentation in computational vision can hardly be overemphasized. It is quite disconcerting to see so much research being done today, with the ultimate goal seemingly only that of mathematical sophistication. Such work disregards what visual perception is and defies being tested through relevant experiments. Experiments that disregard the richness of the world, using impoverished or artificial testbeds, are also useless. Those who ignore mathematics are not to be commended either.

The most valuable computational vision theories are those continuously subjected to testing. Karl Popper (1982) focused on refutation instead of confirmation, and goes as far as to suggest that the role of the scientist is to falsify theories, not verify them. In that vein, Daugman's experimental work on disproving zero crossings as a complete low-level visual representation is to be commended.

Abstraction, a refuge for many, is perceived as a dangerous weapon by Brooks (1988). He correctly argues that AI researchers partition the problem they work on into two components. The AI component, which they solve, and the non-AI component, which they don't solve. Typically AI "succeeds" by defining the parts of the problem that are unsolved as not being AI. The principal mechanism for this partitioning is abstraction, and it is usually used to factor out all aspects of perception and motor skills. Representation, not direct experience, determines all meaning. But, as Moravec (1983a) argued earlier, those aspects are the very ingredients of intelligent behavior! Brooks further argues that these aspects, which are neglected, are the hard problems solved by intelligent systems, and that the shape of solutions to these problems greatly constrains the correct solutions of the small pieces of intelligence that remain.

To summarize, the theme underlying computational vision has been how the perceptual system can cope with the sheer complexity of the visual task and still display so much robustness and fault tolerance. We have suggested throughout our exposition the concepts of system modularity and integration and the relevance of systemic transformations. We have considered perception as a sequence of transformations, some of them carried out in parallel, whose goal is to capture some invariant aspect of the world surrounding us. Invariance allows us to recognize, so that we can then move and perform safely. Furthermore, as signals are transformed at each processing stage, a view of the visual field gradually emerges. The locality of analysis, characteristic of the early stages of visual processing, gradually loosens up, and a holistic view, predicated upon the preservation of information, finally emerges. Two additional major computational characteristics, distributed computation and active perception, help to make vision feasible. Distributed computation refers to both representation and processing, and an appropriate cytoarchitecture could most likely provide an invariant low-level joint image representation. Distributed computation also discredits atomistic perception, characteristic of reductionist models. As Lewontin (1989) correctly points out for biologists, "the problem is that in the very operation of determining what things are made of we take them to bits, and in ways that destroy the very relations that may be of essence. We murder to dissect." Active and directed perception is much related to mobility, and it is distributed when one considers that the observer has the choice to decide what affordances to be attuned to and when. Adaptation also has a key role in enhancing visual capability. Adaptation innately encodes regularities likely to be observed around us. Recalling that unity resides in lawful changes and transformations, such regularities mostly involve the geometry of image formation and motion. Nature having no way to forecast everything, however, allows enough room for nurturing and learning so the organism can differentiate better.

Active, directed perception and adaptation can be further linked.

According to Popper (1982), passive Darwinism can hardly explain why a well-adapted organism, living in a relatively stable environment, should change or evolve. However, to quote from Popper

> The problem may be soluble if we bring in active Darwinism; that is, something like mind. For in this case, we shall have mutations for exploratory behavior, for seeking other environments, new types of food, new ecological niches: a new habitat. This is what biologists have seen and what they call the habitat selection by the organism. Of course, it can be conjectured that it has its basis in mutations, like everything else. But by actively selecting a new ecological niche, a new way of living, new preferences, a new habitat, the organism exposes itself to a new type of selection pressure, a selection pressure that molds it to fit better into that new habitat, into that new way of life, which it has actively chosen, which it has preferred. In this way, the preference, the active choice made by the organism, turns into a choice of a characteristic kind of selection pressure, and in this way it can become the choice of a characteristic direction in the evolution of the species.

The visual potentials, introduced in Chapter 7 for recognition purposes, can then be made computationally efficient only when active perception and functional adaptation are linked. The developmental issue, characteristic of any organism, is an active research issue for many biologists, and computational vision can draw some useful analogies. As we have already discussed in Section 8.4.8., Edelman considers the development of the visual system as one of the many outgrowths of evolution. The invisible hand facilitating such a development is natural selection. Those visual programs leading to malfunctioning organisms have been lost in evolution because their owners left no progeny. All that is left is the collection of programs that give the appearance of organization simply because they work.

Even as our discussion closes on an integrated theory of computational vision, we wish that one could get hold of that fortunate librarian alluded to by Borges, who might have read the secret solution of an intricate puzzle, such as visual perception. As for future research, what we need most is new "vision," and further conceptual growth and maturation for modeling visual perception.

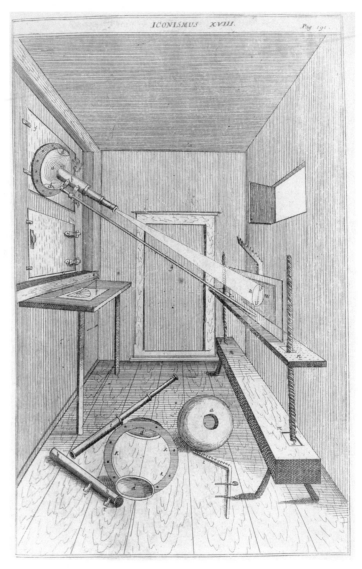

Direction of light into room through camera obscura. (From Zahn, J., Oculus artificialis., 1685–86.) Reprinted courtesy of the National Library of Medicine.

References

For brevity, the following frequently cited sources are given in abbreviated form:

AIJ	Artificial Intelligence Journal
AIM	Artificial Intelligence Magazine
BBS	Behavioral and Brain Sciences
BiolCyb	Biological Cybernetics
CS	Computing Surveys
CVGIP	Computer Vision, Graphics, and Image Processing
CVPR	Conference on Computer Vision and Pattern Recognition
ICASP	International Conference on Acoustics, Speech, and Signal Processing
ICCV	International Conference on Computer Vision
ICNN	International Conference on Neural Networks
ICPR	International Conference on Pattern Recognition
IJCAI	International Conference on Artificial Intelligence
IJRR	International Journal of Robotics Research
JOSA	Journal of Optical Society of America
P-IEEE	Proceedings of the IEEE
PR	Pattern Recognition
SA	Scientific American
SP	Signal Processing
T-IT	IEEE Transactions on Information Theory
T-PAMI	IEEE Transactions on Pattern Analysis and Machine Intelligence
T-SMC	IEEE Transactions on Systems, Man, and Cybernetics
VR	Vision Research

Abbott, A.L., and Ahuja, N. (1988). Surface reconstruction by dynamic integration of focus, camera vergence, and stereo. *ICCV*, Tampa, Florida, 532–543.

Abramovich, I., and Wechsler, H. (1986). Levels of representation for expert systems, *TR*, University of Minnesota.

Abu-Mostafa, Y. (1986). Neural networks for computing. In J.S. Denker, (ed.), *Neural Networks for Computing*. American Institute of Physics **151**, 1–6.

Abu-Mostafa, Y., and St. Jacques, Y. (1985). Information capacity of the Hopfield model. *T-IT* **7**, 1–11.

Acharya, R., Hefferman, P.B., Robb, R., and Wechsler, H. (1987). High-speed 3D imaging of the beating heart using temporal estimation. *CVGIP* **39**, 279–290.

Ackley, D.H., Hinton, G.E., and Sejnowski, T.J. (1985). A learning algorithm for Boltzmann machines. *Cognitive Sciences* **9**, 147–169.

Ada, G.L., and Nossal, G. (1987). The clonal-selection theory. *SA* **August**, 62–69.

Adelson, E.H., and Movshon, J.A. (1983). Phenomenal coherence of moving visual patterns. *Nature* **300**, 523–525.

Ahmed, N., and Rao, K.R. (1975). *Orthogonal Transforms for Digital Signal Processing*. Springer-Verlag, New York.

Aizerman, M.A., Braverman, E.M., and Rozoner, L.I. (1964). Theoretical foundations of the potential function method in pattern recognition learning. *Automation and Remote Control* **25**, 821–837.

Albus, J.S. (1981). *Brains, Behavior, and Robotics*. McGraw-Hill, New York.

Alomoinos, J. (1988). Visual shape computation. *P-IEEE* **76**(8), 899–916.

Alomoinos, J., and Shulman, D. (1989). *Integration of Visual Modules*. Academic Press, Boston.

Amari, S., and Arbib, M.A. (eds.) (1982). *Competition and Cooperation in Neural Nets*. Springer-Verlag, New York.

Anderson, B.D.O., and Moore, J.B. (1979). *Optimal Filtering*. Prentice Hall, Englewood Cliffs, New Jersey.

Anderson, C.H., Burt, P.J., and Van der Wall, G.S. (1985). Change detection and tracking using pyramid transform techniques. *Proc. of SPIE Conf. on Intelligence, Robots and Computer Vision* **579**, 72–78.

Anderson, J.A., and Rosenfeld, E. (eds.) (1988). *Neurocomputing.* MIT Press, Cambridge, Massachusetts.

Anderson, J.A., Silverstein, J.W., Ritz, S.A., and Jones, R.S. (1977). Distinctive features, categorical perception, and probability learning: some application of a neural model. *Psychological Review* **84**, 413–451.

Anderson, J.A., Golden, R.M., and Murphy, G.L. (1986). Concepts in distributed systems. In H. Szu, (ed.), *Proc. of SPIE Conf. on Hybrid and Optical Systems*, ASPIE, Seattle, Washington.

Andress, K.M., and Kak, A. (1988). Evidence accumulation and flow of control in a hierarchical spatial reasoning system. *AIM* **9**(2), 75–94.

Andrews, H.C., and Hunt, B.R. (1977). *Digital Image Restoration.* Prentice Hall, Englewood Cliffs, New Jersey.

Annaratone, M., Arnold, E., Gross, T., Kung, H.T., Lam, M., Menzilcioglu, O., and Webb, J.A. (1987). The WARP computer: architecture, implementation, and performance. *TR*-87-18, Robotics Institute, Carnegie-Mellon University.

Arun, K.S., Huang, T.S., and Blostein, S.D. (1987). Least-squares fitting of two 3D point sets. *T-PAMI* **9**(5), 698–700.

Asada, M., and Tsuji, S. (1987). Shape from projecting a stripe pattern. *CAR–TR–263*, Computer Vision Lab, University of Maryland.

Asada, M., Fukui, Y., and Tsuji, S. (1988). Representing a global map for a mobile robot with relational local maps from sensory data. *ICPR*, 520–524.

Achby, W.R. (1960). *Design for a Brain.* Wiley, New York.

Athale, R.A., Lee, J.N., Robinson, E.L., and Szu, H. (1982). Acousto-optic processors for real-time generation of time-frequency representations. *Optics Letters* **8**, 166–168.

Athale, R.A., Szu, H., and Friedlander, C. (1986). Optical implementation of associative memory with controlled nonlinearity in the correlation domain. *Optics Letters* **11**, 482–484.

Athas, W., and Seitz, C. (1988). Multicomputers: message-passing concurrent computers. *Computer* **21**(8), 9–25.

Ayer, A.J. (1984). *Philosophy in the Twentieth Century.* Vintage Books, New York.

Bajcsy, R. (1988). Active perception. *P-IEEE* **76**(8), 996–1005.

Bajcsy, R., and Lieberman, L. (1976). Texture gradient as a depth cue. *CGIP* **5**, 52–67.

Ballard, D.H. (1986). Cortical connections and parallel processing: structure and function. *BBS* **9**, 67–120.

Ballard, D.H. (1988). Eye fixation and early vision: kinetic depth. ICVV, Tampa, Florida, 524–531.

Ballard, D.H., and Brown, C. (1982). *Computer Vision*. Prentice Hall, Englewood Cliffs, New Jersey.

Ballard, D.H., and Sabah, D. (1983). Viewer independent shape recognition. *T-PAMI* **5**(6), 653–660.

Ballard, D.H., Hinton, G.E., and Sejnowski, T.J. (1983). Parallel vision computation. *Nature* **306**(5938), 21–26.

Bamler, R., and Glunder, H. (1983a). Coherent-optical generation of the Wigner distribution function of real-valued 2-D signals. In *Proc. 10th Int. Optical Computing Conf.*, Cambridge, 117–121.

Bamler, R., and Glunder, H. (1983b). The Wigner distribution of two-dimensional signals: coherent-optical generation and display. *Optica Acta* **30**, 1789–1803.

Barlow, H.B. (1972). Single units and sensation: a neuron doctrine for perceptual psychology? *Perception* **1**, 371–394.

Barnard, S.T., and Thompson, W.B. (1980). Disparity analysis of images. *T-PAMI* **2**(4), 333–340.

Barr, A. (Vol. 1–2), Cohen, P., (Vol. 3), and Feigenbaum, E. (Vol. 1–3) (eds.) (1981). *Handbook of Artificial Intelligence*. Morgan-Kaufman, San Mateo, California.

Barrow, H.G., and Tenenbaum, J.M. (1978). Recovering intrinsic scene characteristics from images. In A.R. Hanson, and E.M. Riseman (eds.), *Computer Vision Systems*. Academic Press, New York.

Bartelt, H. (1980). The Wigner distribution function and its optical production. *Optics Communications* **25**, 32–37.

Bastiaans, M.J. (1978). The Wigner distribution function applied to optical signals and systems. *Optics Communications* **25**, 26–30.

Bastiaans, M.J. (1980). Wigner distribution function and its application to first-order optics. *JOSA* **69**, 1710–1716.

Baum, E.B., and Wilczek, F. (1988). Supervised learning of probability distributions by neural networks. In D. Anderson (ed.), *Neural Information Processing Systems*. American Institute of Physics (AIP), New York, 52–61.

Beck, J. (1983). Textural segmentation, second-order statistics, and textural elements. *BiolCyb* **48**, 125–130.

Beck, J., Sutter, A., and Ivry, R. (1987). Spatial frequency channels and perceptual grouping in texture segregation. *CVGIP* **37**, 299–325.

Besl, P.J., and Jain, R.C. (1985). Three-dimensional object recognition. *CS* **17**(1), 75–145.

Besl, P.J., and Jain, R.C. (1986). Invariant surface characteristics for three-dimensional object recognition in range images. *CVGIP* **33**, 33–80.

Besl, P.J., and Jain, R.C. (1988). Segmentation through variable-order surface fitting. *T-PAMI* **10**(2), 167–192.

Biederman, I. (1987). Recognition-by-components: a theory of human image understanding. *Psychological Review* **94**, 115–147.

Bienenstock, E., Fogelman-Soulie, F., and Weisbuch, G. (eds.) (1986). *Disordered Systems and Biological Organization*. Springer-Verlag, New York.

Binford, T.O. (1982). Survey of model-based image analysis systems. *IJRR* **1**(1), 18–64.

Blake, A. (1989). Comparison of the efficiency of deterministic and stochastic algorithms for visual reconstruction. *T-PAMI* **11**(1), 2–12.

Blake, A., and Zisserman, A. (1987). *Visual Reconstruction*. MIT Press, Cambridge, Massachusetts.

Blakemore, C., and Campbell, F.W. (1969). On the existence of neurons in the human visual system selectively sensitive to the orientation and size of retinal images. *J. Physiology* **203**, 237–260.

Bobrow, D.G., and Stefik, M.J. (1983). The LOOPS Manual. Xerox PARC.

Bobrow, D.G., and Winograd, T. (1977). An overview of KRL. *Cognitive Sciences* **1**, 3–46.

Boudreaux-Bartels, G.F. (1983). *Time-Frequency Signal Processing Algorithms: Analysis and Synthesis Using Wigner Distributions*. Ph.D. Dissertation, Dept. of Electrical Engineering, Rice University.

Bourne, L.E., Jr., and Restle, F. (1959). Mathematical theory of concept identification. *Psychological Review* **66**, 278–296.

Bracewell, R. (1978). *The Fourier Transform and Its Applications*. McGraw-Hill, New York.

Bracewell, R. (1989). The Fourier transform. *SA* **June**, 86–95.

Brachman, R.J. (1978). A structured paradigm for representing knowledge. BBN Report No. 3605, Bolt Beranek & Newman, Cambridge, Massachusetts.

Brachman, R.J. (1979). On the epistemological status of semantic networks. In N.V. Findler (ed.), *Associative Networks: Representation and Use of Knowledge by Computers*. Academic Press, New York, 3–50.

Brady, M. (1982). Computational approaches to image understanding. *CS* **14**(1), 3–72.

Brodatz, P. (1966). *A Photographic Album for Artists and Designers*. Dover, New York.

Brooks, R.A. (1981). Symbolic reasoning among 3-D models and 2-D images. *AIJ* **17**, 285–348.

Brooks, R.A. (1985). Visual map making for a mobile robot. *IEEE Int. Conference on Robotics and Automation*, 819–824.

Brooks, R.A. (1986). A robust layered control system for a mobile robot. *IEEE Journal of Robotics and Automation* **2**, 14–22.

Brooks, R.A. (1988). Intelligence without representation. TR, MIT AI Laboratory.

Brooks, R.A. (1990). A robot being. In P. Dario, P. Aebisher, and G. Sandini (Eds.), *Robots and Biological Systems*. Springer–Verlag, New York.

Bruck, J., and Goodman, J.W. (1987). On the power of neural networks for solving hard problems. *IEEE Conference on Neural Information Processing Systems—Natural and Synthetic*, Denver, Colorado.

Burt, P.J. (1988). Smart sensing within a pyramid vision machine. *P-IEEE* **76**(8), 1006–1015.

Burt, P.J., and Adelson, E.H. (1983). The Laplacian pyramid as a compact image code. *IEEE Trans. on Communications* **31**(4), 532–540.

Caelli, T. (1985). Three processing characteristics of visual texture segmentation. *Spatial Vision* **1**(1), 19–30.

Campbell, F.W., and Robson, J.G. (1968). Application of Fourier analysis to the visibility of gratings. *J. Physiology* **197**, 551–566.

Campbell, F.W., Cooper, G.F., and Enroth-Cugell, C. (1969). The spatial selectivity of the visual cells of the cat. *J. Physiology* **203**, 223–235.

Cantoni, V., and Levialdi, S. (1987). PAPIA. In L. Uhr (ed.), *Parallel Computer Vision*. Academic Press, Boston, 3–14.

Carnevali, P., Coletti, L., and Patarnello, S. (1985). Image processing by simulated annealing. *IBM J. Res. Develop* **29**(6), 569–579.

Carpenter, G.A., Cohen, M.A., and Grossberg, S. (1987). Computing with neural networks. *Science* **235**, 1226–1227.

Casasent, D., and Chang, W.T. (1983). Generalized chord transformation for distortion-invariant optical pattern recognition. *Applied Optics* **22**(14), 2087–2094.

Casasent, D., and Psaltis, D. (1975). Position, rotation and scale invariant optical correlation. *Applied Optics* **15**, 1795–1799.

Castleman, K. (1979). *Digital Image Processing*. Prentice-Hall, Englewood Cliffs, New Jersey.

Caulfield, H.J., and Weinberg, M.H. (1982). Computer recognition of 2-D pattern using generalized matched filters. *Applied Optics* **21**(9).

Cavanagh, P. (1978). Size and position invariance in the visual field. *Perception* **7**, 167–177.

Cavanagh, P. (1984). Image transforms in the visual system. In P. Dodwell and T. Caelli, (eds.) *Figural Synthesis*. Erlbaum, Hillsdale, New Jersey, 185–213.

Cavanagh, P. (1985). Local log polar frequency analysis in the striate cortex as a basis for size and orientation invariance. In D. Rose and V.G. Dobson (eds.), *Models of the Visual Cortex*. Wiley, New York, 85–95.

Cavanagh, P. (1989). How 3D are we? *Vision and 3D Representation* (*ABSTRACTS*), University of Minnesota, May 24–26.

Chakravarty, I., and Freeman, H. (1982). Characteristic views as a basis for 3D object recognition. *Proc. SPIE on Robot Vision* **336**, 37–45.

Chandrasekaran, B. (1985). Generic tasks in knowledge-based reasoning: Characterizing and designing expert systems at the "right" level of abstraction. *Proc. of the IEEE 2nd Int. Conf. on AI Applications*.

Chin, R.T., and Dyer, C.R. (1986). Model-based recognition in robot vision. *CS* **18**(1), 67–108.

Claasen, T.A.C.M., and Mecklenbrauker, W.F.G. (1980). The Wigner distribution—A tool for time-frequency analysis. Parts I–III. *Philips Journal of Research* **35**, 217–250, 276–300, 372–389.

Claasen, T.A.C.M., and Mecklenbrauker, W.F.G. (1984). On the time-frequency discrimination of energy distributions: Can they look sharper than Heisenberg? *ICASP*, San Diego, California.

Clark, M., and Bovik, A.C. (1989). Experiments in segmenting texton patterns using localized spatial filters. *PR* **22**(6), 707–718.

Cohen, F.S. (1986). Markov random fields for image modeling and analysis. In V.B. Desai (ed.), *Modeling and Application of Stochastic Processes*. Kluwer, Norwell, Massachusetts, 243–272.

Cohen, L. (1966). Generalized phase-space distribution functions. *J. Math. Physics* **7**, 781–786.

Cohen, M.A., and Grossberg, S. (1983). Absolute stability of global pattern formation and parallel memory storage by competitive neural networks. *T-SMC* **13**(5), 815–825.

Connell, H.J., and Brady, M. (1987). Generating and generalizing models of visual objects. *AIJ* **31**, 159–183.

Connors, R.W., and Harlow, C.A. (1980). A theoretical comparison of texture algorithms. *T-PAMI* **2**(3), 204–222.

Cook, S.A. (1971). The complexity of theorem proving procedures. *Proc. 3rd Ann. ACM Symp. on Theory of Computing*, 151–158.

Crick, F. (1989). The recent excitement about neural networks. *Nature* **337**(12), 129–130.

Crowder, R.G. (1976). *Principles of Learning and Memory*. Erlbaum, Hillsdale, New Jersey.

Crowley, J.L. (1984). Machine vision: three generations of commercial systems. TR-Robotics Institute, Carnegie-Mellon University.

Crowley, J.L., and Parker, A.C. (1984). A representation for shape based on peaks and ridges in the difference of low-pass transform. *T-PAMI* **6**(2), 156–169.

Cruz-Young, C., and Tam, J.Y. (1985). NEP: an emulation assist processor for parallel associative networks. IBM Palo Alto Scientific Center Report Number G320–3475.

Cutting, J.E. (1986). *Perception with An Eye for Motion*. MIT Press, Cambridge, Massachusetts.

Daugman, J.G. (1983). Six formal properties of 2-D anisotropic visual filters: structural principles and frequency/orientation selectivity. *T-SMC* **13**(5), 882–887.

Daugman, J.G. (1985). Uncertainty relation for resolution in space, spatial frequency, and orientation optimized by two-dimensional visual cortical filters. *JOSA* **2**, 1160–1169.

Daugman, J.G. (1987). Image analysis and compact coding by oriented 2-D Gabor primitives. *SPIE Proceedings*, 758.

Daugman, J.G. (1988a). Complete discrete 2-D Gabor transforms by neural networks for image analysis and compression. *IEEE Trans. Acoustics, Speech and Signal Processing* **36**(7), 1169–1179.

Daugman, J.G. (1988b). Pattern and motion vision without Laplacian zero crossings. *JOSA* **5**(7), 1142–1148.

De Bruijn, N.G. (1967). Uncertainty principles in Fourier analysis. In O. Shisha (ed.), *Inequalities*. Academic Press, New York, 57–71.

DeJong, K. (1988). Learning with genetic algorithms. *Machine Learning* **3**, 121–138.

DeKleer, J. (1986). An assumption-based truth maintenance system (TMS). *AIJ* **28**(2), 127–163.

DeMenthon, D. (1986). Inverse perceptive of a road from a single image. *CAR–TR–210*, Computer Vision Lab, University of Maryland.

DeValois, K.K., DeValois, R.L., and Yund, E.W. (1979). Responses of striate cortex cells to gratings and checkerboard patterns. *J. Physiology* **291**, 483–505.

Deans, S.R. (1981). Hough transform from Radon transform. *T-PAMI* **3**(2), 185–188.

Denker, J.S. (ed.) (1986). *Neural Networks for Computing*. American Institute of Physics (AIP), New York.

Derin, H., and Elliott, H. (1987). Modeling and segmentation of noisy and textured images using Gibbs random fields. *T-PAMI* **9**(1), 39–55.

Dickmans, E.D., and Graefe, V. (1988a). Dynamic monocular machine vision. *Machine Vision and Applications* **1**, 233–240.

Dickmans, E.D., and Graefe, V. (1988b). Applications of dynamic monocular machine vision. *Machine Vision and Applications* **1**, 241–261.

Dijkstra, E.W. (1976). *A Discipline of Programming*. Prentice-Hall, Englewood Cliffs, New Jersey.

Dodwell, P.C. (1983). The Lie transformation group model of visual perception. *Perception and Psychophysics* **34**(1), 1–16.

Downing, D., and Bennett, I. (1988). Multi-transputer based parallel implementation of feature extraction for object recognition. *Proc. of the 8th OCCAM*, Sheffield, UK.

Doyle, J. (1982). A glimpse of truth-maintenance. In P.W. Winston and R.N. Brown (eds.), *Artificial Intelligence: MIT Perspective*. MIT Press, Cambridge, Massachusetts, 119–135.

Dreyfus, H. (1979). *What Computers Can't Do: The Limits of Artificial Intelligence*. Harper and Row, New York.

Duda, R., and Hart, P. (1973). *Pattern Classification and Scene Analysis*. Wiley, New York.

Dudgeon, D.E., and Mersereau, R.M. (1984). *Multidimensional Digital Signal Processing*. Prentice Hall, Englewood Cliffs, New Jersey, 137–138.

Durani, T. (1983). Systolic processor for computing the Wigner distribution. *Defense Electronics*.

Eagleson, R. (1987). Estimating 3D motion parameters from the changing responses of 2D bandpass spatial frequency filters. *IEEE Montreal Technologies Conference: Compint '87*, 102–105.

Edelman, G.M. (1982). Group selection and higher brain function. *Bulletin of the American Academy of Arts and Science*, **XXXVI**, 1.

Edelman, G.M. (1987). *Neural Darwinism. The Theory of Neuronal Group selection*. Basic Books, New York.

Edelman, G.M., and Reeke, G.N. (1982). Selective networks capable of representative transformations, limited generalizations, and associative memory. *Proc. Natl. Acad. Sci.* **79**, 2091–2095.

Elfes, A. (1989). Using occupancy grids for mobile robot perception and navigation. *Computer* **22**(6), 46–57.

Ellman, T. (1989). Exploration-based learning: A survey of programs and perspectives. *CS* **21**(2), 163–221.

Enroth-Cugell, C., and Robson, J.G. (1966). The contrast sensitivity of retinal ganglion cells for the cat. *J. Physiology* **187**, 517–552.

Erman, L.D., Hayes-Roth, F., Lesser, R.V., and Reddy, R.D. (1980). The Hearsay-II speech understanding system: integrating knowledge to resolve uncertainty. *CS* **12**(2), 213–253.

Fahlman, S.E. (1979). *NETL: A System for Representing and Using Real-World Knowledge*. MIT Press, Cambridge, Massachusetts.

Fahlman, S.E., and Hinton, G.E. (1987). Connectionist architectures for artificial intelligence. *Computer* **20**(1), 100–109.

Farah, M.J. (1985). The neurological basis of mental imagery. In S. Pinker (ed.), *Visual Cognition*. MIT Press, Cambridge, Massachusetts, 245–272.

Faugeras, O.D., and Hebert, M. (1986). The representation, recognition, and locating of 3D objects. *IJRR* **5**(3), 27–52.

Faugeras, O.D., and Hebert, M. (1987). The representation, recognition, and positioning of 3D shapes from range data. In T. Kanade (ed.), *Three-Dimensional Machine Vision*. Kluwer, Norwell, Massachusetts, 301–354.

Feldman, J.A., Fanty, M.A., Goddard, N.H., and Lynne, K.J. (1988). Computing with structured connectionist networks. *Comm. of the ACM* **31**(2), 170–187.

Fennema, C.I., and Thompson, W.B. (1979). Velocity determination in scenes containing several moving objects. *CGIP* **9**, 201–315.

Fikes, R.E., and Nilsson, N.J. (1971). STRIPS: A new approach to the application of theorem proving to problem solving. AIJ **2**, 189–208.

Finke, R.A. (1985). Theories relating mental imagery to perception. *Psychological Bulletin* **98**(2), 236–259.

Fischler, M. (1989). Representation and scene modeling problem. *Vision and 3D Representation (ABSTRACTS)*, University of Minnesota, May 24–26.

Fischler, M., and Bolles, R. (1981). Random-sample consensus: a paradigm for model fitting with applications to image analysis and automated cartography. *Comm. of the ACM* **24**(6), 381–395.

Fischler, M., and Firschein, O. (eds.) (1987). *Readings in Computer Vision*. Morgan Kauffman, San Mateo, California.

Flandrin, P. (1984). Some features of time-frequency representations of multicomponent signals. *ICASP*, San Diego, California.

Flandrin, P., and Escudie, B. (1979). Sur une condition necessaire et suffisante de positivite de la representation cojointe en temps en frequence des signaux d'energie finie. *Compte Rendus Academy Sciences* (Paris) **288**, Ser. A, 307–309.

Fleet, D.F., and Jepson, A.D. (1989). Computation of normal velocity from local phase information. *CVPR*, San Diego, California, 379–386.

Forney, G.D. (1973). The Viterbi algorithm. *P-IEEE* **61**, 268–278.

Frisby, J.P. (1980). *Seeing*. Oxford University Press.

Frisby, J.P., and Mayhew, J.E.W. (1979). Does visual texture discrimination precede binocular fusion? *Perception* **8**, 153–156.

Fu, K.S. (Ed.) (1982). *Syntactic Pattern Recognition and Applications*. Prentice-Hall, Englewood Cliffs, New Jersey.

Fukushima, K. (1984). A hierarchical neural network model for associative memory. *BiolCyb* **50**, 105–113.

Gablick, S. (1977). *Progress in Arts*. Rizolli, New York.

Gabor, D. (1946). Theory of communication. *J. IEEE* **93**, 429–459.

Gafni, H., and Zeevi, Y.Y. (1977). A model for separation of spatial and temporal information in the visual system. *BiolCyb* **28**, 73–82.

Gafni, H., and Zeevi, Y.Y. (1979). A model for processing of movement in the visual system. *BiolCyb* **32**, 165–173.

Gagalowicz, A. (1981). A new method for texture field synthesis: some applications to the study of human vision. *T-PAMI* **3**(5), 520–533.

Garey, M.R., and Johnson, D.S. (1979). *Computers and Intractability. A Guide to the Theory of NP-Completeness*. W.H. Freeman, San Francisco.

Garvey, T.D., and Lowrance, J.D. (1983). Evidential reasoning: an implementation for multisensor integration. *TN* 307, AI Center, SRI, Palo Alto, CA.

Garvey, T.D., Lowrance, J.D., and Fischler, M. (1981). An inference technique for integrating knowledge from disparate sources. *IJCAI*, Vancouver, Canada, 319–325.

Gaska, J.P., Foster, K.H., Nagler, M., and Pollen, D.A. (1983). Spatial and temporal frequency selectivity of V2 neurons in the Macaque monkey. *Invest. Ophthalmol. Visual Sci. (Suppl.)* **24**, 228.

Gazzaniga, M.S. (ed.) (1988). *Perspectives in Memory Research*. MIT Press, Cambridge, Massachusetts.

Geisler, W., and Hamilton, D. (1986). Sampling theory analysis of spatial vision. *JOSA* **3**, 62–70.

Gelb, A. (1974). *Applied Optimal Estimation*. MIT Press, Cambridge, Massachusetts.

Geman, S., and Geman, D. (1984). Stochastic relaxation, Gibbs distributions, and the Bayesian restoration of images. *T-PAMI* **6**(6), 721–741.

Georgeff, M.P., and Lansky, A.L. (1987). Reactive reasoning and planning. *Proceedings of the American Association for Artificial Intelligence*, 677–682.

Gibson, J. (1950). *The Perception of the Visual World*. Houghton-Mifflin, Boston, Massachusetts.

Gibson, J. (1966). *The Senses Considered as Perceptual Systems*. Houghton-Mifflin, Boston, Massachusetts.

Gibson, J. (1979). *The Ecological Approach to Visual Perception*. Houghton-Mifflin, Boston, Massachusetts.

Ginsberg, M.L. (1989). Universal planning: An (almost) universally bad idea. *AIM* **10**(4), 40–44.

Ginsburg, A.P. (1980). Specifying relevant spatial information for image evaluation and display design: an explanation of how we see certain objects. *Proceedings of the SID* **21**(3), 219–227.

Gordon, J., and Shortliffe, E. (1984). The Dempster-Shafer theory of evidence. In B.G. Buchanan, and E.H. Shortliffe (eds.), *Rule-Based Expert Systems*. Addison-Wesley, Reading, Massachusetts.

Gordon, M. (1988). Probabilistic and genetic algorithms for document retrieval. *Comm. of the ACM* **31**(10), 1208–1218.

Gould, J.L. (1982). *Ethology*. Norton, New York.

Gould, S.J. (1989). *Wonderful Life: The Burgess Shale and the Nature of History*. W.W. Norton, New York.

Goutsias, J. (1988). Mutually compatible Gibbs images: properties, simulation and identification. *ICASP*, New York.

Grimson, W., and Lozano-Perez, T. (1986). Model-based recognition and localization from sparse range data. In A. Rosenfeld (ed.), *Techniques for 3D Machine Perception*. North-Holland, Amsterdam, Holland, 113–148.

Grossberg, S. (1988). Nonlinear neural networks: Principles, mechanisms, and architectures. *Neural Networks* **1**(1), 17–61.

Gurari, E. (1989). *An Introduction to the Theory of Computation*. Computer Science Press, Rockville, Maryland.

Gurari, E., and Wechsler, H. (1982). On the difficulties involved in the segmentation of pictures. *T-PAMI* **4**(3), 304–306.

Hall, C.F., and Hall, E.L. (1979). A non-linear model for the spatial characteristics of human visual models in image processing. *T-SMC* **7**, 161–170.

Haralick, R. (1986). Glossary of computer vision. Machine Vision International.

Haralick, R., Shanmugam, K., and Dinstein, I. (1973). Textural features for image classification. *T-SMC* **3**(1), 610–621.

Hartline, H.K. (1940). The receptive field of the optic nerve fibers. *J. Physiology* **130**, 690–699.

Hartline, P., Kass, L., and Loop, M. (1978). Merging of modalities in the optic tectum: infrared and visual integration in the rattlesnake. *Science* **199**, 545–548.

Hayes, P.J. (1981). The logic of frames. In B.L. Webber, and N.J. Nilsson (eds.), *Readings in Artificial Intelligence*, Morgan Kaufmann, San Mateo, California.

Hayes-Roth, F., Waterman, D.A., and Lenat, D.B. (1983). *Building Expert Systems*. Addison-Wesley, Reading, Massachusetts.

Hebb, D.O. (1949). *The Organization of Behavior*. Wiley, New York.

Hecht-Nielsen, R. (1986). Artificial neural system technology. TRW, AI Center.

Heeger, D.J. (1987). Optical flow from spatiotemporal filters. *ICCV*, London, England, 181–190.

Helmholtz, H. (1878). The facts of perception. In R. Kahl (ed.), *Selected Writings of Hermann von Helmholtz*. Wesleyan University Press, 1971, 366–407.

Henderson, T.C. (1983). Efficient 3D object representation for industrial vision systems. *T-PAMI* **5**(6), 609–617.

Henderson, T.C., Allen, P., Cox, I., Mitche, A., Durrant-Whyte, H., and Snyder, W. (1987). Workshop on multisensor integration in manufacturing automation. UCCS–87–006, Computer Science, University of Utah.

Hering, E. (1868). *Die Lehre von binocularem sehen*.

Hester, C., and Casasent, D. (1981). Interclass discrimination using synthetic discriminant functions (SDF). *Proc. SPIE on Infrared Technology for Detection and Classification*, **302**.

Hillis, W.D. (1985). *The Connection Machine*. MIT Press, Cambridge, Massachusetts.

Hinton, G.E. (1981). A parallel computation that assigns canonical object-based frames of reference. *IJCAI*, Vancouver, Canada.

Hinton, G.E., and Parsons, L.M. (1988). Scene-based and viewer-centered representations for comparing shapes. *Cognition* **30**, 1–35.

Hintzmann, D.L. (1974). Theoretical implications of the spacing effect. In R.L. Solso (ed.), *Theories in Cognitive Psychology: The Loyola Symposium*. Erlbaum, Hillsdale, New Jersey, 77–99.

Hoffman, D.D., and Richards, W.A. (1985). Parts of recognition. In S. Pinker (ed.), *Visual Recognition*. MIT Press, Cambridge, Massachusetts, 65–96.

Hoffman, W.C. (1977). An informal historical description of the "LTG/NP." *Cahiers de Psychologie* **20**, 139–150.

Holland, J.H. (1975). *Adaptation in Natural and Artificial Systems*. University of Michigan Press.

Holland, J.H., Holyoak, K.F., Nisbett, R.E., and Thagard, P.R. (1986). *Induction: Processes of Inference, Learning, and Discovery*. MIT Press, Cambridge, Massachusetts.

Holland, S.W., Rossol, L., and Ward, M.R. (1979). CONSIGHT-I: A vision-controlled robot system for transferring parts from belt conveyors. In G.G. Dodd, and L. Rosol, (eds.), *Computer Vision and Sensor-Based Robots*. Plenum Press, New York, 81–100.

Hopcroft, J., and Ullman, J. (1979). *Introduction to Automata Theory, Languages and Computation*. Addison-Wesley, Reading, Massachusetts.

Hopfield, J.J. (1982). Neural networks and physical systems with emergent collective computational abilities. *Proc. Natl. Acad. Sci. USA* **79 April**, 2554–2558.

Hopfield, J.J. (1984). Neurons with graded response have collective computational properties like those of two-state neurons. *Proc. Natl. Acad. Sci. USA* **81**, 3058–3092.

Hopfield, J.J., and Tank, D.W. (1986). Computing with neural circuits: a model. *Science* **233**, 625–633.

Hopfield, J.J., Feinstein, D.I., and Palmer, R.G. (1983). "Unlearning" has a stabilizing effect in collective memories. *Nature* **304**, 158–159.

Horn, B.K.P. (1974). Determining lightness from an image. *CGIP* **3**, 277–299.

Horn, B.K.P. (1984). Extended Gaussian images. *P-IEEE*, 1671–1686.

Horn, B.K.P., and Schunk, M. (1981). Determining optical flow. *AIJ* **17**, 185–203.

Horowitz, E., and Sahni, S. (1978). *Fundamentals of Computer Algorithms*. Computer Science Press, Rockville, Maryland.

Hou, S.H. (1987). The fast Hartley transform algorithm. *IEEE Trans on Computers*, C-36(2), 147–156.

Hrechanyk, L.M., and Ballard, D.H. (1982). A connectionist model of form perception. *Proc. of IEEE Workshop on Computer Vision*. Rindge, New Hampshire.

Hu, M.K. (1962). Visual pattern recognition by moment invariants. *T-IT* 8, 179–187.

Huang, T.S., and Blostein, S.D. (1985). Robust algorithms for motion estimated based on two sequential stereo image pairs. *CVPR*, 518–523.

Hubel, D.H., and Wiesel, T.N. (1962). Receptive fields, binocular interaction and functional architecture in the cat's visual cortex. *J. Physiology* **160**, 106–154.

Hughes, R. (1982). *Shock of the New*. Knopf, New York.

Hummel, R.A. (1986). Representations based on zero-crossings in scale-space. *CPVR*, 204–209.

Hummel, R.A. (1987). The scale space formulation of pyramid data structures. In L. Uhr (ed.), *Parallel Computer Vision*. Academic Press, New York, 107–124.

Hummel, R.A., and Zucker, S.W. (1983). On the foundations of relaxation labeling processes. *T-PAMI* **5**(3), 267–287.

Hunt, A.E., and Sanderson, A.C. (1982). Vision-based predictive robotic tracking of a moving target. CMU–R–TR–82–15, The Robotic Institute, Carnegie-Mellon University.

Hurlbert, A.C. (1986). Formal connections between lightness algorithms. *JOSA* **3**, 1684–1693.

Hurlbert, A.C., and Poggio, T. (1988). Synthesizing a color algorithm from examples. *Science* **239**, 482–485.

Huttenlocher, D.P., and Ullman, S. (1987). Object recognition using alignment. *ICCV*, London, England, 102–111.

Huttenlocher, D.P., and Ullman, S. (1988). Recognizing solid objects by alignment. *Proc. Image Understanding Workshop* **2**, Cambridge, Massachusetts, 1114–1124.

Hwang, K., and Briggs, F. (1984). *Computer Architecture and Parallel Processing*. McGraw Hill, New York.

Ikeuchi, K. (1980). Shape from regular patterns. *ICPR*, Miami, Florida, 1032–1039.

Ikeuchi, K., and Kanade, T. (1988). Automatic generation of object recognition programs. *P-IEEE* **76**(8), 1016–1035.

Jackel, L.D., Graf, H.P., Hubbard, W., Denker, J.S., and Anderson, D. (1988). An application of neural net chips: handwritten digit recognition. Bell Labs.

Jacobson, L. (1987). Conjoint Image Representations and Their Applications. Ph.D. Thesis, Univ. of Minnesota.

Jacobson, L., and Wechsler, H. (1982a). The Wigner distribution and its usefulness for 2-D image processing. *ICPR*, Munich, West Germany.

Jacobson, L., and Wechsler, H. (1982b). A paradigm for invariant object recognition of brightness, optical flow and binocular disparity images. *Pattern Recognition Letters* **1**, 61–68.

Jacobson, L., and Wechsler, H. (1983). The composite pseudo-Wigner distribution. *ICASP*, Boston, Massachusetts.

Jacobson, L., and Wechsler, H. (1985). Joint spatial/spatial-frequency representations for image processing. *SPIE/Cambridge Int. Conference on Intelligent Robots and Computer Vision*, Boston, Massachusetts.

Jacobson, L., and Wechsler, H. (1987). Derivation of optical flow using a spatiotemporal-frequency approach. *CVGIP* **38**, 29–65.

Jacobson, L., and Wechsler, H. (1988). Joint spatial/spatial-frequency representation. *SP* **14**, 37–68.

Janez, L. (1986). Visual grouping without low spatial frequencies. *VR* **24**(3), 271–274.

Jenkin, M., and Kolers, P.A. (1986). Some problems with correspondence. RBCV–TR–86–10, Computer Sci. Dept., Univ. of Toronto, Canada.

Jepson, A.D., and Jenkin, M. (1989). The fast computation of disparity from phase differences. *CVPR*, San Diego, California, 386–398.

Johansson, G. (1976). Spatio-temporal differentiation and integration in visual motion perception. *Psychol. Res.* **38**, 379–393.

Julesz, B. (1975). Experiments in the visual perception of texture. *SA* **232**, 34–43.

Julesz, B., and Bergen, J. (1983). Textons, the fundamental elements in preattentive vision and perception of textures. *Bell Systems Technical Journal* **62**(6), 1619–1645.

Kak, A. (1983). Depth perception for robots. TR–EE 83–44, Purdue University.

Kak, A., and Chen, S. (eds.) (1987). *Spatial Reasoning and MultiSensor Fusion Workshop*. Morgan Kauffmann, San Mateo, California.

Kak, A.C., Boyer, K.L., Chen, C.H., Safranek, R.J., and Yang, H.S. (1986). A knowledge-based robotic assembly cell. *IEEE Expert* **1**(1), 63–85.

Kamgar-Parsi, B., and Kamgar-Parsi, B. (1987a). An efficient model of neural networks for optimization. *ICNN*, San Diego, California.

Kamgar-Parsi, B., and Kamgar-Parsi, B. (1987b). Evaluation of quantization errors in computer vision. CAR–TR–316, Center for Automation Research, University of Maryland.

Kamgar-Parsi, B., and Kamgar-Parsi, B. (1989). On problem solving with Hopfield neural networks. *ICNN*, Washington, D.C.

Kamgar-Parsi, B., Kamgar-Parsi, B., and Wechsler, H. (1990). Simultaneous fitting of several planes to point sets using neural networks, *CVGIP* (to appear).

Kanade, T. (1980). Survey. Region segmentation: signal vs. semantics. *CGIP* **13**, 279–297.

Kanatani, K. (1986). Group theoretical methods in image understanding. CAR–TR–214, Center for Automation Research, University of Maryland.

Kanerva, P. (1988). *Sparse Distributed Memory*. MIT Press, Cambridge, Massachusetts.

Kant, K., and Zucker, S. (1984). Trajectory planning in time-varying environment: TPP = PPP + VPP. TR–84–7R, Electrical Engineering, McGill University.

Karp, R.M. (1972). Reducibility among combinational problems. In R.E. Miller and J.W. Thatcher (eds.), *Complexity of Computer Computation*. Plenum Press, New York, 85–103.

Karp, R.M. (1988). In A. Goel (ed.), A report on the AAAI Symposium on parallel models of intelligence. *Sigart* **106**.

Kass, M., Witkin, A., and Terzopoulos, D. (1987). Snakes: active contour models. *ICVV*, London, England, 259–268.

Keller, J.M., Chen, S., and Crownover, R.M. (1989). Texture description and segmentation through fractal geometry. *CVGIP* **45**, 150–166.

Kent, E. (1985). PIPE: pipelined image processing engine. *Journal of Parallel and Distributed Computing* **2**, 50–78.

Kent, E., and Albus, J.S. (1984). Servoed world models as interfaces between robot control systems and sensory data. *Robotica* **2**, 17–25.

Kirkpatrick, S., Gebalt, C.D., and Vecchi, M.P. (1983). Optimization by simulated annealing. *Science* **220**, 671–680.

Kirousis, L., and Papadimitriou, C. (1985). The complexity of recognizing polyhedral scenes. *26th Annual Symposium of Foundations of Computer Science.*

Kjell, B., and Wang, P. (1987). Pyramid linking algorithms for image segmentation of the connection machine. *SPIE* **848**, 234–240.

Klopf, A.H. (1982). *The Hedonistic Neuron. A Theory of Memory, Learning and Intelligence.* Harper and Row, New York.

Klopf, A.H. (1987). Drive-reinforcement learning: a real-time learning mechanism for unsupervised learning. *ICNN*, San Diego, California.

Knudsen, E.I. (1982). Auditory and visual maps of space in the optic tectum of the owl. *J. of Neuroscience* **2**, 1177–1194.

Knudsen, E.I., and Konishi, M. (1978). A neural map of auditory space in the owl. *Science* **200**, 795–797.

Koenderink, J.J. (1988). Scale-time. *BiolCyb* **58**, 159–162.

Koenderink, J.J., and van Doorn, A.J. (1979). The internal representation of solid shape with respect to vision. *BiolCyb* **32**, 211–216.

Koenderink, J.J., and van Doorn, A.J. (1980). Photometric invariants related to solid shape. *Opt. Acta* **27**, 981–996.

Koenderink, J.J., and van Doorn, A.J. (1981) Exterospecific component for the detection of structure and motion in three dimensions. *JOSA* **71**, 953–957.

Koenderink, J.J., and van Doorn, A.J. (1982). The shape of smooth objects and the way contours end. *Perception* **11**, 129–137.

Koenderink, J.J., and van Doorn, A.J. (1986). Dynamic shape. *BiolCyb* **53**, 383–396.

Koenderink, J.J., and van Doorn, A.J. (1987). Representation of local geometry in the visual system. *BiolCyb* **55**, 367–375.

Kohonen, T. (1987). *Self-Organization and Associative-Memories* (2nd ed). Springer-Verlag, New York.

Kolb, B., and Whishaw, I.Q. (1985). *Fundamentals of Human Neuropsychology* (2nd ed.). W.H. Freeman, San Francisco, California.

Kolmogorov, A.N. (1957). On the representation of continuous functions of one variable by superposition of continuous functions of one variable and addition. *AMS Translation* **2**, 55–59.

Koren, I., Mendelson, B., Peleg, I., and Silberman, G. (1988). A data-driven VLSI array for arbitrary algorithms. *Computer* **21**(10), 30–44.

Korn, G., and Korn, T.M. (1968). *Mathematical Handbook for Scientists and Engineers* (2nd ed.). McGraw Hill, New York.

Kosslyn, S.M. (1980). *Image and Mind*. Harvard University Press, Cambridge, Massachusetts.

Kuan, D.T., and Drazovich, R.J. (1986). Model based interpretation of 3-D range data. In A. Rosenfeld (ed.), *Techniques for 3-D Machine Perception*. North-Holland, New York, 219–230.

Kube, P., and Pentland, A. (1988). On the imaging of fractal surfaces, *T-PAMI* **10**(5), 704–707.

Kuffler, S.W. (1952). Neurons in the retina: organization, inhibition and excitation problems. *Symp. Quantitative Biology* **27**, 281–292.

Kuhn, T.S. (1962). *The Structure of Scientific Revolutions*. University of Chicago Press.

Kuipers, B., and Levitt, T. (1988). Navigation and mapping in large-scale space. *AIM* **9**(2), 25–43.

Kulikowski, J.J., Marcelja, S., and Bishop, P.O. (1982). Theory of spatial position and spatial frequency relations in the receptive fields of simple cells in the visual cortex. *BiolCyb* **43**, 187–198.

Kumar, B.V.K.V., and Carroll, C. (1983). Pattern recognition using Wigner distribution function. *IEEE Proc. 10th Int. Optical Computing Conf.*, Cambridge. 130–136.

Kung, H.T. (1982). Why systolic architectures? *Computer* **15**(1), 37–46.

Kung, S., Lo, S.C., Jean, S.N., and Hwang, J.N. (1987). Wavefront array processors—concept to implementation. *Computer* **20**(7), 18–34.

Kunt, M., Benard, M., and Leonardi, R. (1987). Recent results in high-compression image coding. *IEEE Trans. on Circuits and Systems* **34**(14), 1306–1336.

Land, E.H., and McCann, J.J. (1971). Lightness and retinex theory. *JOSA* **61**, 1–11.

Lashley, K.S. (1950). In search of the engram. *Society of Experimental Biology Symposium* No. 4: *Psychological Mechanisms in Animal Behavior.* Cambridge University Press, 454–455, 468–473, 477–480.

Leipnik, R. (1959). Entropy and the uncertainty principle. *Information and Control* **2**, 64–79.

Leipnik, R. (1960). The extended entropy uncertainty principle. *Information and Control* **3**, 18–25.

Levine, M. (1986). *Man and Machine Vision.* McGraw Hill, New York.

Lewontin, R.C. (1989). The science of metamorphis. *The New York Review of Books* **XXXVI**(7).

Lin, Z.C., Huang, T.S., Blostein, S.D., Lee, H., and Margerum, E.A. (1986). Motion estimation from 3-D point sets with and without correspondences. *CVPR*, Miami Beach, Florida, 194–201.

Linsker, R. (1988). Self-organization in a perceptual network. *Computer* **March**, 105–117.

Lippmann, R.P. (1987). An introduction to computing with neural nets. *IEEE ASSP Magazine* **April**, 4–22.

Longuet-Higgins, H.C., and Prazdny, K. (1980). The interpretation of moving retinal images. *Proc. of the Royal Society, London B* **208**, 385–397.

Lowe, D.G. (1985). *Perceptual Organization and Visual Recognition.* Kluwer, Norwell, Massachusetts.

Lumia, R., and Albus, J.S. (1988). Teleoperation and automation for space robotics. *Robotics and Autonomous Systems* **4**(1), 27–34.

Lynch, G. (1986). *Synapses, Circuits, and the Beginning of Memory.* MIT Press, Cambridge, Massachusetts.

Maffei, L., and Fiorentini, A. (1973). The visual cortex as a spatial frequency analyzer. *VR* **13**, 1255–1267.

Maffei, L., and Fiorentini, A. (1976). The unresponsive regions of visual cortical receptive fields. *VR* **16**, 1131–1267.

Makovski, A. (1983). *Medical Imaging Systems.* Prentice Hall, Englewood Cliffs, New Jersey.

Malik, J. (1989). Representing constraints for infering 3D scene structure from monocular cues. *Vision and 3D Representation (ABSTRACTS)*, University of Minnesota, May 24–26.

Mallat, S. (1989). A theory for multiresolution signal decomposition: the wavelet representation. *T-PAMI* **11**(7), 674–693.

Mandelbrot, B.B. (1982). *The Fractal Geometry of Nature*. W.H. Freeman, San Francisco, California.

Marcelja, S. (1980). Mathematical description of the responses of simple cells. *JOSA* **70**, 1297–1300.

Marinovic, N.M., and Eichmann, G. (1985). An expansion of Wigner distribution and its applications. *ICASP*, 1021–1024.

Marr, D. (1982). *Vision*. W.H. Freeman, San Francisco, California.

Marr, D., and Hildreth, E. (1979). Theory of edge detection. AI Memo 518, MIT AI Lab.

Marroquin, J.L. (1984). Surface reconstruction preserving discontinuities. AI Memo 792, MIT AI Lab.

Massone, L., Sandini, G., and Tagliasco, V. (1985). "Form-invariant" topological mapping strategy for 2D shape recognition. *CVGIP* **30**, 169–188.

Matthies, L., Szeliski, R., and Kanade, T. (1987). Kalman filter-based algorithms for estimating depth from image sequences. CMU–CS–87–185, Computer Science, Carnegie–Mellon University.

McClellan, J.H., Parks, T.W., and Rabiner. L.R. (1973). A computer program for designing optimum FIR linear phase digital filters. *IEEE Transactions on Audio and Electroacoustics* **21**(6), 506–526.

McClelland, J.L., Rumelhart, D.E., and the PDP Research Group (eds.) (1986). *Parallel Distributed Processing* (Vols. 1, 2). MIT Press, Cambridge, Massachusetts.

McCune, B.P., and Drazovich, R.J. (1983). Radar with sight and knowledge. *Defense Electronics* **August**.

McLean-Palmer, J., Jones, J., and Palmer, J. (1985). New degrees of freedom in the structure of simple receptive fields. *Investigative Ophthalmology and Visual Science Supplement* (ARVO) **265**.

Mead, C. (1989). *Analog VLSI and Neural Systems*. Addison-Wesley, Reading, Massachusetts.

Meadows, D.M., Johnson, W.O., and Allen, J.B. (1970). Generation of surface contours by Moiré patterns. *Applied Optics* **9**(4), 942–947.

Meer, P., Baugher, E.S., and Rosenfeld, A. (1987). Frequency domain analysis and synthesis of image pyramid generating kernels. *T-PAMI* **9**(4), 512–522.

Meer, P., Wang, S., and Wechsler, H. (1989). Edge detection by associative mapping. *PR* **22**(5), 491–504.

Metropolis, N., Rosenbluth, A.W., Rosenbluth, M.N., Teller, A.H., and Teller, E. (1953). Equation of state calculation by fast computing machines. *J. Chem. Phys.* **21**, 1087–1092.

Michalski, R.S. (1983). A theory and methodology of inductive learning. In R.S. Michalski, J.G. Carbonell, and T.M. Mitchell (eds.), *Machine Learning*. Tioga Publishing, Palo Alto, California, 83–134.

Michalski, R.S. (1989). Two-tiered concept meaning, inferential matching, and conceptual cohesiveness. In S. Vosniadou and A. Orton (eds.), *Similarity and Analogy*. Cambridge University Press.

Mingolla, E., and Todd, J.T. (1986). Perception of solid shape from shading. *BiolCyb* **53**, 137–151.

Minsky, M. (1975). A framework for representing knowledge. In P.W. Winston (ed.), *The Psychology of Computer Vision*. McGraw-Hill, New York, 211–277.

Minsky, M. (1986). *The Society of Mind*. Simon and Schuster, New York.

Minsky, M., and Papert, S. (1987). *Perceptrons* (2nd ed.). MIT Press, Cambridge, Massachusetts.

Mishkin, M., and Appenzeller, T. (1987). The anatomy of memory. *SA* **June**, 80–89.

Mishkin, M., Ungerleider, L.G., and Macko, K.A. (1983). Object vision and spatial vision: two cortical pathways. *Trends in NeuroScience* **6**, 414–437.

Mohideen, S., Cherkassky, V., and Wechsler, H. (1989). Computation of the DAM and numerical stability of the generalized inverse matrix. *ICNN*, Washington, D.C.

Moore, D.J.H. (1972). An approach to the analysis and extraction of pattern features using integral geometry. *T-SMC* **2**, 97–102.

Moore, D.J.H., and Parker, D.J. (1974). Analysis of global pattern features. *PR* **6**, 149–164.

Moravec, H.P. (1983a). Locomotion, vision, and intelligence. The Robotic Institute, Carnegie-Mellon University.

Moravec, H.P. (1983b). The stanford cart and the CMU rover. *P-IEEE* **71**(7), 872–884.

Moravec, H.P. (1987). Sensor fusion in certainty grids for mobile robots. *Annual Research Review*, Robotics Institute, Carnegie-Mellon University.

Mountcastle, V.B. (1978). An organizing principle for cerebral function: the unit module and the distributed system. In G.M. Edelman and V.B. Mountcastle (eds.), *The Mindful Brain: Cortical Organization and the Group-Selective Theory of Higher Brain Function*. MIT Press, Cambridge, Massachusetts.

Movshon, J.A., Thompson, I.D., and Tolhurst, D.J. (1978a). Spatial summation in the receptive fields of simple cells in the cat's striate cortex. *J. Physiology* **283**, 53–77.

Movshon, J.A., Thompson, I.D., and Tolhurst, D.J. (1978b). Receptive field organization of complex cells in the cat's striate cortex. *J. Physiology* **283**, 79–99.

Murray, D., Kashko, A., and Buxton, H. (1986). A parallel approach to the picture restoration algorithm of Geman and Geman on an SIMD machine. *Image and Vision Computing* **4**(3), 133–142.

Nahar, S., Sahni, S., and Shragowitz, E. (1984). Experiments with simulated annealing. TR84, Computer Science, University of Minnesota.

Negahdaripour, S., and Horn, B. (1985). Determining 3D motion of planar objects from image brightness measurements. *IJCAI*, Los Angeles, California, 898–901.

Neisser, U. (1976). *Cognition and Reality*. W.H. Freeman, San Francisco, California.

Neisser, U. (1989). Direct perception and recognition as distinct perceptual systems. *Cognitive Science Society Address*.

Newell, A. (1988). Symposium on "Parallel Models of Intelligence." Ohio State University.

Nii, H.P. (1984). Signal-to-symbol transformation: reasoning in the HASP/SIAP program. *ICASP*, San Diego, California, 39A.3.1–39A.3.4.

Nilsson, N.J. (1965). *Learning Machines*. McGraw Hill, New York.

Nilsson, N.J. (1974). Artificial intelligence. *IFIP* 74, Stockholm, Sweden.

Nilsson, J.J. (1984). SHAKEY the robot. TN 323, SRI AI Center.

Oja, E. (1982). A simplified neuron model as a principal component analyzer. *J. Math. Biology* **15**, 267–273.

Oppenheim, A.V., and Lim, J.S. (1981). The importance of phase in signals. *P-IEEE* **69**, 529–541.

Oppenheim, A.V., Shafer, R.W., and Stockham, Jr., T.G. (1968). Nonlinear filtering of multiplied and convolved signals. *P-IEEE* **56**, 1264–1291.

Papathomas, T., and Julesz, B. (1988). Lie differential operators in animal and machine vision. *Proc. of the Conf. From Pixels to Features*, Bonas, France.

Pavlidis, T. (1977). *Structural Pattern Recognition*. Springer-Verlag, New York.

Peleg, S., Naor, J., Hartley, R., and Avnir, D. (1984). Multiple resolution texture analysis and classification. *T-PAMI* **6**(4), 518–523.

Penrose, R. (1989). *The Emperor's New Mind*. Oxford University Press.

Pentland, A. (1983). Fractal-based descriptions. *IJCAI*, Munich, West Germany, 981–983.

Pentland, A. (1987). A new sense for depth of field. *T-PAMI* **9**(4), 523–531.

Pentland, A. (1988). Shape information from shading: A theory about human perception. *ICCV*, 404–413, Tampa, Florida.

Pinker, S. (1985). Visual cognition: an introduction. In S. Pinker (ed.), *Visual Cognition*. MIT Press, Cambridge, Massachusetts, 1–64.

Platt, J.C., and Barr, A.H. (1988). Constrained differential optimization. In D. Anderson (ed.), *Neural Information Processing Systems*. American Institute of Physics (AIP), New York, 612–621.

Poggio, T. (1985). Early vision: from computational structure to algorithms and parallel hardware. *CVGIP* **31**, 139–155.

Pollen, D.A., and Ronner, S.F. (1981). Phase relationships between adjacent simple cells in the visual cortex. *Science* **212**, 1409–1411.

Pollen, D.A., and Ronner, S.F. (1983). Visual cortical neurons as localized spatial frequency filters. *T-SMC* **13**, 907–915.

Polzleitner, W., and Wechsler, H. (1990). Selective and focused invariant recognition using DAM. *T-PAMI* **12** (to appear).

Popper, K. (1982). The place of mind in nature. In R. Elvee (ed.), *Mind in Nature*. Harper and Row, New York, 31–59.

Porat, M., and Zeevi, Y.Y. (1988). The generalized Gabor scheme of image representation in biological and machine vision. *T-PAMI* **10**(4), 452–468.

Poultis, D., and Farhat, N. (1965). Optical information processing based on an associative-memory model of neural nets with thresholding and feedback. *Optics Letter* **10**(2), 98.

Putnam, H. (1975). The meaning of meaning. In K. Gunderson (ed.), *Language, Mind and Knowledge*. Minnesota Studies in the Philosophy of Science, 7, University, Minnesota, Minneapolis.

Rashid, R.F. (1980). Towards a system for the interpretation of moving light displays. *T-PAMI* **2**(6), 574–581.

Reddi, S.S. (1981). Radial and angular moment invariants for image identification. *T-PAMI* **3**(2), 240–242.

Reed, T., and Wechsler, H. (1990). Segmentation of textured images and Gestalt organization using spatial/spatial-frequency representations. *T-PAMI* **12**(1), 1–12.

Reeves, A. (1984). Parallel computer architectures for image processing. *CVGIP* **25**, 68–88.

Rescorla, R.A., and Wagner A.R. (1972). A theory of Pavlovian conditioning: variations on the effectiveness of reinforcement and non-reinforcement. In A.H. Black and W.F. Prokasy (eds.), *Classical Conditioning II: Current Research and Theory*. Appleton-Century-Crofts, New York.

Rich, E. (1989). *Artificial Intelligence*. McGraw Hill, New York.

Richards, G. (1988). Implementation of back-propagation on a transputer. *Proc. of the 8th OCCAM*, Sheffield, UK.

Richmond, B., Optican, L., and Gawne, T.J. (1988). Neurons use multiple messages encoded in temporally modulated spike trains to represent pictures. In J. Kulikowski (ed.), *Seeing Contour and Colour*. Pergamon Press, London.

Rock, I. (1984). *Perception*. Scientific American, New York.

Rock, I., Wheeler, D., and Tudor, L. (1989). Can we imagine how objects look from other viewpoints? *Cognitive Psychology* **21**, 185–210.

Rodieck, R.W. (1965). Quantitative analysis of cat retinal ganglion cell response to visual stimuli. *VR* **5**, 585–601.

Rosenblatt, F. (1961). *Principles of Neurodydnamics: Perceptrons and the Theory of Brain Mechanism*. Spartan Books, New York.

Rosenfeld, A. (1986). Expert vision systems: some issues. *CVGIP* **34**, 99–117.

Rosenfeld, A., and Thurston, M. (1971). Edge and curve detection for visual scene analysis. *IEEE Trans. Computers* **C-20**, 562–569.

Rosenfeld, A., Hummel, R., and Zucker, S. (1976). Scene labeling by relaxation operations. *T-SMC* **6**(6), 420–433.

Rubin, S.M. (1980). Natural scene recognition using locus search. *CGIP* **13**, 298–333.

Ruck, D.W., Rogers, S.K., Kabrisky, M., and Maybeck, P.S. (1990), Back propagation: A degenerated Kalman filter? *T-PAMI* (under review).

Rumelhart, D.E., Hinton, G.E., and Williams, R.J. (1986). Learning internal representations by error-propagation. In J.L. McClelland and D.E. Rumelhart (eds.), *Parallel Distributed Processing* **1**, MIT Press, Cambridge, Massachusetts, 318–362.

Rusell, B. (1921). *The Analysis of Mind*. MacMillan, New York.

Sachs, M.B., Nachmias, J., and Robson, J.G. (1971). Spatial-frequency channels in human vision. *JOSA* **61**, 1176–1186.

Samet, H. (1989a). *The Design and Analysis of Spatial Data Structures*. Addison-Wesley, Reading, Massachusetts.

Samet, H. (1989b). *Applications of Spatial Data Structures*. Addison-Wesley, Reading, Massachusetts.

Sandini, G., Frigato, C., Grosso, E., and Tistarelli, M. (1990). Multimodal integration in artificial vision. In A. Bejczy (ed.), *Robots with Redundancy*. Springer-Verlag, New York.

Schaefer, D. Ho, P., Boyd, J., and Vallejos, C. (1987). The GAM pyramid. In L. Uhr (Ed.), *Parallel Computer Vision*. Academic Press, Boston, 15–42.

Schank, R.C., and Abelson, R.P. (1977). *Scripts, Plans, Goals, and Understanding*. Erlbaum, Hillsdale, New Jersey.

Schenker, P.S., Wang, K.M., and Cande, E.G. (1981), Fast adaptive algorithms for low-level scene analysis: application of polar exponential growth (PEG) representation to high-speed scale and rotation invariant target segmentation. *SPIE*, **281**, 47–57.

Schoppers, M.J. (1989) In defence of reaction plans as caches. *AIM* **10**(4), 51–60.

Schwartz, E.L. (1977). Spatial mapping in the primate sensory projection: analytic structure and relevance to perception. *BiolCyb* **25**, 181–194.

Schwartz, E.L. (1981). Cortical anatomy, size invariance, and spatial frequency analysis. *Perception* **10**, 455–468.

Schwartz, J.T., Sharir, M., and Hopcroft, J. (1987). *Planning, Geometry, and Complexity of Robot Motion*. Ablex Publishing, Norwood, New Jersey.

Searl, J. (1984). *Minds, Brains, and Science*. Harvard University Press.

Sedgewick, H.A. (1983). Environment-centered representation of spatial layout: available visual information from texture and perspective. In J. Beck, B. Hope and A. Rosenfeld (eds.), *Human and Machine Vision*. Academic Press, New York, 425–458.

Serra, J. (1982). *Image Analysis and Mathematical Morphology*. Academic Press, New York.

Shafer, G. (1976). *A Mathematical Theory of Evidence*. Princeton University Press.

Sharir, M. (1989). Algorithmic motion planning in robotics. *Computer* **22**(3), 9–20.

Sheil, B. (1983). The artificial intelligence tool box. *Proc. of the NYU Symposium on Artificial Intelligence and Business*.

Shepard, R.N. (1983). Ecological constraints on internal representations. James J. Gibson Memorial Lecture, Cornell University.

Shepard, R.N., and Cooper, L.A. (1986). *Mental Images and Their Transformations*. MIT Press, Cambridge, Massachusetts.

Shmuel, A., and Werman, M. (1989). Active vision: 3D from an image sequence. TR-89-13, Computer Science Department, Hebrew University, Israel.

Shortliffe, E., and Buchannan, B. (1984). *Rule-Based Expert Systems*. Addison-Wesley, Reading, Massachusetts.

Simon, H. (1982). *The Science of the Artificial* (2nd ed.). MIT Press, Cambridge, Massachusetts.

Simon, H. (1983). Why should machines learn. In R.S. Michalski, J.G. Carbonell and T.M. Mitchell (eds.). *Machine Learning. An AI Approach*. Tioga Press, Palo Alto, California, 25–38.

Singhal, S., and Wu, L. (1989). Training multilayer perceptrons with the extended Kalman algorithm. In D.S. Touretzky (ed.), *Advances in Neural Network Information Processing*. Morgan Kaufmann, San Mateo, California, 133–140.

Skarda, C.A., and Freeman, W.J. (1987). How brains make chaos in order to make sense of the world. *BBS* **10**, 161–195.

Sklansky, J. (1978). On the Hough technique for curve detection. *IEEE Trans. on Computers* **27**(10), 923–926.

Slamecka, N.J. (1987). The law of frequency. In D.S. Gorfein and R.R. Hoffman (eds.), *Memory and Learning: The Ebbinghaus Centennial Conference*, Erlbaum, Hillsdale, New Jersey.

Sperry, R.W. (1956). The eye and the brain. *SA* **194**(5), 48–52.

Steinbuch, K. (1961). Die Lernmatrix. *Kybernetik* **1**, 36–45.

Steinman, R.M., Levinson, J.Z., Collewijn, H., and van der Steen, J. (1985). Vision in the presence of known natural image motion. *JOSA* **2**(2), 226–233.

Stephanou, H.E., and Lu, S.Y. (1988). Measuring consensus effectiveness by a generalized entropy criterion. *T-PAMI* **10**(4), 544–554.

Stephanou, H.E., and Sage, A.P. (1987). Perspectives on imperfect information processing. *T-SMC* **17**(5), 780–798.

Sternberg, S.R. (1985). An overview of image algebra and related architectures. In S. Levialdi (ed.), *Integrated Technology for Parallel Image Processing*. Academic Press, New York.

Stiles, G.S., and Denq, D.L. (1985). On the effect of noise on the Moore-Penrose generalized inverse associative memory. *T-PAMI* **7**(3), 358–360.

Suh, J.Y., and Van Gucht, D. (1987). Incorporating heuristic information into genetic scarch. *TR*, Department of Computer Science, Indiana University.

Sutton, R.S., and Barto, A.G. (1981). Toward a modern theory of adaptive networks: expectation and prediction. *Psychological Review* **88**, 135–171.

Szu, H., and Hartley, R. (1987). Fast simulated annealing. *Physics Letters A* **122**(3, 4), 157–162.

Szu, H., and Messner, R.P. (1986). Adaptive invariant novelty filters. *P-IEEE* **7**(3), 518–519.

Terzopoulos, D. (1986). Image analysis using multigrid relaxation methods. *T-PAMI* **8**(2), 129–139.

Terzoupoulos, D. (1988). The computation of visible-surface representations. *T-PAMI* **10**(4), 417–438.

Terzoupolos, D., Witkin, A., and Kass, M. (1987). Symmetry-seeking models for 3D object reconstruction. *ICVV*, London, England, 269–276.

Thompson, W.B., Mutch, K.M., and Berzins, V.A. (1984). Analyzing object motion based on optical flow. *ICPR*, 791–794.

Thorpe, C., Hebert, M.H., Kanade, T., and Shafer, S. (1988). Vision and navigation for the Carnegie-Mellon Navlab. *T-PAMI* **10**(3), 362–372.

Toffoli, T., and Margolus, N. (1987). *Cellular Automata Machines*. MIT Press, Cambridge, Massachusetts.

Treisman, A. (1985). Preattentive processing in vision. *CVGIP* **31**(2), 156–177.

Tsotsos, J.K. (1987). A "complexity level" analysis of vision. *ICCV*, London, England, 346–355.

Tsotsos, J.K. (1989). The complexity of perceptual search tasks. *IJCAI*, Detroit, Michigan.

Tsuji, S., and Zheng, J.Y. (1987). Visual path planning by a mobile robot. *IJCAI*, 1127–1130.

Tucker, L., and Robertson, G. (1988). Architecture and applications of the connection machine. *Computer* **21**(8), 26–39.

Turau, V. (1988). A model for a control and monitoring system for an autonomous mobile robot. *Robotics and Autonomous Systems* **4**(1), 41–48.

Turk, M.A., Morgenthaler, D.G., Gremban, K.D., and Marra, M. (1988). VITS—A vision system for autonomous land vehicle navigation. *T-PAMI* **10**(3), 342–361.

Uhr, L. (ed.) (1987). *Parallel Computer Vision*. Academic Press, Boston.

Ullman, J. (1984). *Computational Aspects of VLSI*. Computer Science Press, Rockville, Maryland.

Ullman, S. (1979a). Relaxation and constrained optimization by local processes. *CGIP* **10**, 115–125.

Ullman, S. (1979b). *The Interpretation of Structure from Motion*. MIT Press, Cambridge, Massachusetts.

Ullman, S. (1980). Against direct perception. *BBS* **3**, 373–415.

Ullman, S. (1985). Visual routines. In S. Pinker (ed.), *Visual Cognition*. MIT Press, Cambridge, Massachusetts, 97–160.

Vamos, T. (1977). Industrial objects and machine part recognition. In K.S. Fu (ed.), *Syntactic Pattern Recognition Applications*. Springer-Verlag, New York, 244–265.

Van Essen, D. (1979). Hierarchical organization and functional streams in the visual cortex. *Annual Review of Neuroscience* **2**, 227–263.

Van Essen, D. (1985). Functional organization of primate visual cortex. In A. Peters and E.G. Jones (eds.), *Cerebral Cortex*. Plenum Press, New York, **3**, 259–329.

Van Essen, D., and Maunsell, J.H.R. (1983). Hierarchical organization and functional streams in the visual cortex. *Trends in Neuro Science*.

Veen, A. (1986). Dataflow machine architecture. *CS* **18**(4), 365–397.

Ville, J. (1948). Theorie et applications de la notion de signal analytique. *Cable et Transmission* **2**, 61–74.

Von Neumannn, J. (1958). *The Computer and the Brain*. Yale University Press.

Walker, E.L., Herman, M., and Kanade, T. (1988). A framework for representing and reasoning about 3D objects for vision. *AIM* **9**(2), 47–58.

Waltz, D.L., and Pollack, J.B. (1985). Massively parallel parsing. *Cognitive Science* **9**, 51–74.

Warrington, E.E. (1982). Neuropsychological studies of object recognition. *Phil. Trans. R. Soc. London* **258**, 15–33.

Watson, A.B. (1987a). Human visual representation of spatial imagery. *Perception* **16**, 227–228.

Watson, A.B. (1987b). The cortex transform: rapid comutation of simulated neural images. *CVGIP* **39**, 311–327.

Watson, A.B., and Ahumada, A.J., Jr. (1983). A look at motion in the frequency domain. *SIGRAPH/SIGART Interdisciplinary Workshop, Motion: Representation and Perception*.

Waxman, A.M., and Duncan, J.J. (1986). Binocular image flows. *T-PAMI*, 31–38.

Waxman, A.M., Wu, J., and Bergholm, F. (1988). Convected activation profiles and the measurement of visual motion. *CVPR*, 717–723.

Wechsler, H., and Zimmerman, L. (1988). 2D invariant object recognition using distributed associative memories. *T-PAMI* **10**(6), 811–821.

Wechsler, H., and Zimmerman, L. (1989). Distributed associative memory for bin-picking. *T-PAMI* **11**(8), 814–822.

Wechsler, H., and Zimmerman, L. (1990a). Redundant vision for robotics. In A. Bejczy (ed.), *Robots with Redundancy.* Springer-Verlag, New York.

Wechsler, H., and Zimmerman, L. (1990b). Dynamic vision. In P. Dario, P. Aebisher and G. Sandini (eds.), *Robots and Biological Systems.* Springer-Verlag, New York.

Weiman, C.F.R., and Chaikin, G. (1979). Logarithmic spiral grids for image processing and display. *CGIP* **11**, 197–226.

Weiss, I. (1988). Projective invariants of shapes. *Proc. DARPA Image Understanding Workshop*, 1125–1134.

Wertheimer, M. (1958). Principles of perceptual organization. In D.C. Bardslee and M. Wertheimer (eds.), *Readings in Perception.* Van Nostrand, Princeton, New Jersey, 115–135.

Weszka, J.S. Dyer, C.R., and Rosenfeld, A. (1976). A comparison study of texture measures for terrain classification. *T-SMC* **6**(4), 269–286.

Weyl, H. (1932). *Theory of Groups and Quantum Mechanics.* Dutton, New York.

Whitehead, S.D., and Ballard, D. (1988). A computational model for activity control in autonomous robots. *TR*, Department of Computer Science, University of Rochester, Rochester, New York.

Whitehead, S.D., and Ballard, D. (1989). Reactive behavior, learning, and anticipation. *TR*, Department of Computer Science, University of Rochester, Rochester, New York.

Widrow, B., and Winter, R. (1988). Neural nets for adaptive filtering and adaptive pattern recognition. *Computer* **March**, 25–40.

Wiener, N. (1948). *Cybernetics.* Wiley, New York.

Wigner, E. (1932). On the quantum correction for thermodynamic equilibrium. *Physical Review* **40**, 749–759.

Wigner, E. (1979). Quantum-mechanical distribution function revisited. In W. Yowgram and A. van der Merwe (eds.), *Perspectives in Quantum Theory.* Dover, New York.

Wilensky, R. (1983). *Planning and Understanding. A Computational Approach to Human Reasoning.* Addison-Wesley, Reading, Massachusetts.

Wilson, G.V., and Pawley, G.S. (1988). On the stability of the traveling salesman problem algorithm of Hopfield and Tank. *BiolCyb* **58**, 63–70.

Wilson, H.R., and Bergen, J.R. (1979). A four mechanism model for spatial vision. *VR* **19**, 19–32.

Wilson, R. (1987). Finite prolate spheroidal sequences and their applications I: generation and properties. *T-PAMI* **9**(6), 787–795.

Wilson, R., and Grandlund, G.H. (1984). The uncertainty principle in image processing. *T-PAMI* **6**(6), 758–767.

Winograd, T., and Flores, F. (1987). *Understanding Computers and Cognition*. Addison-Wesley, Reading, Massachusetts.

Winson, J. (1985). *Brain and Psyche: The Biology of the Unconscious*. Doubleday, New York.

Winston, P.H. (1975). Learning structural descriptions from examples. In P.H. Winston (ed.), *The Psychology of Computer Vision*. McGraw-Hill, New York, 157–210.

Winston, P.H. (1980). Learning and reasoning by analogy. *Comm. of the ACM* **23**, 689–703.

Winston, P.H. (1984). *Artificial Intelligence*. Addison-Wesley, Reading, Massachusetts.

Witkin, A., and Tenenbaum, J.M. (1983). On the role of structure in vision. In J. Beck, B. Hope and A. Rosenfeld (eds.), *Human and Machine Vision*. Academic Press, New York, 481–544.

Woods, W.A. (1975). What's in a link? Foundations for semantic networks. In D.G. Bobrow and A.M. Collins (eds.), *Representation and Understanding*. Academic Press, New York, 35–82.

Woodward, P.M. (1953). *Probability and Information Theory with Applications to Radar*. Pergamon Press, London.

Yang, H.S., and Kak, A (1986). Determination of the identity, position and orientation of the topmost object in a file. *CVGIP* **36**, 229–255.

Yeshurun, Y., and Schwartz, E.L. (1989). Cepstral filtering on a columnar image architecture: a fast algorithm for binocular stereo segmentation. *T-PAMI* **11**(7), 759–767.

Yonas, A., Arterberry, M.E., and Granrud, C.E. (1987). Space perception in infancy. *Annals of Child Development* **4**, 1–34.

Zadeh, L.A. (1986). A simple view of the Demster-Shafer theory of evidence and its implication for the rule of combination. *AIM* **7**(2), 85–90.

Zetzsche, C., and Caelli, T. (1989). Invariant pattern recognition using multiple filter image representations. *CVGIP* **45**, 251–262.

Zhou, Y.T., and Chellappa, R. (1988). Stereo matching using a neural network. *ICASSP*, New York.

Zwicke, P.E., and Kiss, I., Jr. (1983). A new implementation of the Mellin transform and its application to radar classification of ships. *T-PAMI* **5**(2), 191–198.

Man standing looks through telescope. (From Descartes, R., Opera Philosophica, 1656.) Reprinted courtesy of the National Library of Medicine.

Author Index

Subject Index